IUTAM/IACM/IABEM Symposium on Advanced Mathematical and Computational Mechanics Aspects of the Boundary Element Method

# IUTAM/IACM/IABEM Symposium on Advanced Mathematical and Computational Mechanics Aspects of the Boundary Element Method

held in Cracow, Poland, 31 May–3 June 1999

*Edited by*

TADEUSZ BURCZYNSKI

*Department for Strength of Materials and Computational Mechanics,*
*Silesian University of Technology, Gliwice, Poland,*
*and Institute of Computer Modelling, Cracow University of Technology, Cracow, Poland*

Springer-Science+Business Media, B.V.

A C.I.P. Catalogue record for this book is available from the Library of Congress.

DOI 10.1007/978-94-015-9793-7

---

*Printed on acid-free paper*

# Contents

# Preface

The IUTAM/IACM/IABEM Symposium on Advanced Mathematical and Computational Mechanics Aspects of the Boundary Element Method was held in Cracow, Poland, from 31st May to 3rd June 1999. The site of the Symposium was the Polish Academy of Arts and Sciences. The Symposium was attended by 60 persons from 17 countries. In addition, several Polish students and research associates participated in the meeting.

The objective of the Symposium was to provide a forum for researchers in the boundary element method and boundary integral formulations in general to present contemporary concepts and techniques leading to the advancements of capabilities and understanding of the mathematical and computational aspects of the method in mechanics. The Symposium put special emphasis on theoretical and numerical issues, as well as new formulations and approaches for special and important fields of mechanics. In the 4-day Symposium a total 45 invited papers were presented. The following topics were covered:

- Singularity and hypersingular formulations,
- Regularity,
- Errors error estimators,
- Adaptive methods,
- Galerkin formulations,
- Coupling of BEM-FEM,
- Non-deterministic (stochastic and fuzzy) BEM formulations,
- Heat conduction, diffusion and radiation,
- Non-linear problems,
- Dynamics and time-depending problems,
- Fracture mechanics,

- Thermoelasticity and poroelasticity,
- Fluid mechanics,
- Aerodynamics and acoustics
- Contact problems,
- Biomechanics,
- Optimization and sensitivity analysis problems,
- Ill posed and inverse problems,
- Identification problems.

The lectures and their discussions clearly showed the remarkable evolution of the Boundary Element Method since the first IUTAM Symposium on the similar topic organized by T.A.Cruse in 1987 in San Antonio (USA). Several new aspects have come into focus since that time.

This volume contains 35 papers. Much to the regret of the Scientific Committee some manuscripts were not submitted. All the papers contained herein have reviewed to the standard of leading scientific journal. The Editor would like to acknowledge the great efforts on behalf of both the authors and reviewers.

The Editor particularly wishes to thank the Bureau of IUTAM, the Executive Council of IACM and IABEM, and the International Scientific Committee. Part of the success of the Symposium is a consequence of the excellent facilities provided by the Polish Academy of Arts and Sciences in Cracow.

The smooth running of the Symposium owes much to the initiative and the organizational skills of B.Ulejczyk, R.Bialecki and E.Majchrzak from Silesian University of Technology and M.Stanuszek and A.Karafiat from Cracow University of Technology.

Finally, the Editor would like to express his gratitude to the sponsoring organizations who have supported the Symposium financially, namely International Union of Theoretical and Applied Mechanics (IUTAM), International Association for Computational Mechanics (IACM), International Association for the Boundary Element Method (IABEM), Committee of Mechanics of Polish Academy of Sciences, Department for Strength of Materials and Computational Mechanics of Silesian University of Technology in Gliwice and Institute of Computer Modelling of Cracow University of Technology.

Gliwice/Cracow, April 2001 Tadeusz Burczynski

# Comittees and Sponsors

**International Scientific Committee**

T.Burczynski, Chairman (Poland)

T.A.Cruse, Co-Chairman (USA)

D.Beskos (Greece)

H.D.Bui (France)

L.B.Freund (USA)

S.Kobayashi (Japan)

G.Kuhn (Germany)

H.A.Mang (Austria)

L.Morino (Italy)

C.Pozrikidis (USA)

F.Rizzo (USA)

**Sponsors**

International Union of Theoretical and Applied Mechanics (IUTAM)

International Association for Computational Mechanics (IACM)

International Association for the Boundary Element Method (IABEM)

Committee of Mechanics of Polish Academy of Sciences

Department for Strength of Materials and Computational Mechanics
Silesian University of Technology, Gliwice, Poland

Institute of Computer Modelling
Cracow University of Technology, Cracow, Poland

# A DUAL RECIPROCITY BOUNDARY ELEMENT FORMULATION FOR TRANSIENT NON-LINEAR CONDUCTION-RADIATION PROBLEMS

Jutta Blobner
*Institute of Applied Mechanics, University of Erlangen, Nuremberg,*
*Egerlandstr. 5, D-91058 Erlangen, Germany*
blobner@ltm.uni-erlangen.de

Günther Kuhn
*Institute of Applied Mechanics, University of Erlangen, Nuremberg,*
*Egerlandstr. 5, D-91058 Erlangen, Germany*
gkuhn@ltm.uni-erlangen.de

Ryszard A. Białecki
*Institute of Thermal Technology, Silesian Technical University*
*Konarskiego 22, PL-44 101 Gliwice, Poland*
bialecki@itc.ise.polsl.gliwice.pl

Matjaž Hriberšek
*Faculty of Mechnical Engineering, University of Maribor,*
*Smetanowa 17, SI-2000 Maribor, Slovenia*
matjaz.hribersek@uni-mb.si

**Keywords:** Dual Reciprocity Boundary Element Method, transient temperature, heat conduction, heat radiation, non-linear material, non-linear boundary condition

**Abstract** A boundary-only formulation for transient temperature fields in bodies of non-linear material properties and arbitrary non-linear boundary conditions has been developed. The option for self-irradiating boundaries has been included in the formulation. Heat conduction equation has been partially

*T. Burczynski (ed.), IUTAM/IACM/IABEM Symposium on Advanced Mathematical and Computational Mechanics Aspects of the Boundary Element Method*, 1–11.

linearized by Kirchhoff's transformation. The result has been discretized by Dual Reciprocity Boundary Element Method. The integral equation of heat radiation has been discretized by standard boundary element method. The coupling of the resulting two sets of equations has been accomplished by eliminating the radiative heat fluxes arising in both sets. The final set of ordinary differential equations has been solved using Runge-Kutta solver with automatic time step adjustment. Another discussed problem is the coupling of transient conduction in a solid with natural convection in the surrounding fluid by discretizing the fluid itself.

## Introduction

Heat transfer problems of practically arbitrary complexity can be nowadays solved using the available commercial software based on FEM or FDM. This optimistic picture is however overshadowed by an important limiting factor. The development of the codes reflects the state of the fundamental literature on heat transfer where the description of the phenomenon and the associated solution techniques are typically limited to a model of only one heat transfer mode.

In the simplest cases of heat conduction in a solid, a typical approach is to solve the governing differential equation subjected to a set of prescribed boundary conditions. The most straightforward method of acquiring the values of these conditions are the measurements. This option however, is not only expensive and time consuming, but it is also impracticable.

Another, more realistic option is to compute the values of the heat flux and temperature arising in the boundary conditions. These functions result from the convective heat transfer in the surrounding fluid and the irradiation of the boundary by the environment. Thus, a realistic heat conduction problem encompasses mutually interrelated all modes of heat transfer: conduction, radiation and convection. Moreover, all constituent equations are coupled and have to be solved simultaneously. The situation is very much the same for problems originally formulated as pure convection or radiation. A deeper analysis of such models leads finally to a multi mode heat transfer problem.

The paper addresses a problem of transient heat conduction in a solid whose material properties are non-linear. The body exchanges heat with the environment by radiation and convection. The model deals with a situation when a portion of the boundary can irradiate itself, simultaneously exchanging heat by convection. Another discussed problem is the coupling of transient conduction in a solid with natural convection in the surrounding fluid.

# 1. FORMULATION OF THE CONDUCTION-RADIATION PROBLEM

## 1.1. HEAT CONDUCTION

A transient temperature field $T$ in a 2D domain $\Omega$ is considered. The domain is homogeneous, material properties are temperature dependent, the material is isotropic. The governing equation describing the temperature field has a form

$$\nabla[k(T)\nabla T(\mathbf{x},t)] = c(T)\rho(T)\dot{T}(\mathbf{x},t) \qquad \mathbf{x} \in \Omega,\ t > 0, \tag{1.1}$$

where $t$ denotes the time, $\mathbf{x}$ is the vector co-ordinate of a current point, $k$, $c$, $\rho$ are heat conductivity, specific heat capacity and density, respectively, and dot stands for the temporal derivative.

The initial condition has a standard form. Different kinds of boundary conditions may be defined on separate portions of the boundary $\Gamma$ of the domain. The set of possible boundary conditions encompasses prescribed temperature (Dirichlet condition), prescribed heat flux (Neumann condition) and a general, temperature dependent non-linear Neumann condition covering practically all remaining cases. The latter can be written in form of a heat balance on the interface between solid and the surrounding fluid

$$q(\mathbf{x},t) = q^{\mathrm{c}}(\mathbf{x},t) + q^{\mathrm{r}}(\mathbf{x},t) \qquad \mathbf{x} \in \Gamma,\ t \geq 0, \tag{1.2}$$

where $q$, $q^{\mathrm{c}}$ and $q^{\mathrm{r}}$ stand for the conductive, convective and radiative heat fluxes, respectively. Convective or radiative heat fluxes may vanish, so that condition (1.2) covers the situation of heat transfer by (non-linear) convection, radiation and the simultaneous energy transport by these two modes.

The conductive heat flux is defined as

$$q = -k\frac{\partial T}{\partial n} \quad , \tag{1.3}$$

with $n$ standing for the outward normal to the boundary. The convective heat flux can be generally written in a form of a non-linear Robin condition as

$$q^{\mathrm{c}}(\mathbf{x},t) = h[T(\mathbf{x},t)][T(\mathbf{x},t) - T_{\mathrm{f}}(\mathbf{x},t)] \quad , \tag{1.4}$$

where $T_{\mathrm{f}}$ is the temperature of the fluid exchanging heat by convection with the boundary and $h$ is the (temperature dependent) convective heat transfer coefficient.

## 1.2. HEAT RADIATION

The radiating walls are grey, diffusive emitters and reflectors of the radiant energy. The heat conducting solid is opaque to radiation. The medium surrounding the solid is transparent to radiation. Due to the high speed of the propagation of electromagnetic waves, the transient components of the radiative transfer are neglected.

The uncoupled radiative heat flux $q^{\mathrm{r}}$ is a non-linear function of local and ambient temperatures. For grey surfaces the radiative heat flux can be written as

$$q^{\mathrm{r}}(\mathbf{x},t) = \sigma\epsilon_{\mathrm{b-e}}[T(\mathbf{x},t)][T^4(\mathbf{x},t) - T_{\mathrm{e}}^4(\mathbf{x},t)] \quad , \tag{1.5}$$

with $\sigma$ standing for the Stefan-Boltzmann constant. $\epsilon_{\mathrm{b-e}}$ being the radiative heat interchange factor between boundary and environment. $T_{\mathrm{e}}$ denotes the temperature of an external surface exchanging heat by radiation with the point under consideration.

The coupled radiative heat flux is a solution of the governing equation of radiation. Cavities open to the environment should be closed by introducing a fictitious, black ($\epsilon = 1$) surface $\Gamma_{\mathrm{f}}$ having the temperature of the environment $T_{\mathrm{e}}$ [1]. The governing equation of heat radiation reads then

$$\begin{aligned} q^{\mathrm{r}}(\mathbf{x}) + \epsilon(\mathbf{x})e_{\mathrm{b}}[T(\mathbf{x})] &= \epsilon(\mathbf{x})\int_{\Gamma_{\mathrm{r}}}\left\{e_{\mathrm{b}}[T(\mathbf{y})] + \frac{1-\epsilon(\mathbf{y})}{\epsilon(\mathbf{y})}q^{\mathrm{r}}(\mathbf{y})\right\}K(\mathbf{x},\mathbf{y})\,\mathrm{d}\Gamma_{\mathrm{r}}(\mathbf{y}) \\ &+ \epsilon(\mathbf{x})\int_{\Gamma_{\mathrm{f}}} e_{\mathrm{b}}[T_{\mathrm{e}}(\mathbf{y})]K(\mathbf{x},\mathbf{y})\,\mathrm{d}\Gamma_{\mathrm{f}}(\mathbf{y}) \quad , \end{aligned} \tag{1.6}$$

where $e_{\mathrm{b}}$ is the blackbody emissive power. In the case of a grey surface, it is related to the temperature by the fourth power law, i.e. $e_{\mathrm{b}} = \sigma T^4$. $\mathbf{x}$, $\mathbf{y}$ are observation and current point, respectively, $\epsilon$ is the emissivity. $\Gamma_{\mathrm{r}}$ denotes the real surface of the cavity. The shape of the fictitious surface $\Gamma_{\mathrm{f}}$ is arbitrary, but it should be convex with respect to the cavity. Kernel function $K(\mathbf{x},\mathbf{y})$ depends solely on the geometry and is defined as

$$K(\mathbf{x},\mathbf{y}) = \frac{\cos\phi_x \cos\phi_y}{2|\mathbf{x}-\mathbf{y}|} \quad , \tag{1.7}$$

with $|\mathbf{x}-\mathbf{y}|$ denoting the distance between observation and current point, $\phi_x, \phi_y$ are the angles made by the line connecting these points and the normals at points $\mathbf{x}$ and $\mathbf{y}$, respectively.

# 2. DISCRETIZATION

## 2.1. CONDUCTION

Heat conduction equation is discretized using BEM with Dual Reciprocity applied to transform the domain integrals into a finite sum of boundary integrals. The original technique to account for the non-linearity of the material properties has been described in our earlier work [2]. The resulting set of non-linear ordinary differential equations is expressed in terms of vectors $\boldsymbol{\psi}$, $\dot{\boldsymbol{\psi}}$ and $\mathbf{q}$ storing the nodal values of the Kirchhoff transform, its temporal derivative and the conductive flux, respectively. Using matrix notation the set can be written as

$$\mathbf{H}\boldsymbol{\psi} - \mathbf{G}\,\mathbf{q} = \mathbf{C}\mathbf{A}^{-1}\dot{\boldsymbol{\psi}} \quad , \tag{1.8}$$

where $\mathbf{H}$ and $\mathbf{G}$ are standard influence matrices and the capacity matrix $\mathbf{C}$ is defined as

$$\mathbf{C} = (\mathbf{H}\,\hat{\boldsymbol{\Psi}} - \mathbf{G}\,\hat{\mathbf{Q}})\,\mathbf{F}^{-1} \quad , \tag{1.9}$$

with matrices $\hat{\boldsymbol{\Psi}}$ and $\hat{\mathbf{Q}}$ being assemblies of column vectors $\hat{\boldsymbol{\psi}}^j$ and $\hat{\mathbf{q}}^j$ storing the nodal values of known Dual Reciprocity interpolation functions [2]. It should be noted that despite the non-linearity of the problem at hand, matrices $\mathbf{H}$, $\mathbf{G}$, and $\mathbf{C}$ are constant and depend only on the geometry. The entire material non-linearity is concentrated in the diagonal matrix $\mathbf{A}$ storing the nodal values of the heat diffusivity.

## 2.2. RADIATION

The discretization of the heat radiation equation has been described elsewhere [1, 3]. Thus, the details of this procedure will not be repeated here. The main features of this technique are

- The entire radiating surface is subdivided into boundary elements.
- Distributions of both the radiative heat flux and the blackbody emissive power within elements are approximated using locally based shape functions.
- The geometry of boundary elements is approximated using similar functions. Due to the complexities caused by the presence of shadow zones, only linear shape functions have been used.
- The set of algebraic equations has been generated using nodal collocation.

The result of the discretization is a set of linear equations having a form

$$\mathbf{B}\mathbf{e}_{\mathrm{b}} + \mathbf{D}\mathbf{q}^{\mathrm{r}} = \mathbf{c} \quad , \tag{1.10}$$

where $\mathbf{B}$ and $\mathbf{D}$ are constant influence matrices depending on geometry and emissivities at collocation points. $\mathbf{c}$ is a constant vector arising only in the case of an open cavity. It stores the product of appropriate columns of the radiation influence matrix **B** and the blackbody emissive power of the environment. Vectors $\mathbf{e}_b$ and $\mathbf{q}^r$ store the nodal values of the blackbody emissive powers and the radiative heat fluxes, respectively.

## 3. SOLUTION

The final set of equations is of mixed differential-algebraic type and consists of three systems

- non-linear ordinary differential equations (1.8), describing the conduction in the solid,
- system of algebraic equations (1.10) modeling the mutual irradiation of concave portions of the boundary and the influence of the environment radiation,
- the discretized version of the heat balance on the radiating boundaries (1.2) defining the coupling between the first two sets. Additional non-linear coupling conditions are due to the continuity of temperatures at the radiating surface. The latter leads to strongly non-linear equation, as the relationship between the blackbody emissive power and the temperature is described by the fourth power law.

In the first step of the solution procedure the dimensionality of eq. (1.8) is reduced by the number of unknown conductive fluxes using the concept of static condensation. In the next step the substitution of algebraic equations (1.10) and (1.2) into the reduced set of ordinary differential equations is carried out. As a result, the model of the conjugate heat transfer is transformed to a set of ordinary differential equations. The latter is implicit in temporal derivatives and thus difficult to handle by a time stepping procedure. This can be remedied by solving the set with respect to the temporal derivative. The manipulations carried out to form the final set of differential equations from the original differential-algebraic system, consist of matrix multiplications and some variants of Gaussian condensation. Therefore, although the problem is strongly non-linear, the generation of the final set of equations does not require any iterations.

The resulting set of ordinary differential equations is finally solved using an explicit, self adaptive Runge-Kutta solver. Thus, neither linear, nor non-linear algebraic equations are to be solved. Moreover, the length of the time step is determined internally ensuring stability and robustness of the procedure. These features have been confirmed by some benchmark tests [2] and several practical examples. The next section describes briefly one of the solved problems.

**Example; Coupled radiation problem**

The transient temperature field in a concave profile, which is cooled down to room temperature after the process of heat treatment, is sought for. Fig. 1 shows the geometry, prescribed boundary conditions and applied meshing.

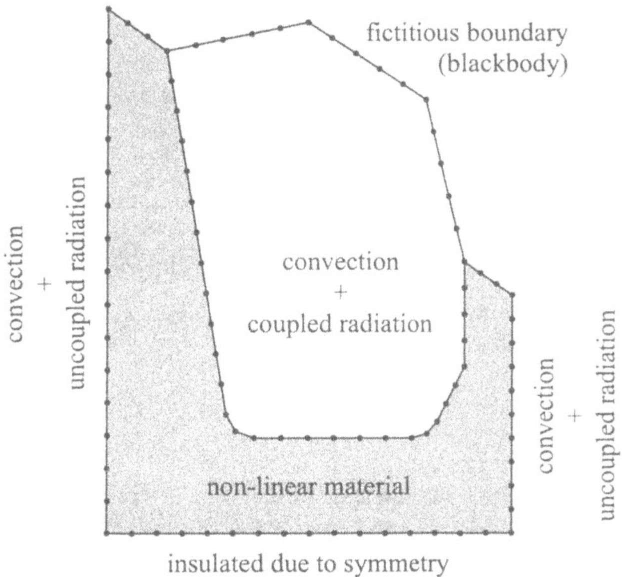


*Figure 1* Geometry, boundary conditions and meshing

The insulation at the lower boundary results from the symmetry condition, whereas all other boundaries may exchange heat with the environment by radiation and convection. Here, convective heat transfer is modeled by a linear Robin boundary condition. The surrounding fluid is assumed to be transparent to radiation. At the concave part of the boundary, coupled radiation has been taken into account. Therefore, a fictitious black boundary has been introduced. Heat transfer coefficient and material properties depend on temperature. Details on modeling can be found in [2].

The upper part of Fig. 2 depicts isotherms of resulting temperatures in the profile after $t = 6\,\mathrm{s}$ and after $t = 300\,\mathrm{s}$. The graphs in the lower part of Fig. 2 show the computed heat fluxes along the perimeter of the concave boundary between points A and B.

The convective heat flux is much lower than the coupled radiative heat flux. Thus, neglecting radiation would lead to large errors. The jumps of the coupled radiative heat flux are due to the geometric corners. To

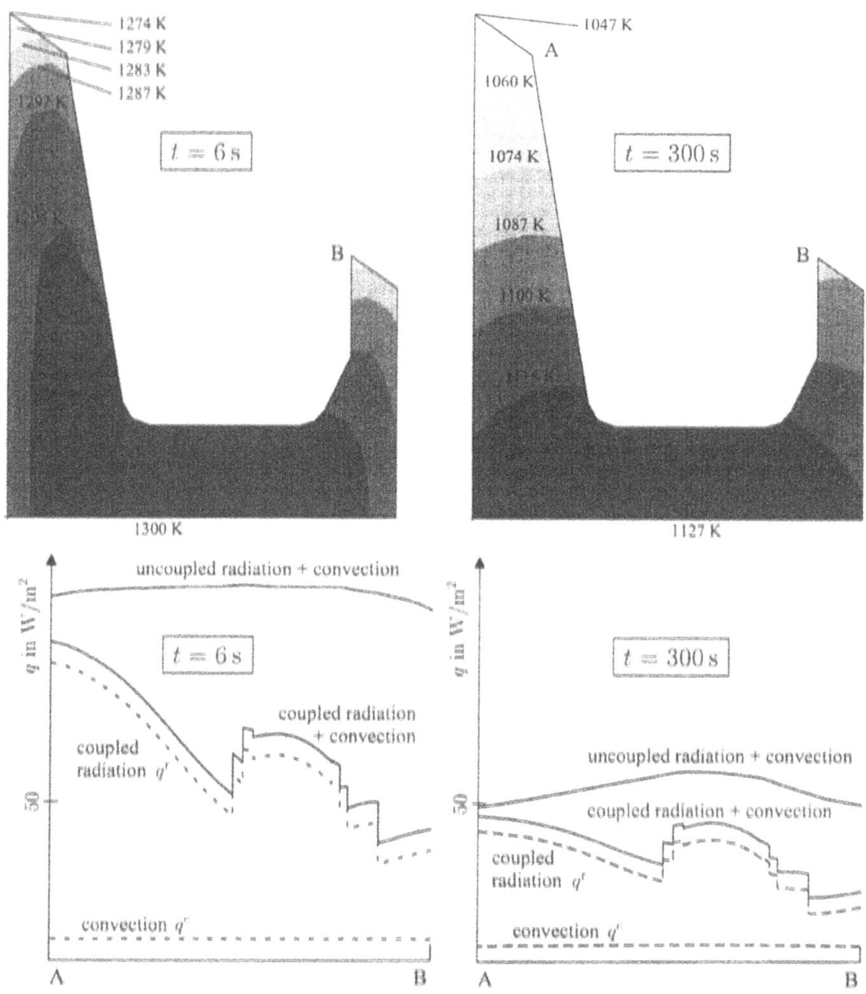


*Figure 2* Top: isotherms for $t = 6\,\mathrm{s}$ (left) and $t = 300\,\mathrm{s}$ (right)
Bottom: heat fluxes $q, q^c, q^r$ along the concave part of the boundary for $t = 6\,\mathrm{s}$ (left) and $t = 300\,\mathrm{s}$ (right)

illustrate the influence of coupled radiation, the sum of the radiative and convective heat fluxes is shown if uncoupled radiation is assumed, *i.e.* if the radiative heat flux is calculated from the conventional Stefan-Boltzmann law. It can be seen that the simplified model leads to significant errors.

# 4. ALTERNATIVE MODELING OF THE SURROUNDING FLUID

In order to model convective heat transfer with a surrounding fluid in a more precise way than by the Robin boundary condition, the fluid itself has to be discretized. The viscous fluid flow can be described by the Boundary Domain Integral Method (BDIM) [4] and implicitly coupled with the DRBEM, which characterizes heat transfer in the solid part [5]. In the fluid part, the non-linear set of transient equations for flow kinematics, vorticity transport and heat transport is solved with respect to unknown velocities, vorticities and temperatures.

The coupling of both methods combine boundary-only discretization of the solid and both boundary and internal discretization of fluid regions. In order to applying efficiently iterative solvers, the sub domain technique is used both in solid and in fluid parts of the computational domain.

**Example; Conjugate heat transfer**

Conjugate natural convection in a square cavity with conducting side wall is selected for the test case problem. The cavity is heated at the left, while the right side is kept at initial temperature, all other boundaries are insulated. The solid is subdivided into two sub domains and discretized by boundary elements only, since domain integrals are transformed to boundary ones by DRBEM. For the fluid, a number of macro elements are introduced and BDIM is applied. Fig. 3 depicts geometry, boundary conditions, material parameters and applied meshing.

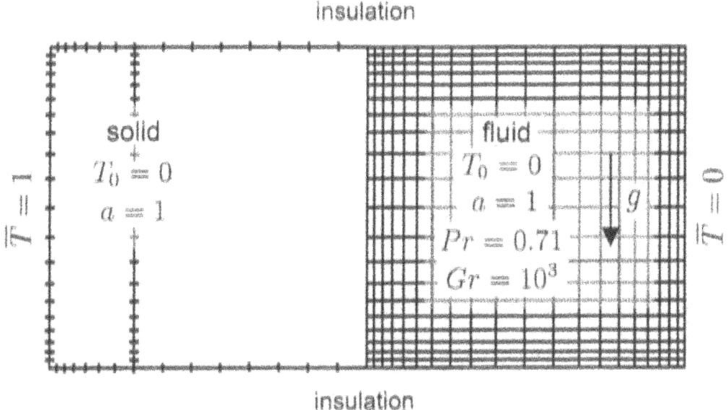


*Figure 3* Presentation of conjugate heat transfer problem

Fig. 4 shows the resulting distributions of temperature in solid and fluid (upper part) of velocities (lower left) and the vorticities (lower right) in the fluid for a selected Fourier number.

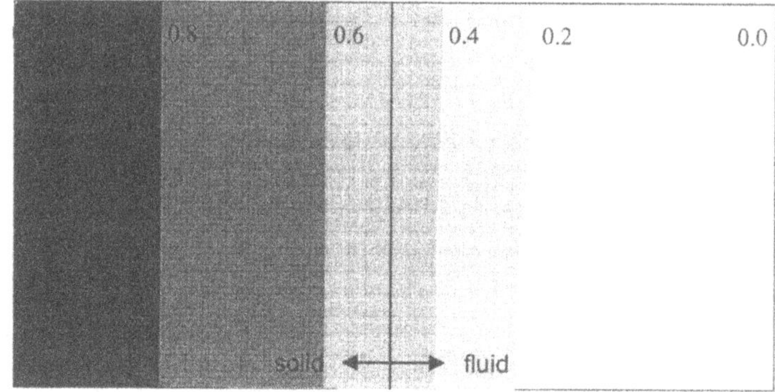


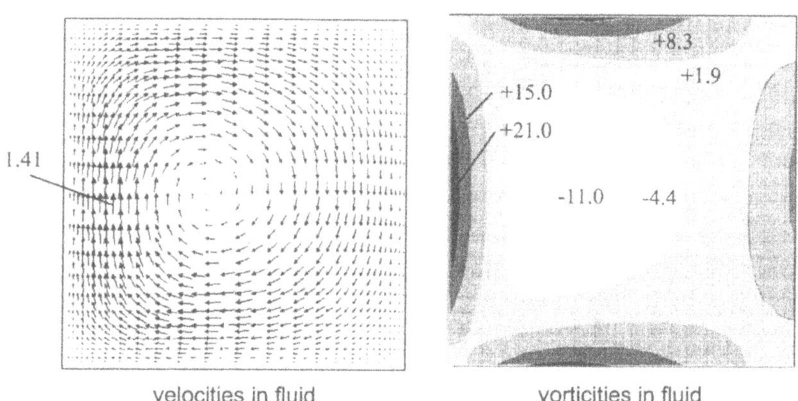


*Figure 4* Temperatures, velocities and vorticities for $Fo = 0.75$

## 5. CONCLUSIONS

The Boundary Element Technique has been used to solve coupled heat transfer problems involving transient conduction, radiation, and natural convection.

Two procedures of coupling single heat transfer modes have been developed and tested. The philosophy behind the first approach is to use autonomous codes to model single heat transfer modes. The information about the coupling condition at the interface is then transferred to the solver *via* disk files. In the course of the time stepping procedure the matrices arising in the final set of the ordinary differential equations were generated only once. The approach leads to an explicit time stepping procedure.

This technique has been applied to transient, non-linear material conduction in a solid containing self irradiating portions of the boundary. It has proved to be robust and accurate.

The second approach was to treat the adjacent domains, with the energy transport ruled by different heat transfer modes, as substructures of the entire domain. The iterative solution is then accomplished within one code and encompasses the entire computational domain.

This strategy has been applied to a problem of energy transfer in a viscous fluid exchanging heat with a heat conducting massive wall.

It has been shown that BEM is capable of solving strongly non-linear conjugate heat transfer problems. So far coupling of only two heat transfer modes have been considered. The logical step of the further research will be to develop an algorithm capable of dealing with simultaneous heat transfer by conduction, convection and radiation.

## Acknowledgments

The work of J. Blobner has been financed by the 'Deutsche Forschungsgemeinschaft' under grant Ku 360/12, and the work of R.A. Białecki has been financed by the Polish Committee for Scientific Research under grant 8 T10B 014 11.

## References

[1] Białecki R.A. and Grela L.: 'Practical aspects of developing heat radiation BEM code', in C.A. Brebbia and A. Kassab (eds.), *Proceedings of the 9th International Conference on Boundary Element Technology*, Computational Mechanics Publications, 29-39 (1994).

[2] Blobner J., Białecki R.A., and Kuhn G.: Transient non-linear heat conduction-radiation problems; A Dual Reciprocity formulation. *Int. J. Num. Meth. Eng.* (1999) 49, 1865-1882.

[3] Białecki R.A.: *Solving Heat Radiation Problems Using the Boundary Element Method*, Computational Mechanics Publications, Southampton and Boston, 1993.

[4] Hriberšek M., and Kuhn G.: Conjugate heat transfer by Boundary Domain Integral Method. *Eng. Anal. Bound. Elem.* (2000) 24 (4), 297-305.

[5] Blobner J., Hriberšek M., and Kuhn G.: Dual Reciprocity BEM - BDIM technique for conjugate heat transfer computations. *Comput. Meth. Appl. Mech. Eng.* (in press).

# A GENERAL ALGORITHM FOR THE DIRECT NUMERICAL EVALUATION OF ELEMENT INTEGRALS IN THE 2D GALERKIN BEM

Marc Bonnet[(a)], Massimo Guiggiani[(b)]

*(a) Laboratoire de Mécanique des Solides (CNRS UMR 7649), Ecole Polytechnique, France*

*(b) Dipartimento di Matematica, Università di Siena, Via del Capitano 15, 53100 Siena, Italy*

bonnet@lms.polytechnique.fr, guiggiani@unisi.it

**Keywords:** Symmetric Galerkin BEM, singular integration

**Abstract** This paper deals with a direct approach for the evaluation of singular element integrals arising from the discretization of hypersingular boundary integral equations (BIEs) in symmetric Galerkin form, for two-dimensional problems.

## 1. INTRODUCTION

The direct evaluation of element integrals arising from the discretization of strongly singular or hypersingular boundary integral equations (BIEs) in collocation form, summarized in [4], is well-established and widely used.

In this paper, the direct approach is developed for hypersingular BIEs in weighted-residual form, i.e. for Galerkin boundary element methods (GBEMs). In line with previous works, the limiting form of the integral identity as a small parameter $\varepsilon$ vanishes must be derived, and this is again done *after* discretization. References [1, 6] implement, in a different way, the same general principles.

The present investigation differs from previous ones in that GBEMs involve *double* element integrals. Hence, two cases of potentially singular integrations arise: coincident (double integral on twice the same element) and adjacent (double integral on a pair of touching elements). The algorithm presented in this communication deals with both the coincident and adjacent cases, for 2D problems. Indeed, it appears that

*T. Burczynski (ed.), IUTAM/IACM/IABEM Symposium on Advanced Mathematical and Computational Mechanics Aspects of the Boundary Element Method*, 13–23.

they must be considered together: each one generates a potentially infinite term, which cancel each other. Double integrals are treated as a whole (i.e. not as inner singular integrals followed by outer nonsingular ones), through the introduction of suitable coordinate transformations in the *two-dimensional* space of intrinsic coordinates.

The usual direct approach to collocation BEM does not resort to prior regularization [4]. Similarly, the method presented here is not based on any prior analytical manipulation, like the double integration by parts on the hypersingular kernel which is used in other contributions, see e.g. [2].

The proposed algorithm is in particular applicable to symmetric GBEMs, which are under intense investigation (see the survey paper [3]), and is devised so as to define in that case a perfectly symmetric integration procedure, even when the numerical quadrature is not exact. However, its definition is valid whether the GBEM formulation considered is symmetric or not, the key features being the *hypersingularity* of the kernels involved and the *double* integration.

## 2. INTEGRAL STATEMENTS

Let $\Omega$ be a 2D bounded domain with boundary $\Gamma = \partial\Omega$ (possibly with a finite number of corners). With $\mathbf{n}(\mathbf{y})$ and $\mathbf{t}(\mathbf{y})$ we indicate, respectively, the normal and tangent unit vectors at $\mathbf{y} \in \Gamma$ (Figure 1).

The starting point is the 3rd Green identity for $\varphi$ and $q = \partial\varphi/\partial n$ on the punctured domain $\Omega_\varepsilon$, with boundary $\Gamma - e_\varepsilon + s_\varepsilon$ (Figure 1). In this case we deal with the hypersingular BIE

$$\int_{\Gamma - e_\varepsilon + s_\varepsilon} [V(\mathbf{x},\mathbf{y})\varphi(\mathbf{y}) - W(\mathbf{x},\mathbf{y})q(\mathbf{y})]\, ds_y = 0. \tag{1}$$

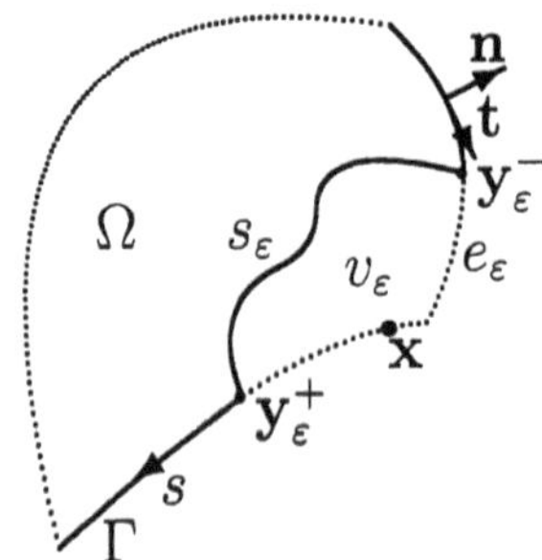


*Figure 1* Exclusion of the singular point $\mathbf{x}$ by a vanishing neighbourhood $v_\varepsilon$.

We assume for simplicity that the free-plane fundamental solution for the Laplace equation is being used, i.e. $G(\mathbf{x},\mathbf{y}) = -1/(2\pi)\ln r$ and hence

$$\begin{aligned} W(\mathbf{x},\mathbf{y}) &= -G_{,i}(\mathbf{x},\mathbf{y})n_i(\mathbf{x}) = \frac{1}{2\pi}\frac{r_i n_i(\mathbf{x})}{r^2}, \\ V(\mathbf{x},\mathbf{y}) &= W_{,j}n_j(\mathbf{y}) = -G_{,ij}(\mathbf{x},\mathbf{y})n_i(\mathbf{x})n_j(\mathbf{y}) = O(r^{-2}), \end{aligned} \tag{2}$$

where $\mathbf{r} = \mathbf{y} - \mathbf{x}$, $r = |\mathbf{r}|$ and $r_i$ is the $i$-th Cartesian component of $\mathbf{r}$.

Let $\psi(\mathbf{x})$ be any sufficiently regular function defined on $\Gamma$. *Before* taking the limit for $\varepsilon \to 0$, we can use $\psi(\mathbf{x})$ to weigh the integral identity (1):

$$\int_\Gamma \psi(\mathbf{x}) \left\{ \int_{\Gamma - e_\varepsilon + s_\varepsilon} [V(\mathbf{x},\mathbf{y})\varphi(\mathbf{y}) - W(\mathbf{x},\mathbf{y})q(\mathbf{y})]\, ds_y \right\} ds_x = 0. \tag{3}$$

As usual the density function $\varphi$ is assumed to be $C^{(1,\alpha)}$ at $\mathbf{x}$. For the selected fundamental solution the integrals on $s_\varepsilon$ can be evaluated analytically. This step is not influenced by the presence of the weight function $\psi$ and has to be performed for the collocation BEM as well.

In principle any shape may be used for $v_\varepsilon$, since the overall result will be independent of this shape. Therefore, it is just more convenient to select a circular shape or, more precisely, $|y_\varepsilon^+ - \mathbf{x}| = |y_\varepsilon^- - \mathbf{x}| = \varepsilon$, thereby avoiding a $\ln\varepsilon$ contribution from $\left[G_{,i}(\mathbf{x},\mathbf{y})\right]_{y_\varepsilon^-}^{y_\varepsilon^+}$. The direct SGBEM formulation is thus sought as the limiting form as $\varepsilon \to 0$ of the identity

$$\begin{aligned} 0 = \int_\Gamma \psi(\mathbf{x}) &\left\{ \frac{1}{2}q(\mathbf{x}) + \varphi(\mathbf{x})t_i(\mathbf{x})\left[G_{,i}(\mathbf{x},\mathbf{y})\right]_{y_\varepsilon^-(\mathbf{x})}^{y_\varepsilon^+(\mathbf{x})} \right. \\ &\left. + \int_{\Gamma - e_\varepsilon} \left[V(\mathbf{x},\mathbf{y})\varphi(\mathbf{y}) - W(\mathbf{x},\mathbf{y})q(\mathbf{y})\right] ds_y \right\} ds_x + o(1), \end{aligned} \tag{4}$$

where $o(1)$ accounts for all vanishing contributions. It should be noted that within curly braces we have precisely the hypersingular boundary integral equation as obtained for the collocation BEM by Guiggiani et al [5].

## 3. PRELIMINARY DEFINITIONS

Let the boundary $\Gamma$ be modeled by (curvilinear) boundary elements and let $\mathbf{y} \in E$ and $\mathbf{x} \in \tilde{E}$, $E$ and $\tilde{E}$ being two such elements. In the Galerkin BEM (regardless of its symmetry) we have to deal with double integrals on $\tilde{E} \times E$.

If the two elements are *disjoint*, that is, do not share a common endpoint, the double integration is performed using ordinary means since the integrand is continuous.

On the other hand, singularities in the integrand function arise either when the two elements share one common endpoint (i.e., they are *adjacent*), or when they are *coincident* (i.e., $E = \tilde{E}$). In these cases the appropriate setting is

$$\int_{\tilde{E}} \int_{E - e_\varepsilon(\mathbf{x})} \{\cdots\} ds_y ds_x,$$

where, although not strictly necessary, we can use the following restricted definition for $e_\varepsilon(\mathbf{x})$

$$e_\varepsilon(\mathbf{x}) = \{\mathbf{y} \in E : |\mathbf{y} - \mathbf{x}| \leq \varepsilon, \text{ with } \mathbf{x} \in \tilde{E}\}, \tag{5}$$

Each boundary element is analytically defined by means of suitable parametric equations. Typically the Cartesian coordinates of a point $\mathbf{y}$ of $E$ are given by $\mathbf{y} = \mathbf{y}(\xi) = \sum_{p=1}^{N_e} N_p(\xi)\mathbf{z}^p$, where $-1 \leq \xi \leq 1$ is the parameter (or intrinsic coordinate), $N_p(\xi)$ are cardinal shape functions and $\mathbf{z}^p$ are the geometric nodes. We will denote by $\mathbf{a}(\xi)$ and $\mathbf{t}(\xi)$, respectively, the natural and normalized tangent vector to an element:

$$\mathbf{a}(\xi) = \frac{d\mathbf{y}}{d\xi} = \sum_{p=1}^{N_e} N_p'(\xi)\mathbf{z}^p, \qquad \text{so that} \qquad \mathbf{t}(\xi) = \frac{\mathbf{a}(\xi)}{|\mathbf{a}(\xi)|} = \frac{\mathbf{a}(\xi)}{a(\xi)}, \tag{6}$$

and by $\mathbf{m}(\xi)$ and $\mathbf{n}(\xi)$ the natural and normalized normal vectors:

$$\mathbf{m}(\xi) = \mathbf{e}_3 \wedge \mathbf{a}(\xi), \qquad \mathbf{n}(\xi) = \frac{\mathbf{m}(\xi)}{|\mathbf{m}(\xi)|} = \frac{\mathbf{m}(\xi)}{a(\xi)}. \tag{7}$$

Moreover we have $\mathbf{n}(\mathbf{y})ds_y = \mathbf{n}(\xi)J(\xi)d\xi = \mathbf{m}(\xi)d\xi$. Also define

$$\mathbf{b}(\xi) = \frac{d\mathbf{a}}{d\xi} = \frac{d^2\mathbf{y}}{d\xi^2} = \sum_{p=1}^{N_e} N_p''(\xi)\mathbf{z}^p. \tag{8}$$

## 4. COINCIDENT ELEMENTS

Let $\mathbf{y}$ and $\mathbf{x}$ belong to the same boundary element $E = \tilde{E}$. We will consider the hypersingular integral

$$I_C = \int_E \int_{E - e_\varepsilon(\mathbf{x})} V_{ij}(\mathbf{y} - \mathbf{x}) n_i(\mathbf{x}) n_j(\mathbf{y}) \psi(\mathbf{x}) \varphi(\mathbf{y}) ds_y ds_x, \tag{9}$$

where $V_{ij}(\mathbf{y} - \mathbf{x}) = -G_{,ij}(\mathbf{y} - \mathbf{x})$ and

$$e_\varepsilon(\mathbf{x}) = \{\mathbf{y} \in E : |\mathbf{y} - \mathbf{x}| \leq \varepsilon, \text{ with } \mathbf{x} \in E\}. \tag{10}$$

The distance vector $\mathbf{r}$ now becomes

$$\mathbf{r} = \mathbf{y}(\xi) - \mathbf{x}(\eta) = \sum_{p=1}^{N_e} \left[N_p(\xi) - N_p(\eta)\right] \mathbf{z}^p, \tag{11}$$

and $(r = 0) \Leftrightarrow (\mathbf{y} = \mathbf{x}) \Leftrightarrow (\xi = \eta)$. Therefore, in the parameter space we have to integrate over the square $(\eta, \xi) \in [-1,1] \times [-1,1]$ minus a (narrow, non uniform) strip across the $\xi = \eta$ diagonal. The strip is the image of $E \times e_\varepsilon(\mathbf{x})$.

The double integration (9) will be treated by subdividing the square $(\eta, \xi) \in [-1,1] \times [-1,1]$ into two triangular regions, labeled (1) and (2). In each region a new pair of coordinates $(u, v)$ is introduced according to the following scheme

$$\begin{aligned}
&\text{region (1):} \quad \begin{cases} \eta = u(1-v) - v \\ \xi = u(1-v) + v \end{cases} \quad \text{with } u \in [-1,1],\ v \in [0,1], \\
&\text{region (2):} \quad \begin{cases} \eta = u(1+v) - v \\ \xi = u(1+v) + v \end{cases} \quad \text{with } u \in [-1,1],\ v \in [-1,0].
\end{aligned} \tag{12}$$

It should be noted that in both cases if $v = 0$ we have $\xi = \eta = u$. Another feature of these coordinate transformations is that $u$ and $v$ vary between fixed values, that is the range of variation of each coordinate does not depend on the other one. Also useful are the relations $d\xi d\eta = 2(1 \mp v) du dv$.

In terms of the new coordinates, the distance vector $\mathbf{r}$ is

$$\mathbf{r} = \mathbf{r}(u, v) = \sum_{p=1}^{N_e} \left[N_p(\xi(u,v)) - N_p(\eta(u,v))\right] \mathbf{z}^p. \tag{13}$$

We are also interested in the Taylor expansion of the function $\mathbf{r}(u, v)$ as a function of $v$ and near $v = 0$

$$\begin{aligned}
\mathbf{r}(u, v) &= 2\mathbf{a}(u)\, v \mp \frac{1}{2} 4u\mathbf{b}(u)\, v^2 + o(v^2) \\
&= 2v\left[\mathbf{a}(u) \mp uv\mathbf{b}(u) + o(v)\right] = 2v\hat{\mathbf{r}}(u, v),
\end{aligned} \tag{14}$$

where, since $u = \xi = \eta$ when $v = 0$, $\mathbf{a}$ and $\mathbf{b}$ are precisely the functions defined in equations (6) and (8).

Similarly, the distance $r = |\mathbf{r}|$ has the following Taylor expansion

$$r = 2va(u)\left[\pm 1 - uvc(u) + o(v)\right] = \pm 2v\hat{r}(u, v), \tag{15}$$

where $a(u) = |\mathbf{a}(u)|$ and

$$c(u) = \frac{\mathbf{a} \cdot \mathbf{b}}{\mathbf{a} \cdot \mathbf{a}}(u) = \frac{a'}{a}(u). \tag{16}$$

The integrand function in equation (9) as a function of $u$ and $v$ becomes

$$V_{ij}(\mathbf{r})n_i(\mathbf{x})n_j(\mathbf{y})\psi(\mathbf{x})\varphi(\mathbf{y})ds_y ds_x = \frac{1}{v^2}F(u,v)dudv \tag{17}$$

where $F(u,v)$ is, in each triangular region, a regular function. We will use $F^{(1)}(u,v)$ and $F^{(2)}(u,v)$ to refer to the restrictions of $F(u,v)$ on subregion (1) and (2), respectively.

The direct algorithm (Guiggiani et al. [5]) relies on a two-term Taylor expansion of $F(u,v)$ around $v=0$

$$F(u,v) = F(u,0) + \frac{\partial F}{\partial v}(u,0)\, v + O(v^2). \tag{18}$$

In the present case (Laplace equation) the two terms are found to be:

$$\begin{aligned} F(u,0) &= \frac{1}{4\pi}\psi(u)\varphi(u), \\ \frac{\partial F}{\partial v}(u,0) &= \frac{1}{4\pi}\Big\{\big[\psi\varphi' - \psi'\varphi\big](u) \mp \big[u\psi\varphi\big]'(u)\Big\}. \end{aligned} \tag{19}$$

It is worth noting that $F(1,0)=0$ if $\psi(1)\varphi(1)=0$, and $F(-1,0)=0$ if $\psi(-1)\varphi(-1)=0$.

The boundary of the exclusion neighbourhood $e_\varepsilon(\mathbf{x})$ is defined by condition (10), which together with expansion (15), leads to

$$\varepsilon = \pm 2v\hat{r}(u,v) = 2va(u)\big[\pm 1 - uvc(u) + o(v)\big], \tag{20}$$

which, upon reversion, gives

$$\begin{aligned} v_\varepsilon &= \pm\alpha(\varepsilon,u) = \pm\big[\varepsilon\beta(u) + \varepsilon^2\gamma(u) + o(\varepsilon^2)\big] \\ &= \pm\frac{\varepsilon}{2a(u)}\left[1 + \frac{uc(u)}{2a(u)}\varepsilon + o(\varepsilon)\right], \end{aligned} \tag{21}$$

where

$$\beta(u) = \frac{1}{2a(u)} \qquad \text{and} \qquad \gamma(u) = \frac{uc(u)}{(2a(u))^2}. \tag{22}$$

According to equations (17) and (21), the hypersingular double integral (9) becomes in terms of the coordinates $u$ and $v$

$$\begin{aligned} I_C &= \int_E \int_{E-e_\varepsilon(\mathbf{x})} V_{ij}(\mathbf{y}-\mathbf{x})n_i(\mathbf{x})n_j(\mathbf{y})\psi(\mathbf{x})\varphi(\mathbf{y})ds_y ds_x \\ &= \int_{-1}^{1}\left\{\int_{-1}^{-\alpha(\varepsilon,u)} \frac{F^{(2)}(u,v)}{v^2}dv + \int_{\alpha(\varepsilon,u)}^{1} \frac{F^{(1)}(u,v)}{v^2}dv\right\}du. \end{aligned} \tag{23}$$

Following the direct algorithm, the first two terms of expansion (18) are added and subtracted in (23), where $F^{(1)}(u,0) = F^{(2)}(u,0)$. As typical in the direct method, all the potentially singular integrals are now trivial functions of $v$ and can be integrated analytically.

Summing up, the following form of the double integral (23) is obtained

$$
\begin{aligned}
I_C = &\int_{-1}^{1}\int_{-1}^{0}\frac{1}{v^2}\Big\{F^{(2)}(u,v) - \Big[F^{(2)}(u,0) + \frac{\partial F^{(2)}}{\partial v}(u,0)\,v\Big]\Big\}dvdu \\
&+ \int_{-1}^{1}\int_{0}^{1}\frac{1}{v^2}\Big\{F^{(1)}(u,v) - \Big[F^{(1)}(u,0) + \frac{\partial F^{(1)}}{\partial v}(u,0)\,v\Big]\Big\}dvdu \\
&+ 2\int_{-1}^{1}F(u,0)\Big[\frac{1}{\varepsilon\beta(u)} - \frac{\gamma(u)}{\beta(u)^2} - 1\Big]du \\
&+ \int_{-1}^{1}\Big[\frac{\partial F^{(2)}}{\partial v}(u,0) - \frac{\partial F^{(1)}}{\partial v}(u,0)\Big]\big[\ln|\varepsilon\beta(u)|\big]du + O(\varepsilon). \qquad (24)
\end{aligned}
$$

The distinctive feature of this expression is that all divergent terms are of $\varepsilon$-type. Of course they will eventually disappear.

# 5. ADJACENT ELEMENTS

Let $E$ and $\tilde{E}$ be two *adjacent* boundary elements, with $\mathbf{y} \in E$ and $\mathbf{x} \in \tilde{E}$. By adjacent elements we mean two elements having a common endpoint. The singularity occurs when $\mathbf{y} = \mathbf{x}$ which may happen either when $(\eta,\xi) = (-1,1)$ or when $(\eta,\xi) = (1,-1)$, depending on which element comes first on the boundary $\Gamma$.

As above, we will consider the integral

$$
I_A = \int_{\tilde{E}}\int_{E-e_\varepsilon(\mathbf{x})} V_{ij}(\mathbf{y}-\mathbf{x})\tilde{n}_i(\mathbf{x})n_j(\mathbf{y})\tilde{\psi}(\mathbf{x})\varphi(\mathbf{y})ds_y ds_x, \qquad (25)
$$

where $V_{ij}(\mathbf{y}-\mathbf{x}) = -G_{,ij}(\mathbf{y}-\mathbf{x})$. The definition of $e_\varepsilon(\mathbf{x})$ was given in (5).

The distance vector $\mathbf{r}$ is such that $(r = 0) \Leftrightarrow (\mathbf{y} = \mathbf{x}) \Leftrightarrow (\xi\eta = -1)$. To make the distinction clearer, quantities defined on $\tilde{E}$ are also marked by a tilde. Therefore, in the parameter space we have to integrate over the square $(\eta,\xi) \in [-1,1] \times [-1,1]$ minus a small region around either vertex $(\eta,\xi) = (1,-1)$ or vertex $(-1,1)$. This region is the image of $\tilde{E} \times e_\varepsilon(\mathbf{x})$.

The double integration (25) will be treated by subdividing again the square $(\eta,\xi) \in [-1,1] \times [-1,1]$ into two triangular regions, labeled (1) and (2). Moreover, the mappings $(\eta,\xi) \Leftrightarrow (u,v)$ defined in (12) could be used in this case as well.

**Singularity at** $(\eta,\xi)=(1,-1)$. In this case the integral over region (1) is not singular and we will consider in detail only the integral $I_A^{(2)}$ over region (2).

If the mappings defined in (12) are employed, the singularity would occur at $v=-1$. Therefore, it is merely a matter of convenience to shift the $v$ coordinate in (12) so that the singular point is defined by $v=0$ and thus obtaining $d\xi d\eta = 2v du dv$.

As before, we are interested in the Taylor expansion of the function $\mathbf{r}(u,v)=\mathbf{y}(\xi(u,v))-\mathbf{x}(\eta(u,v))$ as a function of $v$ near $v=0$

$$\begin{aligned}\mathbf{r}(u,v) &= [(u+1)\mathbf{a}(-1)-(u-1)\tilde{\mathbf{a}}(+1)]\,v+o(v)\\ &= 2v\mathbf{s}(u)+o(v)=2v\bar{\mathbf{r}}(u,v),\end{aligned} \tag{26}$$

where $\mathbf{a}(-1)=\mathbf{a}$ and $\tilde{\mathbf{a}}(1)=\tilde{\mathbf{a}}$ are the natural tangent vectors (6) to the elements $E$ and $\tilde{E}$ at their common endpoint. Equation (26) defines the vector $\mathbf{s}(u)$. The distance $r=|\mathbf{r}|$ has the expansion

$$r = 2vs(u)+o(v)=2v\bar{r}(u,v), \tag{27}$$

with $s(u)=|\mathbf{s}(u)|$ and $\bar{r}(u,v)=|\bar{\mathbf{r}}(u,v)|$.

The integrand function in (25) as a function of $u$ and $v$ becomes

$$V_{ij}(\mathbf{r})\tilde{n}_i(\mathbf{x})n_j(\mathbf{y})\tilde{\psi}(\mathbf{x})\varphi(\mathbf{y})ds_y ds_x = \frac{1}{v}Q(u,v)\,du dv, \tag{28}$$

where $Q(u,v)$ is a regular function.

Here a first-order Taylor expansion of $Q(u,v)$ around $v=0$ is sufficient: $Q(u,v)=Q(u,0)+o(1)$, where the term $Q(u,0)$ is easily found to be

$$Q(u,0)=\frac{1}{2}V_{ij}(\bar{\mathbf{r}}(u,0))\tilde{m}_i(1)m_j(-1)\tilde{\psi}(1)\varphi(-1). \tag{29}$$

It is worth noting that $Q(u,0)\equiv 0$ whenever $\tilde{\psi}(1)\varphi(-1)=0$. This observation has a strong relevance in the selection of suitable boundary elements in the Galerkin BEM with hypersingular kernels.

The boundary of the exclusion neighbourhood $e_\varepsilon(\mathbf{x})$ is defined by condition (5), which together with expansion (27), leads to

$$v_\varepsilon = \varepsilon\bar{\beta}(u)+o(\varepsilon)=\frac{\varepsilon}{2s(u)}+o(\varepsilon), \tag{30}$$

where $\bar{\beta}(u)=1/(2s(u))$.

According to equations (28) and (30), the singular integral (25) over adjacent elements when $E$ follows $\tilde{E}$ becomes

$$I_A = \int_{\tilde{E}} \int_{E - e_\varepsilon(\mathbf{x})} V_{ij}(\mathbf{y}-\mathbf{x}) \tilde{n}_i(\mathbf{x}) n_j(\mathbf{y}) \tilde{\psi}(\mathbf{x}) \varphi(\mathbf{y}) ds_y ds_x = I_A^{(1)} + I_A^{(2)}$$

$$= I_A^{(1)} + \int_{-1}^{1} \int_{\varepsilon/(2s(u))}^{1} \frac{Q(u,v)}{v} du dv + O(\varepsilon). \quad (31)$$

Following the direct method, the first term of expansion of $Q$ is added and subtracted in (31). Moreover, expansion (30) is considered and all singular integrals are performed analytically. The final result is

$$I_A^{(2)} = \int_0^1 \left\{ \int_{-1}^{1} \frac{Q(u,v) - Q(u,0)}{v} dv + Q(u,0) \ln |2s(u)| \right\} du$$

$$- \ln |\varepsilon| \int_{-1}^{1} Q(u,0) du + O(\varepsilon). \quad (32)$$

As in equation (24), the divergent term is of $\varepsilon$-type.

**Singularity at** $(\eta, \xi) = (-1, 1)$. This time, the singularity in the integral (25) (with $e_\varepsilon$ still defined in (5)) occurs when $(\eta, \xi) = (-1, 1)$ (i.e., $\tilde{E}$ following $E$ on $\gamma$). The integral over region (2) is then nonsingular, and essentially the same treatment as above is applied to the integral over region (1).

## 6. EVALUATION OF THE SINGULAR FREE TERM

Let us go back to the starting identity (4) for the direct SGBEM. It can be shown that the following result holds

$$\int_{\tilde{E}} \psi(\mathbf{x}) \varphi(\mathbf{x}) t_i(\mathbf{x}) G_{,i}(\mathbf{x}, \mathbf{y}_\varepsilon^+(\mathbf{x})) ds_x - \int_{\tilde{E}} \psi(\mathbf{x}) \varphi(\mathbf{x}) t_i(\mathbf{x}) G_{,i}(\mathbf{x}, \mathbf{y}_\varepsilon^-(\mathbf{x})) ds_x$$

$$= -\frac{1}{\pi\varepsilon} \int_{-1}^{1} \psi(u) \varphi(u) a(u) du + \frac{1}{2\pi} [\psi(1)\varphi(1) + \psi(-1)\varphi(-1)]$$

$$- \frac{\psi(1)\varphi(1)}{4\pi} \left[ 1 + \frac{\theta^+ \cos\theta^+}{\sin\theta^+} \right] - \frac{\psi(-1)\varphi(-1)}{4\pi} \left[ 1 + \frac{\theta^- \cos\theta^-}{\sin\theta^-} \right] + o(1) \quad (33)$$

## 7. A NUMERICAL TEST

A MATLAB program has been written to test the present direct integration method. To demonstrate its accuracy, we present a comparison between numerical and exact (analytical) values of double singular integrals of the form:

$$\int_{E+\tilde{E}} \int_{E+\tilde{E}-e_\varepsilon} W(\mathbf{x}, \mathbf{y})\psi(\mathbf{x})\varphi(\mathbf{y}) ds_y ds_x \tag{34}$$

where $E, \tilde{E}$ are straight boundary elements meeting at their common endpoint with an angle $\theta$ (figure 2) and $\psi$ and $\varphi$ are taken as the same piecewise linear 'hat function':

$$(\psi, \varphi)(\xi) = \frac{1+\xi}{2} \quad (\text{on } E) \qquad \text{and} \quad (\psi, \varphi)(\xi) = \frac{1-\xi}{2} \quad (\text{on } \tilde{E})$$

Integral (34) is scale-independent (this stems from the fact that the kernel $W(\mathbf{x}, \mathbf{y})$ is homogeneous of degree -2 with respect to the position vector $\mathbf{y} - \mathbf{x}$), and hence depends only on $\theta$ and the element length ratio $\eta = |\tilde{E}|/|E|$). The exact value (in the limiting case as $\varepsilon \to 0$) of (34) is found to be:

$$\begin{aligned} &- [2 + (\eta + \eta^{-1}) \cos\theta] \ln(2\cos\theta + \eta + \eta^{-1}) + (\eta - \eta^{-1}) \cos\theta \ln\eta \\ &\qquad - 2\sin\theta \Big[\eta \tan^{-1} \frac{\sin\theta}{(\cos\theta + eta)} + \eta^{-1} \tan^{-1} \frac{\sin\theta}{(\cos\theta + \eta^{-1})}\Big] \end{aligned}$$

The relative error, for various corner angles $\theta_k = k\pi/12$, $0 \leq k \leq 6$ (i.e. $0 \leq \theta_k \pi/2$), are obtained as follows:

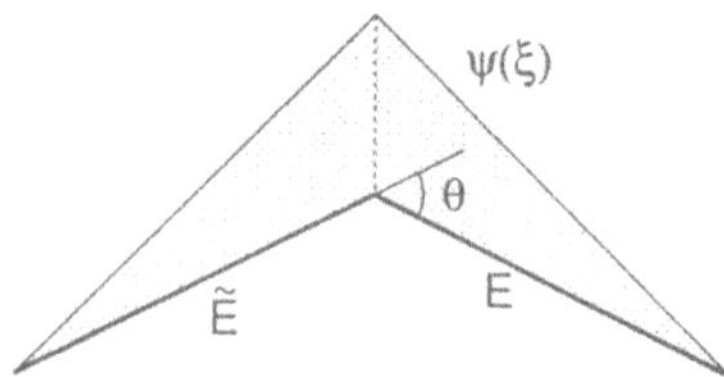


*Figure 2* Adjacent straight boundary elements $E, \tilde{E}$ and the 'hat function'.

| | $\theta\,(^0)$ | $n_G = 4$ | $n_G = 6$ | $n_G = 8$ | $n_G = 10$ |
|---|---|---|---|---|---|
| $\eta = 1$ | 0 | $3.56\,10^{-06}$ | $4.98\,10^{-09}$ | $5.96\,10^{-12}$ | $2.86\,10^{-14}$ |
| | 15 | $3.14\,10^{-06}$ | $3.38\,10^{-09}$ | $2.64\,10^{-12}$ | $2.30\,10^{-14}$ |
| | 30 | $2.52\,10^{-06}$ | $6.49\,10^{-10}$ | $-6.48\,10^{-13}$ | $2.74\,10^{-14}$ |
| | 45 | $-5.83\,10^{-06}$ | $-1.30\,10^{-08}$ | $-2.28\,10^{-11}$ | $-1.30\,10^{-14}$ |
| | 60 | $-1.65\,10^{-04}$ | $-1.20\,10^{-06}$ | $-8.03\,10^{-09}$ | $-5.09\,10^{-11}$ |
| | 75 | $-1.50\,10^{-03}$ | $-2.88\,10^{-05}$ | $-5.01\,10^{-07}$ | $-8.22\,10^{-09}$ |
| | 90 | $-9.28\,10^{-03}$ | $-3.96\,10^{-04}$ | $-1.53\,10^{-05}$ | $-5.58\,10^{-07}$ |
| $\eta = \sqrt{8}$ | 0 | $-1.82\,10^{-04}$ | $-1.12\,10^{-06}$ | $-5.96\,10^{-09}$ | $-2.78\,10^{-11}$ |
| | 15 | $4.55\,10^{-05}$ | $1.59\,10^{-06}$ | $9.97\,10^{-09}$ | $7.78\,10^{-12}$ |
| | 30 | $4.55\,10^{-04}$ | $-1.46\,10^{-06}$ | $-3.60\,10^{-08}$ | $2.18\,10^{-10}$ |
| | 45 | $1.65\,10^{-04}$ | $-1.33\,10^{-05}$ | $2.52\,10^{-07}$ | $-2.67\,10^{-09}$ |
| | 60 | $-2.43\,10^{-03}$ | $5.56\,10^{-05}$ | $-1.07\,10^{-06}$ | $1.74\,10^{-08}$ |
| | 75 | $-9.67\,10^{-03}$ | $4.31\,10^{-04}$ | $-1.63\,10^{-05}$ | $5.40\,10^{-07}$ |
| | 90 | $-2.37\,10^{-02}$ | $8.60\,10^{-04}$ | $2.90\,10^{-05}$ | $-7.60\,10^{-06}$ |

# References


[1] ANDRÄ, H., SCHNACK, E. Integration of singular Galerkin-type boundary element integrals for 3D elasticity problems. *Numerische Mathematik*, **76**, 143–165 (1997).

[2] BONNET, M. Regularized direct and indirect symmetric variational BIE formulations for three-dimensional elasticity. *Engng. Anal. with Bound. Elem.*, **15**, 93–102 (1995).

[3] BONNET, M., MAIER, G., POLIZZOTTO, C. On symmetric galerkin boundary element method. *Appl. Mech. Rev.*, **51**, 669–704 (1998).

[4] GUIGGIANI, M. Formulation and numerical treatment of boundary integral equations with hypersingular kernels. In V. Sladek (ed.), *Singular Integrals in Boundary Element Methods*, chap. 3, pp. 85–124. Comp. Mech. Publ., Southampton (1998).

[5] GUIGGIANI, M., KRISHNASAMY, G., RUDOLPHI, T. J., RIZZO, F. J. A general algorithm for the numerical solution of hypersingular boundary integral equations. *ASME J. Appl. Mech.*, **59**, 604–614 (1992).

[6] SAUTER, S. A., LAGE, C. Transformatio of hypersingular integrals and black-box cubature. Tech. Rep. 97–15, University of Kiel, Germany (1973).

# APPLICATIONS OF BOUNDARY INTEGRAL EQUATIONS FOR SOLVING SOME IDENTIFICATION PROBLEMS IN ELASTICITY

Huy Duong BUI
*Electricité de France, Division R & D ,*
*92140 Clamart, France*
*Laboratoire de Mécanique des Solides (CNRS UMR 7649),*
*Ecole Polytechnique, 91128 Palaiseau, France*
bui@lms.polytechnique.fr

Andrei CONSTANTINESCU
*Laboratoire de Mécanique des Solides (CNRS UMR 7649),*
*Ecole Polytechnique, 91128 Palaiseau, France*
constant@lms.polytechnique.fr

**Keywords:** elasticity, identification, anisotropy, observation equation

**Abstract** The identification of distributed elastic moduli of damaged materials or crack like defects in elastic media are inverse problems known as generalized elastic tomography. It consists of recovering the damaged zone or the crack in a 3D body using mechanical overdetermined boundary data. For 21 distributed unknowns which are small perturbations of elastic isotropy, a linear system of rank 5 may be derived directly from the observation equations which involves both domains and boundary integrals, with actual mechanical fields and the proposed adjoint fields. It is found that the generalization of Calderon's method in elasto-statics provided a linear system of rank 5, hence identification problems for small symmetry up to 5 elastic moduli fields could be solved. Finally, the problem of identification of a plane crack in 3D elastic body illustrates the ability of the observation equation method to provide closed form solution for the identification of the crack plane and the crack geometry.

*T. Burczynski (ed.), IUTAM/IACM/IABEM Symposium on Advanced Mathematical and Computational Mechanics Aspects of the Boundary Element Method,* 25–35.

## Introduction

The identification of spatially distributed elastic moduli is an inverse problem generally denoted as generalized elastic tomography. In a certain number of cases the origin of distributed elastic moduli is damage or microcracking. This also implies that the interior distribution is a priori unknown and generally not measurable by direct methods. One can only imagine to know certain quantities like applied forces or displacements on the boundary of the body and to identify the distribution of interior of elastic moduli from these measurements.

This problem has been discussed in the last years in a series of papers generally devoted to the linearized identification problem and a series of results have been obtained. Thus, the linearized identification problem in elasticity has been solved for the isotropic case (2 unknowns : Young and Poisson moduli fields) by Ikehata (1990), Nakamura and Uhlmann (1991). The case of general anisotropy has been discussed by Constantinescu (1994) under the assumption that the orientation angles of the material symmetries and the eigentensors are known a priori. Theses works generalize to elasticity previous results obtained for inverse electricty problems, Calderon (1980), Kohn and Vogelius (1984).

The case of general anisotropie present a series of difficulties in both electricity and elasticity. Kohn and Vogelius (1987) discussed an example of non uniqueness in anisotropic conductivity problem and Constantinescu (1994) made similar analyses for anisotropic elasticity.

The present paper is first devoted to such problems by presenting a series of integrals equations which can be used for solving elastic identification problems. The method considered here is based on the observation equation. As a starting point, we shall consider the observation equation established initially for the electrostatic linear identification problem by Calderon (1980). This equation conducts to different formulations of the identification problem and to uniqueness results in the isotropic and anisotropic cases. Finally the paper is devoted to plane crack identification in 3D elastic body. Most works in the literature are concerned with the control of the crack and the use of classical boundary integral equations with singular Green function kernels. These methods are used for numerical purposes. Here we shall rather discuss the observable equation method using appropriate adjoint fields which conducts to a closed form solution of the inverse problem, recently given by Andrieux et al (1997).

# 1. THE DIRECT AND THE INVERSE PROBLEM

Let us consider an elastic body $\Omega$ under the hypothesis of small strains and rotations. The displacement, strain and respectively stress fields, denoted by $\boldsymbol{u}, \boldsymbol{\epsilon}, \boldsymbol{\sigma}$ are subject, considering the absence of a body force, to the following set of equations:

$$\begin{aligned} \boldsymbol{\varepsilon}(\boldsymbol{u}) &= \frac{1}{2}(\nabla + \nabla^T)\boldsymbol{u} \\ \boldsymbol{\sigma} &= \boldsymbol{L} : \boldsymbol{\varepsilon} \\ \text{div } \boldsymbol{\sigma} &= 0 \end{aligned} \tag{1}$$

where $\boldsymbol{L}$ represents the forth order tensor of elastic moduli.

The *direct elasticity problem* consists of computing a solution of the system of partial differential equations (1) with known elastic moduli $\boldsymbol{L}$ and given one of the following boundary conditions on $\partial\Omega$: imposed displacements $\boldsymbol{u}|_{\partial\Omega}$ or imposed tractions $\boldsymbol{\sigma n}|_{\partial\Omega} = \boldsymbol{T}$. The pairs of corresponding boundary conditions $(\boldsymbol{u}, \boldsymbol{T})$ can be more generally described in terms of the Dirichlet-to-Neumann data map:

$$\Lambda_{\boldsymbol{L}} : \boldsymbol{u} \longrightarrow \Lambda_{\boldsymbol{L}}(\boldsymbol{u}) = \boldsymbol{T} \tag{2}$$

which maps a given boundary displacement in the corresponding boundary traction.

The *inverse elasticity problem* seeks to determine the *unknown* elastic moduli $\boldsymbol{L}$ from the partial knowledge of the Dirichlet-to-Neumann data map $\Lambda_{\boldsymbol{L}}$. This means from a series of overdetermined boundary conditions, i.e. simultaneously known displacements $\boldsymbol{u}$ and tractions $\boldsymbol{T}$.

Let us now suppose that the elastic moduli $\boldsymbol{L}$ can be expressed as:

$$\boldsymbol{L}(\boldsymbol{x}) = \boldsymbol{L}_0 + \delta\boldsymbol{L}(\boldsymbol{x}) \qquad \forall \boldsymbol{x} \in \Omega$$

where $\boldsymbol{L}_0$ is a homogenous distribution of isotropic elastic moduli and $\delta\boldsymbol{L}(\boldsymbol{x})$ is a small perturbation of $\boldsymbol{L}_0$ representing a heterogeneous distribution of anisotropic elastic moduli.

The *linearized identification problem* seeks to determine only the anisotropic perturbation $\delta\boldsymbol{L}(\boldsymbol{x})$ from the partial knowledge of the Dirichlet-to-Neumann data map.

# 2. THE OBSERVATION EQUATION

The derivation of an explicit observation equation using the boundary data $(\boldsymbol{u}, \boldsymbol{T})$ is a classical result (Ikehata (1990), Nakamura and Uhlmann (1991), Bui (1993)) and a straightforward generalization of the technique used previously in electricity (Calderon (1980), Friedmann and Vogelius (1989)).

The observation equation is based on the variational formulation of the direct problem (1) and the use of special choosen adjoint test functions.

Let us consider the adjoint elastic fields $\boldsymbol{u}_0^*$, elastic solution on $\Omega$ with elastic moduli $\boldsymbol{L}_0$ and boundary conditions given by the traction vector $\boldsymbol{T}^*$.

For the same given traction vector $\boldsymbol{T}$ on $\partial\Omega$, let us denote two actual displacement fields:

- the *perturbed* solution $\boldsymbol{u}$, computed with elastic moduli $\boldsymbol{L} = \boldsymbol{L}_0 + \delta\boldsymbol{L}$, and
- the *unperturbed* solution $\boldsymbol{u}_0$ computed with unperturbed elastic moduli $\boldsymbol{L}_0$

For the linearized inverse problem, where the unknown is the perturbation $\delta\boldsymbol{L}$, one can write two observation equations corresponding to the actual displacements fields. Thus, the nonlinear observation equation is given by:

$$\begin{aligned} -\int_\Omega \nabla\boldsymbol{u}_0^* : \delta\boldsymbol{L} : \nabla\boldsymbol{u}\, d\Omega &= \int_{\partial\Omega} (\boldsymbol{u}^* \cdot \boldsymbol{T} - \boldsymbol{u} \cdot \boldsymbol{T}^*)\, dS \qquad (3) \\ &= \int_{\partial\Omega} (\boldsymbol{u}^* \cdot \Lambda_{\boldsymbol{L}}(\boldsymbol{u}) - \boldsymbol{u} \cdot \Lambda_{\boldsymbol{L}}(\boldsymbol{u}^*))\, dS := \mathcal{D} \end{aligned}$$

while the linearized observation equation is similarly expressed as:

$$\begin{aligned} -\int_\Omega \nabla\boldsymbol{u}_0^* : \delta\boldsymbol{L} : \nabla\boldsymbol{u}_0\, d\Omega &= \int_{\partial\Omega} (\boldsymbol{u}^* \cdot \boldsymbol{T} - \boldsymbol{u} \cdot \boldsymbol{T}^*)\, dS \qquad (4) \\ &= \int_{\partial\Omega} (\boldsymbol{u}^* \cdot \Lambda_{\boldsymbol{L}}(\boldsymbol{u}) - \boldsymbol{u} \cdot \Lambda_{\boldsymbol{L}}(\boldsymbol{u}^*))\, dS := \mathcal{D} \end{aligned}$$

The right hand side $\mathcal{D}$ represents the boundary data and is in both cases the same. Its value is known from the true boundary data on the actual solid $(\boldsymbol{u}, \boldsymbol{T})$ and from the computed boundary values of the choosen adjoint fields on the unperturbed solid $(\boldsymbol{u}^*, \boldsymbol{T}^*)$.

From the general form of the observation equation it is obvious that a convenient method for computing field distributions is provided by the *boundary integrals equations.*

## 3. THE LOADING TYPES AND THE PARAMETERS

In the sequel, we shall adress the identification problem as expressed by the linearized observation equation (4).

One can remark that, the equation (4) alone is not sufficient for the identification of the anisotropic moduli tensor $\delta \boldsymbol{L}(x)$ with the 21 components $\delta L_{ijkl}(\boldsymbol{x})$. The identification depends on the different "types" of mechanical loadings provided during the experiments on the body and creating the data set: $\boldsymbol{Y} = \{\mathcal{D}_m\}_{m\in M}$.

In order to obtain identifiability, one important task is to provide data, rich enough to span the whole space of the 21 unknown functions $\boldsymbol{X} = \{\delta L_{ijkl}(\boldsymbol{x})\}$, or in other terms to construct a one-to-one mapping $\boldsymbol{X} \longrightarrow \boldsymbol{Y}$.

As it will be shown in the next sections on different examples, this task is not a straightforward one.

## 4. AN EXAMPLE OF NON-UNIQUENESS

The identifiability of the perturbation of elastic moduli $\delta \boldsymbol{L}$ from the complete knowledge of the Dirichlet-to-Neumann data map, has been proven in the case of an isotropic perturbation in a series of papers (Ikehata (1990), Nakamura and Uhlmann (1991)). Constantinescu (1994,1995) studied the case where 3 Euler angles associated with the material symmetries and 6 second order eigentensors $\boldsymbol{N}_k$, $k = 1,6$ are known (with unit norm $tr \boldsymbol{N}_k \boldsymbol{N} = 1$) and conjectured the identifiability of 6 Kelvin's eigenelastic moduli $c_k$, $\boldsymbol{L} = c_k \boldsymbol{N}_k \otimes \boldsymbol{N}_k$, from the Dirichlet-to-Neumann map. The complet anisotropic case, without apriori knowledge of the anisotropy or of the spatial distribution is to our knowledge still an open problem as presented in the sequel.

Let us suppose that a symmetric and positive definite tensor of elastic moduli $\boldsymbol{L}(\boldsymbol{x})$ is given. Is it possible to find another tensor of elastic moduli $\boldsymbol{L}'(\boldsymbol{y})$ over the same domain $\Omega$, such that the Dirichlet-to-Neumann maps coincide $\Lambda_{\boldsymbol{L}} = \Lambda_{\boldsymbol{L}'}$, for both elastic bodies ?

Based on a result on the conductivity problem by Kohn and Vogelius (1987), Constantinescu (1994) gave in certain cases a positive answer to this nonuniqueness question.

Let us choose a one-to-one differentiable map on $\Omega$, $\boldsymbol{\psi} : \boldsymbol{x} \rightarrow \boldsymbol{y} = \boldsymbol{\psi}(\boldsymbol{x})$ such that $\boldsymbol{\psi}(\boldsymbol{x}) = \boldsymbol{x}$ on the boundary $\partial\Omega$.

For an actual displacement field $\boldsymbol{u}(\boldsymbol{x})$ corresponding to a solution with elastic moduli $\boldsymbol{L}$, let us define the displacement field $\boldsymbol{u}'(\boldsymbol{y})$ as the "parallel transport" of the actual field:

$$\boldsymbol{u}'(\boldsymbol{y}) = \boldsymbol{u}(\boldsymbol{\psi}^{-1}(\boldsymbol{y})) \tag{5}$$

If we define the elastic moduli $\boldsymbol{L}'$ as:

$$\boldsymbol{L}'_{ijkl}(\boldsymbol{x}) = |\det(\boldsymbol{\psi})|^{-1} \boldsymbol{L}_{inkm}(\boldsymbol{x})\, \boldsymbol{\psi}_{n,j}(\boldsymbol{x})\, \boldsymbol{\psi}_{m,l}(\boldsymbol{x}) \tag{6}$$

we remark that $\boldsymbol{u}'$ is an elastic solution with moduli $\boldsymbol{L}'$. Moreover $\boldsymbol{u}$ and $\boldsymbol{u}'$ have the same boundary displacement, the same traction vector $\boldsymbol{T}$ and will create the same deformation energy.

It can be checked that $\boldsymbol{L}'$ is positive definite, because $\boldsymbol{L}$ is assumed positive definite and $\det(\boldsymbol{\psi}) = 0$. However, the symmetries conditions:

$$L'_{ijkl} = L'_{jikl} = L'_{klij} \tag{7}$$

have yet to be satisfied in order to obtain a complete elasticity tensor.

A closer look at the preceding system of equations (7) shows that, for a given $\boldsymbol{L}$, there are more equations than unknowns. For example, in the linear case, there are 18 linearized equations for determining the vector field $\boldsymbol{\psi}(\boldsymbol{x}) - \boldsymbol{x}$, which vanishes on the boundary. The linearized solution of (7) is generally $\boldsymbol{\psi}(\boldsymbol{x}) - \boldsymbol{x} = 0$, except for some particular $\boldsymbol{L}$. The non-uniqueness in this identification problem corresponds to some particular choice of both $\boldsymbol{L}$ and $\boldsymbol{\psi}$ satisfying (7).

In what follows, we shall restrict ourself to the case of a homogeneous isotropic $\boldsymbol{L}_0$, but we shall accept that the perturbation $\delta\boldsymbol{L}(\boldsymbol{x}) = \boldsymbol{L}(\boldsymbol{x}) - \boldsymbol{L}_0$ may be anisotropic and heterogenous.

# 5. THE LINEARIZED INVERSE PROBLEM

## 5.1. GENERALIZATION OF CALDERON'S METHOD

In order to find the unknown perturbation of the elastic moduli $\delta\boldsymbol{L}(\boldsymbol{x})$, we will assume that $\delta\boldsymbol{L}(\boldsymbol{x}) = \boldsymbol{0}$ on $\partial\Omega$, and apply the technique outlined by Calderon (1980) in electricty or Ikehata (1990) in isotropic elasticity.

Let us introduce the complex vectors $\boldsymbol{Z}, \boldsymbol{Z}^* \in \boldsymbol{C}^3$, by:

$$\boldsymbol{Z} = \frac{1}{2}(\boldsymbol{\zeta} + i\boldsymbol{\zeta}^{\perp}) \qquad\qquad \boldsymbol{Z}^* = \frac{1}{2}(\boldsymbol{\zeta} - i\boldsymbol{\zeta}^{\perp}) \tag{8}$$

where $\boldsymbol{\zeta} = (\zeta_1, \zeta_2, \zeta_3) \in \boldsymbol{R}^3$ and $\boldsymbol{\zeta}^{\perp}$ is a vector perpendicular to $\boldsymbol{\zeta}$ and having the same norm.

The elastic field $\boldsymbol{u}_0$ is introduced using the scalar harmonic field $B_0(\boldsymbol{x}) := \exp(-i\boldsymbol{Z}\cdot\boldsymbol{x})$ as follows:

$$\begin{aligned} \boldsymbol{u}_0 &:= \nabla B_0(\boldsymbol{x}), & \boldsymbol{\varepsilon}_0 &= -\boldsymbol{Z}\otimes\boldsymbol{Z}\exp(-i\boldsymbol{Z}\cdot\boldsymbol{x}), \\ \boldsymbol{\sigma}_0 &:= 2\mu_0\nabla\nabla B_0(\boldsymbol{x}), & \boldsymbol{T} &= -2\mu_0\boldsymbol{Z}(\boldsymbol{Z}\cdot\boldsymbol{n})\exp(-i\boldsymbol{Z}\cdot\boldsymbol{x}) \end{aligned} \tag{9}$$

where $\boldsymbol{n}$ is the unit outward normal on the boundary $\partial\Omega$ and $\mu_0$ is the shear moduli corresponding to $\boldsymbol{L}_0$.

Similarly the adjoint elastic field $\boldsymbol{u}_0^*$ is derived from the scalar harmonic field $B_0^*(\boldsymbol{x}) := \exp(-i\boldsymbol{Z}^* \cdot \boldsymbol{x})$;

$$\begin{aligned} \boldsymbol{u}_0^* &:= \nabla B_0^*(\boldsymbol{x}), & \boldsymbol{\varepsilon}_0^* &= -\boldsymbol{Z}^* \otimes \boldsymbol{Z}^* \exp(-i\boldsymbol{Z}^* \cdot \boldsymbol{x}), \\ \boldsymbol{\sigma}_0^* &= 2\mu_0 \nabla\nabla B_0^*(\boldsymbol{x}), & \boldsymbol{T}^* &= -2\mu_0 \boldsymbol{Z}^*(\boldsymbol{Z}^* \cdot \boldsymbol{n}) \exp(-i\boldsymbol{Z}^* \cdot \boldsymbol{x}) \end{aligned} \tag{10}$$

The linear observation equation (4) becomes:

$$\begin{aligned} \int_\Omega (\boldsymbol{Z} \otimes \boldsymbol{Z}) : \delta \boldsymbol{L}(\boldsymbol{x}) &: (\boldsymbol{Z}^* \otimes \boldsymbol{Z}^*) \exp(-i\boldsymbol{\zeta} \cdot \boldsymbol{x}) d\Omega_{\boldsymbol{x}} \qquad (11) \\ &= \int_{\partial\Omega} (\boldsymbol{u}^* \cdot \boldsymbol{T} - \boldsymbol{u} \cdot \boldsymbol{T}^*)\, dS := \mathcal{D}(\boldsymbol{\zeta}, \boldsymbol{\zeta}^\perp) \end{aligned}$$

The right hand side of (11) is the data $D(\boldsymbol{\zeta}, \boldsymbol{\zeta}^\perp)$.

Let us extend formally the definition of $\delta \boldsymbol{L}(\boldsymbol{x})$ on the whole space $\boldsymbol{R}^3$ by letting $\delta \boldsymbol{L}(\boldsymbol{x}) = \boldsymbol{0}, \boldsymbol{x} \in \boldsymbol{R}^3 \setminus \Omega$ and introduce the Fourier transform $\mathcal{F}[\delta \boldsymbol{L}](\boldsymbol{\zeta})$ of $\delta \boldsymbol{L}(\boldsymbol{x})$. Then equation (11) becomes:

$$(\boldsymbol{Z} \otimes \boldsymbol{Z}) : \mathcal{F}[\delta \boldsymbol{L}](\boldsymbol{\zeta}) : (\boldsymbol{Z}^* \otimes \boldsymbol{Z}^*) = \int_{\partial\Omega} (\boldsymbol{u}^* \cdot \boldsymbol{T} - \boldsymbol{u} \cdot \boldsymbol{T}^*)\, dS \tag{12}$$

## 5.2. THE LOADING TYPES AND PARAMETERS

In order to construct a linear system for the Fourier transform $\mathcal{F}[\delta \boldsymbol{L}](\boldsymbol{\zeta})$, let us choose a series of $\boldsymbol{\zeta}^\perp$ in the the plane perpendicular to $\boldsymbol{\zeta}$. Their position is completely determined by an angle in this plan denoted by $\phi$. An "experiment" and the corresponding adjoint field are now completly characterized by $\boldsymbol{\zeta}$ and $\phi$. Using this notation one can write equation (12) in the following form:

$$a_{ij}(\boldsymbol{\zeta};\phi) \mathcal{F}[\delta \boldsymbol{L}]_{ijhk}(\boldsymbol{\zeta}) b_{hk}(\boldsymbol{\zeta};\phi) = \mathcal{D}(\boldsymbol{\zeta};\phi). \tag{13}$$

with $a_{ij}(\boldsymbol{\zeta};\phi)$ and $b_{hk}(\boldsymbol{\zeta};\phi)$ polynomes of degree 4 in $\cos(\phi)$.

Choosing different "types" of experiments with different angles $\phi_\alpha, \alpha = 1, 2, 3, \ldots$ creates the linear system for the Fourier transform:

$$a_{ij}(\boldsymbol{\zeta};\phi_\alpha) \mathcal{F}[\delta \boldsymbol{L}]_{ijhk}(\boldsymbol{\zeta}) b_{hk}(\boldsymbol{\zeta};\phi_\alpha) = \mathcal{D}(\boldsymbol{z};\phi_\alpha) \alpha = 1, 2, 3, \ldots.$$

The general anisotropic case corresponds to 21 elastic coefficients which can be expressed by: 6 eigenelastic moduli $c_k$ corresponding to the eigentensors $\boldsymbol{N}_k$, $k = 1, \ldots, 6$, (subject to the normalization condition $\mathrm{tr}\boldsymbol{N}_k \otimes \boldsymbol{N}_k = 1$), 6 traces $\mathrm{tr}\boldsymbol{N}_k$, 6 traces $\mathrm{tr}\boldsymbol{N}_k \otimes \boldsymbol{N}_k \otimes \boldsymbol{N}_k$ and 3 Euler angles of the material symmetry, Cowin and Mehrabadi (1990). If the

Euler angles of the material symmetry and the 6 eigentensors $\boldsymbol{N}_k$ are known, the variations $\delta c_k(\boldsymbol{z})$ are determined by the $6 \times 6$ linear systems

$$\sum_{k=1}^{6} \mathcal{F}\delta c_k(\boldsymbol{\zeta})(\boldsymbol{Z}\cdot\boldsymbol{N}_k\cdot\boldsymbol{Z})(\boldsymbol{Z}^*\cdot\boldsymbol{N}_k\cdot\boldsymbol{Z}^*) = \mathcal{D}(\zeta;\phi_l), l = 1,6 \qquad (14)$$

Unfortunately, the construction (14) cannot determine more than 5 unknowns,because it can be proved that the rank of the linear system is only 5. This is due to the fact that $\boldsymbol{s}_0^* := 2\mu_0\nabla\nabla B_0^*(\boldsymbol{x})$, is a deviatoric field and thus it does not span the space of symetric second order tensors. To obtain higher rank of the linear system, more adjoint fields with non zero traces are needed. Even with the incomplete adjoint fields, the equation (14) may ensure the identifiability of some elastic moduli or orientation angles but not all, for hexagonal (5), cubic (3) and isotropic (2) symmetries. For tetragonal (6), and trigonal (6), we need complementary informations (for example one orientation angle). The identification of the orthotropic symmetry (9) is possible if the orientation angles of the material symmetry are known. Finally, the identification of the general triclinic elastic perturbation (21 unknowns), by the present method, is still an open problem, since we need new appropriate adjoint fields to complete those given above.

## 6. PLANAR CRACK IDENTIFICATION IN ELASTOSTATICS

In this section, we present the analytical solution to the inverse problem of the identification of a planar crack in 3D elasticity, given recently by Andrieux et al (1997) who derived their solution using the observation equation. The crack opening displacement discontinuity $[\![\boldsymbol{u}]\!]$ satisfies the integral equation over the unknown crack $S$ with the normal $\boldsymbol{N}$ to one crack face:

$$\begin{aligned}\int_S [\![\boldsymbol{u}]\!]\cdot\boldsymbol{L}_0:\boldsymbol{\varepsilon}(\boldsymbol{v})\cdot\boldsymbol{N}dS &= \int_{\partial\Omega}(\boldsymbol{T}(\boldsymbol{w})\cdot\boldsymbol{v} - \boldsymbol{w}\cdot\boldsymbol{L}_0:\boldsymbol{\varepsilon}(\boldsymbol{v})\cdot\boldsymbol{n})dS \qquad (15)\\ &= \mathcal{D}(\boldsymbol{v};\boldsymbol{w},\boldsymbol{T}(\boldsymbol{w})),\end{aligned}$$

where $\boldsymbol{v}(\boldsymbol{x})$ is the adjoint field corresponding to the *uncracked* body, hence a twice differentiable field. The right hand side $\mathcal{D}$ is a linear functional of $\boldsymbol{v}$, with coefficients depending on the data pair $(\boldsymbol{w},\boldsymbol{T}(\boldsymbol{w}))$. The observation equation (15) is very similar to (3). By choosing appropriately the adjoint fields, one can determine the normal $\boldsymbol{n}$, the crack plane position and what is more important, the crack geometry. The normal is determined by the adjoint fields $\boldsymbol{v}$, linear functions of the coordinates such that the corresponding stresses are constant, defined by:

$$v_k^{ij} = (\boldsymbol{L}^{-1})_{klmn}(\delta_{im}\delta_{jn} + \delta_{in}\delta_{jm})x_l \tag{16}$$

with all indexes taking values in the set $\{1, 2, 3\}$.

Two data sets $(\boldsymbol{w}_1, \boldsymbol{T}(\boldsymbol{w}_1))$ and $(\boldsymbol{w}_2, \boldsymbol{T}(\boldsymbol{w}_2))$ are required for identifying the normal $\boldsymbol{N}$ on the crack plane. The proof, more technical than difficult, can be found in Andrieux et al (1997). Once the normal is determined, a change of coordinates permits to define the axes $Ox_1x_2$ parallel to the crack plane. The position of the crack plane defined as $x_3 = c$ is then determined by the adjoint field $\boldsymbol{v}(\boldsymbol{x})$, a polynome of second order of $x_1$ and $x_2$ corresponding to the bending solution in elasticity.

Let us sketch the determination of the crack geometry with more details. The adjoint fields are defined by the following equations. One introduces the index $\boldsymbol{\zeta} = (\zeta_1, \zeta_2, 0)$ running over $\boldsymbol{R}^2$ and the family of complex vector fields $\boldsymbol{Z}$ and its conjugate $\boldsymbol{Z}^*$ belonging to the complex vectorial space $\boldsymbol{C}^3$ defined by:

$$\boldsymbol{Z} = (\boldsymbol{\zeta} + i|\boldsymbol{\zeta}|\boldsymbol{N}) \qquad\qquad \boldsymbol{Z}^* = (\boldsymbol{\zeta} - i|\boldsymbol{\zeta}|\boldsymbol{N}) \tag{17}$$

Two kinds of adjoint fields are considered :

$$\boldsymbol{v}^+(\boldsymbol{x}, \boldsymbol{\zeta}) = \nabla \exp(-i\boldsymbol{Z} \cdot \boldsymbol{x}) + \nabla \exp(-i\boldsymbol{Z}^* \cdot \boldsymbol{x}), \tag{18}$$

$$\boldsymbol{v}^-(\boldsymbol{x}, \cdot) = \nabla \exp(-i\boldsymbol{Z} \cdot \boldsymbol{x}) - \nabla \exp(-i\boldsymbol{Z}^* \cdot \boldsymbol{x}). \tag{19}$$

Equation (15) can be splitted into 2 equations for the Fourier transforms of the normal and tangential components of the discontinuity $\boldsymbol{D}(\boldsymbol{\zeta}) := [\![\boldsymbol{u}]\!](\boldsymbol{\zeta})$ .

$$\mathcal{F}[D_3](\boldsymbol{\zeta}) = \frac{1+\nu}{2E|\boldsymbol{\zeta}|^2}\mathcal{D}(\boldsymbol{v}^+(\boldsymbol{\zeta}); \boldsymbol{w}, \boldsymbol{T}(\boldsymbol{w})), \tag{20}$$

$$\zeta_1\mathcal{F}[D_1](\boldsymbol{\zeta}) + \zeta_2\mathcal{F}[D_2](\boldsymbol{\zeta}) = \frac{1+\nu}{2iE|\boldsymbol{\zeta}|}\mathcal{D}(\boldsymbol{v}^-(\boldsymbol{\zeta}); \boldsymbol{w}, \boldsymbol{T}(\boldsymbol{w})). \tag{21}$$

where $E, \nu$ are respectively the Young modulus and the Poisson ratio.

Choosing the indexes $\boldsymbol{\zeta} = (\zeta_1, 0, 0)$, $\boldsymbol{\zeta} = (0, \zeta_2, 0)$, we obtain from equation (21) the partial Fourier transforms of the tangential components of the discontinuity. And finally, the crack discontinuity, hence the crack itself, is explicitly known from the boundary data $(\boldsymbol{w}, \boldsymbol{T}(\boldsymbol{w}))$ as proven in Andrieux et al. (1997).

## Acknowledgments

This paper was partially supported by EC Contract no. ERBIC15 CT97 0706.

# References

[Alessandrini1985] Alessandrini, G. On the Identification of the Leading Coefficient of an Elliptic Equation. *Bolletino U.M.I., Annalisi funzionale et applicazioni, Serie VI*, IV-C:87–111, 1985.

[Alessandrini1988] Alessandrini, G. Stable Determination of Conductivity by Boundary Measurements. *Appl. Analysis*, 27:153–172, 1988.

[Andrieux and Ben Abda1994] Andrieux S. and Ben Abda A. Determination of planar cracks by external measurements via reciprocity map functionals: identifiability results and direct location methods. in H.D.Bui et al., editors *Inverse Problems in Engineering*, Balkema, Rotterdam/ Brookfield, 1994.

[Andrieux1992] Andrieux S. Fonctionnelles d'écart la réciprocité généralisée et identification.. *C. R. Acad. Sci. Paris*, I,315, 1992.

[Andrieux and Ben Abda and Bui1997] Andrieux, S. and Ben Abda, A. and Bui, H.D. On the Identification of Planar Crack in Elasticity Via Reciprocity Gap Concept. *C. R. Acad. Sci. Paris*, I,324:1431–1438, 1997.

[Bonnet and Constantinescu1993] Bonnet, M. and Constantinescu A. Quelques remarques sur l'identification de modules élastiques l'aide de mesures sur la frontire. *Actes du 11e Congres francais de Mecanique*, Villeneuve d'Ascq, 6-10 Sept. 1993, Presse Univ. de Lille, 1993.

[Bui1994] Bui, H.D. Introduction aux problmes inverses en mecanique des materiaux. Eyrolles, Paris, 1993 (English Version: Inverse Problems in the Mechanics of Materials. An Introduction, CRC Press, Boca Raton, 1994; Japanese Version: Shokabo, Tokyo, 1994).

[Calderon1980] Calderon, A. On an Inverse Boundary Value Problem. *Seminar on numerical analysis and its application to continuum physics*, Soc. Brasiliera de Matematica, Rio de Janeiro, pp. 65-73, 1980.

[Constantinescu1994] Constantinescu A. Sur l'identification des modules elastiques. PhD Thesis, Ecole Polytechnique, Palaiseau, France, 1994.

[Constantinescu1995a] Constantinescu A. On the identification of elastic moduli from displacement-force boundary measurements. *Int. J. of Inverse Problems in Engineering*, 1:293–315, 1995

[Cowin and Mehrababdi1990] Cowin, S.C. and Mehrabadi, M. Eigentensors of Linear Elastic Materials. *Quart. J. Mech. Appl. Math* , 43:1, 1990.

[Friedman and Vogelius1989] Friedman, A. and Vogelius, M. Determining Cracks by Boundary Measurements. *Indiana Univ. Math. J.* , 48(3), 1989.

[Grediac1992] Grediac M. and Toukourou, C. and Vautrin, A. Inverse Problem in Mechanics of Structures: A New Approach Based on Displacement Field Processing. Bui H.D. et Tanaka M., editors *In Inverse Problems in Engineering Mechanics*, IUTAM Symposium, Tokyo, Springer Verlag, 1992

[Ikehata1990] Ikehata, M. Inversion for the Linearized Problem for an Inverse Boundary Value Problem in Elastic Prospection. *SIAM J. Appl. Math.*, 50(6):1635, 1990.

[Isaacson and Isaacson1989] Isaacson, D. and Isaacson E.L. Comments on Calderon's Paper "On an Inverse Boundary Value Problem". *Math. Compt.* , 52:553, 1989.

[Kohn and Vogelius1984] Kohn, R,V. and Vogelius M. Determining Conductivity by Boundary Measurements. *Comm. Pure Appl. Math.*, 27:289–298, 1984.

[Kohn and Vogelius1987] Kohn, R,V. and Vogelius M. Relaxation of a Variational Method for Impedance Computed Tomography. *Comm. Pure Appl. Math.*, 15:745–777, 1987.

[Lions1968] Lions, J.L. *Contrôle optimal de systèmes gouvernés par des équations aux dérivées partielles.* Dunod, Paris, 1968

[Nakamura and Uhlmann1991] Nakamura, G. and Uhlmann, G. Uniqueness for Identifying Lamé Moduli by Dirichlet to Neumann Map. Yamaguti M. editor, *Inverse problems in Engineering Sciences*, ICM-90 Satellite Conference Proceedings, Springer Verlag, Tokyo, 1991.

[Sabatier1987] Sabatier, P. *Basic Methods of Tomography and Inverse Problems.* Malvern Physics Series, Adam Hilger, Bristol, 1987.

[Sun and Uhlmann1991] Sun, Z. and Uhlmann, G. *Generic Uniqueness for Determined Inverse Problems in 2D.* Yamaguti M. editor, Inverse problems in Engineering Sciences, ICM-90 Satellite Conference Proceedings, Springer Verlag, Tokyo, 1991.

# EVOLUTIONARY BEM COMPUTATION IN SHAPE OPTIMIZATION PROBLEMS

T. Burczynski[(1,2)], W.Beluch[(1)], A. Dlugosz[(1)], G.Kokot[(1)], W.Kus[(1)], and P.Orantek[(1)]

*(1) Department for Strength of Materials and Computational Mechanics, Silesian University of Technology, Konarskiego 18a, 44-100 Gliwice, Poland*

*(2) Institute of Computer Modelling, Cracow University of Technology. ul. Warszawska 24, 31-155 Krakow, Poland*

**Keywords:** Boundary Element, Evolutionary Algorithm, Shape Optimization

**Abstract** The aim of the paper is to develop of the coupling of the boundary element method (BEM) and evolutionary algorithms (EA) to shape optimization problems in applied sciences and engineering. New approaches of the evolutionary BEM computation in optimization are proposed to: (i) shape optimization of structures under statical and dynamical loading, (ii) shape optimization of structures under thermomechanical loading, (iii) shape optimization of cracked structures for criteria expressed by stress intensity factors, and (iv) shape optimization of elasto-plastic structures. Several numerical examples for optimization of 2-D structures are presented.

## 1. INTRODUCTION

The aim of the paper is to develop an application of the boundary element method and evolutionary algorithms to optimization problems. This work is an extension of previous papers in which the coupling of genetic/evolutionary algorithms and the boundary element method (BEM) has been used in shape optimization of elastic bodies [9], generalized shape optimization (shape + topology optimization) [8], optimization of cracked structures [4], shape optimization of thermoelastic [5] and elasto-plastic [7] structures and in inverse problems [3].

Evolutionary algorithms can be considered as modified and generalized classical genetic algorithm [11] in which populations of chromosomes are coded by *floating point representation* and the new modified crossover and mutation operations are introduced. The evolutionary algorithm

*T. Burczynski (ed.), IUTAM/IACM/IABEM Symposium on Advanced Mathematical and Computational Mechanics Aspects of the Boundary Element Method, 37–49.*

starts with a population of randomly generated chromosomes from a feasible solution domain. These chromosomes, which have the vector or matrix structure, evolve towards better solutions by applying genetic operators such as selection, mutation and crossover. After applying genetic operators, the new population has a better fitness. The probability of crossover and mutation does not have to be constant as in classical genetic algorithms and can change during the evolutionary process. An objective function (fitness function) with constraints plays the role of the environment to distinguish between good and bad chromosomes and select the better solution.

In order to evaluate the fitness function the direct boundary or boundary-initial value problems should be solved. In the paper the BEM [10] is proposed to solve the direct problem. Such selection has some advantages. One of the main features is the fact that the boundary element mesh generation in shape optimization does not cause difficulties because boundary element discretization is able to adopt itself to a new configuration without major distortion of the boundary element nets. To increase the efficiency and accuracy it is convenient to implement an adaptive grid refinement and incorporate an automatic generator of boundary element meshes.

In the present paper the new approaches of the evolutionary BEM computation in optimization are proposed: (i) shape optimization of structures under statical and dynamical loading, (ii) shape optimization of structures under thermo-mechanical loading, (iii) shape optimization of cracked structures, and (iv) shape optimization of elasto-plastic structures.

## 2. FORMULATION OF THE PROBLEM

A two dimensional (2-D) body, which occupies a domain $\Omega$ bounded by a boundary $\Gamma$, is considered. Boundary conditions in the form of the displacement and traction fields are prescribed. In the case of dynamical problems initial conditions are also prescribed. One should find an optimal shape of the boundary $\Gamma$ by minimizing an objective functional:

$$\min_{x} Jo(\mathbf{x}) \tag{1}$$

with constraints:

$$J_{\alpha}(\mathbf{x}) = 0, \quad \alpha = 1,2,...,A \quad and \quad J_{\beta}(\mathbf{x}) \geq 0, \beta = 1,2,...,B \tag{2}$$

where: $\mathbf{x}$ is a vector of shape design parameters.

The problem can be solved by using conventional optimization procedures, based on sensitivity information. The boundary element method was applied in several shape optimization problems in which minimization procedures used sensitivity analysis [1,2,6]. Unfortunately, approaches based on sensitivity information have some drawbacks: an objective function must be continuous, shape variation of the boundary defect should be regular, a hessian of the objective function should be positive definite, there is a large probability of convergence to a local optimum, computation starts from a single point narrowing the search domain and a choice of the starting point may influence the convergence.

Because of this difficulties there is a demand for checking other methods, free from the restrictions mentioned above. The genetic algorithms belong to these methods. Genetic algorithms are stochastic algorithms whose search methods model natural phenomena: genetic inheritance and Darvinian strife for survival. Classical genetic algorithms are based only on the fitness value information and coded chromosomes. They work on populations of solutions and use binary operators of crossover and mutation and the probability of operators is constant.

Evolutionary algorithms are modified and generalized classical genetic algorithms in which populations of chromosomes are not coded binary and *floating point representation* is used. They use modified crossover (*simple, arithmetical and heuristic*) and mutation (*uniform, boundary and non-uniform*) operations. The selection is performed in the form of the *ranking selection* or the *tournament selection* and the probability of operators can be variable.

A chromosome, which represents the design vector can be expressed in the form

$$\mathbf{x} = < x_1, x_2, ....x_i, ...x_N >, \quad i = 1, 2, ..N \tag{3}$$

where restrictions on genes $x_j$ are introduced in the form:

$$x_{iL} \leq x_i \leq x_{iR} \tag{4}$$

Genes $x_j$ represent geometry and shape of the body. Values of genes $x_j$ belong to the space of real or natural numbers.

The crossover operation swaps some chromosomes of the selected parents in order to create an offspring. Simple, arithmetical and heuristic crossover operators are applied.

*Simple crossover*

This operator needs two parents and produces two offspring. The simple crossover may produce an offspring outside the design space. To avoid this, a parameter $\alpha \in [0, 1]$ is applied. For randomly generated crossing parameter i it works as follows (chromosomes $\mathbf{x_1}$, $\mathbf{x_2}$ are parents in the vector form):

$$\mathbf{x_1} =< x_1, x_2, ...., x_i, ...x_N > \quad and \quad \mathbf{x_2} =< y_1, y_2, ...., y_i, ...y_N > \tag{5}$$

$$\begin{gathered} '\mathbf{x_1} =< x_1, ...., x_i, y_{i+1}\alpha + x_{i+1}(1-\alpha), ..., y_{N\alpha} + x_N(1-\alpha) > \\ and \\ '\mathbf{x_2} =< y_1, ...., y_i, x_{i+1}\alpha + y_{i+1}(1-\alpha), ..., x_N\alpha + y_N(1-\alpha) > \end{gathered} \tag{6}$$

$$\begin{gathered} '\mathbf{x_1} =< x_1, ...., x_i, y_{i+1}\alpha + x_{i+1}(1-\alpha), ..., y_{N\alpha} + x_N(1-\alpha) > \\ and \\ '\mathbf{x_2} =< y_1, ...., y_i, x_{i+1}\alpha + y_{i+1}(1-\alpha), ..., x_N\alpha + y_N(1-\alpha) > \end{gathered} \tag{7}$$

*Arithmetical crossover*

This operator produces two offsprings which are a linear combination of two parents

$$'\mathbf{x_1} = \alpha x_1 + (1-\alpha)x_2 \quad 'x_1 = \alpha x_2 + (1-\alpha)x_1 \tag{8}$$

*Heuristic crossover*

This operator produces a single offspring from two parents:

$$'\mathbf{x_3} = r(x_2 - x_1) + x_2 \tag{9}$$

where: $r$ is a random value from the range [0,1] and $J(x_2) \leq J(x_1)$.

Mutation operators such as uniform mutation, boundary mutation, non-uniform mutation and special gradient mutation are used:

$$\mathbf{x_1} =< x_1, x_2, ...., x_i, ...x_N > \quad '\mathbf{x_1} =< x_1, x_2, ....,' x_i, ...x_N > \tag{10}$$

*Uniform mutation*

Offspring are allowed to move freely within the feasible domain and the gene $'x_i$ takes any arbitrary value from the range $[x_{iL}, x_{iR}]$.

*Boundary mutation*

The chromosome can take only boundary values of the design space, $'\mathbf{x_i} = x_{iL}$ *or* $'\mathbf{x_i} = x_{iR}$. The boundary mutation works very well when the solution lies either on or near the boundary of the feasible search space.

*Non-uniform mutation*

This mutation operator depends on generation number t and is used to tune of the system

$$'\mathbf{x_i} = \begin{cases} x_i + \Delta(t, x_{iR} - x_i) & if \quad a \quad random \quad digit \quad is \quad 0 \\ x_i - \Delta(t, x_i - x_{iL}) & if \quad a \quad random \quad digit \quad is \quad 1 \end{cases} \tag{11}$$

where: the function $\Delta$ takes value from the range [0,y].

The objective function given by (1) plays the role of a fitness function. Evolutionary optimization is performed by using the evolutionary algorithm which minimizes the fitness function (1) with respect to shape parameters. The constraints (2) are taken into account by the penalty function method.

The flow chart of the proposed approach of the evolutionary optimization using the boundary element method is presented in Fig. 1.

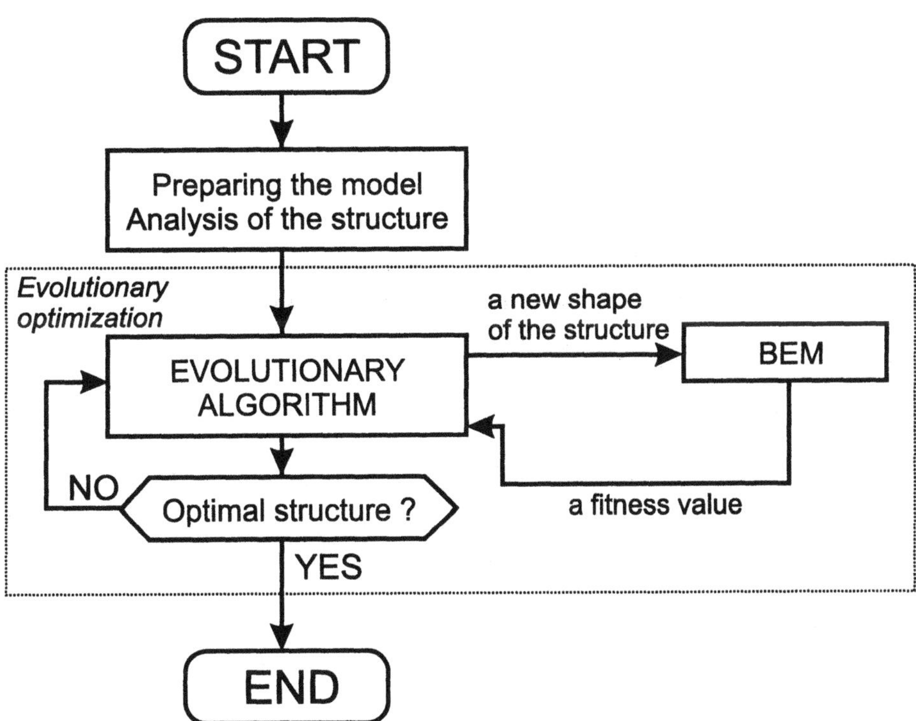


*Figure 1* Block diagram of the evolutionary optimization

Depending on the problem the various BEM approaches will be applied in evolutionary optimization.

## 3. SHAPE OPTIMIZATION OF ELASTIC STRUCTURES UNDER STATICAL AND DYNAMICAL LOADING

The problem of shape optimization is formulated as minimization of the following functionals:

$$J_0 = \int_\Omega \Psi(\sigma, \varepsilon, u) \mathrm{d}\Omega + \int_\Gamma \varphi(u, p) \mathrm{d}\Gamma \quad or \quad J_0 = \int_\Omega \mathrm{d}\Omega \tag{12}$$

with constraints $J_\alpha$ and $J_\beta$ (2) imposed on boundary stresses and boundary displacements.

The displacement field $\mathbf{u}$ is computed by solving a boundary-initial value problem of elastodynamics by the dual reciprocity boundary element method (DRBEM). The vector boundary integral equation of DRBEM used in the paper has the form:

$$\mathbf{cu} + \int_\Gamma \mathbf{P}^* \mathbf{u} \mathrm{d}\Gamma - \int_\Gamma \mathbf{U}^* \mathbf{p} \mathrm{d}\Gamma = \rho \left\{ -\mathbf{c}\Psi^k + \int_\Gamma \mathbf{U}^* \Sigma^k d\Gamma - \int_\Gamma \mathbf{P}^* \Psi^k d\Gamma \right\} \ddot{\alpha}^k \tag{13}$$

where: $\rho$ is a mass density, $\mathbf{U}^*$ and $\mathbf{P}^*$ are fundamental solutions of elastostatics, $\Psi^k$ and $\Sigma^k$ are pseudo-displacement and -traction fields, respectively, generated by $f^k$, where $\mathbf{u}(\mathbf{z}) = \alpha^k f^k(\mathbf{z}), \mathbf{z} \in \boldsymbol{\Omega}$.

In the case of elastostatics the right side of eq. (12) equals zero. NURBS or B-splines are used to modelling the geometry of the structures. Co-ordinates of control points are design variables and play the role of genes.

*Numerical example 1.*

A 2-D cantilever beam subjected to a constant point force is considered (Fig.2a). The fitness function is to minimize the compliance with the constraint condition in the form of the volume constraint. The optimal shape and topology are presented in Fig.2b. In order to generate voids in the domain the coupling the BEM and the bubble method has been applied [8].

*Numerical example 2.*

Shape optimization problem of a dynamical loaded structures (Fig.3a and b) is considered for criterion of minimum volume with stress and displacement constraints. The dot lines denote boundaries that undergo shape optimization. The final optimal shapes of the structures is shown in Fig.3c.

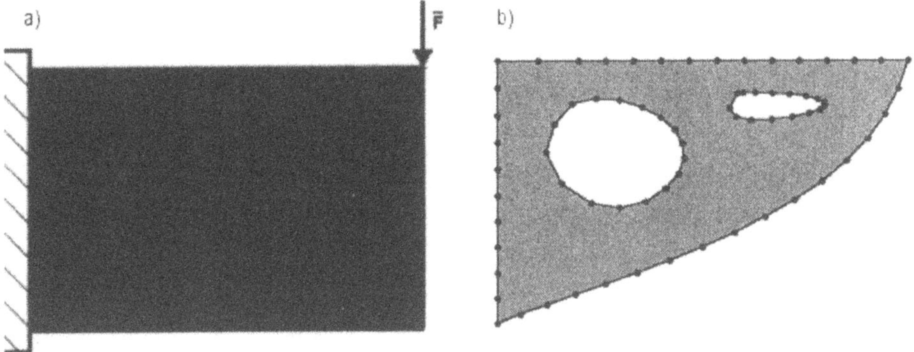


*Figure 2* The evolutionary optimization of the cantilever beam

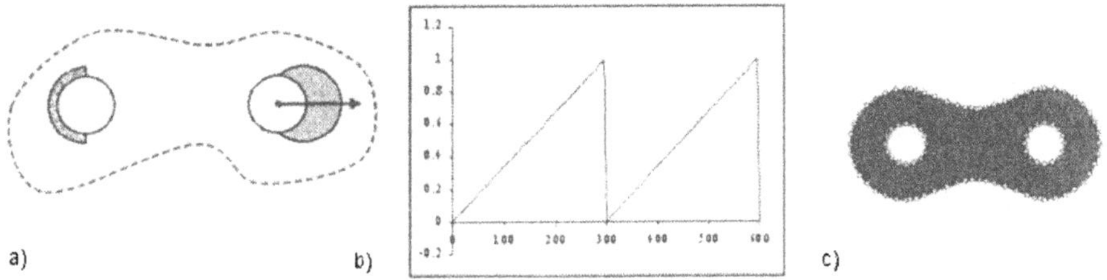


*Figure 3* Shape optimization the structure: (a) structure with boundaries undergoing shape optimization, (b) dynamical load acting upon structure, (c) optimal shape of structure

# 4. SHAPE OPTIMIZATION OF STRUCTURES UNDER THERMOMECHANICAL LOADING

In the shape optimization of thermoelastic structures a criterion of minimum equivalent stresses on the boundary is proposed:

$$J_0 = \int_{\Gamma} \left( \frac{\sigma_{eq}}{\sigma_0} \right)^n \mathrm{d}\Gamma \qquad (14)$$

where: $\sigma_{eq}$ is the equivalent von Mises stress and $\sigma_0$ is an admissible value of stresses.

In order to evaluate the fitness function (1.14) the direct boundary-value problem of linear, homogeneous, isotropic, steady-state thermoelasticity theory is solved. The boundary only integral equation for thermoelastic problem is obtained

$$\mathbf{cu} + \int_{\Gamma} \mathbf{P}^* \mathbf{u} d\Gamma = \int_{\Gamma} \mathbf{U}^* \mathbf{p} d\Gamma + \int_{\Gamma} (\mathbf{P}T - \mathbf{Q}q) \, d\Gamma \tag{15}$$

in which needed boundary temperature T and its normal derivative q are calculated by solving the boundary integral equation for the heat conduction problem:

$$\mathbf{cT} + \int_{\Gamma} \mathbf{Q}^* \mathbf{T} d\Gamma = \int_{\Gamma} \mathbf{T}^* \mathbf{q} d\Gamma \tag{16}$$

where: $\mathbf{Q}^*$ and $\mathbf{T}^*$ are fundamental solutions for the Laplace equation, $\mathbf{P}$ and $\mathbf{Q}$ are known co-ordinate functions.

*Numerical example 3.*

The plane K structure (Fig.4a) is optimized for the criterion (1.14). The Bezier curves are used to model the optimized boundary. Parameters of evolutionary algorithms are: number of chromosomes - 200; number of generations - 500. The optimal shape of the structure with conditions: $T_1 = 100^o C$, $T_2 = 0^o C$, $p = 500 kN/m$ is presented in Figure 4b.

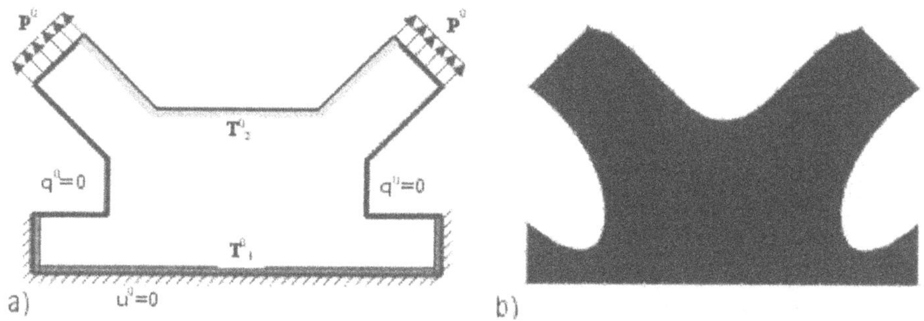


*Figure 4* The plane K structure under thermomechanical loading: a) before, b) after shape optimization

## 5. SHAPE OPTIMIZATION OF CRACKED STRUCTURES

In the shape optimization of elastic cracked structures three kinds of the criteria can be proposed:

(i) minimization of the maximum crack opening (MCO):

$$J_o \equiv MCO = max \ll u \gg = max(u^+ - u^-), \tag{17}$$

where: $u = \sqrt{u_i u_i}$ and $u^+$, $u^-$ are the displacement values of the coincident nodes laying on the opposite sides of the crack,
(ii) minimization of the reduced J-integral in the form:

$$J_0 \equiv J_{red} = \left(J_1^2 + J_2^2\right)^{1/2} \tag{18}$$

where: $J_1$ and $J_2$ are Rice J-integrals for 1 and 2 tip of the crack,
(iii) minimization of the reduced stress intensity factor in the form:

$$J_0 \equiv K_{red} = \sum_{i=1}^{n} w_i K_i \tag{19}$$

where: $K_i$ - stress intensity factors, $w_i$ - weight factors ($\Sigma w_i$=1), $n = 4$ for one internal crack (2-D problems). Co-ordinates of control points of the NURBS are used in modelling the optimal shape of the boundary.

The direct problem is solved by using the dual boundary element method (DBEM). Two vector boundary integral equation of DBEM are formulated: - for displacements

$$\mathbf{cu} + \int_{\Gamma} \mathbf{P}^* \mathbf{u} d\Gamma = \int_{\Gamma} \mathbf{U}^* d\Gamma \tag{20}$$

and for tractions

$$\frac{1}{2}\mathbf{p} + \mathbf{n} \int_{\Gamma} \mathbf{S}^* \mathbf{u} d\Gamma = n \int \mathbf{D}^* \mathbf{p} d\Gamma \tag{21}$$

where: $\mathbf{P}^*$, $\mathbf{U}^*$, $\mathbf{S}^*$ and $\mathbf{D}^*$ are fundamental solutions of elastostatics, $\mathbf{n}$ an unit outward normal vector.

*Numerical example 4.*

The boundary of a plate contained two cracks (Fig.5a) is optimized. The criterion of minimum $K_{red}$ is applied. Parameters of evolutionary algorithms are: number of generations - 1500; population - 100. Constraints on the equivalent von Mises stresses are imposed on the boundary. Optimal solution has been found in 1280 generation (Fig.5b, Table 1).

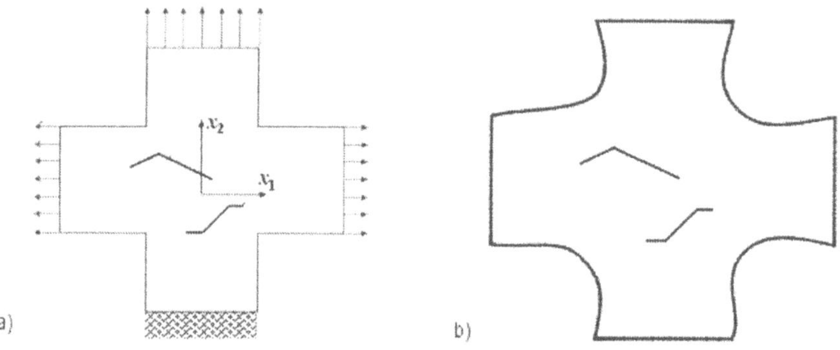


*Figure 5* Plate with two cracks: a) before, b) after optimization

*Table 1* Results of shape optimization for the plate with two cracks

| *Criteria* | *Before* | *After* |
|---|---|---|
| $K_I1, K_{II}1$ | 5.2456, 1.5965 | 4.6088, 1.2586 |
| $K_I2, K_{II}2$ | 4.3746, 3.1095 | 3.42266, 2.5185 |
| $K_I3, K_{II}3$ | 6.0637, 0.4374 | 5.4322, 0.6571 |
| $K_I4, K_{II}4$ | 4.4595, 0.1934 | 4.1093, 0.5211 |
| $K_{red}$ | 4.2679 | 3.9072 |
| $J1, J2$ | 15.6422, 15.0033 | 11.8879, 9.4548 |
| $J3, J4$ | 19.2496, 10.3772 | 15.5943, 8.936 |
| $MCO1, MCO2$ | 1.3821, 0.8829 | 1.1099, 0.8001 |
| $\sigma_{eq\,max}$ | 75.6664 | 51.2198 |

# 6. SHAPE OPTIMIZATION OF ELASTOPLASTIC STRUCTURES

The problem of shape optimization is formulated as minimization of the plastic areas which can be obtained using following functional :

$$\min_{\mathbf{x}} F(\mathbf{x}) = \int_{\Omega} \left(\frac{\sigma_a}{\sigma_0}\right)^n d\Omega \tag{22}$$

where :

$$\sigma_a = \begin{cases} \sigma & \text{when} \quad \sigma_{eq} \geq \sigma_{\text{p}} \\ 0 & \text{when} \quad \sigma_{eq} < \sigma_{\text{p}} \end{cases} \tag{23}$$

$\sigma_{eq}$ is a Huber-von Mises equivalent stress, $\sigma_0$ is a reference stress and $\sigma_p$ yield stress Constrains on maximum area of the structure and genes values are used. The displacements and stresses are computed using integral equations:

$$\mathbf{cu} = \int_{\Gamma} \mathbf{U}^* \mathbf{p} \mathrm{d}\Gamma - \int_{\Gamma} \mathbf{P}^* \mathbf{u} \mathrm{d}\Gamma + \int_{\Omega} \mathbf{U}^* \mathbf{b} \mathrm{d}\Omega + \int_{\Omega} \mathbf{T}^* \varepsilon^{\mathrm{p}} \mathrm{d}\Omega \tag{24}$$

$$\sigma(y) = \int_{\Gamma} \mathbf{D}^* \mathbf{p} \mathrm{d}\Gamma - \int_{\Gamma} \mathbf{S}^* \mathbf{u} \mathrm{d}\Gamma + \int_{\Omega} \mathbf{D}^* \mathbf{b} \mathrm{d}\Omega + \int_{\Omega} \mathbf{T}^* \varepsilon^{p} \mathrm{d}\Omega + \mathbf{I}^{\sigma} \varepsilon^{p}(y) \quad y \in \Omega \tag{25}$$

where **u** are displacements, **p** tractions, **b** body loads, $\varepsilon_p$– plastic strains $\sigma(x)$ stress value inside body, **U***, **P***, **T***, **D*** and **S*** are fundamental solutions and $I^{\sigma}$ is a coefficient.

*Numerical example 5.*

A K type structure is considered. In Fig.6a and 6b the structure after first and 275 generation are shown. The gray color is used to indicate the plastic areas.

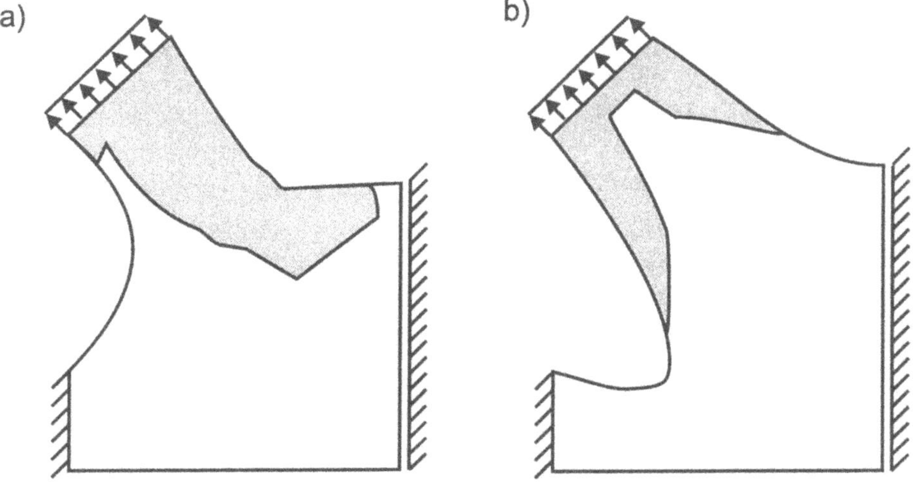


*Figure 6* The evolutionary optimization of the K type structure: a) before, b) after optimization.

# 7. CONCLUDING REMARKS

Several numerical examples of shape optimization. have been presented. Coupling of the boundary element method and the evolutionary methods appear to be very convenient and useful in such problems.

One disadvantage of the evolutionary methods is the time consuming calculation because in order to achieve a satisfactory solution one should produce many generations.

The original feature of the paper is the formulation and solution of the evolutionary BEM optimization of structures being under mechanical, thermomechanical and dynamical loading.

## Acknowledgments

This research was carried out in the framework of the KBN grant 8T11F00316.

## References

[1] T. Burczyński, Applications of BEM in sensitivity analysis and optimization *Computational Mechanics*, **13**, No. 1/2 (1993)

[2] T. Burczyński, Recent Advances in Boundary Element Approach to Design Sensitivity Analysis - Survey, Chapter 1. *Design Sensitivity Analysis* (Eds. M.Kleiber and T.Hisada), Atlanta Technology Publications, Atlanta (1993).

[3] T. Burczyński, W.Beluch, A.Długosz, P.Orantek and M.Nowakowski, Evolutionary methods in inverse problems of engineering mechanics, Proc. ISIP 2000, Nagano 2000.

[4] T. Burczyński, W. Beluch, G. Kokot, Optimization of cracked structures using boundary elements and evolutionary computation, In: *Boundary Element Techniques* (ed. M.H.Aliabadi), London (1999).

[5] T. Burczyński, A. Długosz, Evolutionary Optimization in Thermoelastic Problems using the Boundary Element Method. Proc. Symposium of the International Association for Boundary Element Methods, Brescia Italy 2000.

[6] T. Burczyński, J.H. Kane, and C. Balakrishna, Comparison of shape design sensitivity analysis via material derivative-adjoint variable and implicit differentiation techniques for 3-D and 2-D curved boundary elements, *Computer Methods in Applied Mechanics and Engineering*, **142**, 1997, pp.89-109 (1997).

[7] T. Burczyński and W. Kuś, Evolutionary methods in shape optimization of elastoplastic structures 33rd Solid Mechanics, Zakopane (2000).

[8] T. Burczyński, G. Kokot, The evolutionary optimization using genetic algorithms and boundary elements, Proc. 3rd World Congress on Structural and Multidisciplinary Opimization, Buffalo, New York, USA (1999).

[9] E. Kita, H. Tani, Shape optimization of continuum structures by genetic algorithm and boundary element method, *Engineering Analysis with Boundary Elements*, Vol. **19** (1977).

[10] M. Kleiber (ed.) *Handbook of Computational Mechanics,* Springer (1998).

[11] Z. Michalewicz, *Genetic Algorithms + Data Structures = Evolutionary Programs.* Springer-Verlag, AI Series, New York (1992).

# A REGULARIZED DIRECT SYMMETRIC GALERKIN BIE FORMULATION FOR THREE-DIMENSIONAL ELASTOPLASTICITY

Bertrand Burgardt[a], Marc Bonnet[b], Anh Le Van[a]

*(a) Laboratoire Mécanique et Matériaux, Ecole Centrale de Nantes, France*

*(b) Laboratoire de Mecanique des Solides (CNRS UMR 7649), Ecole Polytechnique, France*

Bertrand.Burgardt@ec-nantes.fr, bonnet@lms.polytechnique.fr, Anh.Le-Van@ec-nantes.fr

**Abstract** This paper deals with a symmetric regularized variational boundary/domain formulation for quasi-static 3D elastoplasticity, which is shown to express the stationarity of a certain energy functional. An implicit constitutive integration scheme is implemented. The global consistent tangent operator associated with the non-linear algebraic system of equations is shown to be symmetric.

**Keywords:** Symmetric Galerkin BEM, elastoplasticity, implicit constitutive integration

## 1. THE QUASI-STATIC ELASTOPLASTIC MODEL

Mechanical problems with material or geometrical non-linearities can be solved with BEM. In that case, the boundary integrals are supplemented with domain integrals containing plastic strains treated as unknown distributions of initial strains. Non-linear domain-BEM formulations are attractive in some specific situations like infinite media, fracture mechanics problems,... This paper addresses the formulation and numerical implementation of a symmetric Galerkin boundary integral equation (SGBIE) method for solving three-dimensional small-strain quasi-static elastic-plastic problems.

Consider a homogeneous body occupying the open domain $\Omega$, referred to a Cartesian orthogonal system. Its piecewise smooth boundary $S$ is split into two disjoint open subsets $S_u$ and $S_t$, over which histories of boundary data (displacement $\boldsymbol{u}^0(\cdot,t)$ and traction $\boldsymbol{t}^0(\cdot,t)$, respectively) are prescribed in a quasi-static manner, $t$ denoting a montonically in-

*T. Burczynski (ed.), IUTAM/IACM/IABEM Symposium on Advanced Mathematical and Computational Mechanics Aspects of the Boundary Element Method,* 51–61.

creasing time-like parameter hereafter referred to simply as 'time'. Let $\Omega_p$ denote the potentially plastic region, i.e. an open bounded subset of $\Omega$ outside of which plastic stains $\boldsymbol{\varepsilon}^p$ and internal variables $\boldsymbol{\alpha}$ are assumed to vanish.

Assuming small displacements and strains, the quasi-static evolution of the considered solid from the initial state is governed (see [9]) by (i) mechanical field equations (equilibrum, compatibility) and boundary conditions, (ii) state laws (generalized Hooke's laws) and (iii) evolution laws (plastic flow rules, consistency condition). The latter equations imply that any numerical solution procedure should involve a time-stepping scheme. In this paper, a single-step integration method [21], namely the well-known backward difference method, is used; it consists in solving an increment problem for every time step with enforcement of the constitutive equations at the step end. This procedure is nowadays quite popular and some interesting results about consistency, convergence and stability are known [18]. After discretization in time, the equations of the evolutive problem are:

$$\text{(field equations)} \qquad \begin{aligned} \operatorname{div} \boldsymbol{\sigma}_{n+1} &= \mathbf{0} \\ \boldsymbol{\varepsilon}_{n+1} = \tfrac{1}{2}(\boldsymbol{\nabla} + \boldsymbol{\nabla}^T)\boldsymbol{u}_{n+1} &\equiv \boldsymbol{\nabla}^s \boldsymbol{u}_{n+1} \end{aligned} \tag{1}$$

$$\text{(boundary cond.)} \qquad \begin{aligned} \boldsymbol{u}_{n+1} &= \boldsymbol{u}^0_{n+1} \quad \text{on } S_u \\ \boldsymbol{\sigma}_{n+1}.\boldsymbol{n} &= \boldsymbol{t}^0_{n+1} \quad \text{on } S_t \end{aligned} \tag{2}$$

$$\text{(state laws)} \qquad \begin{aligned} \boldsymbol{\sigma}_{n+1} &= \boldsymbol{C} : (\boldsymbol{\varepsilon}_{n+1} - \boldsymbol{\varepsilon}^p_{n+1}) \\ \boldsymbol{q}_{n+1} &= \frac{\partial \Theta}{\partial \boldsymbol{\alpha}} \end{aligned} \tag{3}$$

$$\text{(evolution laws)} \qquad \begin{aligned} \boldsymbol{\varepsilon}^p_{n+1} &= \boldsymbol{\varepsilon}^p_n + \lambda_{n+1} \frac{\partial \phi}{\partial \boldsymbol{\sigma}}(\boldsymbol{\sigma}_{n+1}, \boldsymbol{q}_{n+1}) \\ \boldsymbol{\alpha}_{n+1} &= \boldsymbol{\alpha}_n - \lambda_{n+1} \frac{\partial \phi}{\partial \boldsymbol{q}}(\boldsymbol{\sigma}_{n+1}, \boldsymbol{q}_{n+1}) \\ \text{with } &\lambda_{n+1}\, \phi(\boldsymbol{\sigma}_{n+1}, \boldsymbol{q}_{n+1}) = 0 \\ &\lambda_{n+1} \geq 0 \quad \phi(\boldsymbol{\sigma}_{n+1}, \boldsymbol{q}_{n+1}) \leq 0 \end{aligned} \tag{4}$$

where $\boldsymbol{\sigma}$ and $\boldsymbol{\varepsilon}$ denote stresses and total strains, $\boldsymbol{n}$ the unit normal to $S$ outward to $\Omega$, $\boldsymbol{C}$ the elastic stiffness tensor, $\boldsymbol{q}$ the thermodynamical force associated to $\boldsymbol{\alpha}$, $\Theta$ the hardening potential and $\phi$ the yield surface.

This set of relations can be solved by a FEM scheme (based on variational principles in either displacement [20], mixed [19] or Hu-Washizu [21] forms), but also by means of boundary/domain formulations.

## 2. VARIATIONAL INTEGRAL FORMULATION

Boundary element methods for quasi-static elastoplasticity based on variational principles were proposed in the late eighties [12, 14] and have since seen several implementations for two-dimensional problems [3]. Weakly singular SGBIE formulations for three-dimensional elasticity are expounded in [2] (as stationarity equations of the elastic potential energy for displacement fields in elastic equilibrium) and [10]. The former formulation is here extended to small-strain elastoplasticity.

Let $\varsigma_{n+1} = (\boldsymbol{u}_{n+1}, \boldsymbol{\sigma}_{n+1}, \boldsymbol{\varepsilon}_{n+1}, \boldsymbol{\varepsilon}^p_{n+1}, \boldsymbol{q}_{n+1}, \boldsymbol{\alpha}_{n+1})$ denote the solution to the governing relations at time $t_{n+1}$. The following representation formula holds for any displacement solving the field equations (1) and Hooke's law:

$$u_k(\boldsymbol{x}) = \int_S \tilde{t}_i(\boldsymbol{y})\ U_i^k(\boldsymbol{y}-\boldsymbol{x})\,\mathrm{d}S_y - \int_S \Sigma_{ij}^k(\boldsymbol{y}-\boldsymbol{x})n_j(\boldsymbol{y})\ \tilde{u}_i(\boldsymbol{y})\,\mathrm{d}S_y + \int_\Omega \Sigma_{ij}^k(\boldsymbol{y}-\boldsymbol{x})\ \varepsilon_{ij}^p(\boldsymbol{y})\,\mathrm{d}V_y \quad (5)$$

where $\tilde{\boldsymbol{u}}, \tilde{\boldsymbol{t}}$ are the boundary displacement and traction associated with $\boldsymbol{u}$, respectively, $\boldsymbol{U}(\boldsymbol{r})$ and $\boldsymbol{\Sigma}(\boldsymbol{r}) = \boldsymbol{C} : \boldsymbol{\nabla}\boldsymbol{U}(\boldsymbol{r})$ the Kelvin fundamental displacement and elastic stress, respectively; strains and stresses at $\boldsymbol{x}$ are then obtained upon differentiation of the above formula with respect to $\boldsymbol{x}$, considering $\boldsymbol{\varepsilon}^p$ as initial strains.

Following [2] and [16], the variational equation on the submanifold $\chi^\sigma_{n+1} = ((\tilde{\boldsymbol{u}}, \tilde{\boldsymbol{t}}), \boldsymbol{\sigma}, \boldsymbol{\varepsilon}^p, \boldsymbol{q}, \boldsymbol{\alpha})_{n+1}$ of $\varsigma_{n+1}$.

$$\begin{aligned}
&\int_{S_t} \tilde{\boldsymbol{u}}.[\boldsymbol{C}:\boldsymbol{\nabla}^s\delta\boldsymbol{u}\ ].\boldsymbol{n}\,\mathrm{d}S + \int_{S_u} \tilde{\boldsymbol{u}}^0.[\boldsymbol{C}:\boldsymbol{\nabla}^s\delta\boldsymbol{u}].\boldsymbol{n}\,\mathrm{d}S \\
&\quad - \int_{S_t} \tilde{\boldsymbol{t}}^0.\gamma_0\delta\boldsymbol{u}\,\mathrm{d}S - \int_{S_u} \tilde{\boldsymbol{t}}.\gamma_0\delta\boldsymbol{u}\,\mathrm{d}S - \int_{\Omega_p} \nabla\delta\boldsymbol{u}:\boldsymbol{C}:\boldsymbol{\varepsilon}^p\,\mathrm{d}V \\
&\quad + \int_{\Omega_p} \delta\boldsymbol{\alpha}.[\frac{\partial\Theta}{\partial\boldsymbol{\alpha}}(\boldsymbol{\alpha}) - \boldsymbol{q}]\,\mathrm{d}V + \int_{\Omega_p} \delta\boldsymbol{\varepsilon}^p:[\boldsymbol{\sigma} - \boldsymbol{C}:(\boldsymbol{\nabla}^s\boldsymbol{u} - \boldsymbol{\varepsilon}^p)]\,\mathrm{d}V \\
&\quad - \int_{\Omega_p} \delta\boldsymbol{\sigma}:[\boldsymbol{\varepsilon}^p - \boldsymbol{\varepsilon}^p_n - \lambda\ \partial_{\boldsymbol{\sigma}}\phi]\,\mathrm{d}V - \int_{\Omega_p} \delta\lambda\ \phi\,\mathrm{d}V \\
&\quad + \int_{\Omega_p} \delta\boldsymbol{q}.[-\boldsymbol{\alpha} + \boldsymbol{\alpha}_n - \lambda\ \partial_{\boldsymbol{q}}\phi]\,\mathrm{d}V = 0
\end{aligned} \quad (6)$$

holds true for any displacement variation $\delta\boldsymbol{u}$ solving the homogeneous elastic equilibrium field equation, i.e. of the form

$$\delta u_k(\boldsymbol{y}) = \int_{S_u} \delta\tilde{t}_i(\boldsymbol{x})U_i^k(x-y)\,\mathrm{d}S_x - \int_{S_t} \Sigma_{ij}^k(x-y)n_j(\boldsymbol{x})\delta\tilde{u}_i(\boldsymbol{x})\,\mathrm{d}S_x \quad (7)$$

(with $\delta\tilde{\boldsymbol{u}} \in \tilde{H}^{1/2}(S_t)$ and $\delta\tilde{\boldsymbol{t}} \in H^{-1/2}(S_u)$), stress variations $\delta\boldsymbol{\sigma} \in L^2(\Omega)$, plastic strain variations $\delta\boldsymbol{\varepsilon}^p \in L^2(\Omega)$ and variations $\delta\boldsymbol{\alpha} \in L^2(\Omega)$, $\delta\boldsymbol{q} \in L^2(\Omega)$, $\delta\lambda \in \{\gamma \in L^2(\Omega),\ \gamma \geq 0\}$. From eq. (6) on, $(\tilde{\boldsymbol{u}}, \tilde{\boldsymbol{t}}, \boldsymbol{\sigma}, \boldsymbol{\varepsilon}^p, \boldsymbol{q}, \boldsymbol{\alpha})$ are implicitly taken at $t = t_{n+1}$ and the index $n+1$ is omitted for brevity.

The substitution of (7) in (6) mandates a careful treatment, because the traces of $\delta\boldsymbol{u}$ and $[\boldsymbol{C} : \boldsymbol{\nabla}^s\delta\boldsymbol{u}].\boldsymbol{n}$ on $S$ involve non-integrable kernels. To overcome this difficulty, the following definition of the hypersingular boundary operator on $S$ is used [8]

$$\int_S \boldsymbol{\phi}.(\boldsymbol{C}:\boldsymbol{\nabla}^s\delta\boldsymbol{u}).\boldsymbol{n}\,\mathrm{d}S = \int_\Omega \boldsymbol{\nabla}^s(\gamma_0^-\boldsymbol{\phi}):\boldsymbol{C}:\boldsymbol{\nabla}^s\delta\boldsymbol{u}\,\mathrm{d}V \quad \forall\boldsymbol{\phi} \in L^2(S) \tag{8}$$

for $\delta\boldsymbol{u}$ satisfying $\operatorname{div}(\boldsymbol{C}:\boldsymbol{\nabla}^s\delta\boldsymbol{u}) = 0$ in $\Omega$ ($\gamma_0^-$: extension operator around $S$). By taking into account the following decomposition of the hypersingular kernel $C_{ijab}\Sigma^a_{kl,b}(r)$ [13]

$$C_{ijab}\Sigma^a_{kl,b}(r) = -e_{jfq}e_{lhs}B_{iqks,fh}(r) - C_{ijkl}\delta(0)$$
$$B_{iqks}(r) = \frac{G}{4\pi(1-\nu)r}[\,\delta_{qs}r_{,i}r_{,k} + \delta_{ik}\delta_{qs} - 2\nu\delta_{is}\delta_{kq} - (1-\nu)\delta_{iq}\delta_{ks}\,]$$

and applying Stokes theorem to reduce the singularity of kernels in both boundary and domain integral operators, a fully regularized variational formulation is finally obtained from (6):

$$\begin{aligned}
&a_{uu}(\delta\tilde{\boldsymbol{u}}, \tilde{\boldsymbol{u}}) + a_{ut}(\delta\tilde{\boldsymbol{u}}, \tilde{\boldsymbol{t}}) && = \ell_u(\delta\tilde{\boldsymbol{u}}) + a_{u\varepsilon}(\delta\tilde{\boldsymbol{u}}, \boldsymbol{\varepsilon}^p)\\
&a_{tu}(\delta\tilde{\boldsymbol{t}}, \tilde{\boldsymbol{u}}) + a_{tt}(\delta\tilde{\boldsymbol{t}}, \tilde{\boldsymbol{t}}) && = \ell_t(\delta\tilde{\boldsymbol{t}}) + a_{t\varepsilon}(\delta\tilde{\boldsymbol{t}}, \boldsymbol{\varepsilon}^p)\\
&\textstyle\int_{\Omega_p} \delta\boldsymbol{\varepsilon}^p:\boldsymbol{\sigma}\,\mathrm{d}V + a^{\boldsymbol{\sigma}}_{\varepsilon\varepsilon}(\delta\boldsymbol{\varepsilon}^p, \boldsymbol{\varepsilon}^p) && = \ell_\varepsilon(\delta\boldsymbol{\varepsilon}^p) + a_{\varepsilon u}(\delta\boldsymbol{\varepsilon}^p, \tilde{\boldsymbol{u}}) + a_{\varepsilon t}(\delta\boldsymbol{\varepsilon}^p, \tilde{\boldsymbol{t}})\\
& && \quad + \textit{Evolution laws in weak form}
\end{aligned} \tag{9}$$

with some properties of symmetry and sign-definiteness proved in [16]. The terms corresponding to the purely elastic response of the body can be found in [2] or [10], while the new ones are given by:

$$a_{u\varepsilon}(\delta\tilde{\boldsymbol{u}},\boldsymbol{\varepsilon}^p) = -\int_{S_t}\int_{\Omega_p} R_s\delta\tilde{u}_k(\boldsymbol{x})e_{jfq}B_{iqks,f}(r)(\boldsymbol{y}-\boldsymbol{x})\varepsilon^p_{ij}(\boldsymbol{y})\,\mathrm{d}S_y\,\mathrm{d}S_x$$

$$a_{t\varepsilon}(\delta\tilde{\boldsymbol{t}},\boldsymbol{\varepsilon}^p) = \int_{S_u}\int_{\Omega_p} \delta\tilde{t}_k(\boldsymbol{x})\Sigma^k_{ij}(\boldsymbol{y}-\boldsymbol{x})\varepsilon^p_{ij}(\boldsymbol{y})\,\mathrm{d}V_y\,\mathrm{d}S_x$$

$$\begin{aligned}a^{\sigma}_{\varepsilon\varepsilon}(\delta\boldsymbol{\varepsilon}^p,\boldsymbol{\varepsilon}^p) = &\int_{\Omega_p}\int_{\Omega_p}(e_{qjf}\delta\varepsilon^p_{ij,f})(\boldsymbol{x})B_{iqks}(\boldsymbol{y}-\boldsymbol{x})(e_{shl}\varepsilon^p_{kl,h})(\boldsymbol{y})\,\mathrm{d}V_y\,\mathrm{d}V_x\\ &-\int_{\Omega_p}\int_{S_p}(e_{qjf}\delta\varepsilon^p_{ij,f})(\boldsymbol{x})B_{iqks}(\boldsymbol{y}-\boldsymbol{x})(e_{shl}n_h\varepsilon^p_{kl})(\boldsymbol{y})\,\mathrm{d}S_y\,\mathrm{d}V_x\\ &-\int_{\Omega_p}\int_{S_p}(e_{qjf}\varepsilon^p_{ij,f})(\boldsymbol{x})B_{iqks}(\boldsymbol{y}-\boldsymbol{x})(e_{shl}n_h\delta\varepsilon^p_{kl})(\boldsymbol{y})\,\mathrm{d}S_y\,\mathrm{d}V_x\\ &+\int_{S_p}\int_{S_p}(e_{qfj}\delta\varepsilon^p_{ij}n_f)(\boldsymbol{x})B_{iqks}(\boldsymbol{y}-\boldsymbol{x})(e_{shl}\varepsilon^p_{kl}n_h)(\boldsymbol{y})\,\mathrm{d}S_y\,\mathrm{d}S_x\end{aligned}$$

Following [16], this set of variational equations is found to express the stationarity of the functional $L(\boldsymbol{\chi}^{\sigma})$ defined by

$$\begin{aligned}L(\boldsymbol{\chi}^{\sigma}) = &\frac{1}{2}a_{uu}(\tilde{\boldsymbol{u}},\tilde{\boldsymbol{u}}) + \frac{1}{2}a_{tt}(\tilde{\boldsymbol{t}},\tilde{\boldsymbol{t}}) + \frac{1}{2}a^{\sigma}_{\varepsilon\varepsilon}(\boldsymbol{\varepsilon}^p,\boldsymbol{\varepsilon}^p)\\ &+ a_{ut}(\tilde{\boldsymbol{u}},\tilde{\boldsymbol{t}}) - a_{u\varepsilon}(\tilde{\boldsymbol{u}},\boldsymbol{\varepsilon}^p) - a_{t\varepsilon}(\tilde{\boldsymbol{t}},\boldsymbol{\varepsilon}^p) - \ell_u(\tilde{\boldsymbol{u}}) - \ell_t(\tilde{\boldsymbol{t}}) - \ell_\varepsilon(\boldsymbol{\varepsilon}^p)\\ &+ \int_{\Omega_p}\boldsymbol{\sigma}:(\boldsymbol{\varepsilon}^p-\boldsymbol{\varepsilon}^p_n)\,\mathrm{d}V + \int_{\Omega_p}[\,\Theta(\boldsymbol{\alpha}) - \boldsymbol{q}.(\boldsymbol{\alpha}-\boldsymbol{\alpha}_n)\,]\,\mathrm{d}V\\ &- \int_{\Omega_p}\lambda\,\phi(\boldsymbol{\sigma},\boldsymbol{q})\,\mathrm{d}V\end{aligned}$$

Note that $L(\boldsymbol{\chi}^{\sigma})$ does not involve the total strain $\boldsymbol{\varepsilon}$, so that the present formulation might be termed an *assumed-stress method.* Some important results about consistency, stability and convergence of the associated numerical scheme has been established in [7].

Performing the change of variable $\boldsymbol{\sigma} = \boldsymbol{C} : (\boldsymbol{\varepsilon} - \boldsymbol{\varepsilon}^p)$ would yield an *assumed-strain method*, commonly used in FEM and also introduced for collocation BEM in [4]. Here, this manipulation changes $L(\boldsymbol{\chi}^{\sigma})$ into a new functional $L(\boldsymbol{\chi}^{\varepsilon})$ where $\boldsymbol{\chi}^{\varepsilon} = ((\tilde{\boldsymbol{u}},\tilde{\boldsymbol{t}}),\boldsymbol{\varepsilon},\boldsymbol{\varepsilon}^p,\boldsymbol{q},\boldsymbol{\alpha})$. The resulting assumed-strain formulation, which will be retained in the following sections, has obviously the same properties as the assumed-stress one.

# 3. LOCAL INTEGRATION

The inelastic behaviour of the body is here assumed to be described by a Von-Mises yield criterion associated to a linear isotropic and kinematic hardening rule respectively caracterized by two positive constants $h$ and $H$ (see [21] for more general classes of materials). The evolution laws resulting of the stationarity conditions of $L(\chi^\varepsilon)$ are written in a weak form as:

$$\begin{aligned}
&\int_{\Omega_p} \delta\varepsilon : \boldsymbol{C} : (\boldsymbol{\varepsilon}^p - \boldsymbol{\varepsilon}^p_n)\,\mathrm{d}V = \int_{\Omega_p} \delta\varepsilon : \boldsymbol{C} : \lambda\hat{\boldsymbol{n}}\,\mathrm{d}V \\
&\int_{\Omega_p} \delta\boldsymbol{\beta} : (\boldsymbol{\beta} - \boldsymbol{\beta}_n)\,\mathrm{d}V = \int_{\Omega_p} \delta\boldsymbol{\beta} : \frac{2H}{3}\lambda\hat{\boldsymbol{n}}\,\mathrm{d}V \\
&\int_{\Omega_p} \delta p.(p - p_n)\,\mathrm{d}V = \int_{\Omega_p} \delta p.\sqrt{\frac{2}{3}}\lambda\,\mathrm{d}V \\
&\int_{\Omega_p} \delta\lambda\phi\,\mathrm{d}V = 0
\end{aligned} \tag{10}$$

with the classical notations

$$\begin{aligned}
&\phi(\boldsymbol{\sigma} - \boldsymbol{\beta}, p) = \| \boldsymbol{s} - \boldsymbol{\beta} \| - \sqrt{\frac{2}{3}}(\sigma_0 + hp) \\
&\hat{\boldsymbol{n}} = \frac{\xi}{\| \xi \|} \quad \text{with} \quad \xi = \boldsymbol{s} - \boldsymbol{\beta} \\
&\boldsymbol{s} = \boldsymbol{s}_n + 2G(\boldsymbol{e} - \boldsymbol{e}_n - \lambda\hat{\boldsymbol{n}})
\end{aligned} \tag{11}$$

where $\boldsymbol{\beta}$ is the center of the elastic domain, $p$ the cumulated plastic strain, $\boldsymbol{s}$ and $\boldsymbol{e}$ the deviatoric stress strain, respectively.

These set of relations are solved by considering the *strain-driven Return Mapping Algorithm (RMA)*, widely used in FEM [21]. The RMA is based on a predictor-corrector scheme and allows to construct an intermediate configuration consistent with the normality rules and the Hooke's law. If $\lambda = 0$ the material remains elastic, otherwise the internal variables must be updated. $\lambda$ results from the consistency condition at $t_{n+1}$, which corresponds to the projection onto the yield surface of a trial stress defined by $\boldsymbol{\xi}^E$, and is given by:

$$\begin{aligned}
&\boldsymbol{\xi}^E = \boldsymbol{s}_n + 2G(\boldsymbol{e} - \boldsymbol{e}_n) - \boldsymbol{\beta}_n\,, \quad \hat{\boldsymbol{n}} = \frac{\boldsymbol{\xi}^E}{\| \boldsymbol{\xi}^E \|} \quad \text{and} \\
&\lambda = \frac{3}{2(3G + H + h)}\langle \phi(\boldsymbol{\xi}^E, p_n) \rangle
\end{aligned} \tag{12}$$

A similar scheme corresponding to a *stress-driven RMA* is described in [19] but the plastic multiplier $\lambda$ can not be evaluated in the case of perfect plasticity.

# 4. NUMERICAL IMPLEMENTATION

The numerical solution procedure is based upon a discretization of eqs. (9) and (10) in space. As pointed out in [15], the usual boundary element interpolation of unknown displacements and tractions on the boundary (which model the elastic structural behaviour) must be supplemented with a domain interpolation of strains (the potentially plastic region being cut into 'cells') when non-linear material behaviour is present. Similar considerations arise in collocation BEM as well [4, 6].

$$\begin{aligned}
&\boldsymbol{y}(\boldsymbol{x}) = \boldsymbol{N}(\boldsymbol{x}).\boldsymbol{Y}\ , \quad \boldsymbol{\varepsilon}(\boldsymbol{x}) = \boldsymbol{N}(\boldsymbol{x}).\boldsymbol{E}\ , \quad \boldsymbol{\varepsilon}^p(\boldsymbol{x}) = \boldsymbol{N}(\boldsymbol{x}).\boldsymbol{E}^p \\
&\boldsymbol{b}(\boldsymbol{x}) = \boldsymbol{N}(\boldsymbol{x}).\boldsymbol{B}\ , \quad p(\boldsymbol{x}) = \boldsymbol{N}(\boldsymbol{x}).\boldsymbol{P} \\
&\text{and} \quad \lambda(\boldsymbol{x}) = max(\boldsymbol{N}(\boldsymbol{x}).\boldsymbol{\Lambda}, 0)
\end{aligned} \tag{13}$$

where $(\boldsymbol{E}, \boldsymbol{E}^p, \boldsymbol{B}, \boldsymbol{P}, \boldsymbol{\Lambda})$ denotes the vector of domain nodal unknowns and $\boldsymbol{Y}$ the vector of boundary nodal densities. In this way we consider a *continuous* interpolation of the unknown boundary densities $\boldsymbol{y} = (\tilde{\boldsymbol{u}}, \tilde{\boldsymbol{t}})$, the total strain $\boldsymbol{\varepsilon}$, the plastic strain $\boldsymbol{\varepsilon}$ and the internal variables $\boldsymbol{\beta}$ and $p$, contrary to [16] or [11]. In the last cited paper, the authors used compatible plastic strain field in order to avoid domain integral calculation.

A Galerkin scheme is then applied to equations (9) and to the time-discretized plasic strain flow rule. The evolutions of $\boldsymbol{\beta}$, $p$ and the consistency condition are performed at the *nodes* of the domain mesh, which amounts to use Dirac measures at the nodes of $\Omega_p$ for the test functions in eqs. (10). The resulting non-linear algebraic system is written as

$$\boldsymbol{K}.\boldsymbol{Y}_{n+1} = \boldsymbol{F}_{n+1} + \boldsymbol{Q}.\boldsymbol{E}^p_{n+1} \tag{14}$$

$$\boldsymbol{M}.\boldsymbol{E}_{n+1} = \boldsymbol{H}_{n+1} + \boldsymbol{Q}^t.\boldsymbol{Y}_{n+1} - \boldsymbol{Z}.\boldsymbol{E}^p_{n+1} \tag{15}$$

$$\boldsymbol{M}.\boldsymbol{E}^p_{n+1} = \boldsymbol{M}.\boldsymbol{E}^p_n + \boldsymbol{\Psi}_{\boldsymbol{\varepsilon}}(\boldsymbol{E}^p_{n+1}, \boldsymbol{E}_n, \boldsymbol{E}^p_n) \tag{16}$$

*+ Evolution of internal variables and consistency condition*

with the notations

$$\boldsymbol{\Psi}_{\boldsymbol{\varepsilon}}(\boldsymbol{E}_{n+1}) = 2G \int_{\Omega_p} \boldsymbol{N}^t(\boldsymbol{x}) \lambda_{n+1}(\boldsymbol{x})\ \hat{\boldsymbol{n}}_{n+1}\, \mathrm{d}V \tag{17}$$

$$\boldsymbol{M} = \int_{\Omega_p} \boldsymbol{N}^t(\boldsymbol{x}) : \boldsymbol{C} : \boldsymbol{N}(\boldsymbol{x})\, \mathrm{d}V \tag{18}$$

Solving eq. (14) for $\boldsymbol{Y}_{n+1}$ and substituting the result into eqs. (15,16) yields a non-linear system of equations for $\boldsymbol{E}_{n+1}$ and $\boldsymbol{E}^p_{n+1}$ characterizing the material plastic constitutive behaviour:

$$\boldsymbol{M}.\boldsymbol{E}_{n+1} = \boldsymbol{G}_{n+1} - \boldsymbol{\Xi}.\boldsymbol{E}^p_{n+1} \tag{19}$$

$$\boldsymbol{M}.\boldsymbol{E}^p_{n+1} = \boldsymbol{M}.\boldsymbol{E}^p_n + \boldsymbol{\Psi}_{\boldsymbol{\varepsilon}}(\boldsymbol{E}_{n+1}, \boldsymbol{E}_n, \boldsymbol{E}^p_n) \tag{20}$$

which formally involves a nonlinear vector function $\mathcal{A}_\varepsilon$ whose tangent operator is the global consistent tangent operator (CTO) [20]:

$$\mathcal{A}_\varepsilon \begin{pmatrix} \boldsymbol{E}^p_{n+1} \\ \boldsymbol{E}_{n+1} \end{pmatrix} = \begin{pmatrix} \boldsymbol{u}^0_{n+1} \\ \boldsymbol{t}^0_{n+1} \end{pmatrix} \qquad \partial\mathcal{A}_\varepsilon = \begin{bmatrix} \Xi & \boldsymbol{M} \\ \boldsymbol{M} & \boldsymbol{\Gamma}(\boldsymbol{E}_{n+1}) \end{bmatrix}$$

$$\boldsymbol{\Gamma}(\boldsymbol{E}_{n+1}) \equiv \boldsymbol{M}.\frac{\partial \boldsymbol{E}^p_{n+1}}{\partial \boldsymbol{E}_{n+1}} = 2G \int_{\Omega_p} \boldsymbol{N}^t(\boldsymbol{x}) \frac{\partial(\lambda_{n+1}\hat{\boldsymbol{n}}_{n+1})}{\partial \boldsymbol{\varepsilon}_{n+1}} \boldsymbol{N}(\boldsymbol{x})\, \mathrm{d}V$$

The global CTO $\partial\mathcal{A}_\varepsilon$ is *symmetric* but from a numerical point of view, this formulation leads to an algebraic system of large size. Unfortunately neither $\boldsymbol{\Gamma}$ nor $\Xi$ are invertible [5, 14]. It is more interesting to eliminate the plastic strain vector $\boldsymbol{E}^p_{n+1}$ in order to get a non-linear equation for $\boldsymbol{E}_{n+1}$. Equations (19,20) then become (with $\boldsymbol{E} \equiv \boldsymbol{E}_{n+1}$)):

$$\boldsymbol{R}_\varepsilon(\boldsymbol{E}) \equiv \boldsymbol{E} + \boldsymbol{M}^{-1}\Xi\boldsymbol{M}^{-1}(\boldsymbol{\Psi}_\varepsilon(\boldsymbol{E}) + \boldsymbol{M}\boldsymbol{E}^p_n) - \boldsymbol{M}^{-1}\boldsymbol{G}_{n+1} = 0 \quad (21)$$

This equation is solved with a *Newton-Raphson method* and the associated consistent iterative scheme reads:

$$[\boldsymbol{I} + 2G\boldsymbol{M}^{-1}\Xi\boldsymbol{M}^{-1}\boldsymbol{\Gamma}]\Delta\boldsymbol{E}^i = -\boldsymbol{R}_\varepsilon(\boldsymbol{E}^i) \qquad \boldsymbol{E}^{i+1} = \boldsymbol{E}^i + \Delta\boldsymbol{E}^i \quad (22)$$

where $\boldsymbol{I} + 2G\boldsymbol{M}^{-1}\Xi\boldsymbol{M}^{-1}\boldsymbol{\Gamma}$, the global CTO corresponding to (22), is seen to be non-symmetric and fully populated because of $\Xi$. This presentation encompasses the treatment proposed in [11] which corresponds to a modified Newton-Raphson scheme with a tangent operator equal to $\boldsymbol{I}$.

Concerning the numerical evaluation of the integral operators, since the present formulation involves integrable kernels, classical Gauss quadrature rules are used for the inner integration (after removing of the singularity by Duffy's coordinates) and the outer integration (the inner potential being regular). This method is easy to implement but numerical quadrature errors entail slight loss of symmetry becaue the inner and outer integrations are not treated symmetrically [17].

## 5. NUMERICAL RESULTS

As a test example, the elastoplastic torsion problem for a cylindrical body (radius $r$ = 100 mm, height $h$ = 50 mm) is considered. A displacement-controlled twisting motion is prescribed on the plane ends. Perfect plasticity is considered; the material properties are $E$ = 200000 MPa, $\nu = 0.3$ and $\sigma_0 = 240$ MPa. The numerical model is made of 8 27-noded hexaedral cells and 48 9-noded quadrilateral boundary elements (Fig. 1).
The results are compared to the analytical solution and the load-displacement curve is plotted in Fig. 2.

*Figure 1* Discretized BE-CE model. Boundary and internal mesh

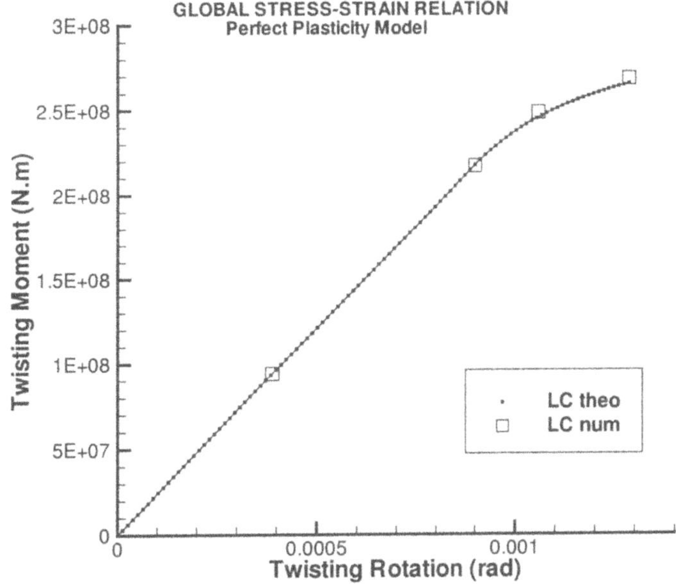


*Figure 2* Twisting moment-Angular rotation curve

# 6. CONCLUSION

A BEM-based strain formulation to quasi-static elastoplasticity evolutive analysis has been presented for general standard materials. The associated iterative scheme admits a global symmetric CTO. Some numerical aspects have now to be improved: quadrature rules taking into account the symmetry of the integration domains and parallel computing. Extensions to damage or finite strain models can be performed in the same way.

## References

[1] M. Bonnet. *Boundary integral equation methods in solids and fluids.* Wiley, 1999.

[2] M. Bonnet. Regularized direct and indirect symmetric variational bie formulations for three-dimensional elasticity. *Eng. Anal. with Bound. Elem.*, 15:93–102, 1995.

[3] M. Bonnet, G. Maier, and C. Polizzotto. Symmetric galerkin boundary element method. *Appl. Mech. Review*, 51:669–704, 1998.

[4] M. Bonnet and S. Mukherjee. Implicit bem formulations for usual and sensitivity problems in elastoplasticity using the consistent tangent operator concept. *Int. J. Solids Struct.*, 33:4461–4480, 1996.

[5] H.D. Bui and K. Dangvan. Sur le problème aux limites en vitesse des contraintes du solide élasto-plastique. *Int. J. Solids Struct.*, 6:183–193, 1970.

[6] B. Burgardt and A. Levan. A fully regularized boundary formulation for three-dimensional elastoplastic problems. In C. Brebbia, A. Kassab, and M. Choppa, editors, *Boundary Elements 20*, pages 95–103. Computational Mechanics Publications, 1998.

[7] C. Comi and G. Maier. Extremum, convergence and stability properties of the finite-increment problem in elastic-plastic boundary element method. *Int. J. Solids Struct.*, 29:249–270, 1992.

[8] M. Costabel. Boundary integral operators on lipschitz domains : elementary results. *SIAM J. Math. Anal.*, 19:613–626, 1988.

[9] J. Lemaitre and J.C. Chaboche. *Mechanics of solid materials.* Cambridge, 1994.

[10] S. Li, M. E. Mear, and L. Xiao. Symmetric weak-form integral equation method for three-dimensional fracture analysis. *Comput. Meth. in Appl. Mech. Engng.*, 51:669–704, 1998.

[11] G. Maier, S. Miccoli, G. Novati, and U. Perego. Symmetric galerkin boundary element method in plasticity and gradient plasticity. *Comput. Mech.*, 17:115–129, 1995.

[12] G. Maier and C. Polizzotto. A galerkin approach to boundary element elastoplastic analysis. *Comput. Meth. in Appl. Mech. Eng.*, 60:175–194, 1987.

[13] J.C. Nedelec. Integral equations with non-integrble kernels. *Integral equations and operator theory*, 5:562–572, 1982.

[14] C. Polizzotto. An energy approach to the boundary element method. part ii : Elastic-plastics solids. *Comput. Meth. in Appl. Mech. Engng.*, 69:263–276, 1988.

[15] C. Polizzotto, C. Panzeca, and M. Zito. A consistent boundary/interior element method for evolutive elastic-plastic structural analysis. In J.H. Kane, G. Maier, N. Tosaka, and S.N. Atluri, editors, *Advances in Boundary Element Techniques*, pages 349–369. Springer Verlag, 1993.

[16] C. Polizzotto and M. Zito. A step-wise variational approach to elastic-plastic analysis by boundary-interior elements. *Int. J. of Plasticity*, 10:81–101, 1994.

[17] S. A. Sauter and C. Lage. Transformation of hypersingular integrals and black-box cubature. Technical Report 97-17, Universität Kiel, 1997. Extended version.

[18] J. C. Simo and S. Govindjee. Non-linear b-stability and symmetry preserving return mapping algorithms for plasticity and viscoplasticity. *Int. J. for Num. Meth. in Engng.*, 31:151–176, 1991.

[19] J. C. Simo, J. G. Kennedy, and R. L. Taylor. Complementary mixed finite element formulations for elastoplasticity. *Comput. Meth. in Appl. Mech. Engng.*, 74:177–206, 1989.

[20] J. C. Simo and R. L. Taylor. Consistent tangent operators for rate independent elastoplasticity. *Comput. Meth. in Appl. Mech. Engng.*, 48:101–118, 1985.

[21] J.C. Simo and T.J.R. Hughes. *Computational Inelasticity*, volume 7 of *Interdisciplinary Applied Mathematics*. Springer Verlag, 1998.

# IMPLEMENTATION OF A SYMMETRIC GALERKIN BOUNDARY ELEMENT METHOD IN QUASI-BRITTLE FRACTURE MECHANICS

Angelo Carini
*Department of Civil Engineering, University of Brescia*
*via Branze 38, 25123 Brescia, Italy*
carini@ing.unibs.it

Alberto Salvadori
*Department of Civil Engineering, University of Brescia*
*via Branze 38, 25123 Brescia, Italy*
alberto@ing.unibs.it

**Keywords:** Variational formulations, Symmetric Galerkin BEM, cohesive fracture mechanics

**Abstract** In the context of cohesive fracture mechanics, a non-associated elastic-plastic based cohesive law is considered in modeling quasi-brittle materials. The structural problem is stated by means of boundary integral equations. In the framework of the symmetric Galerkin formulation, the non associated nature of the cohesive law implies unsymmetry of the relevant boundary operator. To overcome such a drawback, the problem is transformed into an extended equivalent one governed by a symmetric integral operator. It is shown that the new problem admits of a min-max characterization.

## 1. INTRODUCTION

A widely accepted idealized interpretation of fracture processes in quasi-brittle structures (specifically, concrete structures) rests on the "cohesive crack" model. This model is characterized by the following features: the domain $\Omega$, where the problem is defined, contains a discontinuity locus, say $\Gamma_q$, across which displacements discontinuities (or "jumps") $\mathbf{w}$ may occur; along $\Gamma_q$, tractions $\mathbf{q}$ are related to relative dis-

*T. Burczynski (ed.), IUTAM/IACM/IABEM Symposium on Advanced Mathematical and Computational Mechanics Aspects of the Boundary Element Method*, 63–73.

placements $\mathbf{w}$ by an "interface constitutive law" which exhibits, after a peak, a softening (unstable) behavior up to the vanishing of strength; in the domain $\Omega$, the material behaviour is taken as linear elastic; deformations are "small" in the sense that equilibrium relations are not influenced by configuration changes and kinematic compatibility equations are linear. At a certain stage of the structural response to a given loading history, the locus $\Gamma_q$ generally encompasses a portion ("process zone" or "craze") where the two faces interact by tractions $\mathbf{q}$.

In recent years, some plasticity-based interface models have been developed to describe the initiation and propagation of fracture under combined normal and shear stresses. In this context, the present paper is intended to provide a general variational formulation based on the Symmetric Galerkin Boundary Integral Method. After a brief illustration of elastic-plastic based interface models (Section 2), in Section 3 both the linear elastic and cohesive fracture mechanics problems are formulated in terms of symmetric Galerkin boundary integral equations. Such an approach provides in a natural way an equivalent variational formulation for the linear elastic fracture mechanics case. Due to the non-associativity of the elastic-plastic based interface model, a classical variational formulation cannot be given for cohesive fracture mechanics.

Therefore, the problem is transformed, under some additional conditions, into a new "extended" one, (Section 4) which is equivalent to a min-max problem in the original unknowns and in a new set of unknowns.

## 2. ELASTIC PLASTIC BASED INTERFACE MODELS

It has been shown [1], in double edge notched fracture specimens subjected to mixed mode loading, that the failure mode is path-dependent; as a consequence, the cohesive shear stress versus cohesive slip could not be taken as a material property [2].

Shear tractions generate crack slip and, at the same time, some induced normal crack opening (dilatancy). Experimental results [3] indicate that the higher the compressive stress, the smaller the dilatancy; furthermore for a given compressive stress, the rate of normal crack opening decreases with increasing cumulative tangential displacements.

The theory of plasticity provides an appealing framework to describe such characteristic [4], [5]. Every plasticity-based interface model, can be described by the well-known plasticity type constitutive relations [6]:

$$\dot{\mathbf{q}} = \mathbf{D}\,(\dot{\mathbf{w}} - \dot{\mathbf{w}}^{cr}) \tag{1}$$

$$\dot{\chi} = \mathbf{h}(\mathbf{q}, \chi)\dot{\lambda} \tag{2}$$

$$\dot{\mathbf{w}}^{cr} = \frac{\partial \psi}{\partial \mathbf{q}}\dot{\lambda} = \mathbf{m}\dot{\lambda} \tag{3}$$

$$\dot{\phi} = \left(\frac{\partial \phi}{\partial \mathbf{q}}\right)^T \dot{\mathbf{q}} + \left(\frac{\partial \phi}{\partial \chi}\right)^T \dot{\chi} = \mathbf{n}^T\dot{\mathbf{q}} - h\dot{\lambda} \tag{4}$$

$$\begin{cases} \dot{\lambda} \geq 0, \quad \dot{\phi} \leq 0, \quad \dot{\phi}\dot{\lambda} = 0 & \text{if } \phi = 0 \\ \qquad\quad \dot{\lambda} = 0 & \text{if } \phi < 0 \end{cases} \tag{5}$$

where

$$h = -\left(\frac{\partial \phi}{\partial \chi}\right)^T \mathbf{h}(\mathbf{q}, \chi) \tag{6}$$

Eq. (1) defines the elastic part of the interface law, $\mathbf{D}$ being the elastic stiffness matrix (usually diagonal), and states the additivity of the elastic and fracturing displacements. Eq. (2) expresses the hardening-softening laws which govern the evolution of the plastic variables, while eq. (3) expresses the non-associated flow law ($\psi \neq \phi$). Eq. (4) defines the "failure surface" in the traction space and eq. (5) expresses the loading-unloading criterion.

## 3. SYMMETRIC GALERKIN BOUNDARY INTEGRAL FORMULATION

Reference is made to an isotropic homogeneous solid in a cartesian reference system, with domain $\Omega$ and smooth boundary $\Gamma$. Quasi-static external tractions $\bar{\mathbf{p}}$ on $\Gamma_p$, displacements $\bar{\mathbf{u}}$ on $\Gamma_u$ are the imposed mixed boundary conditions. The constitutive law in $\Omega$ is assumed linear elastic, with Young modulus $E$ and Poisson coefficient $\nu$. The domain $\Omega$ contains a discontinuity locus $\Gamma_q$ for the displacement field, in order to model a single fracture ("discrete crack model"). Within $\Gamma_q$, the cohesive law relating tractions $\mathbf{q}$ to the crack opening $\mathbf{w}$ is taken as irreversible, non associated elastic-plastic based, as shortly depicted in the previous section. Assuming small strains and displacements, the solid response to boundary conditions and domain forces $\bar{\mathbf{b}}$ over $\Omega$ is analyzed.

The traditional direct boundary integral equations rest on Green's functions (kernels) (gathered in matrices $\mathbf{G}^{uu}$ and $\mathbf{G}^{pu}$) which represent components $u_i$ ($i = 1, 2, 3$) of the displacement vector $\mathbf{u}$ in a point $\mathbf{x}$ and components $p_i$ of the tractions vector $\mathbf{p}$ on a surface of normal $\mathbf{n}(\mathbf{x})$ at a point $\mathbf{x}$ due to a static discontinuity source, i.e. to a unit force concentrated in space (point $\boldsymbol{\xi}$) and acting on the unbounded elastic space $\Omega_\infty$ in direction $\omega$ ($\omega$ = 1, 2, 3). Beside these kinds of sources, the symmetric Galerkin boundary integral formulation makes also use of kinematic discontinuity sources. Namely, Green's functions (collected in matrices $\mathbf{G}^{up}$ and $\mathbf{G}^{pp}$) which describe displacement components $u_i$ and traction components $p_i$ (on a surface of normal $\mathbf{n}(\mathbf{x})$) at a point $\mathbf{x}$ due to a unit relative displacement concentrated in space at a given point $\boldsymbol{\xi}$ (crossing a surface with normal $\mathbf{n}(\boldsymbol{\xi})$) and acting in $\Omega_\infty$ in direction $\omega$ ($\omega$ = 1, 2, 3). The expressions of the introduced kernels were determined first by Kelvin and Gebbia and can be found, for instance, in [7].

By effect superposition, the *integral equations* governing the incremental cohesive fracture mechanics problem, can be written as (for details see e.g. [8], [9]) :

$$\begin{bmatrix} \int_{\Gamma_u} \mathbf{G}^{uu}[.]\,d\boldsymbol{\xi} & -\fint_{\Gamma_p} \mathbf{G}^{up}[.]\,d\boldsymbol{\xi} & \fint_{\Gamma_q} \mathbf{G}^{up}[.]\,d\boldsymbol{\xi} \\ -\fint_{\Gamma_u} \mathbf{G}^{pu}[.]\,d\boldsymbol{\xi} & \ddashint_{\Gamma_p} \mathbf{G}^{pp}[.]\,d\boldsymbol{\xi} & -\ddashint_{\Gamma_q} \mathbf{G}^{pp}[.]\,d\boldsymbol{\xi} \\ \fint_{\Gamma_u} \mathbf{G}^{pu}[.]\,d\boldsymbol{\xi} & -\ddashint_{\Gamma_p} \mathbf{G}^{pp}[.]\,d\boldsymbol{\xi} & \ddashint_{\Gamma_q} \mathbf{G}^{pp}[.]\,d\boldsymbol{\xi} \end{bmatrix} \begin{bmatrix} \dot{\mathbf{p}} \\ \dot{\mathbf{u}} \\ \dot{\mathbf{w}} \end{bmatrix} =$$

$$= \begin{bmatrix} \dot{\mathbf{f}}^u \\ \dot{\mathbf{f}}^p \\ \dot{\mathbf{f}}^q + \dot{\mathbf{q}} \end{bmatrix} \begin{matrix} \text{on } \Gamma_u \\ \text{on } \Gamma_p \\ \text{on } \Gamma_q \end{matrix} \tag{7}$$

Here $\dot{\mathbf{p}}$, $\dot{\mathbf{u}}$ and $\dot{\mathbf{w}}$ are the unknown traction rate vector on $\Gamma_u$, the unknown displacement rate vector on $\Gamma_p$ and the unknown displacement discontinuity rate vector along $\Gamma_q$. The vector-valued functions $\dot{\mathbf{f}}$, that gather all data (i.e. $\dot{\bar{p}}_i$, $\dot{\bar{u}}_i$, $\dot{\bar{b}}_i$), are as follows:

$$\dot{\mathbf{f}}^u(\mathbf{x}) \equiv \frac{1}{2}\dot{\bar{\mathbf{u}}} - \int_{\Gamma_p} \mathbf{G}^{uu}\dot{\bar{\mathbf{p}}}\,d\boldsymbol{\xi} + \fint_{\Gamma_u} \mathbf{G}^{up}\dot{\bar{\mathbf{u}}}\,d\boldsymbol{\xi} - \int_{\Omega} \mathbf{G}^{uu}\dot{\bar{\mathbf{b}}}\,d\boldsymbol{\xi} \tag{8}$$

$$\dot{\mathbf{f}}^p(\mathbf{x}) \equiv -\frac{1}{2}\dot{\bar{\mathbf{p}}} + \fint_{\Gamma_p} \mathbf{G}^{pu}\dot{\bar{\mathbf{p}}}\,d\boldsymbol{\xi} - \ddashint_{\Gamma_u} \mathbf{G}^{pp}\dot{\bar{\mathbf{u}}}\,d\boldsymbol{\xi} + \fint_{\Omega} \mathbf{G}^{pu}\dot{\bar{\mathbf{b}}}\,d\boldsymbol{\xi} \tag{9}$$

$$\dot{\mathbf{f}}^q(\mathbf{x}) \equiv -\fint_{\Gamma_p} \mathbf{G}^{pu}\dot{\bar{\mathbf{p}}}\,d\boldsymbol{\xi} + \ddashint_{\Gamma_u} \mathbf{G}^{pp}\dot{\bar{\mathbf{u}}}\,d\boldsymbol{\xi} - \fint_{\Omega} \mathbf{G}^{pu}\dot{\bar{\mathbf{b}}}\,d\boldsymbol{\xi} \tag{10}$$

In the above equations, "integrals" concerning the strongly singular kernels $\mathbf{G}^{up}$ and $\mathbf{G}^{pu}$ are understood as Cauchy principal values (and marked by $\fint$), while "integrals" involving the hypersingular kernel $\mathbf{G}^{pp}$ are understood as Hadamard finite-parts (and marked by $\ddashint$). Integral equations (7) can be written in the following more compact form:

$$\mathbf{A}\dot{\mathbf{y}} = \dot{\mathbf{f}} \tag{11}$$

with each term naturally defined by comparison.

As a consequence of a well-known reciprocity relationship for the Green functions (see e.g. [10]), the *linear elastic fracture mechanics problem* (described by eqs. (7) when $\dot{\mathbf{q}} = \mathbf{0}$) admits of a variational formulation.

To this aim, consider the following bilinear form defined over the boundary $\Gamma$ ( $\mathrm{d}\mathbf{x}$ stands for $\mathrm{d}\Gamma(\mathbf{x})$ or $\mathrm{d}\Omega(\mathbf{x})$):

$$\langle \mathbf{g}, \mathbf{y}' \rangle \equiv \int_{\Gamma} \mathbf{g}^T \mathbf{y}' \, \mathrm{d}\mathbf{x} \qquad \text{with } \mathbf{g} \in R(\mathbf{A}) \text{ and } \mathbf{y}' \in D(\mathbf{A}) \tag{12}$$

with $\mathbf{g}$ and $\mathbf{y}'$ regular enough to confer a meaning to the integrations. $R(\mathbf{A})$ and $D(\mathbf{A})$ stands for the range and the domain of the operator $\mathbf{A}$, respectively. When $\mathbf{g} = \mathbf{A}\mathbf{y}$ is taken, eq. (12) defines the bilinear form associated to the operator $\mathbf{A}$:

$$\langle \mathbf{A}\mathbf{y}, \mathbf{y}' \rangle = \int_{\Gamma} (\mathbf{A}\mathbf{y})^T \mathbf{y}' \, \mathrm{d}\mathbf{x} \qquad \text{with } \mathbf{y},\ \mathbf{y}' \in D(\mathbf{A}) \tag{13}$$

It is possible to prove that the integral operator $\mathbf{A}$ is symmetric with respect to the bilinear form (13); namely

$$\langle \mathbf{A}\mathbf{y}, \mathbf{y}' \rangle = \langle \mathbf{A}\mathbf{y}', \mathbf{y} \rangle \qquad \forall \quad \mathbf{y}, \mathbf{y}' \in D(\mathbf{A}) \tag{14}$$

As a consequence, displacement rate vector $\dot{\mathbf{u}}$ on the free boundary $\Gamma_p$, traction rate vector $\dot{\mathbf{p}}$ on the constrained boundary $\Gamma_u$ and crack opening displacement rate vector $\dot{\mathbf{w}}$ on the boundary $\Gamma_q$ (which solve the linear elastic fracture problem in rates) are characterized by the stationarity of the following quadratic functional:

$$\begin{aligned}
\Psi[\dot{\mathbf{u}}, \dot{\mathbf{p}}, \dot{\mathbf{w}}] &= \frac{1}{2}\langle \mathbf{A}\dot{\mathbf{y}}, \dot{\mathbf{y}} \rangle - \langle \dot{\mathbf{f}}, \dot{\mathbf{y}} \rangle \\
&= \frac{1}{2}\int_{\Gamma_u} \dot{\mathbf{p}}(\mathbf{x}) \int_{\Gamma_u} \mathbf{G}^{uu}(\mathbf{x}-\boldsymbol{\xi})\dot{\mathbf{p}}(\boldsymbol{\xi})\,\mathrm{d}\boldsymbol{\xi}\,\mathrm{d}\mathbf{x} - \frac{1}{2}\int_{\Gamma_u} \dot{\mathbf{p}}(\mathbf{x}) \fint_{\Gamma_p} \mathbf{G}^{up}(\mathbf{x}-\boldsymbol{\xi})\dot{\mathbf{u}}(\boldsymbol{\xi})\,\mathrm{d}\boldsymbol{\xi}\,\mathrm{d}\mathbf{x} \\
&+\frac{1}{2}\int_{\Gamma_u} \dot{\mathbf{p}}(\mathbf{x}) \fint_{\Gamma_q} \mathbf{G}^{up}(\mathbf{x}-\boldsymbol{\xi})\dot{\mathbf{w}}(\boldsymbol{\xi})\,\mathrm{d}\boldsymbol{\xi}\,\mathrm{d}\mathbf{x} - \frac{1}{2}\int_{\Gamma_p} \dot{\mathbf{u}}(\mathbf{x}) \fint_{\Gamma_u} \mathbf{G}^{pu}(\mathbf{x}-\boldsymbol{\xi})\dot{\mathbf{p}}(\boldsymbol{\xi})\,\mathrm{d}\boldsymbol{\xi}\,\mathrm{d}\mathbf{x} \\
&+\frac{1}{2}\int_{\Gamma_p} \dot{\mathbf{u}}(\mathbf{x}) \mathop{=\!\!\!\!\!\!\int}_{\Gamma_p} \mathbf{G}^{pp}(\mathbf{x}-\boldsymbol{\xi})\dot{\mathbf{u}}(\boldsymbol{\xi})\,\mathrm{d}\boldsymbol{\xi}\,\mathrm{d}\mathbf{x} - \frac{1}{2}\int_{\Gamma_p} \dot{\mathbf{u}}(\mathbf{x}) \mathop{=\!\!\!\!\!\!\int}_{\Gamma_q} \mathbf{G}^{pp}(\mathbf{x}-\boldsymbol{\xi})\dot{\mathbf{w}}(\boldsymbol{\xi})\,\mathrm{d}\boldsymbol{\xi}\,\mathrm{d}\mathbf{x} \\
&+\frac{1}{2}\int_{\Gamma_q} \dot{\mathbf{w}}(\mathbf{x}) \fint_{\Gamma_u} \mathbf{G}^{pu}(\mathbf{x}-\boldsymbol{\xi})\dot{\mathbf{p}}(\boldsymbol{\xi})\,\mathrm{d}\boldsymbol{\xi}\,\mathrm{d}\mathbf{x} - \frac{1}{2}\int_{\Gamma_q} \dot{\mathbf{w}}(\mathbf{x}) \mathop{=\!\!\!\!\!\!\int}_{\Gamma_p} \mathbf{G}^{pp}(\mathbf{x}-\boldsymbol{\xi})\dot{\mathbf{u}}(\boldsymbol{\xi})\,\mathrm{d}\boldsymbol{\xi}\,\mathrm{d}\mathbf{x} \\
&+\frac{1}{2}\int_{\Gamma_q} \dot{\mathbf{w}}(\mathbf{x}) \mathop{=\!\!\!\!\!\!\int}_{\Gamma_q} \mathbf{G}^{pp}(\mathbf{x}-\boldsymbol{\xi})\dot{\mathbf{w}}(\boldsymbol{\xi})\,\mathrm{d}\boldsymbol{\xi}\,\mathrm{d}\mathbf{x} - \int_{\Gamma_u} \dot{\mathbf{p}}(\mathbf{x})\,\dot{\mathbf{f}}^{u}(\mathbf{x})\,\mathrm{d}\mathbf{x} - \int_{\Gamma_p} \dot{\mathbf{u}}(\mathbf{x})\,\dot{\mathbf{f}}^{p}(\mathbf{x})\,\mathrm{d}\mathbf{x} \\
&- \int_{\Gamma_q} \dot{\mathbf{w}}(\mathbf{x})\,\dot{\mathbf{f}}^{q}(\mathbf{x})\,\mathrm{d}\mathbf{x}
\end{aligned} \tag{15}$$

The stationary point is a *saddle-point* (minimum with respect to $\dot{\mathbf{p}}$ and maximum with respect to both $\dot{\mathbf{u}}$ and $\dot{\mathbf{w}}$ ).

Considering now a cohesive fracture mechanics problem, the constitutive law along the fracture process zone of $\Gamma_q$ must be added to the equation (7) to make the problem complete. For the sake of brevity, the fracture process zone is assumed as coincident with the whole $\Gamma_q$. By means of eqs. (1)-(5), the problem (7) may be rewritten in the following compact form:

$$\begin{cases} \mathbf{L}\dot{\mathbf{\Lambda}} = \dot{\mathbf{\Phi}} \\ \dot{\lambda} \geq 0; \quad \dot{\phi} \leq 0; \quad \dot{\phi}\dot{\lambda} = 0 \quad \text{if } \phi = 0 \\ \dot{\lambda} = 0 \qquad\qquad\qquad\qquad\quad \text{if } \phi < 0 \end{cases} \tag{16}$$

where

$$\mathbf{L} := \begin{bmatrix} \int_{\Gamma_u} \mathbf{G}^{uu}[.]\,d\boldsymbol{\xi} & -\fint_{\Gamma_p} \mathbf{G}^{up}[.]\,d\boldsymbol{\xi} & \fint_{\Gamma_q} \mathbf{G}^{up}[.]\,d\boldsymbol{\xi} & \mathbf{0} \\ -\fint_{\Gamma_u} \mathbf{G}^{pu}[.]\,d\boldsymbol{\xi} & \fint_{\Gamma_p} \mathbf{G}^{pp}[.]\,d\boldsymbol{\xi} & -\fint_{\Gamma_q} \mathbf{G}^{pp}[.]\,d\boldsymbol{\xi} & \mathbf{0} \\ \fint_{\Gamma_u} \mathbf{G}^{pu}[.]\,d\boldsymbol{\xi} & -\fint_{\Gamma_p} \mathbf{G}^{pp}[.]\,d\boldsymbol{\xi} & \fint_{\Gamma_q} \mathbf{G}^{pp}[.]\,d\boldsymbol{\xi} - \mathbf{D} & \mathbf{Dm} \\ \mathbf{0} & \mathbf{0} & \mathbf{n}^T\mathbf{D} & S \end{bmatrix} \begin{matrix} \text{on } \Gamma_u \\ \text{on } \Gamma_p \\ \text{on } \Gamma_q \\ \text{on } \Gamma_q \end{matrix} \tag{17}$$

$$\dot{\mathbf{\Lambda}} = \begin{bmatrix} \dot{\mathbf{p}} \\ \dot{\mathbf{u}} \\ \dot{\mathbf{w}} \\ \dot{\lambda} \end{bmatrix} \begin{matrix} \text{on } \Gamma_u \\ \text{on } \Gamma_p \\ \text{on } \Gamma_q \\ \text{on } \Gamma_q \end{matrix}, \quad \dot{\mathbf{\Phi}} = \begin{bmatrix} \dot{\mathbf{f}}^u \\ \dot{\mathbf{f}}^p \\ \dot{\mathbf{f}}^q \\ \dot{\phi} \end{bmatrix} \begin{matrix} \text{on } \Gamma_u \\ \text{on } \Gamma_p \\ \text{on } \Gamma_q \\ \text{on } \Gamma_q \end{matrix} \tag{18}$$

$$S = -h - \mathbf{n}^T\mathbf{Dm} \tag{19}$$

The definition (17) immediately shows that non-associativity implies unsymmetry of the integral operator with respect to the bilinear form (13). Accordingly, a variational statement in a "classical sense" cannot be given.

The negative definiteness of $S$ is a sufficient condition for existence and uniqueness of the incremental response at the constitutive level (see e.g. [11]). In the case of hardening ($h > 0$) and associated flow-rule, the positive definiteness of $\mathbf{D}$ guarantees the negative definiteness of $S$. In the general case, suitable conditions must be satisfied to guarantee a sign definiteness property for $S$. They can be easily stated when $\mathbf{D}$ is taken as diagonal, what is usual in cohesive fracture mechanics models [12].

## 4. EXTENDED INCREMENTAL PROBLEM

For the sake of clarity, let us expand the operator $\mathbf{L}$ in such a form:

$$\mathbf{L} := \begin{bmatrix} \mathbf{B} & \mathbf{M} \\ \mathbf{N}^T & S \end{bmatrix} \tag{20}$$

where:

$$\mathbf{N} := \begin{bmatrix} \mathbf{0} \\ \mathbf{0} \\ \mathbf{Dn} \end{bmatrix}, \quad \mathbf{M} := \begin{bmatrix} \mathbf{0} \\ \mathbf{0} \\ \mathbf{Dm} \end{bmatrix}, \quad \mathbf{B} := \mathbf{A} - \begin{bmatrix} 0 & 0 & 0 \\ 0 & 0 & 0 \\ 0 & 0 & \mathbf{D} \end{bmatrix} \tag{21}$$

and $\mathbf{A}$ defined in eq. (11). Let the problem (called "extended" with respect to the given problem (16))

$$\begin{cases} \begin{bmatrix} \mathbf{SYM[L]} & \mathbf{SKW[L]} \\ -\mathbf{SKW[L]} & -\mathbf{SYM[L]} \end{bmatrix} \begin{bmatrix} \dot{\mathbf{\Lambda}} \\ \dot{\tilde{\mathbf{\Lambda}}} \end{bmatrix} = \begin{bmatrix} \dot{\mathbf{\Phi}} \\ \dot{\tilde{\mathbf{\Phi}}} \end{bmatrix} \\ \dot{\lambda} \geq 0; \quad \dot{\phi} \leq 0; \quad \dot{\phi}\dot{\lambda} = 0 \quad \text{if } \phi = 0 \\ \dot{\lambda} = 0 \quad \text{if } \phi < 0 \\ \dot{\tilde{\lambda}} \geq 0; \quad \dot{\tilde{\phi}} \geq 0; \quad \dot{\tilde{\phi}}\dot{\tilde{\lambda}} = 0 \quad \text{if } \tilde{\phi} = 0 \\ \dot{\tilde{\lambda}} = 0 \quad \text{if } \tilde{\phi} > 0 \end{cases} \tag{22}$$

be considered, having set

$$\dot{\tilde{\mathbf{\Lambda}}} = \begin{bmatrix} \dot{\tilde{\mathbf{p}}} \\ \dot{\tilde{\mathbf{u}}} \\ \dot{\tilde{\mathbf{w}}} \\ \dot{\tilde{\lambda}} \end{bmatrix} \begin{matrix} \text{on } \Gamma_u \\ \text{on } \Gamma_p \\ \text{on } \Gamma_q \\ \text{on } \Gamma_q \end{matrix}, \quad \dot{\tilde{\mathbf{\Phi}}} = \begin{bmatrix} -\dot{\mathbf{f}}^u \\ -\dot{\mathbf{f}}^p \\ -\dot{\mathbf{f}}^q \\ \dot{\tilde{\phi}} \end{bmatrix} \begin{matrix} \text{on } \Gamma_u \\ \text{on } \Gamma_p \\ \text{on } \Gamma_q \\ \text{on } \Gamma_q \end{matrix} \tag{23}$$

Let $\mathcal{G}$ indicate the following Green's function:

$$\int_{\Gamma_q} \mathcal{G}(\mathbf{x}, \boldsymbol{\xi})(\cdot) \mathrm{d}\Gamma \equiv \mathbf{M}^T \mathbf{B}^{-1} \mathbf{N}(\cdot) \tag{24}$$

As shown in [13] for the non-associated flow theory of plasticity, if the symmetric part of the operator $\int_{\Gamma_q} \mathcal{G}(\mathbf{x}, \boldsymbol{\xi})(\cdot)\mathrm{d}\Gamma - S$ is positive definite with respect to the classical bilinear form, that is

$$\int_{\Gamma_q} \int_{\Gamma_q} \mu(\mathbf{x}) \mathcal{G}(\mathbf{x}, \boldsymbol{\xi}) \mu(\boldsymbol{\xi}) \mathrm{d}\Gamma_\xi \mathrm{d}\Gamma_x - \int_{\Gamma_q} \mu(\mathbf{x}) S(\mathbf{x}) \mu(\mathbf{x}) \mathrm{d}\Gamma_x > 0 \quad \forall \mu(\mathbf{x}) \neq 0 \tag{25}$$

the following propositions hold:

**a.** *Every solution of the problem (22) is such that* $\dot{\tilde{\mathbf{\Lambda}}} \equiv \dot{\mathbf{\Lambda}}$*;*

**b.** *Every solution* $\dot{\mathbf{\Lambda}}$ *of the given problem (16), is also a solution for the extended problem (22) and vice-versa.*

**Proof:** After the condensation of the unknowns $\dot{\mathbf{p}}, \dot{\mathbf{u}}, \dot{\mathbf{w}}, \dot{\tilde{\mathbf{p}}}, \dot{\tilde{\mathbf{u}}}, \dot{\tilde{\mathbf{w}}}$, the system (22) becomes:

$$\begin{cases} \dot{\phi} = \mathbf{N}^T\mathbf{B}^{-1}\left(\dot{\mathbf{f}}^z - \frac{1}{2}\mathbf{M}(\dot{\lambda} + \dot{\tilde{\lambda}})\right) - \frac{1}{2}\mathbf{M}^T\mathbf{B}^{-1}\mathbf{N}(\dot{\lambda} - \dot{\tilde{\lambda}}) + S\dot{\lambda} \\ \dot{\tilde{\phi}} = -\mathbf{N}^T\mathbf{B}^{-1}\left(\dot{\mathbf{f}}^z - \frac{1}{2}\mathbf{M}(\dot{\lambda} + \dot{\tilde{\lambda}})\right) - \frac{1}{2}\mathbf{M}^T\mathbf{B}^{-1}\mathbf{N}(\dot{\lambda} - \dot{\tilde{\lambda}}) - S\dot{\tilde{\lambda}} \\ \dot{\lambda} \geq 0; \quad \dot{\phi} \leq 0; \quad \dot{\phi}\dot{\lambda} = 0 \quad \text{if } \phi = 0 \\ \dot{\lambda} = 0 \qquad\qquad\qquad\qquad\quad \text{if } \phi < 0 \\ \dot{\tilde{\lambda}} \geq 0; \quad \dot{\tilde{\phi}} \geq 0; \quad \dot{\tilde{\phi}}\dot{\tilde{\lambda}} = 0 \quad \text{if } \tilde{\phi} = 0 \\ \dot{\tilde{\lambda}} = 0 \qquad\qquad\qquad\qquad\quad \text{if } \tilde{\phi} > 0 \end{cases} \tag{26}$$

where

$$\dot{\mathbf{f}}^z = \begin{bmatrix} \dot{\mathbf{f}}^u \\ \dot{\mathbf{f}}^p \\ \dot{\mathbf{f}}^q \end{bmatrix} \tag{27}$$

The sign constraints in the system (26) imply the following inequality:

$$\int_{\Gamma_q} (\dot{\phi} + \dot{\tilde{\phi}})(\dot{\lambda} - \dot{\tilde{\lambda}}) \mathrm{d}\Gamma \geq 0 \tag{28}$$

which, using the first two equations of (26), becomes:

$$\begin{aligned} &\int_{\Gamma_q} (\dot{\lambda}(\mathbf{x}) - \dot{\tilde{\lambda}}(\mathbf{x})) \int_{\Gamma_q} \mathcal{G}(\mathbf{x}, \boldsymbol{\xi})(\dot{\lambda}(\boldsymbol{\xi}) - \dot{\tilde{\lambda}}(\boldsymbol{\xi}))\mathrm{d}\Gamma_\xi \mathrm{d}\Gamma_x \\ &\quad - \int_{\Gamma_q} (\dot{\lambda}(\mathbf{x}) - \dot{\tilde{\lambda}}(\mathbf{x}))S(\mathbf{x})(\dot{\lambda}(\mathbf{x}) - \dot{\tilde{\lambda}}(\mathbf{x}))\mathrm{d}\Gamma_x \leq 0 \end{aligned} \tag{29}$$

The strict positive definiteness of the symmetric part of the operator $\int_{\Gamma_q} \mathcal{G}(\mathbf{x}, \boldsymbol{\xi})(\cdot)\mathrm{d}\Gamma_\xi - S$ implies that inequality (29) is satisfied if and only if $(\dot{\lambda} - \dot{\tilde{\lambda}}) = 0$, which is the crucial part of the thesis. The remaining part follows immediately. ||

Defining $\mathbf{s} = \frac{1}{2}(\mathbf{n} + \mathbf{m})$, and $\mathbf{r} = \frac{1}{2}(\mathbf{n} - \mathbf{m})$, let's consider the functional:

$$\begin{aligned} F[\dot{\mathbf{u}}, \dot{\mathbf{p}}, \dot{\mathbf{w}}, \dot{\lambda}, \dot{\tilde{\mathbf{u}}}, \dot{\tilde{\mathbf{p}}}, \dot{\tilde{\mathbf{w}}}, \dot{\tilde{\lambda}}] &= \Psi[\dot{\mathbf{u}}, \dot{\mathbf{p}}, \dot{\mathbf{w}}] - \Psi[\dot{\tilde{\mathbf{u}}}, \dot{\tilde{\mathbf{p}}}, \dot{\tilde{\mathbf{w}}}] \\ &+ \frac{1}{2}\int_{\Gamma_q} -\dot{\mathbf{w}}^T\,\mathbf{D}\dot{\mathbf{w}} + \dot{\lambda}\,S\dot{\lambda} + \dot{\tilde{\mathbf{w}}}^T\,\mathbf{D}\dot{\tilde{\mathbf{w}}} - \dot{\tilde{\lambda}}\,S\dot{\tilde{\lambda}}\,\mathrm{d}\mathbf{x} \\ &+ \int_{\Gamma_q} \dot{\mathbf{w}}^T\,\mathbf{D}\mathbf{s}\dot{\lambda} + \dot{\tilde{\mathbf{w}}}^T\,\mathbf{D}\mathbf{r}\dot{\lambda} - \dot{\mathbf{w}}^T\,\mathbf{D}\mathbf{r}\dot{\tilde{\lambda}} - \dot{\tilde{\mathbf{w}}}^T\,\mathbf{D}\mathbf{s}\dot{\tilde{\lambda}}\,\mathrm{d}\mathbf{x} \end{aligned} \tag{30}$$

The following variational statement can be proved:

**c.** *Displacement rate vector* $\dot{\mathbf{u}}, \dot{\tilde{\mathbf{u}}}$ *on* $\Gamma_p$, *traction rate vector* $\dot{\mathbf{p}}, \dot{\tilde{\mathbf{p}}}$ *on* $\Gamma_u$, *displacement discontinuity rate vector* $\dot{\mathbf{w}}, \dot{\tilde{\mathbf{w}}}$ *on* $\Gamma_q$ *and plastic multiplier rate vector* $\dot{\lambda}, \dot{\tilde{\lambda}}$ *on* $\Gamma_q$, *represent the solution of the extended rate problem (22) if and only if they solve the following min-max problem:*

$$\min_{\dot{\mathbf{p}},\dot{\lambda},\dot{\tilde{\mathbf{u}}},\dot{\tilde{\mathbf{w}}}} \ \max_{\dot{\mathbf{u}},\dot{\mathbf{w}},\dot{\tilde{\mathbf{p}}},\dot{\tilde{\lambda}}} \left\{ F[\dot{\mathbf{u}}, \dot{\mathbf{p}}, \dot{\mathbf{w}}, \dot{\lambda}, \dot{\tilde{\mathbf{u}}}, \dot{\tilde{\mathbf{p}}}, \dot{\tilde{\mathbf{w}}}, \dot{\tilde{\lambda}}] \ \middle| \ \dot{\lambda} \geq 0 \ , \ \dot{\tilde{\lambda}} \geq 0 \right\} \tag{31}$$

**Proof:** Consider the augmented unconstrained functional:

$$L[\dot{\mathbf{u}}, \dot{\mathbf{p}}, \dot{\mathbf{w}}, \dot{\lambda}, \eta, l, \dot{\tilde{\mathbf{u}}}, \dot{\tilde{\mathbf{p}}}, \dot{\tilde{\mathbf{w}}}, \dot{\tilde{\lambda}}, \tilde{\eta}, \tilde{l}] = F[\dot{\mathbf{u}}, \dot{\mathbf{p}}, \dot{\mathbf{w}}, \dot{\lambda}, \dot{\tilde{\mathbf{u}}}, \dot{\tilde{\mathbf{p}}}, \dot{\tilde{\mathbf{w}}}, \dot{\tilde{\lambda}}] +$$
$$\int_{\Gamma_q} \eta(\dot{\lambda} - l^2) d\Gamma + \int_{\Gamma_q} \tilde{\eta}(\dot{\tilde{\lambda}} - \tilde{l}^{\,2}) d\Gamma \tag{32}$$

The inequalities $\dot{\lambda} \geq 0$ and $\dot{\tilde{\lambda}} \geq 0$ have been eliminated by the introduction of the Lagrangian multipliers $\eta$ and $\tilde{\eta}$ after their transformation in equalities through the introduction of the new variables $l$ and $\tilde{l}$.

The stationarity condition $\delta^{(1)} L = 0$ leads to the following equation system:

$$\left\{ \begin{array}{l} \mathbf{A}' \, \dot{\mathbf{y}} + \mathbf{B}' \, \dot{\mathbf{w}} - \dot{\mathbf{f}}^y = \mathbf{0} \\ -\mathbf{A}' \, \dot{\tilde{\mathbf{y}}} - \mathbf{B}' \, \dot{\tilde{\mathbf{w}}} + \dot{\mathbf{f}}^y = \mathbf{0} \\ \mathbf{D}\mathbf{s}\dot{\lambda} - \mathbf{D}\mathbf{r}\dot{\tilde{\lambda}} - \mathbf{D}\dot{\mathbf{w}} + \mathbf{B}'^{\,T} \, \dot{\mathbf{y}} + \mathbf{C}' \, \dot{\mathbf{w}} - \dot{\mathbf{f}}^q = \mathbf{0} \\ \mathbf{D}\mathbf{r}\dot{\lambda} - \mathbf{D}\mathbf{s}\dot{\tilde{\lambda}} + \mathbf{D}\dot{\tilde{\mathbf{w}}} - \mathbf{B}'^{\,T} \, \dot{\tilde{\mathbf{y}}} - \mathbf{C}' \, \dot{\tilde{\mathbf{w}}} + \dot{\mathbf{f}}^q = \mathbf{0} \\ -\mathbf{s}^T\mathbf{D}\dot{\mathbf{w}} - \mathbf{r}^T\mathbf{D}\dot{\tilde{\mathbf{w}}} - S\dot{\lambda} - \eta = 0 \\ \mathbf{r}^T\mathbf{D}\dot{\mathbf{w}} + \mathbf{s}^T\mathbf{D}\dot{\tilde{\mathbf{w}}} + S\dot{\tilde{\lambda}} - \tilde{\eta} = 0 \\ \eta l = 0 \\ \tilde{\eta}\tilde{l} = 0 \\ \dot{\lambda} - l^2 = 0 \\ \dot{\tilde{\lambda}} - \tilde{l}^{\,2} = 0 \end{array} \right. \tag{33}$$

where:

$$\mathbf{A}' \ := \left[ \begin{array}{cc} \int_{\Gamma_u} \mathbf{G}^{uu}[.] \, d\boldsymbol{\xi} & -\fint_{\Gamma_p} \mathbf{G}^{up}[.] \, d\boldsymbol{\xi} \\ -\fint_{\Gamma_u} \mathbf{G}^{pu}[.] \, d\boldsymbol{\xi} & \fint\!\!\!\!=_{\Gamma_p} \mathbf{G}^{pp}[.] \, d\boldsymbol{\xi} \end{array} \right] \begin{array}{l} \text{on } \Gamma_u \\ \text{on } \Gamma_p \end{array} \tag{34}$$

$$\mathbf{B}' \ := \left[ \begin{array}{c} \fint_{\Gamma_q} \mathbf{G}^{up}[.] \, d\boldsymbol{\xi} \\ -\fint\!\!\!\!=_{\Gamma_q} \mathbf{G}^{pp}[.] \, d\boldsymbol{\xi} \end{array} \right] \begin{array}{l} \text{on } \Gamma_u \\ \text{on } \Gamma_p \end{array} \tag{35}$$

$$\mathbf{C}' \ := \fint\!\!\!\!=_{\Gamma_q} \mathbf{G}^{pp}[.] \, d\boldsymbol{\xi} \quad \text{on } \Gamma_q \tag{36}$$

$$\dot{\mathbf{y}} = \left[ \begin{array}{c} \dot{\mathbf{p}} \\ \dot{\mathbf{u}} \end{array} \right] \begin{array}{l} \text{on } \Gamma_u \\ \text{on } \Gamma_p \end{array} , \quad \dot{\tilde{\mathbf{y}}} = \left[ \begin{array}{c} \dot{\tilde{\mathbf{p}}} \\ \dot{\tilde{\mathbf{u}}} \end{array} \right] \begin{array}{l} \text{on } \Gamma_u \\ \text{on } \Gamma_p \end{array} , \quad \dot{\mathbf{f}}^y = \left[ \begin{array}{c} \dot{\mathbf{f}}^u \\ \dot{\mathbf{f}}^p \end{array} \right] \begin{array}{l} \text{on } \Gamma_u \\ \text{on } \Gamma_p \end{array} \tag{37}$$

Moreover, imposing the positiveness of the second variation $\delta^{(2)}L$ with respect to $l$ and its negativeness with respect to $\tilde{l}$, the following inequalities come out:

$$\eta \leq 0 \quad ; \quad \tilde{\eta} \geq 0 \tag{38}$$

It is straightforward to recognize that eqs. (33)-(38) coincide with the extended problem (22).

# 5. CLOSING REMARKS

The present formulation can be used to perform the bifurcation analysis and, as for elastic-plastic problem discretized with the finite element method [14], can be used to define Prager's generalized variables.

The cohesive crack fracture mechanics problem over a time interval can be transformed into an approximate holonomic problem through a suitable change of the complementarity condition [15]. A min-max formulation characterizing the complete evolution of the body response may be given.

The Bramble-Pasciak technique [16] may be used to transform the global saddle-point formulation into a minimum one. Extremal formulations are particularly suitable for finding numerical solutions of the problem through direct solution procedures associated with optimization techniques (like the conjugate gradient method). The value of the functional during the descent process may be used as a measure of the convergence and the value of the functional at the solution may be used to evaluate the approximation error.

Developments in these context are in progress.

# References

[1] Nooru-Mohamed, M.B. and van Mier, J.G.M., Fracture of concrete under Mixed Mode Loading, in *Fracture of concrete and rock, recent development*, S.P. Shah, S.E. Schwartz and B. Barr eds., Elsevier, Amsterdam, 458-467 (1989).

[2] Guo, Z.K., Kobayashi, A.S. and Hawkins, N.M., Mixed Modes I and II Concrete Fracture: An experimental analysis, *J. Appl. Mech.*, ASME, **61**, 815-821 (1994).

[3] Amadei, B., Sture, S., Saeb, S. and Atkinson, R.H., *An evaluation of masonry joint shear strength in existing buildings*, report to NSF, Department of Civil, Environmental and Architectural Engineering, University of Colorado, Boulder, 1989.

[4] Lotfi, H. and Shing, P., Interface model applied to fracture of masonry structures, *J. Struct. Engrg.*, ASCE, **120(1)**, 63-80 (1994).

[5] Carol, I., Prat, P.C. and Lopez, C.M., Normal/shear cracking model: application to discrete crack analysis, *J. Engrg. Mech.*, ASCE, **123(8)**, 1-9 (1997).

[6] Maugin, G.A., *The thermomechanics of plasticity and fracture* , Cambridge University Press, New York, 1992.

[7] Carini, A., De Donato, O., Fundamental solutions for linear viscoelastic continua, *Int. J. Solids Structures*, **29**, 2989-3009 (1992).

[8] Cruse, T.A., *Boundary Element Analysis in Computational Fracture Mechanics*, Kluwer Academic, Dordrecht, The Netherlands, 1988.

[9] Cen, Z., and Maier, G., Bifurcations and instabilities in fracture of cohesive-softening structures: a boundary element analysis, *Fatigue Fract. Engng. Mater. Struct.*, **15**, 911-928 (1992).

[10] Sirtori, S., Maier, G., Novati, G., and Miccoli, S., A Galerkin symmetric boundary-element method in elasticity: formulation and implementation, *Int. J. Numer. Methods Eng.*, **35**, 255-282 (1992).

[11] Maier, G., and Hueckel,T. Non associated and coupled flow rules of elastoplasticity for rock-like materials, *Int. J. Rock Mech. Min. Sci. & Geomech. Abstr.*, **16**, 77-92,(1979)

[12] Salvadori A. Quasi Brittle Fracture Mechanics by Cohesive Crack Models and Symmetric Galerkin Boundary Element Method, PhD Thesis, Politecnico di Milano, Milano, (1999)

[13] Carini, A. and Genna, F., Variational formulations in the non-associated flow theory of plasticity, (submitted).

[14] Comi, C. and Perego, U., A unified approach for variationally consistent finite elements in elastoplasticity, *Comp. Meth. in Appl. Mech. and Engng.*, **121**, 323-344 (1995).

[15] Pandolfi, A. and Carini, A., Some extremum properties of finite-step solutions in elastoplasticity, *Int. J. Solids and Structures*, **36**, 185-218 (1999).

[16] Bramble, J.H. and Pasciak, J.E., A preconditioning technique for indefinite systems resulting from mixed approximations of elliptic problems, *Mathematics of Computation*, **50**, 1-17 (1988).

# ON THE TRUE AND SPURIOUS EIGENSOLUTIONS USING CIRCULANTS FOR REAL-PART DUAL BEM

J. T. Chen, S. R. Kuo and Y. C. Cheng
*Department of Harbor and River Engineering, Taiwan Ocean University, Keelung, Taiwan*

**Abstract** It has been found recently that the multiple reciprocity method (MRM), the real-part BEM and the imaginary-part BEM result in spurious eigenvalues for eigenproblems. In this paper, a circular domain is considered as a demonstrative example. Based on the dual framework of real-part BEM, the true and spurious eigenvalues can be separated by using the singular value decomposition technique (SVD). To understand why the spurious eigenvalues occur, analytical derivation by discretizing the circular boundary into a discrete system is employed and results in a circulants. By using the SVD updating terms, the true eigensolutions can be extracted by merging the two influence matrices in dual BEM. The spurious eigensolutions can be filtered out by using the SVD updating documents where the other two influence matrices are combined.

**Keywords:** real-part BEM; spurious eigensolution;SVD updating technique

## 1. INTRODUCTION

It is well known that fictitious frequency occurs when the singular integral equation or hypersingular integral equation is used alone to solve the exterior acoustic problems (radiation or scattering cases). The reason has been clearly understood that the nonunique solution results from the zero in the denominator [1]. Although some researchers called this irregular value " spurious frequency " [2, 3], we will distinguish the differences between fictitious and spurious solutions in this paper.

For interior eigenproblems, complex-valued boundary integral formulation [4] has been used to determine the eigensolutions. To avoid the complex-valued computation, multiple reciprocity method [5], real-part [6], and imaginary-part [7] formulations have been tried to solve the same problem. However, spurious eigensolutions are embedded in the

*T. Burczynski (ed.), IUTAM/IACM/IABEM Symposium on Advanced Mathematical and Computational Mechanics Aspects of the Boundary Element Method*, 75–85.

above three methods. To overcome this difficulty, the dual formulation in conjunction with the SVD technique [5, 6, 8] is a novel method to extract the true solutions. This technique has been successfully applied to rod [9], beam [10] and cavity problems using dual MRM [5] or real-part dual BEM [6]. A unified method to filter out the spurious solutions is not trivial.

In this paper, we employ the real-part dual BEM to solve the acoustic problem of a circular cavity. After assembling the dual equations, the singular value decomposition (SVD) technique is employed to extract the true and spurious eigenvalues for two-dimensional cavities. The spurious eigensolutions are analytically predicted in the discrete system of circulants and are compared with those obtained by using the real-part dual BEM program, DUALREAL. Finally, the true eigenvalues for a circular cavity are derived analytically by approaching the discrete system to continuous system using the analytical properties of circulants [11]. Two approaches, SVD updating terms and updating documents, are employed to extract the true and spurious solutions respectively. Three results, analytical solution, discrete system solution using circulants and numerical solution using BEM, are compared with each other.

## 2. DUAL INTEGRAL FORMULATION FOR A TWO-DIMENSIONAL INTERIOR ACOUSTIC PROBLEMS

The governing equation for an interior acoustic problem is the Helmholtz equation as follows:

$$(\nabla^2 + k^2)u(x_1, x_2) = 0, \quad (x_1, x_2) \in D,$$

where $\nabla^2$ is the Laplacian operator, $D$ is the domain of the cavity and $k$ is the wave number, which is angular frequency over the speed of sound. The boundary conditions can be either the Neumann or Dirichlet type.

Based on the dual formulation [12], the dual equations for the boundary points are

$$\pi u(x) = C.P.V. \int_B T(s,x)u(s)dB(s) - R.P.V. \int_B U(s,x)t(s)dB(s), \quad x \in B \quad (1)$$

$$\pi t(x) = H.P.V. \int_B M(s,x)u(s)dB(s) - C.P.V. \int_B L(s,x)t(s)dB(s), \quad x \in B \quad (2)$$

where $R.P.V.$, $C.P.V.$ and $H.P.V.$ denote the Riemann principal value, the Cauchy principal value, and the Hadamard principal value, $t(s) = \frac{\partial u(s)}{\partial n_s}$, $B$ denotes the boundary enclosing $D$ and the explicit forms of the four kernels, $U, T, L$ and $M$, can be found in [12].

## 3. CIRCULANT MATRICES FOR INTERIOR PROBLEMS USING THE REAL-PART DUAL BEM

By using the two sets of bases, $J_m(kx)$ and $Y_m(kx)$, we can decompose the two-dimensional real-part kernel functions into

$$U(s,x) = \begin{cases} U^i(\theta,0) = \sum_{m=-\infty}^{\infty} \frac{\pi}{2} Y_m(kR) J_m(k\rho) cos(k\theta), & R > \rho \\ U^e(\theta,0) = \sum_{m=-\infty}^{\infty} \frac{\pi}{2} Y_m(k\rho) J_m(kR) cos(k\theta), & R < \rho \end{cases} \quad (3)$$

$$T(s,x) = \begin{cases} T^i(\theta,0) = \sum_{m=-\infty}^{\infty} \frac{\pi k}{2} Y'_m(kR) J_m(k\rho) cos(k\theta), & R > \rho \\ T^e(\theta,0) = \sum_{m=-\infty}^{\infty} \frac{\pi k}{2} Y_m(k\rho) J'_m(kR) cos(k\theta), & R < \rho \end{cases} \quad (4)$$

$$L(s,x) = \begin{cases} L^i(\theta,0) = \sum_{m=-\infty}^{\infty} \frac{\pi k}{2} Y_m(kR) J'_m(k\rho) cos(k\theta), & R > \rho \\ L^e(\theta,0) = \sum_{m=-\infty}^{\infty} \frac{\pi k}{2} Y'_m(k\rho) J_m(kR) cos(k\theta), & R < \rho \end{cases} \quad (5)$$

$$M(s,x) = \begin{cases} M^i(\theta,0) = \sum_{m=-\infty}^{\infty} \frac{\pi k^2}{2} Y'_m(kR) J'_m(k\rho) cos(k\theta), R > \rho \\ M^e(\theta,0) = \sum_{m=-\infty}^{\infty} \frac{\pi k^2}{2} Y'_m(k\rho) J'_m(kR) cos(k\theta), R < \rho \end{cases} \quad (6)$$

where the superscripts "$i$" and "$e$" denotes interior and exterior domain, $J_m$ and $Y_m$ are the m-th order Bessel functions of the first and second kinds, respectively, $x = (\rho, 0)$ and $s = (R, \theta)$ in polar coordinate. By discretizing $2N$ constant elements on a circular boundary, we have the influence matrix in a form of the circulants as shown belows.

$$[K] = \begin{bmatrix} k_0 & k_1 & k_2 & \cdots & k_{2N-2} & k_{2N-1} \\ k_{2N-1} & k_0 & k_1 & \cdots & k_{2N-3} & k_{2N-2} \\ k_{2N-2} & k_{2N-1} & k_0 & \cdots & k_{2N-4} & k_{2N-3} \\ \vdots & \vdots & \vdots & \ddots & \vdots & \vdots \\ k_1 & k_2 & k_3 & \cdots & k_{2N-1} & k_0 \end{bmatrix} \quad (7)$$

where

$$k_m = \int_{(m-\frac{1}{2})\Delta\theta}^{(m+\frac{1}{2})\Delta\theta} F^e(\theta,0)\rho \, d\theta \approx F^e(\theta_m,0)\rho\,\Delta\theta, \quad m = 0,1,2,\cdots,2N-1 \quad (8)$$

in which $[K]$ can be the influence matrices of $U$, $T$, $L$ or $M$ , $F^e$ can be $U^e$, $T^e$, $L^e$ or $M^e$ kernels as shown in Eqs.(3)-(6), $\theta_m = m\Delta\theta$ and $\Delta\theta = \frac{2\pi}{2N}$ . It is interesting to find that all the matrices in Eq.(7) are symmetric circulants. By using the properties of circulants for the matrices, we have the determinants,

$$det[U] = \lambda_0(\lambda_1\lambda_2\cdots\lambda_{N-1})^2\lambda_N \tag{9}$$

$$det[L] = \mu_0(\mu_1\mu_2\cdots\mu_{N-1})^2\mu_N \tag{10}$$

$$det[T] = \nu_0(\nu_1\nu_2\cdots\nu_{N-1})^2\nu_N \tag{11}$$

$$det[M] = \kappa_0(\kappa_1\kappa_2\cdots\kappa_{N-1})^2\kappa_N \tag{12}$$

where

$$\lambda_\ell = \pi^2\rho Y_\ell(k\rho)J_\ell(k\rho),\ \ell = 0, \pm 1, \cdots, \pm(N-1), N. \tag{13}$$

$$\mu_\ell = \pi^2 k\rho Y_\ell'(k\rho)J_\ell(k\rho),\ \ell = 0, \pm 1, \cdots, \pm(N-1), N. \tag{14}$$

$$\nu_\ell = \pi^2 k\rho Y_\ell(k\rho)J_\ell'(k\rho),\ \ell = 0, \pm 1, \cdots, \pm(N-1), N. \tag{15}$$

$$\kappa_\ell = \pi^2 k^2\rho Y_\ell'(k\rho)J_\ell'(k\rho),\ \ell = 0, \pm 1, \cdots, \pm(N-1), N. \tag{16}$$

## 4. METHODS TO EXTRACT THE TRUE EIGENSOLUTIONS — SVD UPDATING TERMS

Since real-part BEM loses imaginary information, we can reconstruct the independent equation by differentiation. This results in dual formulation in order to extract the true eigensolutions. To obtain an overdetermined system, we can combine $[U]$ and $[L]$ matrices by using updating terms,

$$[C] = \begin{bmatrix} U \\ L \end{bmatrix}_{4N\times 2N} \tag{17}$$

for the Dirichlet problem. Since the eigensolution is nontrival, the rank of $[C]$ must be smaller than $2N$. Therefore, the $2N$ singular values for $[C]$ matrix must have at least one zero value. Based on the equivalence between the SVD technique and the least-squares method in mathematical essence, we have

$$\begin{bmatrix} U^T & L^T \end{bmatrix}\begin{bmatrix} U \\ L \end{bmatrix} = [U^2 + L^2] \tag{18}$$

since $[U]$ and $[L]$ are symmetric. For a circular cavity, we have

$$[U] = [\Phi] \begin{bmatrix} \ddots & & \\ & \lambda_\ell & \\ & & \ddots \end{bmatrix} [\Phi^T] \tag{19}$$

$$[L] = [\Phi] \begin{bmatrix} \ddots & & \\ & \mu_\ell & \\ & & \ddots \end{bmatrix} [\Phi^T] \tag{20}$$

where $[\Phi]$ is a modal matrix. By substituting Eq.(19) and (20) into Eq.(18), we can obtain

$$\left[C^T\right][C] = [\Phi] \begin{bmatrix} \ddots & & \\ & (\lambda_\ell^2 + \mu_\ell^2) & \\ & & \ddots \end{bmatrix} [\Phi^T] \tag{21}$$

Therefore, we have the singular values of $\sqrt{\lambda_\ell^2 + \mu_\ell^2}$, $\ell = 0, \pm 1, \pm 2,$ $\pm(N-1), N$. By plotting the minimum singular value for $C$ versus $k$, only true eigenvalues have dips. The results will be elaborated on later.

## 5. METHODS TO FILTER OUT THE SPURIOUS EIGENSOLUTIONS — SVD UPDATING DOCUMENTS

Based on the dual formulation, the $[U]$ and $[T]$ matrices have the same spurious eigenvalues. This results in spurious eigensolutions. In order to extract the spurious eigenvalues, we can combine the $\left[U^T\right]$ and $\left[T^T\right]$ matrices by using updating documents,

$$[D] = \begin{bmatrix} U^T \\ T^T \end{bmatrix}_{4N \times 2N} \tag{22}$$

Similarly, we have

$$\begin{bmatrix} U & T \end{bmatrix} \begin{bmatrix} U^T \\ T^T \end{bmatrix} = \left[U^2 + T^2\right] \tag{23}$$

since $[U]$ and $[T]$ are symmetric. According to Eqs.(13)-(16), the spurious eigenvalues are embedded in the transposes of $[U]$ and $[T]$ matrices. The singular values for $[D]$ must have at least one zero value. To determine the singular values for $[D]$, we have

$$[D^T]\,[D] = [\Phi] \begin{bmatrix} \ddots & & \\ & (\lambda_\ell^2 + \nu_\ell^2) & \\ & & \ddots \end{bmatrix} [\Phi^T] \tag{24}$$

By plotting the minimum singular values of $\sqrt{\lambda_\ell^2 + \nu_\ell^2}, \ell = 0, \pm 1, \pm 2,$ $\pm\,(N-1), N$ versus $k$ , we can extract the spurious eigenvalues where dips occur.

## 6. NUMERICAL RESULTS AND DISCUSSIONS

A circular cavity with a radius ($\rho = 1\,m$) subjected to the Dirichlet boundary condition ($u = 0, \rho = 1$) is considered. In this case, an analytical solution is available as follows:
eigenequation: $J_m(k_{mn}) = 0,\ m, n = 0, 1, 2, 3 \cdots$ ;
eigenmode : $u(a, \theta) = J_m(k_{mn}a)e^{im\theta}, 0 < a < \rho, 0 < \theta < 2\pi$ .

Twenty constant elements with $N$=10 are adopted in the discrete system. Since two alternatives, the $UT$ or $LM$ equation, can be used to collocate on the circular boundary, two results from the $UT$ and $LM$ methods can be obtained. Fig.1 shows the minimum singular value versus $k$ using the $UT$ method. The true eigenvalues contaminated by the spurious eigenvalues can be obtained as shown in Fig.1 by considering the near zero minimum singular values if only the $UT$ equation is chosen. The true eigenvalues occur at the positions of zeros for $J_m(k_{mn}\rho)$ while the spurious eigenvalues occur at the positions of zeros for $Y_m(k_{mn}\rho)$. Fig.2 shows the minimum singular value versus $k$ only using the $LM$ equation. In a similar way, the true eigenvalues contaminated by spurious eigenvalues can be obtained as shown in Fig.2 by considering the near zero minimum singular values if the $LM$ equation is chosen alone. The true eigenvalues occur at the positions of zeros for $J_m(k_{mn}\rho)$ while the spurious eigenvalues occur at the positions of zeros for $Y_m'(k_{mn}\rho)$. It is interesting to find that no spurious eigenvalues occur as shown in Fig.3 when the $U$ and $L$ matrices are combined as shown in Eq.(18). To extract the spurious eigensolutions, we combine the $U^T$ and $T^T$ matrices as shown in Eq.(23). The minimum singular value versus $k$ is shown in Fig.4, it is found that dips occur only at the positions of spurious eigenvalues.

# 7. CONCLUSIONS

The real-part dual BEM in conjunction with the SVD technique using circulants has been applied to determine the true and spurious eigensolutions of a circular cavity. The spurious eigenvalues have been successfully extracted for discrete system. The true eigenvalues obtained by the real-part dual BEM also match very well the exact solutions. Two approaches, SVD updating terms and updating documents, were proposed to extract the true eigensolutions and to filter out the spurious eigensolutions by using the dual formulation. Three solutions, analytical solution, discrete system solution using circulants and numerical solution using real-part BEM program, are found to be in good agreement.

## References

[1] Chen, J. T. (1998) On fictitious frequencies using dual series representation, *Mechanics Research Communications* **25**(5), 529-534.

[2] Gennaretti, M., Giordani, A., and Morino, L. (1999) A third-order boundary element method for exterior acoustics with applications to scatering by rigid and elastic shells, *J. Sound and Vibration* **225**(5), 699-722.

[3] Schroeder, W., and Wolff, I. (1994) The origin of spurious modes in numerical solutions of electromagnetic field eigenvalue problems, *IEEE Transaction on Microwave Theory and Techniques* **42**(4), 644-653.

[4] Yeih, W., Chen, J. T., Chen, K. H., and Wong, F. C. (1997) A study on the multiple reciprocity method and complex-valued formulation for the Helmholtz equation, *Advances in Engineering Software* **29**(1), 7-12.

[5] Chen, J.T., Huang, C. X., and Wong, F. C. (2000) Determination of spurious eigenvalues and multiplicities of true eigenvalues in the dual multiple reciprocity method using the singular value decomposition technique, *J. Sound and Vibration*, **230**(2), 219-230.

[6] Chen, J.T., Huang, C.X., and Chen, K.H. (1999) Determination of spurious eigenvalues and multiplicities of true eigenvalues using the real-part dual BEM, *Comp. Mech.* **24**(1), 41-51.

[7] Chen, J.T., Kuo, S.R. and Chen, K.H.(1999) A nonsingular integral formulation for the Helmholtz eigenproblems of a circular domain, *J. Chinese Institute of Engineers*, **22**(6), 729-739.

[8] Golub, G.H., and Van Loan, C.F. (1989) *Matrix Computations*, 2nd edition, The Johns Hopkins University Press, Baltimore.

[9] Yeih, W., Chang, J.R., Chang, C.M., and Chen, J.T. (1999) Applications of dual MRM for determining the natural frequencies and natural modes of a rod using the singular value decomposition method, *Advances in Engineering Software* **30**(7), 459-468.

[10] Yeih, W., Chen, J.T., and Chang, C.M. (1999) Applications of dual MRM for determining the natural frequencies and natural modes of an Euler-Bernoulli beam using the singular value decomposition method, *Engng. Anal. Bound. Elem.* **23**, 339-360.

[11] Goldberg, J.L. (1991) Matrix Theory with Applications, McGraw-Hill, New York.

[12] Chen, J.T. and Chen, K.H. (1998) Dual integral formulation for determining the acoustic modes of a two-dimensional cavity with a degenerate boundary, *Engng. Anal. Bound. Elem.* **21**(2), 105-116.

[13] Chen, K.H., Chen, J.T., and Liou, D.Y. (1998) Dual boundary element analysis for an acoustic cavity with an incomplete partition, *Chinese J. Mech.* **14**(2), 1-14 (in Chinese).

[14] De Mey, G. (1977) A simplified integral equation method for the calculation of the eigenvalues of Helmholtz equation, *Int. J. Numer. Meth. Engng.* **11**, 1340-1342.

[15] Kamiya, N., Ando, E. and Nogae, K. (1996) A New Complex-valued formulation and eigenvalue analysis of the Helmholtz Equation by Boundary Element Method, *Advances in Engineering Software* **26**, 219-227.

[16] Liou, D.Y., Chen, J.T., and Chen, K.H. (1999) A new method for determining the acoustic modes of a two-dimensional sound field, *J. Chinese Inst. Civ. Hydr. Engng.* **11**(2), 89-100 (in Chinese).

[17] Nowak, A.J., and Neves, A.C., eds. (1994) *Multiple Reciprocity Boundary Element Method*, Southampton: Comp. Mech. Publ..

[18] Tai, G.R.G., and Shaw R.P. (1974) Helmholtz equation eigenvalues and eigenmodes for arbitrary domains, *J. Acou. Soc. Amer.* **56**, 796-804.

[19] Chen, J.T. (1998) Recent Development of Dual BEM in Acoustic Problems, Keynote lecture, *Proceedings of the 4th World Congress on Computational Mechanics*, E. Onate and S. R. Idelsohn (eds), Argentina, p.106.

[20] Chen, J.T., and Hong, H.K. (1999) Review of dual integral representations with emphasis on hypersingular integrals and divergent series, *Trans. ASME, Appl. Mech. Rev.* **52**(1), 17-33.

[21] Chen, J.T. and Wong, F.C. (1997) Analytical derivations for one-dimensional eigenproblems using dual BEM and MRM, *Engng. Anal. Bound. Elem.* **20**(1), 25-33.

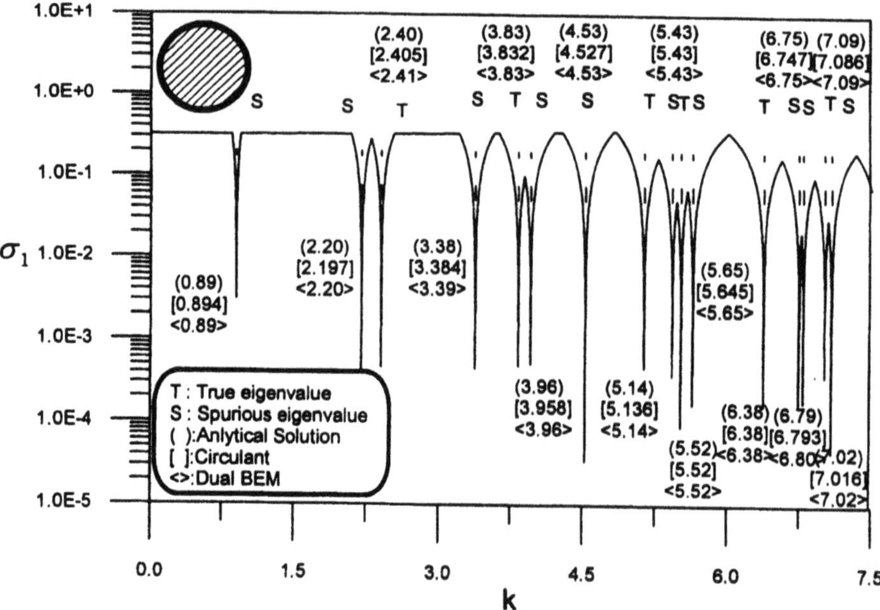


*Fig 1.* The first minimum singular value for different wave numbers using the [U] circulant of real-part dual BEM for the Dirichlet problem (u=0).

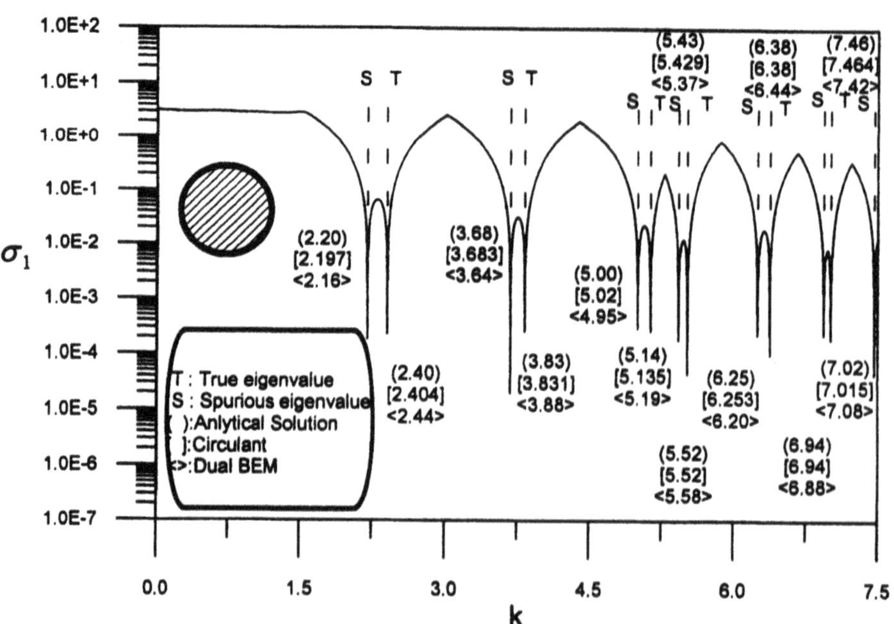


*Fig 2.* The first minimum singular value for different wave numbers using the [L] circulant of real-part dual BEM for the Dirichlet problem (u=0).

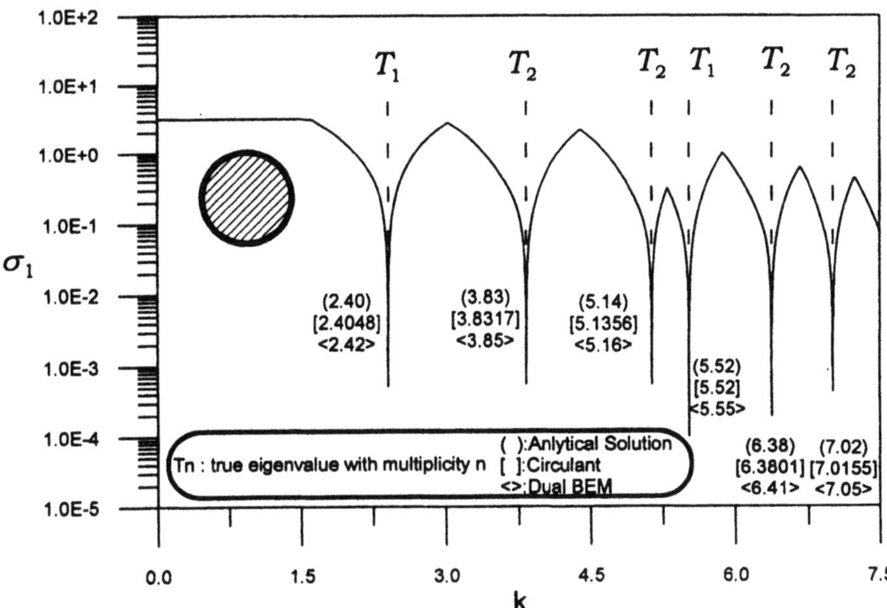


*Fig 3.* The first minimum singular value for different wave numbers using the $\begin{bmatrix} U \\ L \end{bmatrix}$ circulant of real-part dual BEM for the Dirichlet problem (u=0).

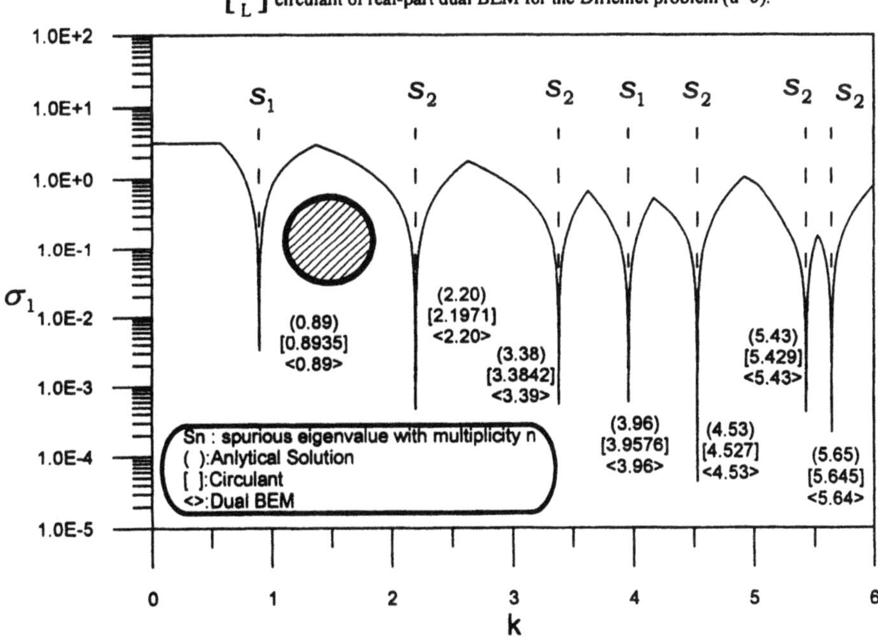


*Fig 4.* The first minimum singular value for different wave numbers using the $\begin{bmatrix} U^T \\ T^T \end{bmatrix}$ circulant of real-part dual BEM.

# HYPERSINGULAR FORMULATION FOR 3-D FRACTURE MECHANICS. A SIMPLE NUMERICAL APPROACH

J. DOMINGUEZ AND M.P. ARIZA

*Escuela Superior de Ingenieros, Universidad de Sevilla,*
*Camino de los Descubrimientos s/n, 41092-Sevilla, Spain*

**Keywords:** boundary elements, fracture mechanics, hypersingular boundary integral equation.

**Abstract** A boundary element approach based on the traction boundary integral equation for elastic problems is presented. The hypersingular and strongly singular integrals appearing in the integral equation are analytically transformed to yield regular or weakly singular integrals which can be numerically evaluated. Regularization of the boundary integrals is done prior to discretization and this process does not require any change of coordinates. In order to satisfy the continuity requirements over the primary variables, collocation points are shifted towards the interior of the elements which are standard continuous quadratic elements. Numerical results obtained for a 3-D fracture mechanics problem show the simplicity and accuracy of the approach.

## 1. INTRODUCTION

There are two main difficulties behind the use of Hypersingular Boundary Elements: One is the continuity conditions that must be satisfied by the density functions; and the other, the integration of the strongly singular and hypersingular kernels obtained by differentiation of the classical ones.

In the present paper a simple strategy to circunvent the first difficulty is proposed. It is an extension to three dimensions of the idea introduced by Gallego and Dominguez (1996) for two-dimensional problems. The elements are continuous with the classical representation of the variables; i.e., with most of the nodes at the contour of the elements. However, collocation is done at points inside the elements where the $C^{1,\alpha}$ continuity requirement is satisfied. Using this strategy the difficul-

*T. Burczynski (ed.), IUTAM/IACM/IABEM Symposium on Advanced Mathematical and Computational Mechanics Aspects of the Boundary Element Method,* 87–97.

ties produced by the continuity requirements at collocation points are avoided and the number of nodal unknowns remains the same as in the classical formulation.

Nevertheless, the main objective of this paper is the integration of hypersingular and strongly singular kernels. The method presented is on the line of the general procedures for evaluation of the hypersingular and strongly singular integrals appearing in the traction BIE.

Guiggiani *et al.* (1992) and Mantic (1994) presented an approach based on regularization by using Taylor's expansion of the displacements about the source point as proposed by Aliabadi *et al.* (1985). In this method, the regularization is done after discretization and the integration is performed in a general way which requires a complicated transformation of coordinates.

In the current paper, the classical integral representation is differentiated and taken to the boundary to obtain a hypersingular BIE in terms of displacements and tractions at the boundary collocation point. The hypersingular and strongly singular integrals are regularized by subtracting two terms of the Taylor's series expansion of the displacement and one term of the traction expansion, at collocation point. All the remaining integrals containing hypersingular and strongly singular kernels are transformed into regular or weakly singular integrals by analytical procedures.

The method is general for boundary surfaces and boundary elements of any shape. The regularization is done prior to any boundary discretization. All the subsequent integrations are performed analytically in a direct way without any change of coordinates. The expressions obtained for the traction BIE are only in terms of regular or weakly singular integrals, which can be evaluated by a standard Gaussian quadrature.

The limiting process shows explicitely all the unbounded terms which finally cancel out. Hadamard Finite Part and Cauchy Principal Value concepts are not necessary.

## 2. TRACTION BIE IN ELASTICITY

The classical displacement integral representation for an internal point $\mathbf{y}$ of an elastic three-dimensional body $\Omega$ bounded by a regular surface $\Gamma$ with unit outward normal $\mathbf{n}(\mathbf{x})$ under zero body forces conditions can be written as:

$$u_l(\mathbf{y}) + \int_\Gamma p^*_{lk}(\mathbf{x},\mathbf{y})u_k(\mathbf{x})d\Gamma - \int_\Gamma u^*_{lk}(\mathbf{x},\mathbf{y})p_k(\mathbf{x})d\Gamma = 0 \qquad (1)$$

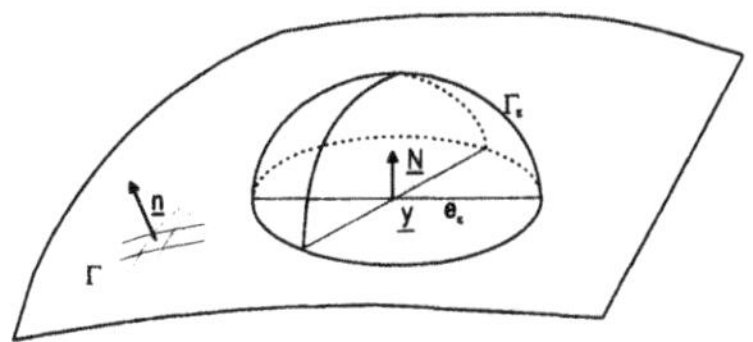

*Figure 1* Vanishing contour around a boundary point.

for l, $k = 1, 2, 3$; where $u_k$ and $p_k$ stand for the $k$ component of displacement and traction vectors, respectively, and $u^*_{lk}$, $p^*_{lk}$ are the fundamental solution displacement and traction tensors, respectively.

The internal stresses can be obtained by differentiation of displacement at point $\mathbf{y}$ and introducing the corresponding strains into the stress-strain relationship. The traction vector components at point $\mathbf{y}$ over a surface with unit outward normal $\mathbf{N}$ is

$$p_l(\mathbf{y}) = \sigma_{lm}(\mathbf{y}) N_m(\mathbf{y}) \tag{2}$$

After carrying out the derivatives of equation (1), using Hooke's law and equation (2), the integral representation for the traction components is obtained.

$$p_l(\mathbf{y}) + \int_\Gamma s^*_{lmk}(\mathbf{x},\mathbf{y}) N_m(\mathbf{y}) u_k(\mathbf{x}) d\Gamma - \int_\Gamma d^*_{lmk}(\mathbf{x},\mathbf{y}) N_m(\mathbf{y}) p_k(x) d\Gamma = 0 \tag{3}$$

where $d^*_{lmk}$ and $s^*_{lmk}$ are linear combinations of derivatives of $u^*_{lk}$ and $p^*_{lk}$, respectively.

The kernel $d^*_{lmk}$ has a strong singularity of order $r^{-2}$ while $s^*_{lmk}$ is hypersingular of order $r^{-3}$, as $r \to 0$.

A limit to the boundary process is carried out to obtain the integral representation for a smooth boundary point $\mathbf{y}$. Using the modified geometry of Figure 1, the following expression can be written:

$$p_l(\mathbf{y}) + \lim_{\varepsilon \to o^+} \left\{ \int_{(\Gamma - e_\varepsilon) + \Gamma_\varepsilon} \left[ s^*_{lmk}(\mathbf{x},\mathbf{y}) N_m(\mathbf{y}) u_k(\mathbf{x}) - d^*_{lmk}(\mathbf{x},\mathbf{y}) N_m(\mathbf{y}) p_k(\mathbf{x}) \right] d\Gamma \right\} = 0 \tag{4}$$

Several aspects related to the above traction representation can be pointed out: (1) the displacement must be differentiable at $\mathbf{y}$ with its first derivatives satisfying a Hölder condition, $u_k \in C^{1,\alpha}$; (2) At non smooth boundary points it is not possible to obtain an integral representation of the tractions only in terms of boundary displacements and tractions following the limit to the boundary process. However, such a process can be followed for the displacement derivatives and after inverting the $\mathbf{c}$ matrix obtain a representation of the tractions by linear combination of the displacement derivatives representation; (3) The surface $\Gamma_\varepsilon$ used to remove point $\mathbf{y}$ from the boundary may have any shape provided it is a regular surface. A spherical surface has been selected because of its simplicity.

Taking into account that $u_k \in C^{1,\alpha}$, at $\mathbf{y}$, it can be expanded as

$$u_k(\mathbf{x}) = u_k(\mathbf{y}) + u_{k,h}(\mathbf{y})(x_h - y_h) + O(r^{1+\alpha}) \tag{5}$$

and the tractions

$$p_k(\mathbf{x}) = \sigma_{kh}(\mathbf{x})n_h(\mathbf{x}) = \sigma_{kh}(\mathbf{y})n_h(\mathbf{x}) + O(r^\alpha) \tag{6}$$

Notice that this particular expansion of $p_k(\mathbf{x})$ around $\mathbf{y}$ with $\mathbf{n}(\mathbf{x})$ as a variable, is required for the surface $\Gamma_\varepsilon$, where $-1 \le \mathbf{n}(\mathbf{x}) \le 1$, but not for $\Gamma - e_\varepsilon$ where $\mathbf{n}(\mathbf{x}) \to \mathbf{N}(\mathbf{y})$, as $r \to 0$.

By using expansions (5) and (6), equation (4) can be written as

$$\begin{aligned}
&p_l(\mathbf{y}) + \lim_{\varepsilon \to o^+} \left\{ \int_{\Gamma - e_\varepsilon} \{ s^*_{lmk} N_m \left[ u_k(\mathbf{x}) - u_k(\mathbf{y}) - u_{k,h}(\mathbf{y})(x_h - y_h) \right] - \right. \\
&d^*_{lmk} N_m \left[ p_k(\mathbf{x}) - p_k(\mathbf{y}) \right] \} d\Gamma + \\
&\int_{\Gamma_\varepsilon} \{ s^*_{lmk} N_m \left[ u_k(\mathbf{x}) - u_k(\mathbf{y}) - u_{k,h}(\mathbf{y})(x_h - y_h) \right] - \\
&d^*_{lmk} N_m \left[ p_k(\mathbf{x}) - \sigma_{kh}(\mathbf{y}) n_h(\mathbf{x}) \right] \} d\Gamma + \\
&u_k(\mathbf{y}) \int_{\Gamma - e_\varepsilon} s^*_{lmk} N_m d\Gamma + u_k(\mathbf{y}) \int_{\Gamma_\varepsilon} s^*_{lmk} N_m d\Gamma + \\
&u_{k,h}(\mathbf{y}) \int_{\Gamma - e_\varepsilon} s^*_{lmk} N_m (x_h - y_h) d\Gamma + u_{k,h}(\mathbf{y}) \int_{\Gamma_\varepsilon} s^*_{lmk} N_m (x_h - y_h) d\Gamma - \\
&\left. p_k(\mathbf{y}) \int_{\Gamma - e_\varepsilon} d^*_{lmk} N_m d\Gamma - \sigma_{kh}(\mathbf{y}) \int_{\Gamma_\varepsilon} d^*_{lmk} N_m n_h d\Gamma \right\} = 0
\end{aligned} \tag{7}$$

or

$$p_l(\mathbf{y}) + T_1 + T_2 + T_3 + T_4 + T_5 + T_6 + T_7 + T_8 = 0 \tag{8}$$

The gradient components of $u_k$ at $\mathbf{y}$, i.e., $u_{k,h}(\mathbf{y})$ should be understood as components of the tangential gradient in the terms $T_1$ and $T_5$ corresponding to the expanssion of $u_k(\mathbf{x})$ over the surface $\Gamma - e_\varepsilon$.

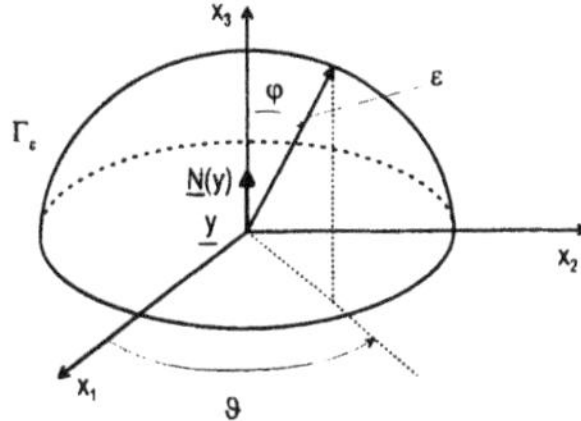


*Figure 2* Local coordinates system.

Since $s^*_{lmk}$ is hypersingular of order $r^{-3}$, and $d^*_{lmk}$ is singular of order $r^{-2}$, as $r \to 0$, the kernels in the integrals $T_1$ and $T_2$ are weakly singular. Therefore, $T_1$ can be numerically computed over the boundary $\Gamma$ and $T_2$=0.

To evaluate $T_3$ to $T_8$ the local system of coordinates in Figure 2 is used again. Consider first the integrals over the spherical surface $\Gamma_\varepsilon$ ($T_4$, $T_6$ and $T_8$). For these integrals (Figure 2) $r = \varepsilon$, $N_1 = N_2 = 0$, $N_3 = 1$, $\partial r/\partial n = 1$, $r_{,3} = n_3 = \cos\varphi$ and $d\Gamma = r^2 \sin\varphi \, d\varphi \, d\theta$. By substitution of these simple geometrical relations the integrals over $\Gamma_\varepsilon$ can be analytically evaluated (see details in Dominguez *et al.*, 2000).

$$T_4 = \lim_{\varepsilon \to o^+} \left\{ u_k(\mathbf{y}) \int_{\Gamma_\varepsilon} s^*_{lmk}(\mathbf{x}, \mathbf{y}) N_m d\Gamma \right\} = u_k(\mathbf{y}) \frac{\mu\,(\upsilon - 2)}{4\,(1-\upsilon)} \delta_{lk} \lim_{\varepsilon \to o} \frac{1}{\varepsilon} \quad (9)$$

$$T_6 = \lim_{\varepsilon \to o^+} \left\{ u_{k,h}(\mathbf{y}) \int_{\Gamma_\varepsilon} s^*_{lmk}(\mathbf{x}, \mathbf{y}) N_m (x_h - y_h) d\Gamma \right\} =$$
$$\frac{\mu\,(5\upsilon - 7)}{30\,(1-\upsilon)} u_{k,h} N_m (\delta_{lk}\delta_{hm} + \delta_{km}\delta_{hl}) \quad (10)$$

$$T_8 = -\lim_{\varepsilon \to o^+} \left\{ \sigma_{kh}(\mathbf{y}) \int_{\Gamma_\varepsilon} d^*_{lmk}(\mathbf{x}, \mathbf{y}) N_m n_h d\Gamma \right\} =$$
$$-\frac{4 - 5\upsilon}{30\,(1-\upsilon)} u_{k,h} N_m (\delta_{lk}\delta_{hm} + \delta_{km}\delta_{hl}) \quad (11)$$

for l = 1, 2 and

$$T_4 = u_k(\mathbf{y}) \frac{-2\mu}{4\,(1-\upsilon)} \delta_{lk} \lim_{\varepsilon \to o} \frac{1}{\varepsilon} \quad (12)$$

$$T_6 = \frac{-\mu}{1-\upsilon}\left[\frac{2+10\upsilon}{30}u_{k,h}\delta_{kh}\delta_{lm} + \frac{14-10\upsilon}{30}u_{k,h}\delta_{lk}\delta_{hm}\right]N_m \quad (13)$$

$$T_8 = -\frac{1}{1-\upsilon}\left[\frac{5\upsilon-1}{30}\sigma_{kh}\delta_{kh}\delta_{lm} + \frac{8-10\upsilon}{30}\sigma_{kh}\delta_{lk}\delta_{hm}\right]N_m \quad (14)$$

for $l = 3$.

In both cases, by adding (10) and (11) or (13) and (14), one obtains:

$$T_6 + T_8 = -\frac{1}{2}\sigma_{lm}(\mathbf{y})N_m = -\frac{1}{2}p_l(\mathbf{y}) \quad (15)$$

By using Stoke's theorem and some simple algebra the term $T_3$ can be transformed in a summation of regular and weakly singular surface integrals, line integrals over the contour of the boundary and a boundless term which cancels out the boundless term resulting from $T_4$. Thus,

$$T_3 = \lim_{\varepsilon\to o^+} u_k(\mathbf{y})\int_{\Gamma-e_\varepsilon} s^*_{lmk}N_m d\Gamma =$$
$$u_k(\mathbf{y})\frac{\mu}{4\pi(1-\upsilon)}\left\{I_{lk} + \lim_{\varepsilon\to o^+}\frac{\pi}{\varepsilon}\left[2+\upsilon\left(\delta_{3l}-1\right)\right]\delta_{lk}\right\} \quad (16)$$

with

$$I_{lk} = 3(1-2\upsilon)\int_\Gamma \frac{1}{r^3}\left(\frac{\partial r}{\partial n}N_k r_{,l} + r_{,m}N_m n_k r_{,l}\right)d\Gamma +$$
$$3\upsilon\int_\Gamma \frac{1}{r^3}\left(\frac{\partial r}{\partial n}N_l r_{,k} + r_{,m}N_m n_l r_{,k}\right)d\Gamma + 3(1-\upsilon)\delta_{lk}\int_\Gamma \frac{1}{r^3}\frac{\partial r}{\partial n}r_{,m}N_m d\Gamma +$$
$$15(\upsilon-1)\int_\Gamma \frac{r_{,l}r_{,k}\frac{\partial r}{\partial n}r_{,m}N_m}{r^3}d\Gamma + 3\upsilon\oint_{\partial\Gamma}\frac{\mathbf{r}\times\mathbf{N}}{r^3}r_{,l}r_{,k}d\mathbf{l}+$$
$$(1-2\upsilon)\delta_{lk}\oint_{\partial\Gamma}\frac{\mathbf{r}\times\mathbf{N}}{r^3}d\mathbf{l}+(4\upsilon-1)\oint_{\partial\Gamma}\frac{\mathbf{r}\times\mathbf{e}_k N_l}{r^3}d\mathbf{l}+$$
$$(1-2\upsilon)\oint_{\partial\Gamma}\frac{\mathbf{r}\times\mathbf{e}_l N_k}{r^3}d\mathbf{l} \quad (17)$$

where the line integrals extend over the contour $\partial\Gamma$ of surface $\Gamma$ and are zero when $\Gamma$ is a closed boundary.

In a similar way the term $T_5$ can be transformed by using Stoke's theorem to yield a summation of line and surface integrals which are, at most weakly singular. Thus,

$$T_5 = \lim_{\varepsilon \to o^+} \left\{ u_{k,h}(\mathbf{y}) \int_{\Gamma - e_\varepsilon} s^*_{lmk} N_m (x_h - y_h)\, d\Gamma \right\} =$$
$$u_{k,h}(\mathbf{y}) \frac{\mu}{4\pi\,(1-\upsilon)} J_{lhk} \tag{18}$$

where

$$J_{lhk} = 3\,(1-2\upsilon) \int_\Gamma \frac{r_{,l} r_{,h} (r_{,m} N_m n_k + N_k \frac{\partial r}{\partial n})}{r^2} d\Gamma +$$
$$3\upsilon \int_\Gamma \frac{r_{,k} r_{,h} (r_{,m} N_m n_l + N_l \frac{\partial r}{\partial n})}{r^2} d\Gamma - 15 \int_\Gamma \frac{r_{,l} r_{,k} r_{,h} \frac{\partial r}{\partial n} r_{,m} N_m}{r^2} d\Gamma +$$
$$3\,(1-3\upsilon) \int_\Gamma \frac{r_{,h} \frac{\partial r}{\partial n} (r_{,k} N_l - r_{,l} N_k)}{r^2} d\Gamma + 3\upsilon \int_\Gamma \frac{r_{,l} r_{,h} n_k r_{,m} N_m}{r^2} d\Gamma +$$
$$\upsilon \oint_{\partial\Gamma} \frac{r_{,l} r_{,h}}{r} (\mathbf{e}_k \times \mathbf{N})\, d\mathbf{l} - \upsilon \int_\Gamma \frac{N_h r_{,l} (n_k - N_k n_m N_m)}{r^2} d\Gamma -$$
$$\upsilon N_h N_k \int_\Gamma \frac{n_l r_{,m} N_m}{r^2} d\Gamma - \upsilon N_h N_k \oint_{\partial\Gamma} \frac{1}{r} (\mathbf{e}_l \times \mathbf{N})\, d\mathbf{l} +$$
$$(1-2\upsilon) \int_\Gamma \frac{N_k r_{,h} (n_l - N_l n_m N_m)}{r^2} d\Gamma -$$
$$(1-3\upsilon) \int_\Gamma \frac{N_l r_{,h} (n_k - N_k n_m N_m)}{r^2} d\Gamma + \upsilon N_k N_l \int_\Gamma \frac{n_h r_{,m} N_m}{r^2} d\Gamma +$$

$$\upsilon N_l N_k \oint_{\partial\Gamma} \frac{1}{r} (\mathbf{e}_h \times \mathbf{N})\, d\mathbf{l} + \delta_{kl} 3\upsilon \int_\Gamma \frac{\frac{\partial r}{\partial n} r_{,m} N_m r_{,h}}{r^2} d\Gamma +$$
$$\delta_{kl} \left\{ (1-\upsilon) \int_\Gamma \frac{n_h r_{,m} N_m}{r^2} d\Gamma + (1-\upsilon) \oint_{\partial\Gamma} \frac{1}{r} (\mathbf{e}_h \times \mathbf{N})\, d\mathbf{l} \right\} +$$
$$\delta_{kh} \left\{ \upsilon \int_\Gamma \frac{n_l r_{,m} N_m}{r^2} d\Gamma + \upsilon \oint_{\partial\Gamma} \frac{1}{r} (\mathbf{e}_l \times \mathbf{N})\, d\mathbf{l} \right\} \tag{19}$$

The term $T_7$ can be very easily transformed by a simple use of Stoke's theorem.

$$T_7 = -\lim_{\varepsilon \to o} \left\{ p_k(\mathbf{y}) \int_{\Gamma - e_\varepsilon} d^*_{lmk} N_m d\Gamma \right\} = p_k(\mathbf{y}) \frac{1}{4\pi\,(1-\upsilon)} K_{lk} \tag{20}$$

with

$$K_{lk} = -\frac{1}{2}\int_{\Gamma} \frac{r_{,m}N_m}{r^2}\left[(1-2\upsilon)\,\delta_{lk} + 3r_{,l}r_{,k}\right] d\Gamma -$$
$$\frac{1}{2}(1-2\upsilon)\int_{\Gamma}\frac{1}{r^2}\left[(N_k - n_k)\,r_{,l} - (N_l - n_l)\,r_{,k}\right] d\Gamma -$$
$$\frac{1}{2}\epsilon_{lkj}(1-2\upsilon)\oint_{\partial\Gamma}\frac{1}{r}dx_j \tag{21}$$

where $\epsilon_{lkj}$ is the permutation symbol which is equal to 1 (-1) if the subscripts permutate cyclically (anticyclically).

By subtitution of equations (9) to (21) into equation (8) one obtains a general traction BIE in terms only of regular and weakly singular integrals.

$$\frac{1}{2}p_l(\mathbf{y}) + \int_{\Gamma}\{s^*_{lmk}N_m\left[u_k(\mathbf{x}) - u_k(\mathbf{y}) - u_{k,h}(\mathbf{y})(x_h - y_h)\right] -$$
$$d^*_{lmk}N_m\left[p_k(\mathbf{x}) - p_k(\mathbf{y})\right]\}\, d\Gamma + \tag{22}$$
$$\frac{\mu}{4\pi(1-\upsilon)}\left[u_k(\mathbf{y})I_{lk} + u_{k,h}(\mathbf{y})J_{lhk} + p_k(\mathbf{y})K_{lk}\right] = 0$$

where $I_{lk}$, $J_{lhk}$ and $K_{lk}$ are given by equations (17), (19) and (21), respectively. This BIE is valid for open or closed, plane or curved boundaries and can be discretized in a straightforward manner using a boundary element formulation.

## 3. BOUNDARY ELEMENT FORMULATION

The present BE technique is based on the use of the traction BIE in combination with the classical BIE to obtain a mixed BE approach very suitable for fracture mechanic problems. It has aspects in common with the one called Dual Boundary Element Method by Mi and Aliabadi (1992). In the current approach, the external boundary and any internal boundary different to crack surfaces are discretized into standard quadratic elements where the classical BIE is used. The traction BIE is used for one face of the crack. To avoid the need of writting another equation for points on the other crack surface, the Crack Opening Displacement (COD) is chosen as the basic variable on the crack. In very special cases where the actual displacement components at the crack surfaces, are required one should write the tractions BIE for one surface of the crack and the standard displacement BIE for the other.

The two facts which make the discretization of the tractions BIE different to the discretization of the classical BIE are: (1), the boundary displacement $u_k$ must satisfy the Hölder continuity condition $u_k \in C^{1,\alpha}$, at **y**; and (2), derivatives, at **y**, of the displacement components exist in the boundary integral equation.

Following the idea of Gallego and Dominguez (1996) for two dimensions, the boundary integral equation is discretized into quadratic surface elements with six or nine nodes located at their usual position to represent the geometry and the boundary variables. However, the collocation is not done at the contour nodes; i.e. at $\xi_1$, $\xi_2 = \pm 1$, but at certain points close to the nodes, inside the element.

It should be noticed that elements of this type are continuous $C^o$ since the boundary variables are written as usual in terms of their values at nodes located at the contour of the element. Only the collocation points are shifted to the interior of the element. Tractions and displacements at each collocation point are expressed in terms of nodal values and shape function values at collocation point. The displacement derivatives are expressed in terms of nodal displacements and the shape function derivatives at collocation point.

As a consequence of the collocation strategy, one may have two or more equations for each nodal component, obtained by collocation at as many points as elements contain the node. These equations are added up to yield only one per nodal component. This Multiple Collocation Approach (MCA) is only used for the nodes on one crack surface except for those inside an element and nodes at the crack front where no collocation at all is required ($\Delta u_k = 0$).

Since the regularization process leads to expressions of the boundary integrals with more terms than the original hypersingular and strongly singular integrals, the regular expressions of the integral will only be used over a part of the surface $\Gamma$ close to the collocation point whereas the original expressions of the integrals are used in the rest of the boundary where they are non-singular.

# 4. NUMERICAL EXAMPLES

A penny shaped crack subject to internal pressure and included in an infinite elastic domain is studied. The crack is subject to internal pressure. The properties of the boundless elastic domain are: shear modulus, $\mu = 10^6 Pa$ and Poisson Ratio, $\nu = 0.3$.

Only one surface of the crack is discretized into quadratic nine node quadrilateral elements with those at the crack front being quarter-point elements (Figure 3). The mode-I Stress Intensity Factor (SIF) $K_I$ is evaluated from the Crack Opening Displacement (COD) at the quarter-point nodes located at the first row of nodes from the crack front.

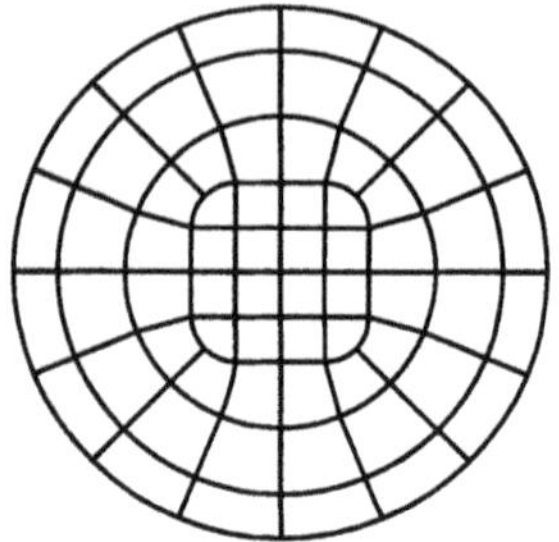

*Figure 3* Boundary element discretization of a penny shaped crack in infinite domain.

$$K_I = \frac{\mu}{2\,(1-\nu)} \Delta u \sqrt{\frac{2\pi}{L}} \tag{23}$$

where L is the element width and $\Delta$u the normal opening displacement.

The SIF evaluated using the BE mesh in Figure 4 is $K_I = 1.989\ \sigma\sqrt{\frac{a}{\pi}}$ which is within a 1% range from the analytical solution ($K_I = 2\ \sigma\sqrt{\frac{a}{\pi}}$) obtained by Sneddon (1946).

## 5. CONCLUSION

A formulation which permits the numerical treatment of traction BIE without special difficulties has been presented in this paper. The hypersingular BIE, is transformed to yield surface integrals which are regular or at most weakly singular, and line integrals over the contour of the boundary. The integration process is completely general and valid for arbitrary shaped cracks.

The regular and weakly singular surface integrals become zero when the integration surface is plane. The line integrals are zero for problems with closed boundaries.

One of the main issues of this paper is to contribute to transform hypersingular formulations of the BEM in something which is as clear, general and easy to handle as the classical formulation. Ideas such as Hadamart Finite Part or Cauchy Principal Values are not needed. All the integrals can be numerically evaluated in an easy way. The boundary of the problem can be curved or plane, closed or open.

Once the integrals have been regularized, the only particular difficulty of HBEM is related to the $C^{1,\alpha}$ continuity requirement over the primary density. A simple BE discretization strategy is adopted to fulfil this regularity condition without increasing the number of equations or introducing more complicated continuous elements.

The generality of the current integration procedure allows for the use of special crack elements such as quarter point. The accuracy of the numerical results obtained is very good.

## Acknowledgments

This work was supported by the Comisión Interministerial de Ciencia y Tecnología of Spain. (PB96-1380 and PB96-1322-C03-01). The financial support is gratefully acknowledged.

## References

[1] Aliabadi M. H., Hall W. S. and Phemister T. G. (1985) "Taylor Expansions for Sin- gular Kernels in Boundary Element Method", Int. J. Num. Meth. Eng., **21**, pp. 2221-2236.

[2] Dominguez J., Ariza M.P. and Gallego R. (2000) "Flux and Traction Boundary Elements without Hypersingular or Strongly Singular Integrals", Int. J. Numer. Meth. Eng., **48**, pp. 111-135.

[3] Gallego R. and Domínguez J. (1996) "Hypersingular BEM for Transient Elastodyna- mics", Int. J. Numer. Meth. Eng., **39**, pp. 1681-1705.

[4] Guiggiani M., Krishnasamy G., Rudolphi T. J. and Rizzo F. J.(1992) "A General Al- gorithm for the Numerical Solution of Hypersingular Boundary Integral Equations", J. Appl. Mech. (ASME), **59**, pp. 604-614.

[5] Mantic V. (1994) "On Computing Boundary Limiting Values of Boundary Integrals with Strongly Singular and Hypersingular Kernels in 3D BEM for Elastostatics", Eng. Anal. with Boundary Elements, **13**, pp. 115-134.

[6] Mi Y. and Aliabadi M. H. (1992) "Dual Boundary Element Method for Three-Dimensional Fracture Mechanics Analysis", Eng. Anal. Boundary Elements, **10**, pp. 161-171.

[7] Sneddon I. N. (1946) "The Distribution of Stress in the Nieghbourhood of a Crack in an Elastic Solid", Proc. Roy. Soc. London, Ser. A, **187**, pp 229-260.

# SHAPE IDENTIFICATION OF 3-D OBSTACLES

Veit Dröge
Friedel Hartmann
*University of Kassel*
*Structural Mechanics*
*D-34125 Kassel*
hartmann@hrz.uni-kassel.de

**Abstract** The aim of the paper is the non-destructive evaluation of elastic media from scattered waves $\boldsymbol{u} = \boldsymbol{u}^{\text{in}} + \boldsymbol{u}^{\text{sc}}$ to determine the shapes and positions of inclusions or cavities in 3-D elastic media. The method works in the frequency domain and is based on the far field properties of the scattered waves.

## 1. SCATTERED WAVES

An incident wave, $\mathcal{L}\boldsymbol{u}^{\text{in}}(\boldsymbol{x}, \omega) = 0$, $\boldsymbol{x} \in R^3$ can be realized by a so-called plane harmonic wave

$$\boldsymbol{u}^{\text{in}}(\boldsymbol{x}, \omega) = \boldsymbol{A}_s^{\text{in}}\, e^{i\boldsymbol{k}_s\boldsymbol{x}} + \boldsymbol{A}_p^{\text{in}}\, e^{i\boldsymbol{k}_p\boldsymbol{x}}$$

where the planes of equal amplitudes are orthogonal to the direction $\boldsymbol{e}_{\text{wave}}$ of propagation. Because of this the wave vectors

$$\boldsymbol{k_s} = \frac{\omega}{c_s}\,\boldsymbol{e}_{\text{wave}}\,, \qquad \boldsymbol{k_p} = \frac{\omega}{c_p}\,\boldsymbol{e}_{\text{wave}}\,,$$

as well as the vectors $\boldsymbol{A}_s, \boldsymbol{A}_p$, which represent the amplitude and directions of the transverse and orthogonal waves respectively

$$\boldsymbol{A}_s \perp \boldsymbol{e}_{\text{wave}}\,, \qquad \boldsymbol{A}_p \,||\, \boldsymbol{e}_{\text{wave}}.$$

are constant.
The total field $\boldsymbol{u}(\boldsymbol{x}, \omega) = \boldsymbol{u}^{\text{in}}(\boldsymbol{x}, \omega) + \boldsymbol{u}^{\text{sc}}(\boldsymbol{x}, \omega)$ consists of the incident wave $\boldsymbol{u}^{\text{in}}(\boldsymbol{x}, \omega)$ and the scattered wave $\boldsymbol{u}^{\text{sc}}(\boldsymbol{x}, \omega)$. The scattered wave must be a homogeneous solution in the exterior, satisfy displacement and traction boundary conditions on the surface $\Gamma = \partial\hat{\Omega}$ of the obstacle

*T. Burczynski (ed.), IUTAM/IACM/IABEM Symposium on Advanced Mathematical and Computational Mechanics Aspects of the Boundary Element Method*, 99–109.

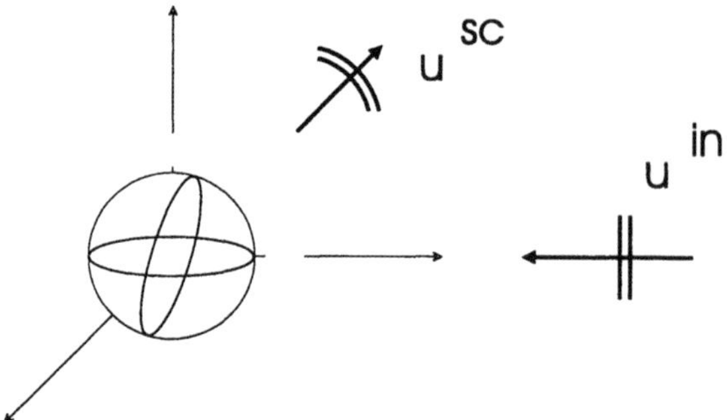


*Figure 1* Fig. 1 Scattering of an elastic wave

and in addition both parts $\boldsymbol{u}_s^{\text{sc}}$ und $\boldsymbol{u}_p^{\text{sc}}$ must satisfy *Sommerfelds radiation condition* at infinity.

## 2. INTEGRAL REPRESENTATIONS

The starting point is the standard integral representation in a domain $\Omega$ which we assume to be bounded by the surface $\Gamma$ of the obstacle and an outer spherical boundary $\Gamma_\infty$ which is the surface of a ball $\Omega_\infty$ which contains the obstacle $\hat{\Omega}$ completely.

$$\mathbf{C}(\boldsymbol{x})\boldsymbol{u}(\boldsymbol{x}) = \iint_\Gamma \mathbf{U}(\boldsymbol{y},\boldsymbol{x})\,\boldsymbol{t}(\boldsymbol{y})\,d\Gamma_y - \oint\!\!\oint_\Gamma \mathbf{T}(\boldsymbol{y},\boldsymbol{x})\,\boldsymbol{u}(\boldsymbol{y})\,d\Gamma_y$$
$$+ \iint_{\Gamma_\infty} \mathbf{U}(\boldsymbol{y},\boldsymbol{x})\,\boldsymbol{t}(\boldsymbol{y})\,d\Gamma_y - \oint\!\!\oint_{\Gamma_\infty} \mathbf{T}(\boldsymbol{y},\boldsymbol{x})\,\boldsymbol{u}(\boldsymbol{y})\,d\Gamma_y\,.$$

Because of $\boldsymbol{u} = \boldsymbol{u}^{\text{sc}} + \boldsymbol{u}^{\text{in}}$ and $\boldsymbol{t} = \boldsymbol{t}^{\text{sc}} + \boldsymbol{t}^{\text{in}}$ this can also be written as

$$\mathbf{C}(\boldsymbol{x})\boldsymbol{u}(\boldsymbol{x}) = \iint_\Gamma \mathbf{U}(\boldsymbol{y},\boldsymbol{x})\,\boldsymbol{t}(\boldsymbol{y})\,d\Gamma_y - \oint\!\!\oint_\Gamma \mathbf{T}(\boldsymbol{y},\boldsymbol{x})\,\boldsymbol{u}(\boldsymbol{y})\,d\Gamma_y$$
$$+ \iint_{\Gamma_\infty} \mathbf{U}(\boldsymbol{y},\boldsymbol{x})\,\boldsymbol{t}^{\text{sc}}(\boldsymbol{y})\,d\Gamma_y - \oint\!\!\oint_{\Gamma_\infty} \mathbf{T}(\boldsymbol{y},\boldsymbol{x})\,\boldsymbol{u}^{\text{sc}}(\boldsymbol{y})\,d\Gamma_y$$
$$+ \iint_{\Gamma_\infty} \mathbf{U}(\boldsymbol{y},\boldsymbol{x})\,\boldsymbol{t}^{\text{in}}(\boldsymbol{y})\,d\Gamma_y - \oint\!\!\oint_{\Gamma_\infty} \mathbf{T}(\boldsymbol{y},\boldsymbol{x})\,\boldsymbol{u}^{\text{in}}(\boldsymbol{y})\,d\Gamma_y\,.$$

Because of the radiation condition the influence of the scattered quantities $\boldsymbol{u}^{\text{sc}}$ and $\boldsymbol{t}^{\text{sc}}$ on the surface $\Gamma_\infty$ is negligible. The integrals of the incident quantities $\boldsymbol{u}^{\text{in}}$ and $\boldsymbol{t}^{\text{in}}$ can be handled as follows: For the incident wave $\boldsymbol{u}^{\text{in}}$ in the domain $\Omega_\infty$ with boundary $\Gamma_\infty$ we have

$$\boldsymbol{u}^{\text{in}}(\boldsymbol{x}) = \iint_{\Gamma_\infty} \mathbf{U}(\boldsymbol{y},\boldsymbol{x})\, \boldsymbol{t}^{\text{in}}(\boldsymbol{y})\, d\Gamma_y - \oint\!\!\!\oint_{\Gamma_\infty} \mathbf{T}(\boldsymbol{y},\boldsymbol{x})\, \boldsymbol{u}^{\text{in}}(\boldsymbol{y})\, d\Gamma_y \tag{1}$$

Hence we obtain for all points inside the domain $\Omega_\infty$ the following integral representation

$$\mathbf{C}(\boldsymbol{x})\boldsymbol{u}(\boldsymbol{x}) = \iint_{\Gamma} \mathbf{U}(\boldsymbol{y},\boldsymbol{x})\, \boldsymbol{t}(\boldsymbol{y})\, d\Gamma_y - \oint\!\!\!\oint_{\Gamma} \mathbf{T}(\boldsymbol{y},\boldsymbol{x})\, \boldsymbol{u}(\boldsymbol{y})\, d\Gamma_y + \boldsymbol{u}^{\text{in}}(\boldsymbol{x})$$

for the solution. Because we can choose the radius of the ball arbitrarily large the equation is valid for any point $\boldsymbol{x} \in R^3$.

Subtracting on both sides the incident wave we obtain an integral representation for the scattered wave alone

$$\begin{aligned}\mathbf{C}(\boldsymbol{x})\boldsymbol{u}^{\text{sc}}(\boldsymbol{x}) = &\iint_{\Gamma} \mathbf{U}(\boldsymbol{y},\boldsymbol{x})\, \boldsymbol{t}^{\text{in}}(\boldsymbol{y})\, d\Gamma_y - \oint\!\!\!\oint_{\Gamma} \mathbf{T}(\boldsymbol{y},\boldsymbol{x})\, \boldsymbol{u}^{\text{in}}(\boldsymbol{y})\, d\Gamma_y \\ &+ \iint_{\Gamma} \mathbf{U}(\boldsymbol{y},\boldsymbol{x})\, \boldsymbol{t}^{\text{sc}}(\boldsymbol{y})\, d\Gamma_y - \oint\!\!\!\oint_{\Gamma} \mathbf{T}(\boldsymbol{y},\boldsymbol{x})\, \boldsymbol{u}^{\text{sc}}(\boldsymbol{y})\, d\Gamma_y\end{aligned}$$

The first two integrals are equivalent with Eq. (1), hence they vanish for any $\boldsymbol{x} \in \Omega$ and we obtain the integral representation for the scattered wave

$$\boldsymbol{C}(\boldsymbol{x})\, \boldsymbol{u}^{sc}(\boldsymbol{x}) = \iint_{\Gamma} \boldsymbol{U}(\boldsymbol{y},\boldsymbol{x})\boldsymbol{t}^{sc}(\boldsymbol{y})\, d\Gamma - \oint\!\!\!\oint_{\Gamma} \boldsymbol{T}(\boldsymbol{y},\boldsymbol{x})\, \boldsymbol{u}^{sc}(\boldsymbol{y})\, d\Gamma\,.$$

## 3. FAR FIELD APPROXIMATIONS

The fundamental solutions are functions of the source point $\boldsymbol{x}$ and the field point $\boldsymbol{y}$. If we let $\boldsymbol{x} = x\,\hat{\boldsymbol{x}}$ where $\hat{\boldsymbol{x}}$ is a unit vector then the distance $r = |\boldsymbol{y} - \boldsymbol{x}|$ can be approximated by $r \approx x - \boldsymbol{x} \bullet \boldsymbol{y}$ so that

$$\frac{e^{ikr}}{r} \approx \frac{e^{ik(x-\hat{\boldsymbol{x}}\cdot\boldsymbol{y})}}{x} \qquad \frac{\partial r}{\partial x_i} \approx \frac{x_i}{x} = \hat{x}_i\,,$$

and hence we obtain for the functions $\psi$, $\chi$ and their derivatives

$$\psi \approx \frac{e^{ik_s r}}{r}, \qquad \chi \approx \frac{e^{ik_s r}}{r} - \frac{{c_s}^2}{{c_p}^2}\frac{e^{ik_p r}}{r},$$

$$\frac{d\psi}{dr} \approx ik_s\frac{e^{ik_s r}}{r}, \qquad \frac{d\chi}{dr} \approx ik_s\frac{e^{ik_s r}}{r} - ik_p\frac{c_{\text{p}}^2}{c_{\text{s}}^2}\frac{e^{ik_p r}}{r}\,.$$

Substituting these results into the previous equations we obtain

$$\mathbf{U}_{lk}^{\infty}(\boldsymbol{y},\boldsymbol{x}) = \frac{1}{4\pi\rho c_{\text{s}}^2}\frac{e^{ik_s x}}{x}\left[\delta_{lk} - \hat{x}_l\hat{x}_k\right] e^{-ik_s\hat{\boldsymbol{x}}\cdot\boldsymbol{y}}$$

$$+\frac{1}{4\pi\rho c_p^2}\frac{e^{ik_p x}}{x}\left[\hat{x}_l\hat{x}_k\right]e^{-ik_p\hat{\boldsymbol{x}}\cdot\boldsymbol{y}}$$

$$\mathbf{T}^{\infty}_{lk}(\boldsymbol{y},\boldsymbol{x}) = \frac{ik_s}{4\pi}\frac{e^{ik_s x}}{x}\left[(\delta_{lk}-2\hat{x}_l\hat{x}_k)(n_j\hat{x}_j)+\hat{x}_k n_l\right]e^{-ik_s\hat{\boldsymbol{x}}\cdot\boldsymbol{y}}$$

$$+\frac{ik_p}{4\pi}\frac{c_s^2}{c_p^2}\frac{e^{ik_p x}}{x}\left[2\hat{x}_l\hat{x}_k n_j\hat{x}_j+\left(\frac{c_p^2}{c_s^2}-2\right)\hat{x}_l n_k\right]e^{-ik_p\hat{\boldsymbol{x}}\cdot\boldsymbol{y}}$$

and for the product with an arbitrary vector $\boldsymbol{f}$

$$\mathbf{U}^{\infty}(\boldsymbol{y},\boldsymbol{x})\,\boldsymbol{f} = \frac{1}{4\pi\rho c_s^2}\frac{e^{ik_s x}}{x}\,\hat{\boldsymbol{x}}\times\left(\boldsymbol{f}\times\hat{\boldsymbol{x}}\right)e^{-ik_s\hat{\boldsymbol{x}}\cdot\boldsymbol{y}}$$

$$+\frac{1}{4\pi\rho c_p^2}\frac{e^{ik_p x}}{x}\,\hat{\boldsymbol{x}}\cdot\left(\boldsymbol{f}\cdot\hat{\boldsymbol{x}}\right)e^{-ik_p\hat{\boldsymbol{x}}\cdot\boldsymbol{y}}$$

$$\mathbf{T}^{\infty}(\boldsymbol{y},\boldsymbol{x})\,\boldsymbol{f} = \frac{ik_s}{4\pi}\frac{e^{ik_s x}}{x}\,\hat{\boldsymbol{x}}\times\left[\left(\boldsymbol{f}\times\hat{\boldsymbol{x}}\right)\left(\boldsymbol{n}\cdot\hat{\boldsymbol{x}}\right)\right.$$

$$\left.+\left(\boldsymbol{n}\times\hat{\boldsymbol{x}}\right)\left(\boldsymbol{f}\cdot\hat{\boldsymbol{x}}\right)\right]e^{-ik_s\hat{\boldsymbol{x}}\cdot\boldsymbol{y}}$$

$$+\frac{ik_p}{4\pi}\frac{c_s^2}{c_p^2}\frac{e^{ik_p x}}{x}\,\hat{\boldsymbol{x}}\cdot\left[2\left(\boldsymbol{f}\cdot\hat{\boldsymbol{x}}\right)\left(\boldsymbol{n}\cdot\hat{\boldsymbol{x}}\right)+\right.$$

$$\left.\left(\frac{c_p^2}{c_s^2}-2\right)\left(\boldsymbol{f}\cdot\boldsymbol{n}\right)\right]e^{-ik_p\hat{\boldsymbol{x}}\cdot\boldsymbol{y}}\;.$$

If $(R,\hat{\boldsymbol{x}})$ denotes the polar coordinates of a point $\boldsymbol{x}$ and if $R$ is large then the far field can be represented by the following asymptotic expansion

$$\boldsymbol{u}^{sc}(R,\boldsymbol{x}) = \boldsymbol{A}_s(\boldsymbol{x})\frac{e^{i\,k_s\,R}}{R}+\boldsymbol{A}_p(\boldsymbol{x})\frac{e^{i\,k_p\,R}}{R}$$

where the factors $\boldsymbol{A}_s$ and $\boldsymbol{A}_p$ are independent of the distance $R$.

$$\boldsymbol{A}_s(\boldsymbol{x}) = \frac{\boldsymbol{x}}{4\pi\rho c_s^2}\int_\Gamma\{ik_s\rho c_s^2[(\boldsymbol{u}\times\boldsymbol{x})(\boldsymbol{n}\cdot\boldsymbol{x}])$$

$$+(\boldsymbol{n}\times\boldsymbol{x})(\boldsymbol{u}\cdot\boldsymbol{x})]+\boldsymbol{t}\times\boldsymbol{x}\}e^{-ik_s\boldsymbol{x}\cdot\boldsymbol{y}}ds_{\boldsymbol{y}}$$

$$\boldsymbol{A}_p(\boldsymbol{x}) = \frac{\boldsymbol{x}}{4\pi\rho c_p^2}\int_\Gamma\{ik_p\rho[2c_s^2(\boldsymbol{u}\cdot\boldsymbol{x})(\boldsymbol{n}\cdot\boldsymbol{x}])+(c_p^2-2c_s^2)(\boldsymbol{u}\cdot\boldsymbol{n})]$$

$$+\boldsymbol{t}\times\boldsymbol{x}\}e^{-ik_p\boldsymbol{x}\cdot\boldsymbol{y}}ds_{\boldsymbol{y}}$$

## 4. ADJOINT PROBLEM

The shape of the $3-D$-obstacle or cavity is found by minimizing the following function with regard to a set of parameters $\boldsymbol{\tau}=(\tau_1,\tau_2,\ldots,\tau_n)$

which represent the position and the shape of the obstacle

$$J(\boldsymbol{u},\boldsymbol{\tau}) = \frac{1}{2}\int_C |\boldsymbol{u}_\tau(\boldsymbol{y}) - \boldsymbol{u}(\boldsymbol{y})|^2\, q(\boldsymbol{y}) ds_{\boldsymbol{y}}\,.$$

To calculate the derivative of the function $J$

$$\frac{d}{d\tau} J(\boldsymbol{u},\boldsymbol{\tau}) = \Re \int_C \frac{\partial g}{\partial \boldsymbol{u}} \frac{\partial \boldsymbol{u}_\tau^{sc}}{\partial \tau}\, ds$$

we need the gradient $\partial \boldsymbol{u}_\tau^{sc}/\partial \boldsymbol{\tau}$. Following the ideas of Haug et. al. [2] and Bonnet [3] the gradient can be found by solving an adjoint problem. In the following we shall assume that the traction vector of the field $\boldsymbol{u}$ vanishes on the surface of the obstacle (cavity).
We consider the differential equation for the scattered field $\boldsymbol{u}^{\mathrm{sc}}$

$$\left[\mu\Delta + (\lambda+\mu)\nabla\nabla + \rho\omega^2\right] \boldsymbol{u}^{\mathrm{sc}}(\boldsymbol{x}) = 0$$

with the boundary condition

$$\boldsymbol{t}(\boldsymbol{x}) = \boldsymbol{t}^{\mathrm{sc}}(\boldsymbol{x}) + \boldsymbol{t}^{\mathrm{in}}(\boldsymbol{x}) = 0, \qquad \boldsymbol{x} \in \Gamma$$

and appropriate radiation conditions. A solution of this problem satisfies

$$\mathcal{A}_\Omega(\boldsymbol{u}^{\mathrm{sc}}, \boldsymbol{w}) = \mathcal{B}_\Omega(\boldsymbol{w}) \qquad \text{for all } \boldsymbol{w} \in H^1(\Omega) \tag{2}$$

where

$$\mathcal{A}_\Omega(\boldsymbol{u}^{\mathrm{sc}}, \boldsymbol{w}) = \iiint_\Omega E(\boldsymbol{u}^{\mathrm{sc}}, \boldsymbol{w})\, d\Omega\,, \qquad \mathcal{B}_\Omega(\boldsymbol{w}) = \iint_\Gamma t(\boldsymbol{u}^{\mathrm{sc}})\boldsymbol{w}^\star\, d\Gamma$$

and

$$E(\boldsymbol{u},\boldsymbol{w}) = \tfrac{1}{2}\mu(\nabla\boldsymbol{u} + \nabla\boldsymbol{u}^T)\cdot(\nabla\boldsymbol{w} + \nabla\boldsymbol{w}^T)^\star + \lambda\, \mathrm{div}\boldsymbol{u}\, \mathrm{div}\boldsymbol{w}^\star - \rho\omega^2 \boldsymbol{u}\, \boldsymbol{w}^\star.$$

denotes the strain energy density.

The property (2) leads to a variational formulation: we are looking for a solution $\boldsymbol{z}$

$$\mathcal{A}_\Omega(\boldsymbol{z}, \boldsymbol{w}) = \iint_C \frac{\partial g}{\partial {\boldsymbol{u}_\tau}^\star} \boldsymbol{w}^\star\, dC \qquad \text{for all } \boldsymbol{w} \in H^1(\Omega). \tag{3}$$

If we let

$$\boldsymbol{w} = \frac{\partial \boldsymbol{u}_\tau^{\mathrm{sc}}}{\partial \tau}\,,$$

we obtain

$$\frac{d}{d\tau} J(\boldsymbol{u};\tau) = \Re\left\{\mathcal{A}_\Omega(\boldsymbol{z}, \frac{\partial \boldsymbol{u}^{\mathrm{sc}}}{\partial \tau})\right\} = \Re\left\{\mathcal{A}_\Omega(\frac{\partial \boldsymbol{u}^{\mathrm{sc}}}{\partial \tau}, \boldsymbol{z})\right\}.$$

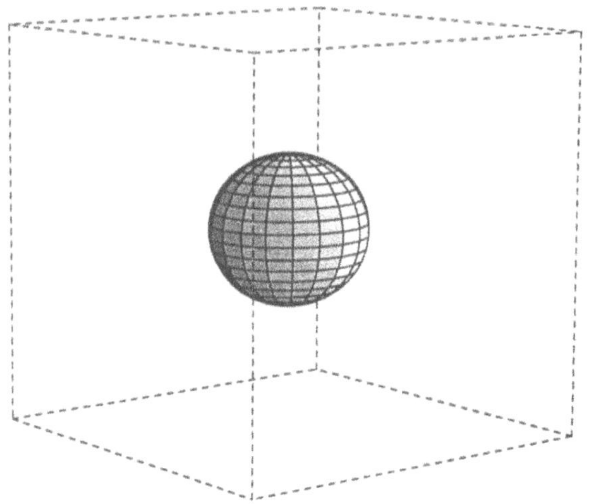

*Figure 2* Fig. 2 Original shape at the start of the iteration

Because the solution $\boldsymbol{z}$ of the adjoint problem does not depend on $\tau$, $\partial \boldsymbol{z}/\partial \tau = 0$, we may write

$$\mathcal{A}_\Omega(\frac{\partial \boldsymbol{u}^{\,\mathrm{sc}}}{\partial \tau}, \boldsymbol{z}) = \iiint_\Omega E(\frac{\partial \boldsymbol{u}^{\,\mathrm{sc}}}{\partial \tau}, \boldsymbol{z})\, d\Omega = \iiint_\Omega \frac{\partial E(\boldsymbol{u}^{\,\mathrm{sc}}, \boldsymbol{z})}{\partial \tau}\, d\Omega\,.$$

The domain $\Omega$ of integration depends on $\tau$ and if we interchange differentiation with integration we therefore obtain the following expression

$$\mathcal{A}_\Omega(\frac{\partial \boldsymbol{u}^{\,\mathrm{sc}}}{\partial \tau}, \boldsymbol{z}) = \underbrace{\frac{d}{d\tau} \iiint_{\Omega_\tau} E(\boldsymbol{u}^{\,\mathrm{sc}}, \boldsymbol{z})\, d\Omega}_{\mathcal{A}_1} - \iint_\Gamma E(\boldsymbol{u}^{\,\mathrm{sc}}, \boldsymbol{z})\Theta\, d\Gamma,$$

where

$$\Theta(\boldsymbol{y}) = \left.\frac{d\boldsymbol{y}(\tau)}{d\tau}\right|_\Gamma \cdot \boldsymbol{n}.$$

Because of (2) the integral in $\mathcal{A}_1$ can be transformed into a surface integral

$$\mathcal{A}_1 = \frac{d}{d\tau} \iint_{\Gamma_\tau} \boldsymbol{t}^{\,\mathrm{sc}} \boldsymbol{z}^\star\, d\Gamma = \iint_\Gamma \left\{ \frac{d}{d\tau}(\boldsymbol{t}^{\,\mathrm{sc}} \boldsymbol{z}^\star) - 2K \boldsymbol{t}^{\,\mathrm{sc}} \boldsymbol{z}^\star \Theta \right\} d\Gamma$$

where the *mean curvature* is

$$K = -\frac{1}{2} \operatorname{div} \boldsymbol{n}.$$

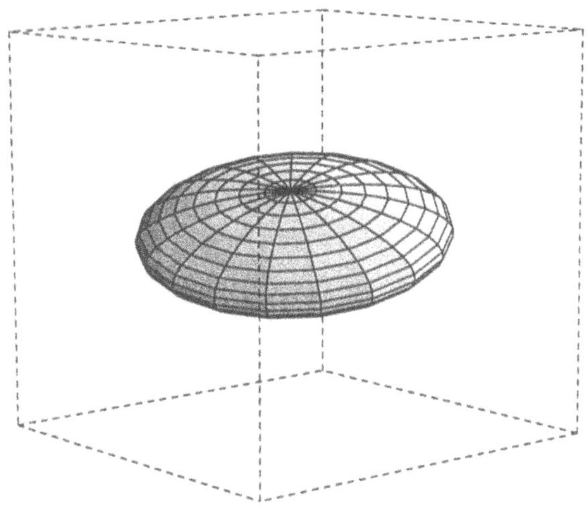

*Figure 3* Fig. 3 Final shape after 50 iterations, $e_J = 5.7 \cdot 10^{-8}$

Because of zero tractions on the surface of the cavity we have for the incident wave

$$\mathcal{A}_1 = \iint_\Gamma \left\{ 2K\boldsymbol{t}^{\text{in}}\boldsymbol{z}^\star\Theta - \frac{d\boldsymbol{t}^{\text{in}}}{d\tau}\boldsymbol{z}^\star - \boldsymbol{t}^{\text{in}}\frac{d\boldsymbol{z}^\star}{d\tau} \right\} d\Gamma$$

and therefore

$$\begin{aligned}\mathcal{A}_1 = \iint_\Gamma &(2Kt_i^{\text{in}}{z_i}^\star\Theta - t_{i,n}^{\text{in}}{z_i}^\star\Theta + \mu(u_{i,j}^{\text{in}} + u_{j,i}^{\text{in}}){z_i}^\star D_j\Theta \\ &+\lambda u_{j,j}^{\text{in}}{z_i}^\star D_i\Theta - t_i^{\text{in}}{z_{i,n}}^\star\Theta)d\Gamma\end{aligned}$$

Here $D_j$ denotes the tangential derivative. To simplify we drop in the following equations the superscript $^{\text{in}}$. The first term in the integral can be written as

$$\begin{aligned}&\iint_\Gamma 2Kt_i^{\text{in}}{z_i}^\star\Theta d\Gamma \\ &= \iint_\Gamma 2K\left[\mu(u_{i,j} + u_{j,i})n_j + \lambda u_{j,j}n_i\right]{z_i}^\star\Theta\, d\Gamma \\ &= -\iint_\Gamma \left\{ D_j\left[\mu(u_{i,j} + u_{j,i}){z_i}^\star\Theta\right] + D_i\left[\lambda u_{j,j}{z_i}^\star\Theta\right] \right\} d\Gamma \\ &= -\iint_\Gamma \left\{ \mu{z_i}^\star\Theta D_j[u_{i,j} + u_{j,i}] + \mu(u_{i,j} + u_{j,i})\Theta D_j{z_i}^\star \right. \\ &\qquad +\mu(u_{i,j} + u_{j,i}){z_i}^\star D_j\Theta \\ &\qquad \left. +\lambda{z_i}^\star\Theta D_i u_{j,j} + \lambda u_{j,j}\Theta D_i{z_i}^\star + \lambda u_{j,j}{z_i}^\star D_i\Theta \right\} d\Gamma\end{aligned}$$

and we so obtain

$$\mathcal{A}_1 = -\iint_\Gamma \left\{\mathcal{A}_2 + \mathcal{A}_3 z_i{}^\star\right\}\Theta\, d\Gamma$$

where

$$\begin{aligned}\mathcal{A}_2 &= t_i z_{i,n}{}^\star + \mu(u_{i,j} + u_{j,i})D_j z_i{}^\star + \lambda u_{j,j} D_i z_i{}^\star\,,\\ \mathcal{A}_3 &= t_{i,n} + \mu D_j[u_{i,j} + u_{j,i}] + \lambda D_i u_{j,j}\,.\end{aligned}$$

Because

$$\begin{aligned}t_i z_{i,n} &= \mu(u_{i,j} + u_{j,i})(z_{i,j}{}^\star - D_j z_i{}^\star) + \lambda u_{j,j}(z_{i,i}{}^\star - D_i z_i{}^\star)\,,\\ t_{i,n} &= \mu u_{i,nn} + \mu n_j u_{j,in} + \lambda n_i u_{j,jn}\,.\end{aligned}$$

we get

$$\mathcal{A}_2 = \mu(u_{i,j}+u_{j,i})z_{i,j}{}^\star + \lambda u_{j,j} z_{i,i}{}^\star = \tfrac{1}{2}\mu(u_{i,j}+u_{j,i})(z_{i,j} + z_{j,i})^\star + \lambda u_{j,j} z_{i,i}{}^\star.$$

and

$$\mathcal{A}_3 = \mu(u_{i,jj} + u_{j,ij}) + \lambda u_{j,ji} = -\rho\omega^2 u_i.$$

Hence

$$\begin{aligned}\mathcal{A}_1 &= -\iint_\Gamma \left\{\tfrac{1}{2}\mu(u_{i,j} + u_{j,i})(z_{i,j} + z_{j,i})^\star + \lambda u_{j,j} z_{i,i}{}^\star - \rho\omega^2 u_i z_i{}^\star\right\}\Theta\, d\Gamma\\ &= -\iint_\Gamma E(\boldsymbol{u}^{\,\mathrm{in}}, \boldsymbol{z})\Theta\, d\Gamma. \qquad (4)\end{aligned}$$

Because the energy term is bilinear we therefore have in the end

$$\frac{d}{d\tau}J(\boldsymbol{u};\tau) = -\Re\left\{\iint_\Gamma E(\boldsymbol{u}, \boldsymbol{z})\Theta\, d\Gamma\right\}.$$

Hence we can calculate the gradient of the functional which is to be minimized by evaluating a surface integral. The difficult part is to evaluate the strain energy density $E$ on the surface $\Gamma$. Using the concept of a tangential derivative $\nabla_S$ we obtain after some intermediate steps the following expression

$$\begin{aligned}E(\boldsymbol{u}, \boldsymbol{z}) = \frac{1}{2}\mu(\nabla_S\, \boldsymbol{u}^S + \nabla_S\, \boldsymbol{u}^{ST})\cdot(\nabla_S\, \boldsymbol{z}^S + \nabla_S\, \boldsymbol{z}^{ST})^\star\\ + \frac{2\lambda\mu}{\lambda + 2\mu}\,\mathrm{div}_S \boldsymbol{u}^S\, \mathrm{div}_S \boldsymbol{z}^{S\star} - \rho\,\omega^2 \boldsymbol{u}\, \boldsymbol{z}^\star.\end{aligned}$$

The density depends on the solution $\boldsymbol{z}$ of the adjoint problem

$$A_\Omega(\boldsymbol{z}, \boldsymbol{w}) = \int_\Gamma \frac{\partial g}{\partial \boldsymbol{u}_\tau}\boldsymbol{w}\, ds \qquad \text{for all } \boldsymbol{w} \in H^1(\Omega)$$

which is the weak form of the following boundary value problem

$$-\left[\mu\Delta + (\lambda+\mu)\nabla\nabla + \rho\omega^2\right] \boldsymbol{z}(\boldsymbol{x}) = \frac{\partial g}{\partial u^\star}\, \delta_C(\boldsymbol{x})$$

with the boundary condition

$$\boldsymbol{t}(\boldsymbol{z}(\boldsymbol{x})) = 0, \qquad \boldsymbol{x} \in \Gamma$$

and a radiation condition as $r \to \infty$.
The symbol $\delta_C$ represents the characteristic function of the surface $C$. For to have in the frequency domain the same set of equations in the direct problem as in the adjoint problem we formulate the equation for the adjoint problem with the conjugated term $\boldsymbol{z}^\star$

$$-\left[\mu\Delta + (\lambda+\mu)\nabla\nabla + \rho\omega^2\right] \boldsymbol{z}(\boldsymbol{x})^\star = \frac{\partial g}{\partial u}\, \delta_C(\boldsymbol{x})$$

with boundary condition

$$\boldsymbol{t}(\boldsymbol{z}(\boldsymbol{x})^*) = 0, \qquad \boldsymbol{x} \in \Gamma$$

Hence the solution of the adjoint problem is

$$\mathbf{C}(\boldsymbol{x})\boldsymbol{z}(\boldsymbol{x})^\star = -\iint_\Gamma \mathbf{T}(\boldsymbol{y},\boldsymbol{x})\, \boldsymbol{z}(\boldsymbol{y})^\star \, d\Gamma_y + \iint_C \mathbf{U}(\boldsymbol{y},\boldsymbol{x})\, \frac{\partial g(\boldsymbol{y})}{\partial u}\, dC_y\,,$$

which is the same integral equation as for the field $\boldsymbol{u}$.

## 5. ITERATIONS

Now the algorithm starts with a first guess and then corrects the parameters by iteratively solving the following set of equations: (1) Find the response of the media to an incident wave

$$\boldsymbol{C}(\boldsymbol{x})\boldsymbol{u}(\boldsymbol{x}) = -\int_\Gamma \boldsymbol{T}(\boldsymbol{y},\boldsymbol{x})\, \boldsymbol{u}(\boldsymbol{y})\, ds_{\boldsymbol{y}} + \boldsymbol{u}^{in}(\boldsymbol{x}) \qquad \boldsymbol{x} \in \Gamma$$

(2) Calculate the amplitudes

$$\begin{aligned}
\boldsymbol{A}_s(\boldsymbol{x}) &= \frac{\boldsymbol{x}}{4\pi\rho c_s^2} \int_\Gamma \{ik_s\rho c_s^2[(\boldsymbol{u}\times\boldsymbol{x})(\boldsymbol{n}\cdot\boldsymbol{x}]) + (\boldsymbol{n}\times\boldsymbol{x})(\boldsymbol{u}\cdot\boldsymbol{x})] \\
&\quad + \boldsymbol{t}\times\boldsymbol{x}\}e^{-ik_s\boldsymbol{x}\cdot\boldsymbol{y}}ds_{\boldsymbol{y}} \\
\boldsymbol{A}_p(\boldsymbol{x}) &= \frac{\boldsymbol{x}}{4\pi\rho c_p^2} \int_\Gamma \{ik_p\rho[2c_s^2(\boldsymbol{u}\cdot\boldsymbol{x})(\boldsymbol{n}\cdot\boldsymbol{x}]) + (c_p^2 - 2c_s^2)(\boldsymbol{u}\cdot\boldsymbol{n})] \\
&\quad + \boldsymbol{t}\times\boldsymbol{x}\}e^{-ik_p\boldsymbol{x}\cdot\boldsymbol{y}}ds_{\boldsymbol{y}}
\end{aligned}$$

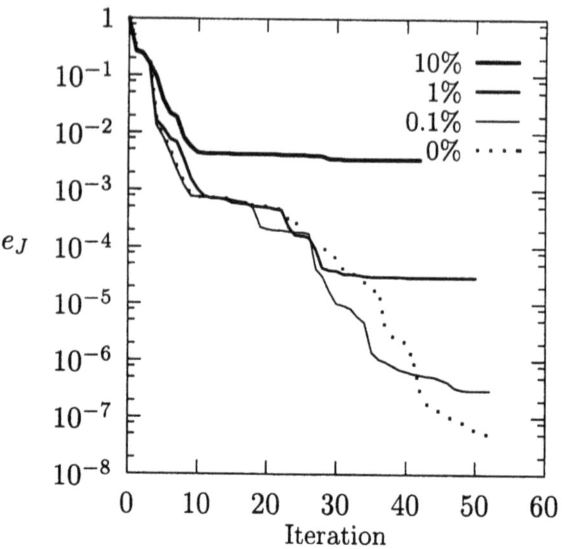


*Figure 4* Fig. 4 Relative error versus number of iterations

(3) Evaluate the function to be minimized

$$J(\boldsymbol{u}) = \frac{1}{2}\int_{\Sigma}\{|\boldsymbol{A}_s(\boldsymbol{y}) - \tilde{\boldsymbol{A}}_s(\boldsymbol{y})|^2 + |\boldsymbol{A}_p(\boldsymbol{y}) - \tilde{\boldsymbol{A}}_s(\boldsymbol{y})|^2 q(\boldsymbol{y}) d\Sigma_{\boldsymbol{y}}$$

(4) Solve the adjoint problem

$$A_\Omega(\boldsymbol{z}, \boldsymbol{w}) = \int_\Gamma \frac{\partial g}{\partial \boldsymbol{u}_\tau} \boldsymbol{w} ds \qquad \text{for all } \boldsymbol{w} \in H^1(\Omega)$$

and (5) calculate the gradient of the function $J$

$$\frac{d}{d\tau_i} J(\boldsymbol{u}, \boldsymbol{\tau}) = -\Re \int_\Gamma [\frac{1}{2}\mu(\nabla_S \boldsymbol{u}^S + \nabla_s \boldsymbol{u}^{ST}) \cdot (\nabla_S \boldsymbol{z}^S + \nabla_S \boldsymbol{z}^{ST})^*]$$
$$+ \frac{2\mu}{\lambda + 2\mu} \text{div}_S \boldsymbol{u}^S \text{div}_S \boldsymbol{z}^{S*} - \rho\omega^2 \boldsymbol{u}\boldsymbol{z}^*] \frac{d\boldsymbol{y}}{d\tau_i} \boldsymbol{n}\, ds_{\boldsymbol{y}}$$

where the subscript $S$ denotes that we must perform differentiation 'within the surface'.

The implementation of the foregoing algorithm into an existing 3-D boundary element code is straightforward. In many test runs we verified the good approximation properties of the proposed method even for configurations with multiple obstacles. Crucial for the success of the method is that the set $\boldsymbol{\tau} = \{\tau_1, \tau_2, \ldots, \tau_n\}$ of parameters is 'complete' in the sense that the shape of the obstacle can be modelled accurately by the parameters $\tau_i$.

## References

[1] T. Burczynski, Applications of BEM in sensitivity analysis and optimization, Comp. Mech. 13, 29-44

[2] E. J. Haug and K. K. Choi and V. Komkov, Design Sensitivity Analysis of Structural Systems, Academic Press, Inc., Orlando, 1986, Mathematics in Science and Engineering, Vol. 177,

[3] M. Bonnet, Shape Identification Using Acoustic Measurements: A Numerical Investigation Using BIE and Shape Differentiation, Inverse Problems in Engineering Mechanics, IUTAM Symposium Tokyo 1992, Eds. M. Tanaka, H. D. Bui, Springer-Verlag, Berlin, 1993,191-200

[4] V. Dröge, Materialfehler in elastischen Medien als inverses Streuproblem, Dissertation University of Kassel 1996

[5] K. Mayer, R. Marklein, K.J. Langenberg, T. Kreutter, Three-dimensional imaging system based on Fourier transform synthetic aperture focusing technique, Ultrasonics 28, 241-255

[6] N. Nishimura, S. Kobayashi, A boundary integral equation method for an inverse problem related to crack detection, Int. J. Num. Meth. Eng. 32, 1371-1387

[7] M. Tanaka, M. Nakamura, S. Hanaoka, Sensitivity analysis of 2-D elastodynamic problems through boundary integral equations, in: Computational Mechanics 95, ed. S.N.Atluri, G. Yagawa, T.A.Cruse, Springer-Verlag Berlin, Vol. 2, 3049-3054

[8] D. J. N. Wall, Uniqueness theorems for the inverse problem of elastodynamic boundary scattering, IMA J. Appl. Math. 44, 221-241

[9] C.F.Ying, R. Truell, Scattering of plane longitudinal wave by a spherical obstacle in an isotropically elastic solid, J. Appl. Phys. 27, 1086-1097

# INTEGRAL TRANSFORM METHODS IN 3-D DYNAMIC FRACTURE MECHANICS

P. Fedelinski
*Department for Strength of Materials*
*and Computational Mechanics,*
*Silesian University of Technology,*
*44-100 Gliwice, ul.Konarskiego 18A, Poland*
fedel@rmt4.kmt.polsl.gliwice.pl

**Keywords:** boundary element method, integral transform methods, dynamic fracture mechanics

**Abstract** The boundary element method (BEM) for three-dimensional dynamic analysis of cracked structures is presented. The solutions are computed in time or frequency domain by using either the Laplace or the Fourier integral transform method. The displacement and the traction boundary integral equations are used in the present approach. The boundary geometry, displacements and tractions are approximated using either constant or quadratic elements. The dynamic stress intensity factors are computed using crack opening displacements. The method is used to analyze the dynamic behaviour of a rectangular bar with an internal square crack and a cracked thick-walled cylinder subjected to internal impact pressure.

## Introduction

The boundary element method (BEM) is widely used in fracture mechanics, because it gives accurate solutions and the modelling of crack growth is simple. A review of boundary element formulations and their applications in fracture mechanics has been presented by Aliabadi (1997). In the present work, three-dimensional crack problems are solved using the dual method combined with integral transform methods. In the dual method, the external boundaries and crack surfaces are discretized. The unknown boundary displacements and tractions are computed directly by the solution of the system of equations. The integral transform methods are chosen because they give very accurate results and the structure of the computer code is relatively simple.

*T. Burczynski (ed.), IUTAM/IACM/IABEM Symposium on Advanced Mathematical and Computational Mechanics Aspects of the Boundary Element Method,* 111–121.

Details of the methods and simple examples of application were presented by Fedelinski (1998a), (1998b), (1998c). In the present paper the displacement and traction boundary integral equations for points along the external boundaries and crack surfaces are presented. The approximation of boundary quantities and the final matrix equation of motion is given. Two numerical examples show possible applications of the methods. The transformed fundamental solutions of elastodynamics are presented in Appendix.

# 1. BOUNDARY INTEGRAL EQUATIONS

Consider a homogeneous, isotropic and linear elastic body containing a crack. Zero body forces and initial conditions are assumed.

The Laplace transform of displacement of a point $\mathbf{x}'$ can be computed from the following displacement boundary integral equation

$$c_{ij}(\mathbf{x}')\bar{u}_j(\mathbf{x}',s) = \int_\Gamma \bar{U}_{ij}(\mathbf{x}',\mathbf{x},s)\bar{t}_j(\mathbf{x},s)d\Gamma(\mathbf{x}) - \fint_\Gamma \bar{T}_{ij}(\mathbf{x}',\mathbf{x},s)\bar{u}_j(\mathbf{x},s)d\Gamma(\mathbf{x}), \quad (1)$$

where $\bar{u}_j(\mathbf{x},s)$, $\bar{t}_j(\mathbf{x},s)$ are the transformed displacements and tractions; $\bar{U}_{ij}(\mathbf{x}',\mathbf{x},s)$, $\bar{T}_{ij}(\mathbf{x}',\mathbf{x},s)$ are the transformed fundamental solutions of elastodynamics, given in Appendix; $c_{ij}(\mathbf{x}')$ is a constant, which depends on the position of the collocation point; $\Gamma$ is the boundary of the body; $\mathbf{x}$ is an integration point, which belongs to the boundary; $s$ is a parameter of the transformation; a bar over a variable denotes the transformed variable; $\fint$ is the Cauchy principal value integral. The summation convention is used for repeated indices.

The general forms of boundary integral equations and the transformed fundamental solutions of elastodynamics are the same for the Laplace and the Fourier transform. Therefore the application of the method will be demonstrated for the Laplace transform.

If the boundaries at coincident collocation points $\mathbf{x}'$ and $\mathbf{x}''$ on opposite crack surfaces are smooth then the displacement equation has the form

$$\frac{1}{2}\bar{u}_i(\mathbf{x}',s) + \frac{1}{2}\bar{u}_i(\mathbf{x}'',s) = \int_\Gamma \bar{U}_{ij}(\mathbf{x}',\mathbf{x},s)\bar{t}_j(\mathbf{x},s)d\Gamma(\mathbf{x}) - \fint_\Gamma \bar{T}_{ij}(\mathbf{x}',\mathbf{x},s)\bar{u}_j(\mathbf{x},s)d\Gamma(\mathbf{x}). \quad (2)$$

The traction boundary integral equation for the coincident points $\mathbf{x}'$ and $\mathbf{x}''$ has the form

$$\frac{1}{2}\bar{t}_j(\mathbf{x}',s) - \frac{1}{2}\bar{t}_j(\mathbf{x}'',s) = n_i(\mathbf{x}')\Big[\ \fint_{\Gamma} \bar{U}_{kij}(\mathbf{x}',\mathbf{x},s)\bar{t}_k(\mathbf{x},s)d\Gamma(\mathbf{x})$$
$$- \fint_{\Gamma} \bar{T}_{kij}(\mathbf{x}',\mathbf{x},s)\bar{u}_k(\mathbf{x},s)d\Gamma(\mathbf{x})\Big], \tag{3}$$

where $n_i(\mathbf{x}')$ is an outward normal unit vector at the collocation point; $\bar{U}_{kij}(\mathbf{x}',\mathbf{x},s)$, $\bar{T}_{kij}(\mathbf{x}',\mathbf{x},s)$ are the transformed fundamental solutions of elastodynamics, given in Appendix; $\fint$ is the Hadamard principal value integral.

## 2. NUMERICAL IMPLEMENTATION

In order to obtain numerical solutions the boundary is divided into boundary elements. Two types of boundary elements are used in the present approach: either constant or quadratic. The constant element has one node situated in the centre of the element. Quadratic continuous elements with 9 nodes, are used for the discretization of the external boundary and quadratic discontinuous for the crack surfaces. The discontinuous elements are used for crack surfaces to satisfy a proper order of continuity of integrand at the collocation point [Guiggiani et al. (1992)]. The boundary geometry, displacements and tractions are interpolated within each element using shape functions.

The displacement equation (1) is applied to nodes which belong to the external boundary. In the dual boundary element method both crack surfaces are discretized. The displacement equation (2) is applied to nodes belonging to one crack surface and the traction equation (3) to nodes on the opposite crack surface. The unknown displacements or tractions on crack surfaces can be computed directly using this method.

The set of boundary equations (1), (2) and (3) for all boundary nodes can be written in a matrix form as follows

$$\bar{\mathbf{H}}\bar{\mathbf{u}} = \bar{\mathbf{G}}\bar{\mathbf{t}}, \tag{4}$$

where $\bar{\mathbf{H}}$, $\bar{\mathbf{G}}$ depend on integrals of transformed fundamental solutions of elastodynamics, interpolating functions and Jacobians; $\bar{\mathbf{u}}$, $\bar{\mathbf{t}}$ contain nodal values of the transformed displacements and tractions. The matrix equation is rearranged according to the boundary conditions to give the equation

$$\bar{\mathbf{A}}\bar{\mathbf{x}} = \bar{\mathbf{B}}\bar{\mathbf{y}}, \tag{5}$$

where $\bar{\mathbf{x}}$ contains unknown transformed displacements and tractions, and $\bar{\mathbf{y}}$ contains prescribed transformed boundary displacements and tractions. The equation can be solved for a particular parameter of the integral transformation. When the Laplace transform is used, the transforms of boundary conditions for simple time variations can be easily calculated analytically. The time-dependent solutions are obtained by solving the equation for a series of parameters and by using a numerical inverse transform. The dependence of the dynamic response of the structure on forcing frequency can be computed by using the Fourier transform.

The transformed fundamental solutions are singular with respect to the distance between the collocation and the integration point. Therefore special methods are used to compute boundary integrals when the collocation point belongs to the integration element. The integral of $\bar{U}_{ij}(\mathbf{x}',\mathbf{x},s)$ is classified as weakly singular while $\bar{T}_{ij}(\mathbf{x}',\mathbf{x},s)$ and $\bar{U}_{kij}(\mathbf{x}',\mathbf{x},s)$ are strongly singular. These integrals are computed using the methods presented by Dominguez (1993).

The integral of $\bar{T}_{kij}(\mathbf{x}',\mathbf{x},s)$ is hypersingular. The fundamental solution of elastostatics is subtracted from the transformed solution and added back. The non-singular difference is integrated numerically while the static fundamental solutions $T_{kij}(\mathbf{x}',\mathbf{x})$ is integrated using the method presented by Guiggiani et al. (1992).

## 3. NUMERICAL EXAMPLES

Two numerical examples are presented to demonstrate possible applications of the method.

The mode-one dynamic stress intensity factors (DSIF) $K_I$ are computed using the crack opening displacements (COD) as follows

$$K_I = \frac{E}{8(1-\nu^2)}\sqrt{\frac{2\pi}{R}}\Delta u, \tag{6}$$

where $E$ is Young's modulus; $\nu$ is Poisson's ratio; $R$ is the distance of coincident points on crack surfaces from the crack front; $\Delta u$ is the difference between the displacements of these points in the direction perpendicular to the crack surface. The displacements of centre nodes of quadratic elements adjacent to the crack front and the centre nodes of the second pair of constant elements, counted from the crack front, are used to compute DSIFs. The time-dependent results are computed by using the solutions for 25 Laplace parameters.

## 3.1. RECTANGULAR BAR WITH AN INTERNAL SQUARE CRACK

A rectangular bar of height $h$ with a square cross section of length $2b \times 2b$ contains an internal square crack of length $2a \times 2a$, as shown in Fig. 1.

The ratios of dimensions are $a/b = 0.5$ and $h/b = 4$. The distance of the crack from the lower constrained end of the bar is $c$. The Poisson's ratio of the material is $\nu = 0.2$. The upper end is subjected to the uniformly distributed impact tension $\sigma_o$ with the Heaviside time dependence. The boundary is divided either into 450 constant elements (250 elements for the external boundary and 100 elements for each crack surface), or into 58 quadratic elements (40 elements for the external boundary and 9 for each crack surface). The dynamic behaviour of the structure is analyzed for three different positions of the crack $c/h = 0.25,\ 0.50,\ 0.75$.

The normalized DSIFs $K_I/K_o$ as functions of normalized time $\tau c_1/h$ and normalized forcing frequency $\omega h/c_1$, where $K_o = \sigma_o\sqrt{\pi a}$, $c_1$ is the velocity of longitudinal wave, are shown in Fig. 2 and 3, respectively. The solutions for constant and quadratic elements are very similar. The position of the crack has a small influence on peak values of DSIFs and first resonant frequency.

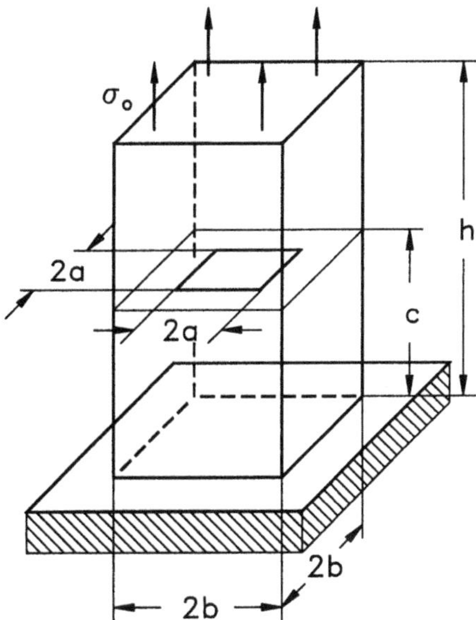


*Figure 1* Rectangular bar with an internal square crack

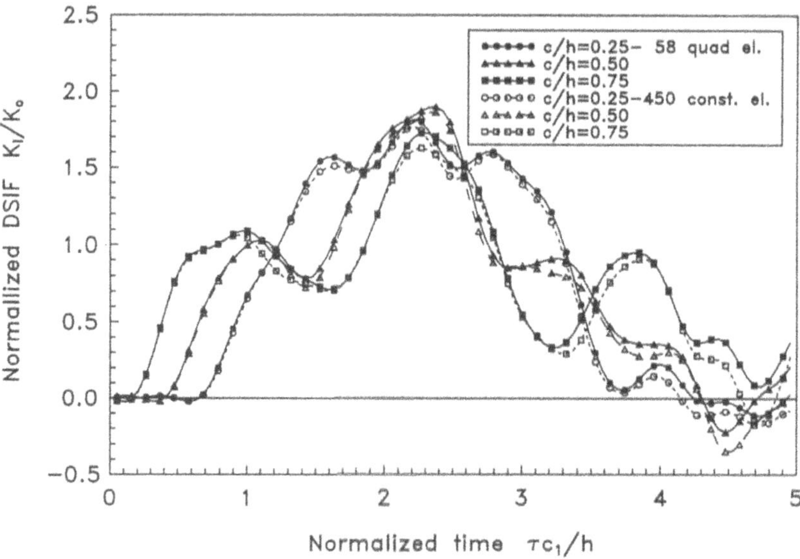


*Figure 2* Normalized DSIF $K_I/K_o$ in the middle of the crack front versus normalized time $\tau c_1/h$ for different positions of the crack in the bar

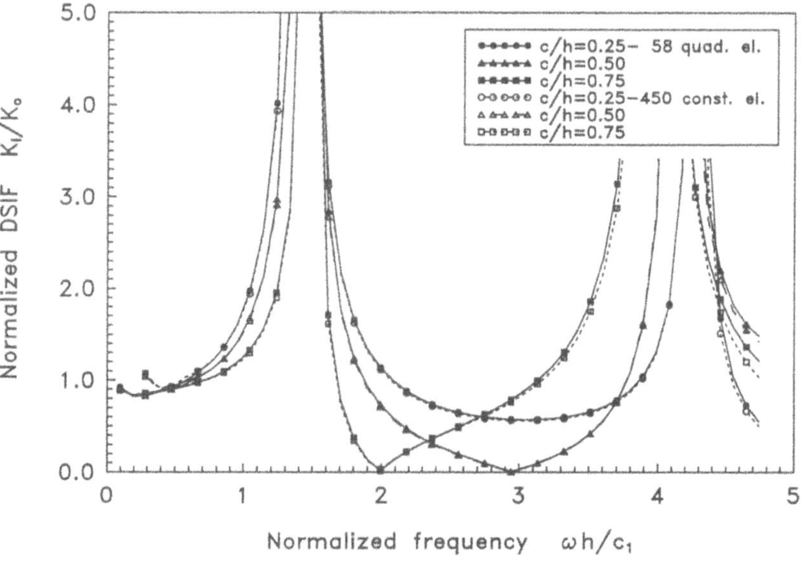


*Figure 3* Normalized DSIF $K_I/K_o$ in the middle of the crack front versus normalized forcing frequency $\omega h/c_1$ for different positions of the crack in the bar

## 3.2. PRESSURIZED CRACKED THICK-WALLED CYLINDER

A thick-walled cylinder with a crack is subjected to internal impact normal pressure $p$, with the Heaviside time-dependence.

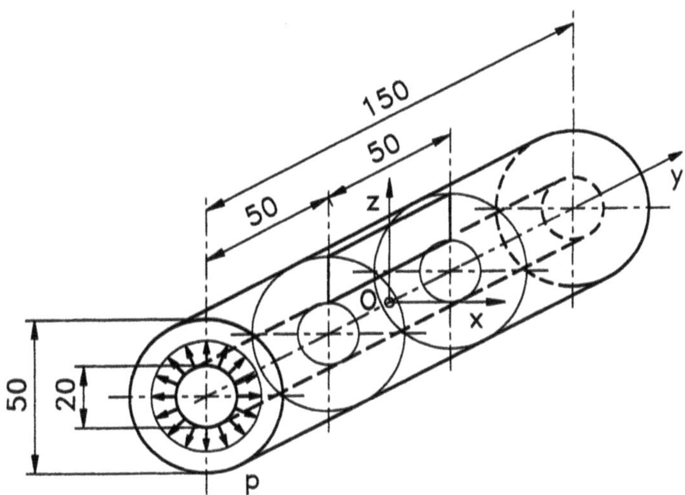


*Figure 4* Cracked thick-walled cylinder subjected to impact internal pressure

The internal radius of the cylinder is $r_1 = 10\ mm$; the external radius is $r_2 = 25\ mm$; the length of the cylinder is $l = 150\ mm$; and the crack length is $2a = 50\ mm$. The dimensions of the cylinder and the coordinate system, are shown in Fig. 4. The material properties of the cylinder are: Young's modulus $E = 0.2 \cdot 10^{12}\ Pa$; Poisson's ratio $\nu = 0.3$; and the mass density $\rho = 8000\ kgm^{-3}$. The cylinder with the crack has two planes of symmetry $yz$ and $xz$ One half of the cylinder is analyzed by constraining the displacements in $y$-direction along the plane $xz$ of symmetry. This part of the cylinder is divided into 606 constant elements.

The normalized DSIFs $K_I/K_o$ are calculated in the middle of the crack front, and plotted in Fig. 5. The normalizing SIF is

$$K_o = \sigma\sqrt{\pi a}, \quad \sigma = \frac{pr_1^2}{r_2^2 - r_1^2}(1 + \frac{r_2^2}{r^2}), \tag{7}$$

where $\sigma$ is the static circumferential stress in the middle of the crack front; $r$ is the distance of the centre of crack front from the axis of the cylinder.

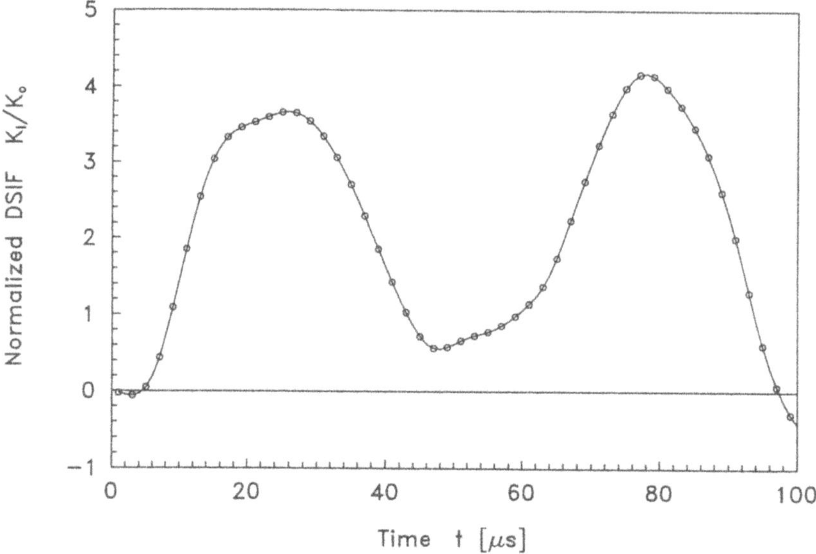


*Figure 5* Normalized DSIF $K_I/K_o$ in the middle of crack front versus time $\tau$ for the cracked pressurized thick-walled cylinder

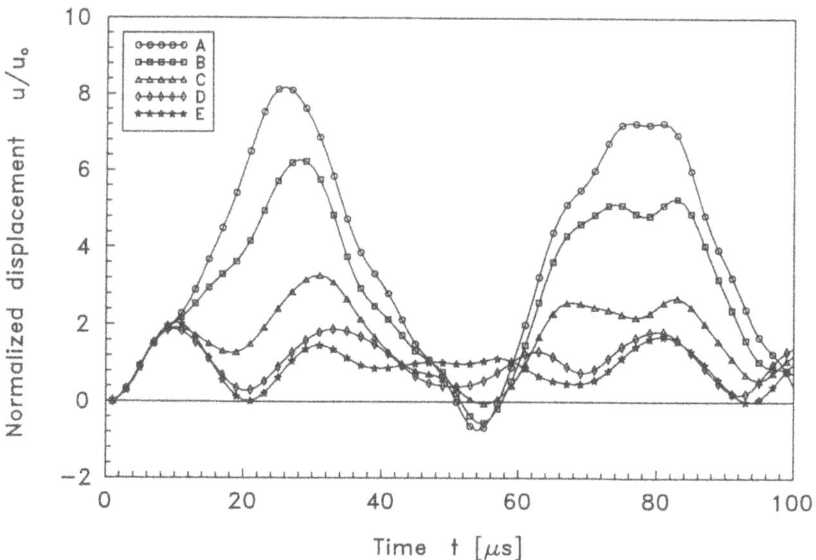


*Figure 6* Normalized displacements $u/u_o$ of points along the external surface versus time $\tau$ for the cracked pressurized thick-walled cylinder

The displacements are computed at 5 points, which have the following $(x, y, z)$ coordinates: $A(25, 2.5, 0)$, $B(25, 17.5, 0)$, $C(25, 32.5, 0)$, $D(25, 47.5, 0)$ and $E(25, 62.5, 0)$. The normalized displacements $u/u_o$ of the external surface, in the radial direction, are shown in Fig. 6. The normalizing displacement $u_o$ is the static displacement

$$u_o = \frac{2r_1^2 r_2 p}{E(r_2^2 - r_1^2)}. \tag{8}$$

## 4. CONCLUSIONS

The boundary element method for the dynamic analysis of three - dimensional structures with cracks is presented. The formulation can be used to compute displacements, strains, stresses and dynamic stress intensity factors (DSIF) as functions of time or forcing frequency by discretizing the boundary of the body. The dynamic stress intensity factors were computed for a rectangular bar with a square internal crack and for a cracked thick-walled cylinder subjected to internal impact pressure.

## Appendix

The transformed fundamental solutions of elastodynamics used in the displacement equation have the forms [Cruse and Rizzo (1968)]:

$$\bar{U}_{ij}(\mathbf{x}', \mathbf{x}, s) = \frac{1}{4\pi\mu}(\psi\delta_{ij} - \chi r_{,i} r_{,j}), \tag{1.A.1}$$

$$\begin{aligned}\bar{T}_{ij}(\mathbf{x}', \mathbf{x}, s) = \frac{1}{4\pi}\Big[(\psi_{,r} - \frac{\chi}{r})(\delta_{ij}\frac{\partial r}{\partial n} + r_{,j} n_i) - 2\frac{\chi}{r}(r_{,i} n_j - 2r_{,i} r_{,j}\frac{\partial r}{\partial n})\\ -2\chi_{,r} r_{,i} r_{,j}\frac{\partial r}{\partial n} + \frac{\lambda}{\mu}(\psi_{,r} - \chi_{,r} - 2\frac{\chi}{r}) r_{,i} n_j\Big],\end{aligned} \tag{1.A.2}$$

where $\mu$ is the shear modulus; $\lambda$ is the Lamé constant; $r$ is the distance between the collocation and the integration point; $n_i$ is a component of the outward normal unit vector at the boundary point; $\delta_{ij}$ is the Kronecker delta; the derivatives of functions $\psi$, $\chi$ with respect to the distance $r$ are denoted by $\psi_{,r}$ and $\chi_{,r}$; an index preceded by a comma, for example $,i$ denotes the derivative with respect to the coordinate $x_i$.

The transformed fundamental solutions of elastodynamics used in the traction equation have the forms [Gaul and Fiedler (1993)]:

$$\begin{aligned}\bar{U}_{kij}(\mathbf{x}', \mathbf{x}, s) = \frac{1}{4\pi}\Big\{\Big[2\frac{\chi}{r} - \frac{\lambda}{\mu}(\psi_{,r} - \chi_{,r} - 2\frac{\chi}{r})\Big]\delta_{ij} r_{,k}\\ -(\psi_{,r} - \frac{\chi}{r})(\delta_{ik} r_{,j} + \delta_{jk} r_{,i}) + 2(\chi_{,r} - 2\frac{\chi}{r}) r_{,i} r_{,j} r_{,k}\Big\},\end{aligned} \tag{1.A.3}$$

$$\bar{T}_{kij}(\mathbf{x}',\mathbf{x},s) = \frac{\mu}{4\pi}\Big\{\frac{\partial r}{\partial n}\Big[4(\chi_{,rr} - 5\frac{\chi_{,r}}{r} + 8\frac{\chi}{r^2})r_{,i}r_{,j}r_{,k}$$
$$-(\psi_{,rr} - \frac{\psi_{,r}}{r} - 3\frac{\chi_{,r}}{r} + 6\frac{\chi}{r^2})(\delta_{ik}r_{,j} + \delta_{jk}r_{,i})$$
$$+2[2\frac{\chi_{,r}}{r} - 4\frac{\chi}{r^2} + \frac{\lambda}{\mu}(\chi_{,rr} + \frac{\chi_{,r}}{r} - 4\frac{\chi}{r^2} - \psi_{,rr} + \frac{\psi_{,r}}{r})]\delta_{ij}r_{,k}\Big]$$
$$+2[2\frac{\chi_{,r}}{r} - 4\frac{\chi}{r^2} + \frac{\lambda}{\mu}(\chi_{,rr} + \frac{\chi_{,r}}{r} - 4\frac{\chi}{r^2} - \psi_{,rr} + \frac{\psi_{,r}}{r})]r_{,i}r_{,j}n_k \quad (1.A.4)$$
$$-(\psi_{,rr} - \frac{\psi_{,r}}{r} - 3\frac{\chi_{,r}}{r} + 6\frac{\chi}{r^2})(r_{,j}n_i + r_{,i}n_j)r_{,k}$$
$$+\Big[4\frac{\chi}{r^2} + 4\frac{\lambda}{\mu}(\frac{\chi_{,r}}{r} + 2\frac{\chi}{r^2} - \frac{\psi_{,r}}{r}) + (\frac{\lambda}{\mu})^2(\chi_{,rr} + 4\frac{\chi_{,r}}{r} + 2\frac{\chi}{r^2} - \psi_{,rr} - 2\frac{\psi_{,r}}{r})\Big]\delta_{ij}n_k$$
$$-2(\frac{\psi_{,r}}{r} - \frac{\chi}{r^2})(\delta_{kj}n_i + \delta_{ki}n_j)\Big\}.$$

The functions $\psi$, $\chi$ and their derivatives are:

$$\psi = -(\frac{c_2}{c_1})^2 e^{-z_1}(1 + z_1)\frac{1}{z_1^2 r} + e^{-z_2}(1 + z_2 + z_2^2)\frac{1}{z_2^2 r}, \quad (1.A.5)$$

$$\psi_{,r} = (\frac{c_2}{c_1})^2 e^{-z_1}(3+3z_1+z_1^2)\frac{1}{z_1^2 r^2} - e^{-z_2}(3+3z_2+2z_2^2+z_2^3)\frac{1}{z_2^2 r^2}, \quad (1.A.6)$$

$$\psi_{,rr} = -(\frac{c_2}{c_1})^2 e^{-z_1}(12 + 12z_1 + 5z_1^2 + z_1^3)\frac{1}{z_1^2 r^3}$$
$$+e^{-z_2}(12 + 12z_2 + 7z_2^2 + 3z_2^3 + z_2^4)\frac{1}{z_2^2 r^3}. \quad (1.A.7)$$

$$\chi = -(\frac{c_2}{c_1})^2 e^{-z_1}(3 + 3z_1 + z_1^2)\frac{1}{z_1^2 r} + e^{-z_2}(3 + 3z_2 + z_2^2)\frac{1}{z_2^2 r}, \quad (1.A.8)$$

$$\chi_{,r} = (\frac{c_2}{c_1})^2 e^{-z_1}(9 + 9z_1 + 4z_1^2 + z_1^3)\frac{1}{z_1^2 r^2}$$
$$-e^{-z_2}(9 + 9z_2 + 4z_2^2 + z_2^3)\frac{1}{z_2^2 r^2}, \quad (1.A.9)$$

$$\chi_{,rr} = -(\frac{c_2}{c_1})^2 e^{-z_1}(36 + 36z_1 + 17z_1^2 + 5z_1^3 + z_1^4)\frac{1}{z_1^2 r^3}$$
$$+e^{-z_2}(36 + 36z_2 + 17z_2^2 + 5z_2^3 + z_2^4)\frac{1}{z_2^2 r^3}, \quad (1.A.10)$$

where $c_1$ is the velocity of longitudinal wave; $c_2$ is the velocity of shear wave; $z_1 = sr/c_1$, $z_2 = sr/c_2$ for the Laplace transforms and $z_1 = -i\omega r/c_1$, $z_2 = -i\omega r/c_2$ for the Fourier transforms.

## References

[1] Aliabadi M.H. (1997) Boundary element formulations in fracture mechanics, *Appl. Mech. Rev.*, **50**, 83-96.

[2] Cruse T.A. and Rizzo, F.J. (1968) A direct formulation and numerical solution of the general transient elastodynamic problem. I, *J. Math. Anal. Appl.*, **22**, 244-259.

[3] Dominguez, J. (1993) *Boundary Elements in Dynamics*, Comp. Mech. Publ., Southampton.

[4] Fedelinski, P. (1998a) The boundary element method in three- dimensional dynamic fracture mechanics - Part I: Theory, in Proc. XXXVII Symposium *Modelling in Mechanics*, *Scient. Pap. Dep. Appl. Mech.*, **6**, 83-88.

[5] Fedelinski, P. (1998b) The boundary element method in three- dimensional dynamic fracture mechanics - Part II: Examples, in Proc. XXXVII Symposium *Modelling in Mechanics*, *Scient. Pap. Dep. Appl. Mech.*, **6**, 89-94.

[6] Fedelinski, P. (1998c) The three-dimensional boundary element method in dynamic analysis of cracks, *Build. Res. J.*, **46**, 257-275.

[7] Gaul, L. and Fiedler, C. (1993) Improved calculation of field variables in the domain based on BEM, *Engng. Anal. Bound. Elem.*, **11**, 257-264.

[8] Guiggiani, M., Krishnasamy, G., Rudolphi, T.J. and Rizzo, F.J. (1992) A general algorithm for the numerical solution of hypersingular boundary integral equations, *Trans. ASME, J. Appl. Mech.*, **59**, 604-614.

# SOLUTION OF PLANE THERMOELASTICITY PROBLEMS OF FRACTURE MECHANICS BY THE DISPLACEMENT AND TEMPERATURE DISCONTINUITY METHOD

Arkadiy Goltsev
*Donetsk State University*
*Universitetskay 24, 83055 Donetsk*
*Ukraine*
goltsev@univ.donetsk.ua

**Keywords:** BEM, discontinuity method, thermoelasticity, fracture mechanics

**Abstract** The displacement discontinuity method is extended on the plane thermoelasticity problems for bodies with cracks. The analytical solution for elementary discontinuities is obtained. The test and illustrated problems are solved.

## 1. INTRODUCTION

It is well known that the plane thermoelasticity crack problems for the infinite medium can be reduced to singular integral equations. The solution may be obtained very easy only for the straight cracks. In the case of arbitrary geometry we should use the numerical methods.

In this paper the displacement discontinuity method described in the monograph of Crouch and Starfield [1] is extended on the plane thermoelasticity problems for bodies with cracks. The base of this method is the analytical solution of thermoelasticity problem for segment with the elementary discontinuity of displacement and temperature. By means of this solution and the BEM the test and illustrated problems are solved.

*T. Burczynski (ed.), IUTAM/IACM/IABEM Symposium on Advanced Mathematical and Computational Mechanics Aspects of the Boundary Element Method,* 123–132.

## 2. ANALYTICAL SOLUTION FOR ELEMENTARY DISCONTINUITIES

Let us consider an infinite plane with a finite segment of length $2a$. The centre of the segment is located in the origin of coordinates $x$, $y$ as shown in Figure 1.

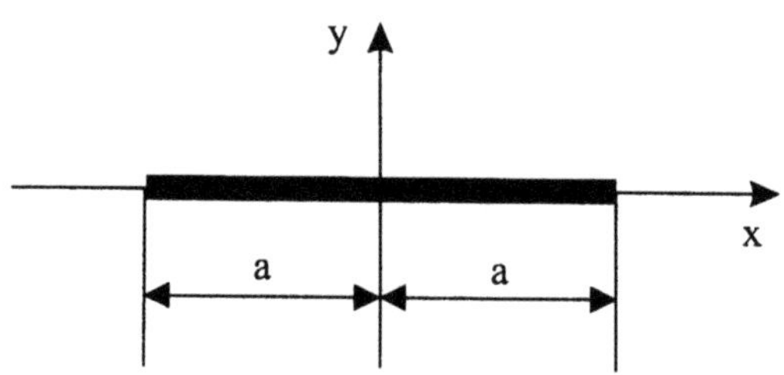


*Figure 1* Finite segment.

It is assumed that the displacements $U$ along axis $x$, $V$ along axis $y$ and temperature $T$ have the constant discontinuities or jumps of their values on the line of the segment ($|x| \leq a$, $y = 0$)

$$[U] = U(x, 0_+) - U(x, 0_-) \tag{1}$$

$$[V] = V(x, 0_+) - V(x, 0_-) \tag{2}$$

$$[T] = T(x, 0_+) - T(x, 0_-) \tag{3}$$

The square brackets mean the value of discontinuity or jump of corresponding function.

In order to obtain the temperature, displacements and stresses in the body the equations of stationary thermoelasticity in the case of the plane strain were used. This is

equation of thermal conductivity

$$\frac{\partial^2 T}{\partial x^2} + \frac{\partial^2 T}{\partial y^2} = 0 \tag{4}$$

equations of equilibrium

$$\frac{\partial \sigma_x}{\partial x} + \frac{\partial \tau_{xy}}{\partial y} = 0, \qquad \frac{\partial \tau_{xy}}{\partial x} + \frac{\partial \sigma_y}{\partial y} = 0 \tag{5}$$

geometrical ratios

$$e_x = \frac{\partial U}{\partial x}, \qquad e_y = \frac{\partial V}{\partial y}, \qquad e_{xy} = \frac{1}{2}\left(\frac{\partial U}{\partial y} + \frac{\partial V}{\partial x}\right) \tag{6}$$

law of thermoelasticity

$$\sigma_x = \frac{E}{1+\nu}\left\{e_x + \frac{\nu}{1-2\nu}\left(e_x + e_y\right)\right\} - \frac{E}{1-2\nu}\alpha_t T \tag{7}$$

$$\sigma_y = \frac{E}{1+\nu}\left\{e_y + \frac{\nu}{1-2\nu}\left(e_x + e_y\right)\right\} - \frac{E}{1-2\nu}\alpha_t T \tag{8}$$

$$\tau_{xy} = \frac{E}{1+\nu} e_{xy} \tag{9}$$

Here $\sigma_x$, $\sigma_y$ and $\tau_{xy}$ are the stresses, $e_x$, $e_y$ and $e_{xy}$ are the strains, $E$ is Young's modulus, $\nu$ is Poisson's ratio and $\alpha_t$ is the coefficient of linear thermal expansion.

The solution of this problem was obtained by means of two-dimensional Fourier transform. The theory of generalized functions was used for determining the derivatives of functions with the jumps. If there is the discontinuity of the function $f(x, y)$ on the line $L$, its derivatives can be determined in the following way:

$$\frac{\partial f}{\partial x} = \left\{\frac{\partial f}{\partial x}\right\}_C + n_1 \delta_L [f] \tag{10}$$

$$\frac{\partial f}{\partial y} = \left\{\frac{\partial f}{\partial y}\right\}_C + n_2 \delta_L [f] \tag{11}$$

where the derivatives in the braces are the classical ones, $n_1$ and $n_2$ are the direction cosines of normal $\vec{n}$ to the line $L$, $\delta_L$ is the two-dimensional Dirac's function defined on the line $L$. The mathematical definition of $\delta$-function is

$$\int_{-\infty}^{+\infty}\int_{-\infty}^{+\infty} \delta_L \left(x - x', y - y'\right) f(x, y)\, dx dy = \int_L f\left(x', y'\right) dL \tag{12}$$

Here $x'$ and $y'$ are the coordinates of the points on the line $L$.

Applying the formula of tow-dimentional Fourier transform

$$\tilde{f}(\xi, \eta) = \frac{1}{2\pi}\int_{-\infty}^{+\infty}\int_{-\infty}^{+\infty} f(x, y) \exp\left(i\left(\xi x + \eta y\right)\right) dx dy \tag{13}$$

to the derivatives (10), (11) we obtain their expressions in the transform space

$$\widetilde{\frac{\partial f}{\partial x}} = (-i\xi)\,\tilde{f} + \frac{1}{2\pi}\int_L n_1\,[f]\exp\left(i\left(\xi x' + \eta y'\right)\right)dL \tag{14}$$

$$\widetilde{\frac{\partial f}{\partial y}} = (-i\eta)\,\tilde{f} + \frac{1}{2\pi}\int_L n_2\,[f]\exp\left(i\left(\xi x' + \eta y'\right)\right)dL \tag{15}$$

where the values with tilde are the transforms, $\xi$ and $\eta$ are the variables in the transform space. It is seen that the transforms of derivatives (14), (15) have the integral parts with the jump of function itself.

Considering now the system of basic equations (4) – (9) in the transform space and taking into account the expressions for derivatives (14), (15) we solve our problem in terms of transforms. The solution is presented in the form of integral represantations as follows:

for temperature

$$\tilde{T} = \frac{1}{2\pi}\int_L K_T\,[T]\exp\left(i\left(\xi x' + \eta y'\right)\right)dL, \qquad K_T = \frac{-i\eta}{\xi^2 + \eta^2} \tag{16}$$

for displacements

$$\tilde{U} = \int_L \left\{D_U^1\,[U] + D_V^1\,[V] + D_T^1\,[T]\right\}[T]\exp\left(i\left(\xi x' + \eta y'\right)\right)dL \tag{17}$$

$$\tilde{V} = \int_L \left\{D_U^2\,[U] + D_V^2\,[V] + D_T^2\,[T]\right\}[T]\exp\left(i\left(\xi x' + \eta y'\right)\right)dL \tag{18}$$

$$D_U^1 = \frac{\nu_0\xi^2 - \nu_1\eta^2}{\Delta_d}(i\eta)\,, \qquad D_U^2 = \frac{\nu_1\xi^2 - \nu_0\eta^2}{\Delta_d}(-i\xi)$$

$$D_V^1 = \frac{\nu_0\xi^2 - \nu_1\eta^2}{\Delta_d}(-i\xi)\,, \qquad D_V^2 = \frac{\nu_2\xi^2 + \nu_1\eta^2}{\Delta_d}(-i\eta)$$

$$\nu_1 = 1 + \nu_0, \qquad \nu_2 = 2 + 3\nu_0, \qquad \nu_0 = \frac{\nu}{1 - 2\nu}$$

$$D_T^1 = \beta_1\frac{\xi\eta}{\Delta\,d}, \qquad D_T^2 = \beta_1\frac{\eta^2}{\Delta\,d}, \qquad \beta_1 = \frac{1+\nu}{1-2\nu}\alpha_t$$

$$\Delta_d = \frac{2\pi(1-\nu)}{1-2\nu}\left(\xi^2+\eta^2\right)^2$$

for stresses

$$\widetilde{\sigma}_x = \int_L \left\{K_U^1[U] + K_V^1[V] + K_T^1[T]\right\}[T]\exp\left(i\left(\xi x' + \eta y'\right)\right)dL \quad (19)$$

$$\widetilde{\sigma}_y = \int_L \left\{K_U^2[U] + K_V^2[V] + K_T^2[T]\right\}[T]\exp\left(i\left(\xi x' + \eta y'\right)\right)dL \quad (20)$$

$$\widetilde{\tau}_{xy} = \int_L \left\{K_U^3[U] + K_V^3[V] + K_T^3[T]\right\}[T]\exp\left(i\left(\xi x' + \eta y'\right)\right)dL \quad (21)$$

$$K_U^1 = \frac{-\xi\eta^3}{\Delta_s}, \qquad K_V^1 = K_U^3 = \frac{\xi^2\eta^2}{\Delta_s}, \qquad K_T^1 = \beta_2\frac{i\eta^3}{\Delta_s}$$

$$K_U^2 = K_V^3 = \frac{-\xi^3\eta}{\Delta_s}, \qquad K_V^2 = \frac{\xi^4}{\Delta_s}, \qquad K_T^2 = \beta_2\frac{i\xi^2\eta}{\Delta_s}$$

$$K_T^3 = \beta_2\frac{-i\xi\eta^2}{\Delta_s}, \qquad \beta_2 = (1+\nu)\,\alpha_t$$

$$\Delta_s = \frac{2\pi\left(1-\nu^2\right)}{E}\left(\xi^2+\eta^2\right)^2$$

Here $L$ is the line of the segment.

Applying the Fourier inversion formula to the expressions (16) – (21) and carrying out the integration on the line of the segment we find our solution. The inversion procedure is described in [2] and includes the principles of generalized functions for taking one improper integral. By analogy with the elasticity problem [1] this solution may be expressed by means of the function $f(x,y)$ and its derivatives in the following way:

for temperature field

$$T = F_T[T], \qquad \frac{\partial T}{\partial x} = F_T^x[T], \qquad \frac{\partial T}{\partial y} = F_T^y[T] \quad (22)$$

$$F_T = -2(1-\nu) f_{,y}, \qquad F_T^x = -2(1-\nu) f_{,xy}$$
$$F_T^y = -2(1-\nu) f_{,yy}$$

for displacements

$$U = F_U^U [U] + F_V^U [V] + F_T^U [T] \tag{23}$$

$$V = F_U^V [U] + F_V^V [V] + F_T^V [T] \tag{24}$$

$$F_T^U = -(1+\nu) \alpha_t y f_{,x}, \qquad F_T^V = -(1+\nu) \alpha_t (y f_{,y} + f)$$

for stresses

$$\sigma_x = P_U^x [U] + P_V^x [V] + P_T^x [T] \tag{25}$$

$$\sigma_y = P_U^y [U] + P_V^y [V] + P_T^y [T] \tag{26}$$

$$\tau_{xy} = Q_U [U] + Q_V [V] + Q_T [T] \tag{27}$$

$$P_T^x = E\alpha_t (y f_{,yy} + 2 f_{,y}), \qquad P_T^y = E\alpha_t y f_{,xx}$$
$$Q_T = -E\alpha_t (y f_{,xy} + f_{,x})$$

The function $f(x,y)$ here is the same one of the elasticity problem [1], that is

$$f(x,y) = -\frac{1}{4\pi(1-\nu)} \left\{ y \left( \arctan\frac{y}{x-a} - \arctan\frac{y}{x+a} \right) - \right.$$

$$\left. -(x-a)\ln\sqrt{(x-a)^2+y^2} + (x+a)\ln\sqrt{(x+a)^2+y^2} \right\} \tag{28}$$

The functions standing before the jumps of the displacements in the expressions (23) – (27) with the accuracy of sign, because of definition of jumps (1) – (3), coincide with the corresponding functions of the elasticity problem. Their values are presented in the monograph [1]. The other functions standing before the jump of the temperature and presented here correspond to the thermoelasticity problem.

The obtained solution may be used for solving the plane thermoelasticity problems of fracture mechanics.

# 3. TEST PROBLEM

To prove the trustworthiness of the method the test problem was examind. This is the straight thermally insulated crack $L$ of length $2l$ in an infinite plane (Figure 2).

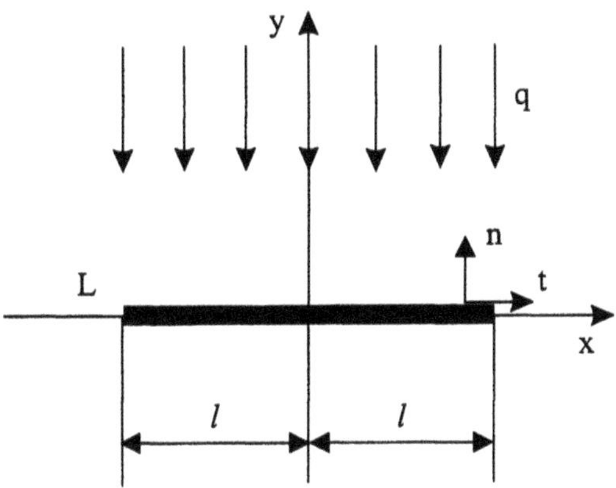


*Figure 2* Straight crack.

A plane is subjected to a remotely uniform heat flow $q$ perpendicular to the line of the crack. It is assumed that the heat does not pass through the crack and the opposite sides of the crack are free from the normal and shear stresses and do not contact each other. Then, the following boundary conditions must be satisfied on the line of the crack:

$$\left.\frac{\partial T}{\partial n}\right|_L = 0, \qquad \sigma_n|_L = 0, \qquad \tau_{nt}|_L = 0 \tag{29}$$

The analytical solution of this problem is well known [3, 4]. The jumps of the temperature and displacements on the line of the crack ($|x| \leq l$, $y = 0$) are defined as follows [3]:

$$[T] = 2q\sqrt{l^2 - x^2}, \quad [U] = \alpha_t\,(1+\nu)\,qx\sqrt{l^2 - x^2}, \quad [V] = 0 \tag{30}$$

The stress intensity factors of mode I ($K_1$) and of mode II ($K_2$) are given by expressions [3]

$$K_1^{\pm} = 0, \qquad K_2^{\pm} = \pm\frac{\alpha_t E q l \sqrt{l}}{4\,(1-\nu)} \tag{31}$$

where the sign plus corresponds to the right tip and the sign minus to the left tip of the crack.

For numerical solution of this problem we use the analytical solution for segment (22) – (27) and technique of the BEM [1]. The following systems of linear algebraic equations are obtained:

$$\sum_{j=1}^{N} A_{ij}\,[T]_j = -q, \qquad \sum_{j=1}^{N} B_{ij}\,[U]_j = \alpha_t\,(1+\nu) \sum_{j=1}^{N} C_{ij}\,[T]_j \tag{32}$$

$$i = 1, \ldots, N$$

$$2\pi A_{ij} = B_{ij} = \frac{2a}{(x_i - x_j)^2 - a^2}, \qquad j \neq 1, \quad j \neq N$$

$$C_{ij} = \ln\left|\frac{x_i - x_j - a}{x_i - x_j + a}\right|, \qquad j \neq 1, \quad j \neq N$$

Here $N$ is the number of the boundary elements of length $2a$.

It should be emphasized that the boundary influence coefficients for the end-elements ($j = 1$, $j = N$) are obtained by means of the integral representations for the heat flow and stresses and analytical behaviour of their jumps. It is the same way as in the elasticity problem [1]. The jumps for the special end-elements on the base of analytical solution (30) are set in the form

$$[T]\,(z) = [T]_{1,N}\sqrt{z/a}, \qquad [U]\,(z) = [U]_{1,N}\sqrt{z/a} \tag{33}$$

where $z$ is the distance from the tip along the crack, the index $1, N$ corresponds to the jumps in the centre of end-elements.

The numerical results for the jumps of temperature and displacement co-ordinate very well with the analytical solution (30). The stress intensity factor was found by calculating the stress near the tip of the crack accoding to the formula

$$K_2 = \tau_{nt}\,(r, 0)\sqrt{2r} \tag{34}$$

This value was compared with the analytical solution (31). For relative distance from the crack tip $10^{-7}$ and $N = 9$ the value of $K_2$ was computed with the error less than 1%.

The numarical investigations show it is enough to take into account only one end-element for calculating stress near by the tip of the crack when we estimate the stress intensity factor. The introduced error in this case is no more than 1%.

## 4. ILLUSTRATED PROBLEM

The thermoelasticity problem for the broken thermally insulated crack containing two links was considered as an illustrated example. The scheme of the loading is shown in Figure 3.

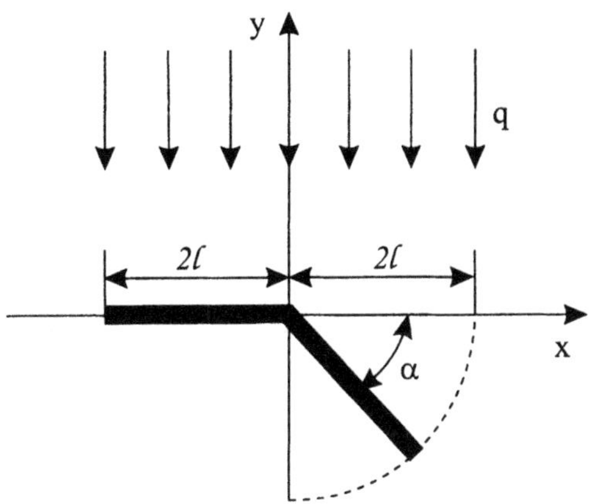


*Figure 3* Broken crack.

The total length of the crack is $4l$. The point of the break is located in the centre of the crack and the angle of the break $\alpha$ is changed from 0 to $-\pi/2$. The uniform heat flow $q$ is perpendicular to the left half of the crack.

When the crack is a straight line, the calculated values of $K_2$ differ from the analytical one (31) no more than 1% for $N = 8$. The dependence of the stress intensity factors from the angle of the break was investigated. The results in the form of graphs are presented in Figure 4 where the values of $K_i^{\pm}$ $(i = 1, 2)$ are given in the relative unit measure with the accuracy of $\alpha_t Eql\sqrt{l}/(1 - \nu)$.

The ciphers 1 and 2 mark the curves for $K_1$ and $K_2$, respectively. As we see, when the crack is broken, the stress intensity factor of mode I does not equal zero because of appearing the jump of the displacement $V$. The curves for the right tip of the crack are shown by solid line and for the left tip by dashed line. The values of $K_2$ at the left tip are negative and the velues of other stress intensity factors are positive.

The graphs in Figure 4 show that the values of stress intensity factor of mode II are constantly reduced with increasing the angle of the break. The values of stress intensity fsctor of mode I are at first increased and then decreased. The maximum value of $K_1$ at the right tip of the crack is reached at the angle that is equal to approximately 70 degrees. When

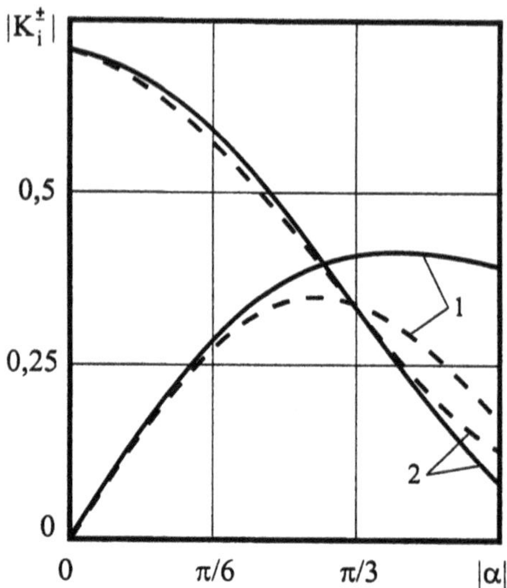


*Figure 4* Curves of stress intensity factors.

the angle of the break is more than 60 degrees, $K_1$ exceed $K_2$ by absolute value at both tips.

The analysis of the results shows that the difference in values between right and left tips of the crack is more considerable for the stress intensity factor of mode I. It is seen from the graphs that the stress intensity factor of mode I at the right tip of the crack is greater than at the left tip within the range $\pi/3 < |\alpha| < \pi/2$. It means that the tips of the crack are not the same in this case. The crack propagation will be started only at the right tip at this circumstances.

The summary of this results is when the crack is broken, it must be considered as a crack of mixed type with taking into account the stress intensity factor of mode II and of mode I also.

## References

[1] Crouch, S.L., and Starfield, A.M. (1983). *Boundary element methods in solid mechanics.* London: George Allen and Unwin.

[2] Shevchenko, V.P. (1977). *Integral transforms in theory of plates and shells* [in Russian]. Donetsk: Donetsk state university.

[3] Panasyuk, V.V., Savruk, M.P., and Datsyshin A.P. (1976). *Distribution of stresses near cracks in plates and shells* [in Russian]. Kiev: Naukova Dumka.

[4] Shi, G.C. (1962). On the singular character of thermal stresses near a crack tip. *J. Appl. Mech.*, 29:587-589.

# INVESTIGATING UNNAMED SINGULARITIES

Massimo Guiggiani
*Dipartimento di Matematica*
*Università di Siena*
*Via del Capitano 15, 53100 Siena, Italy*
guiggiani@unisi.it

Attilio Frangi
*Dipartimento di Ingegneria Strutturale*
*Politecnico di Milano*
*Piazza Leonardo da Vinci 32, 20133 Milano, Italy*
frangi@stru.polimi.it

**Keywords:** Hypersingular integral equations, singular integrals, plate bending, BEM.

**Abstract** Boundary integral equations with extremely singular (i.e., more than hypersingular) kernels would be useful to obtain second and third order derivatives of the primary variable on the boundary. In this paper it is shown how to obtain these boundary integral equations with still unnamed singularities and, moreover, how to efficiently and reliably compute all the singular integrals. This is done by extending in full generality the so called direct approach.

## Introduction

Boundary integral equations (BIE) with strongly singular and hypersingular kernels are now currently employed in many fields of applied mechanics. However, as already stated in [3], boundary integral equations with even higher and still unnamed singularities (i.e., more than hypersingular) are equally well defined and could prove useful in some applications. Since the typical hypersingular boundary integral equation (HBIE) is basically a BIE for $w_{,a}$ (if $w$ is the primary variable), i.e. a gradient BIE, we may call the integral equations for $w_{,ab}$ and for $w_{,abc}$ as second order and third order gradient BIE, respectively, and so on.

*T. Burczynski (ed.), IUTAM/IACM/IABEM Symposium on Advanced Mathematical and Computational Mechanics Aspects of the Boundary Element Method,* 133–143.

Apparently it was in 1994 that Knöpke presented in [6] the first derivation of a second order gradient BIE, that is the identity for the bending moment components of elastic Kirchhoff plates. The regularization was carried out using simple linear solutions for a partial regularization, followed by substantial analytical steps, thus limiting the whole analysis to straight boundaries. Moreover, numerical examples were not presented. The same kind of singularity, but in the context of plane elasticity, was addressed by Watson in [7].

Second and third order gradient BIE's were employed by Frangi and Bonnet in [1] for developing a symmetric Galerkin BEM for Kirchhoff plates. Integration by parts was used to have at most weakly singular kernels. Another paper dealing with supersingularities, that is with what we call here the second-order singular BIE, for potential problems in three-dimensional bodies appeared very recently [5]. The singular integrals are evaluated extending, with substantial work, the direct method (e.g., [2, 3, 4]).

In this paper a systematic direct approach for the evaluation of all singular integrals in the second and third order gradient BIE's is developed. The proposed technique can be seen as the extension to these higher singularities of the *direct method* presented in [3] and reviewed in [4]. The analysis is developed for the bending of thin elastic plates (also called Kirchhoff plates), since this is one of the fields where this kind of BIE's are certainly useful. However, the selection of plate bending is just a matter of definiteness, and does not limitate the generality of the proposed algorithm. Numerical results are presented here for curved (quadratic) elements. In all cases, very accurate results are easily obtained.

## 1. A GENERAL FORMULATION FOR BIE'S

The notation generally employed in Kirchhoff plate theory is here briefly summarized. $\Omega$: mean surface of a thin plate laying on the $x_3 = 0$ plane of the Cartesian reference system $(O; x_1, x_2, x_3)$ ($\mathbf{x} = (x_1, x_2, 0) = (x_1, x_2)$ henceforth); $\Gamma \equiv \partial\Omega$: plate boundary; $E$ and $\nu$: Young modulus and Poisson ratio, respectively; $D = Eh^3/[12(1-\nu^2)]$ flexural rigidity of the plate (of thickness $h$); $\mathbf{n} = (n_1, n_2)$ and $\boldsymbol{\tau} = (\tau_1, \tau_2)$: in-plane normal and tangential versors at $\mathbf{x} \in \Gamma$, respectively; $w$: out-of-plane displacements (deflections); $\phi_N = w_{,m} n_m$, $\phi_T = w_{,m} \tau_m$: normal and tangential slopes, respectively; $m_{mn}$: moment components; $q_m$: shear components $m_N \equiv m_{mn} n_m n_n$, $m_T \equiv m_{mn} n_m \tau_n$, $q = q_m n_m$: normal moments, twisting moments and shears, respectively; $p = p(\mathbf{x})$: vertical pressure acting on the plate mean surface.

A common path leading to the usual boundary element equations begins by enforcing the generalized Betti identity for plates which writes

$$0 = \int_\Gamma \Big[ w^\star(\mathbf{x}) q(\mathbf{x}) - w^\star{}_{,m}(\mathbf{x}) m_{mn}(\mathbf{x}) n_n(\mathbf{x}) + m^\star_{mj}(\mathbf{x}) n_j(\mathbf{x}) w_{,m}(\mathbf{x}) - q^\star(\mathbf{x}) w(\mathbf{x}) \Big] \, \mathrm{d}\ell + \int_\Omega w^\star(\mathbf{x}) p(\mathbf{x}) \, \mathrm{d}\Omega, \quad (1)$$

where $w^\star(\mathbf{x})$ denotes a generic auxiliary field, defined on $\Omega_\infty$ (an ideally infinite plate), representing the static deflection in $\Omega_\infty$ due to a given $p^\star(\mathbf{x})$ such that $p^\star(\mathbf{x}) = 0$ in $\Omega$. Starred quantities flow from $w^\star(\mathbf{x})$ in the same way as unstarred ones flow from $w(\mathbf{x})$.

The classical boundary integral equations can be obtained by setting in turn

$$w^\star = W(\mathbf{y}, \mathbf{x}) = \frac{1}{16\pi D} r^2 \log(r^2/{r_0}^2),$$

thus generating the *displacement equation* (function $W$ expresses vertical displacement at $\mathbf{x}$ due to a concentrated vertical force acting at $\mathbf{y}$) and

$$w^\star = W_{,a}(\mathbf{y}, \mathbf{x}) = \frac{\partial W}{\partial x_a} = -\frac{\partial W}{\partial y_a},$$

with $a = 1, 2$, thus yielding the (first order) *gradient equation* (or hypersingular BIE). These equations contain strong and hyper-singularities which have led to the introduction of different options (analytical or numerical, direct or indirect) for their evaluation.

Boundary integral equations with *more than hypersingular* kernels can be obtained by setting in eqn. (1) either

$$w^\star = W_{,ab}, \qquad \text{or even} \qquad w^\star = W_{,abc}. \quad (2)$$

Owing to their very high order of singularity, the corresponding boundary integral equations cannot be considered as classical ones.

The direct method is here adopted and extended to deal with these new kind of integral equations. They will be henceforth referred to as *second order gradient equation* when employing the first kernel in eq. (2), and *third order gradient equation* when employing the second kernel in (2).

## 1.1. BIE ON THE PUNCTURED DOMAIN

In view of the singular behaviour of some kernel functions for $\mathbf{x} \to \mathbf{y}$ (when $\mathbf{y} \in \Gamma$), a vanishing neighborhood $v_\varepsilon$, of boundary $s_\varepsilon + e_\varepsilon$, is excluded from $\Omega$ as shown in Figure 1. The boundary of the punctured

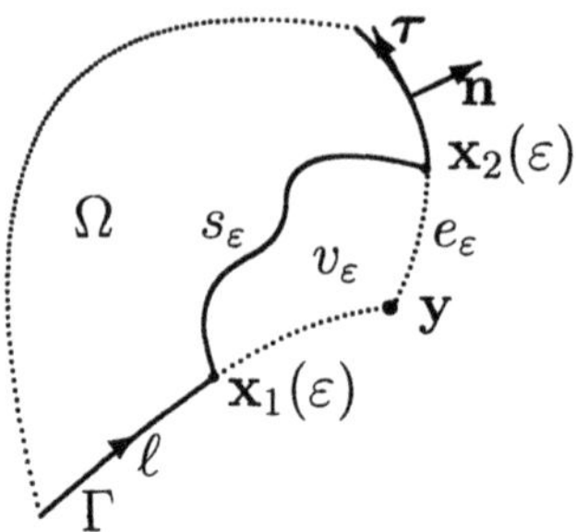


*Figure 1* Exclusion of the singular point $\mathbf{y}$ by a vanishing neighborhood $v_\varepsilon$.

domain $\Omega_\varepsilon = \Omega - v_\varepsilon$ is denoted by $\Gamma_\varepsilon = \Gamma - e_\varepsilon + s_\varepsilon$. The two endpoints of $s_\varepsilon$, ordered according to the positive sense of $\Gamma$, are indicated by $\mathbf{x}_1(\varepsilon)$ and $\mathbf{x}_2(\varepsilon)$. Obviously, $\Omega_\varepsilon$ is just a subdomain of the domain $\Omega$ and no "deformation" of the domain is introduced.

In the following Sections the integral equations are collocated at $\mathbf{y} \in \Gamma$ and the limit expression for $\varepsilon \to 0$ is analyzed, where $\varepsilon$ can be conceived as the radius of the smallest circle centered at $\mathbf{y}$ and containing $v_\varepsilon$. No special constraints are set on the regularity of $\Gamma$ at point $\mathbf{y}$ (i.e., $\mathbf{y}$ can be placed at a corner point or at a point where $\Gamma$ has discontinuous curvature), on the shape of $v_\varepsilon$ [3, 4, 2], as shown in Figure 1, and on the order of singularity of the kernel functions.

The following compact notation is introduced to denote the whole integrand function in eqn. (1)

$$\mathcal{W}(w; \mathbf{y}, \mathbf{x}) = w^\star q - w^\star{}_{,m}\, m_{mn} n_n + m^\star_{mj} n_j w_{,m} - q^\star w, \tag{3}$$

Since the domain integral of pressure $p$ is at most weakly singular, the assumption $p = 0$ is adopted in the sequel for brevity and without loss of generality. On the punctured domain $\Omega_\varepsilon$ the BIE (1) can thus be rewritten as ($p = 0$)

$$0 = \lim_{\varepsilon \to 0} \left\{ \int_{\Gamma - e_\varepsilon} \mathcal{W}(w; \mathbf{y}, \mathbf{x})\, \mathrm{d}\ell + \int_{s_\varepsilon} \mathcal{W}(w; \mathbf{y}, \mathbf{x})\, \mathrm{d}\ell \right\}. \tag{4}$$

Notice that, regardless of the order of singularity, we know that the integral on $\Gamma_\varepsilon = \Gamma - e_\varepsilon + s_\varepsilon$ is identically equal to zero for any $\varepsilon > 0$ and therefore the limit is zero as well [3, 4].

## 1.2. FREE TERM EVALUATION

Let $N$ denotes the *order* of the gradient boundary integral equation, i.e., $N = 2$ for the second order gradient equation and $N = 3$ for the third order gradient equations analyzed herein. To proceed further we assume at this point $C^{N,\alpha}$ continuity for $w$ at $\mathbf{y}$.

*Before* taking the limit, the integral over $\Gamma_\varepsilon = (\Gamma - e_\varepsilon) + s_\varepsilon$ is split into integrals over $\Gamma - e_\varepsilon$ and $s_\varepsilon$ (Figure 1). The latter one is evaluated analytically, even for a non-circular shape of $s_\varepsilon$, yielding

$$\int_{s_\varepsilon} \mathcal{W}(w;\mathbf{y},\mathbf{x})\,\mathrm{d}\ell = \sum_{n=0}^{N} \left\langle \frac{A^{-(N-n)}_{i_1 i_2 \cdots i_n}(\mathbf{y},\mathbf{x})}{r^{N-n}} \right\rangle_{\mathbf{x}_1(\varepsilon)}^{\mathbf{x}_2(\varepsilon)} w_{,i_1 i_2 \cdots i_n}(\mathbf{y}) + \left\langle A^{\log}_{i_1 i_2 \cdots i_N}(\mathbf{y},\mathbf{x}) \log r \right\rangle_{\mathbf{x}_1(\varepsilon)}^{\mathbf{x}_2(\varepsilon)} w_{,i_1 i_2 \cdots i_N}(\mathbf{y}), \quad (5)$$

where $A^{-(N-n)}_{i_1 i_2 \cdots i_n}$ and $A^{\log}_{i_1 i_2 \cdots i_N}$ are regular tensor functions and

$$\left\langle [\cdot](\mathbf{y},\mathbf{x}) \right\rangle_{\mathbf{x}_1(\varepsilon)}^{\mathbf{x}_2(\varepsilon)} = [\cdot](\mathbf{y},\mathbf{x}_2) - [\cdot](\mathbf{y},\mathbf{x}_1).$$

This particular form for indices, which deviates form the choice made throughout the article (where indices are denoted by $i, j \cdots k$ rather than $i_1, i_2 \cdots i_n$), has been adopted in order to introduce a unified notation. Similarly $w_{,i_1 i_2 \cdots i_n}$ denotes the $n^{th}$ order gradient of the plate deflection. For instance, when $N = 3$ and $n = 2$, $A^{-(N-n)}_{i_1 i_2 \cdots i_n} = A^{-1}_{mn}$, and $w_{,i_1 i_2 \cdots i_n} = w_{,mn}$. When $n = 0$ it is assumed that $w_{,i_1 i_2 \cdots i_n} = w$.

As expected, some terms in eqn. (5) are not bounded for $\varepsilon \to 0$, but they will be canceled, as shown in eqn. (10) in the next section, by other terms arising from the integral on $\Gamma - e_\varepsilon$. At the moment, eqn. (5) is simply substituted back into eqn. (4).

## 1.3. SINGULAR INTEGRALS

To deal with the singular integral on $\Gamma - e_\varepsilon$ in eqn. (4) the usual discretization of $\Gamma$ into boundary elements is now taken into account. The arc-length coordinate $\ell$ running along $\Gamma$ (associated to the field point $\mathbf{x}$) is therefore defined by means of a piecewise continuous (i.e., continuous over each boundary element) intrinsic parameter $\xi$ such that $\ell = \ell(\xi)$, with $\xi \in [-1, 1]$ on each boundary element. Therefore, on each element we have that $\mathbf{x} = \mathbf{f}(\xi)$, where the functions $\mathbf{f}$ are usually given in terms of shape functions and nodal values.

The particular (although most difficult) case is analyzed where $\mathbf{y}$ lies between two boundary elements, denoted as $\Gamma^s_1$ and $\Gamma^s_2$. Therefore $\Gamma^s = \Gamma^s_1 \cup \Gamma^s_2$ indicates a part of the boundary strictly containing $\mathbf{y}$.

The evaluation of terms within parentheses in eqn. (4) (see also eqn. (5)) is addressed by means of the following provisions. The integral over $\Gamma^s - e_\varepsilon$ is expressed in terms of the intrinsic coordinate $\xi$. The regular integral over the rest of the boundary $\Gamma - \Gamma^s$ does not need any special treatment and can be evaluated by standard numerical (or analytical) techniques. Thus we have

$$\int_{\Gamma^s - e_\varepsilon} \mathcal{W}(w; \mathbf{y}, \mathbf{x})\, \mathrm{d}\ell = \int_{-1}^{\xi_1(\varepsilon)} \tilde{\mathcal{W}}(\xi) J_1(\xi)\, \mathrm{d}\xi + \int_{\xi_2(\varepsilon)}^{1} \tilde{\mathcal{W}}(\xi) J_2(\xi)\, \mathrm{d}\xi,$$

and eqn. (4) becomes

$$\begin{aligned} \int_{\Gamma - \Gamma^s} \mathcal{W}(w; \mathbf{y}, \mathbf{x})\, \mathrm{d}\ell + \lim_{\varepsilon \to 0} \Bigg\{ \int_{-1}^{\xi_1(\varepsilon)} \tilde{\mathcal{W}}(\xi) J_1(\xi)\, \mathrm{d}\xi + \int_{\xi_2(\varepsilon)}^{1} \tilde{\mathcal{W}}(\xi) J_2(\xi)\, \mathrm{d}\xi \\ + \sum_{n=0}^{N} \left\langle \frac{A_{i_1 i_2 \cdots i_n}^{-(N-n)}(\mathbf{y}, \mathbf{x})}{r^{N-n}} \right\rangle_{\xi_1(\varepsilon)}^{\xi_2(\varepsilon)} w_{,i_1 i_2 \cdots i_n}(\mathbf{y}) \\ + \left\langle A_{i_1 i_2 \cdots i_N}^{\log}(\mathbf{y}, \mathbf{x}) \log r \right\rangle_{\xi_1(\varepsilon)}^{\xi_2(\varepsilon)} w_{,i_1 i_2 \cdots i_N}(\mathbf{y}) \Bigg\} = 0, \end{aligned} \tag{6}$$

where $\tilde{\mathcal{W}}(\xi) = \mathcal{W}\big(w(\xi); \mathbf{y}, \mathbf{x} = \mathbf{f}(\xi)\big)$ is the whole integrand as a function of $\xi$, $r = r(\xi)$ is the distance between the field and source points, $J_\alpha(\xi)$ is the Jacobian for $\ell = \ell(\xi)$ on $\Gamma_\alpha^s$, and $\xi_\alpha(\varepsilon) = \xi\big(\mathbf{x}_\alpha(\varepsilon)\big)$ is, on each element, the intrinsic coordinate corresponding to the endpoints of $e_\varepsilon$. The symbol $\langle \cdot \rangle_{\xi_1(\varepsilon)}^{\xi_2(\varepsilon)}$ has therefore the following meaning

$$\langle \cdot \rangle_{\xi_1(\varepsilon)}^{\xi_2(\varepsilon)} = \langle \cdot \rangle_{\xi(\mathbf{x}_1(\varepsilon))}^{\xi(\mathbf{x}_2(\varepsilon))}$$

**Laurent series.** It is convenient to define in the parameter space of each element around $\mathbf{y}$ the new local variable $\delta$

$$\delta(\xi) = \begin{cases} \xi - 1 & \text{on } \Gamma_1^s, \\ \xi + 1 & \text{on } \Gamma_2^s. \end{cases} \tag{7}$$

Therefore $\mathrm{d}\xi = \mathrm{d}\delta$ and $\delta \in [-2, 0]$ on $\Gamma_1^s$, while $\delta \in [0, 2]$ on $\Gamma_2^s$.

The truncated Laurent expansion $\mathcal{L}[\tilde{\mathcal{W}} J_\alpha]$ with respect to $\delta$ of the singular integrand $\tilde{\mathcal{W}} J_\alpha$ can be evaluated on each element $\Gamma_\alpha^s$ around $\mathbf{y}$ and writes

$$\mathcal{L}[\tilde{\mathcal{W}} J_\alpha](\mathbf{y}, \delta) = \sum_{n=1}^{N+1} \frac{\mathcal{W}_\alpha^{-n}(\mathbf{y})}{\delta^n}. \tag{8}$$

Obviously, $-n$ in the numerator is a superscript, not an exponent. As evident from its definition, the truncated Laurent series $\mathcal{L}[\tilde{\mathcal{W}} J_\alpha]$ only contains the singular terms [3, 4]. According to the roles of eqn. (8) in the following paragraphs, coefficients $\mathcal{W}_\alpha^{-n}(\mathbf{y})$ need be expressed in two different manners.

**Singularity subtraction.** The Laurent expansion (8) is then subtracted and added back to $\tilde{\mathcal{W}}J_\alpha$ in each integral in eqn. (6). On each element $\Gamma^s_\alpha$ the difference $\tilde{\mathcal{W}}J_\alpha - \mathcal{L}[\tilde{\mathcal{W}}J_\alpha]$ is a *regular* function of $\xi$ for $\delta(\xi) \to 0$, and its integral is therefore bounded and easily computable in a numerical way.

**Cancellation of unbounded terms.** All the unbounded terms cancel out in the limit as shown in the following paragraphs with a two step procedure.

First the integrals of the (added back) terms in the truncated Laurent expansion (8) are *analytically* integrated w.r.t. to $\xi$ yielding

$$\sum_{n=1}^{N+1}\left[\mathcal{W}_\alpha^{-n}(\mathbf{y})\int_{(-1)^\alpha}^{\xi_\alpha(\varepsilon)}\frac{\mathrm{d}\xi}{\delta^n}\right] = \left\langle \mathcal{W}_\alpha^{-1}(\mathbf{y})\log|\delta|\right\rangle_{(-1)^\alpha}^{\xi_\alpha(\varepsilon)} - \sum_{n=2}^{N+1}\left\langle \frac{\mathcal{W}_\alpha^{-n}(\mathbf{y})}{n-1}\frac{1}{\delta^{n-1}}\right\rangle_{(-1)^\alpha}^{\xi_\alpha(\varepsilon)}, \tag{9}$$

where $\delta = \delta(\xi)$ according to eqn. (7). Unbounded terms arise, in the limit $\varepsilon \to 0$, for $\xi = \xi_\alpha(\varepsilon)$. It is worth noting that this analytical integration is always a trivial task, regardless of the type of boundary elements employed and of the order of singularity.

Secondly, the contributions of the unbounded free terms arising from the integration over $s_\varepsilon$ in eqn. (5) are expanded themselves in a Laurent series (up to the zero order term, that is one more term with respect to the truncated Laurent expansion $\mathcal{L}[\cdot]$). This is shown to exactly cancel out the unbounded terms in eqn. (9) since

$$\int_{s_\varepsilon}\mathcal{W}(w;\mathbf{y},\mathbf{x})\,\mathrm{d}\ell = \left\langle \mathcal{W}_\alpha^{-1}(\mathbf{y})\log|\delta|\right\rangle_{\xi_1(\varepsilon)}^{\xi_2(\varepsilon)} - \sum_{n=2}^{N+1}\left\langle \frac{\mathcal{W}_\alpha^{-n}(\mathbf{y})}{n-1}\frac{1}{\delta^{n-1}}\right\rangle_{\xi_1(\varepsilon)}^{\xi_2(\varepsilon)} + \sum_{n=0}^{N} a^n_{i_1 i_2\cdots i_n}(\mathbf{y})\, w_{,i_1 i_2\cdots i_n}(\mathbf{y}) + O(\varepsilon),$$

Hence from eqns. (6) (or (5)) and (9) and according to the aforementioned two steps, we obtain the following result, of great relevance in the proposed approach

$$\lim_{\varepsilon\to 0}\left\{\sum_{\alpha=1,2}(-1)^{\alpha+1}\sum_{n=1}^{N+1}\left[\mathcal{W}_{\alpha}^{-n}(\mathbf{y})\int_{(-1)^{\alpha}}^{\xi_{\alpha}(\varepsilon)}\frac{\mathrm{d}\xi}{\delta^{n}}\right]\right.$$

$$\left.+\sum_{n=0}^{N}\left\langle\frac{A_{i_1 i_2\cdots i_n}^{-(N-n)}}{r^{N-n}}\right\rangle_{\xi_1(\varepsilon)}^{\xi_2(\varepsilon)} w_{,i_1 i_2\cdots i_n}(\mathbf{y})+\left\langle A_{i_1 i_2\cdots i_N}^{\log}\log r\right\rangle_{\xi_1(\varepsilon)}^{\xi_2(\varepsilon)} w_{,i_1 i_2\cdots i_N}(\mathbf{y})\right\}$$

$$=\sum_{n=0}^{N}\left[a_{i_1 i_2\cdots i_n}^{n}(\mathbf{y})+b_{i_1 i_2\cdots i_n}^{n}(\mathbf{y})\right] w_{,i_1 i_2\cdots i_n}(\mathbf{y}) \quad (10)$$

where $a_{i_1 i_2\cdots i_n}^{n}(\mathbf{y})$ are the zero order coefficients (i.e., the first non singular terms) of the Laurent series expansion of each free term of eqn. (5) and therefore depend solely on the geometry of the boundary at $\mathbf{y}$. On the other hand, $b_{i_1 i_2\cdots i_n}^{n}(\mathbf{y})w_{,i_1 i_2\cdots i_n}(\mathbf{y})$ stem from the analytical integral over $\Gamma^s$ of the Laurent series of eqn. (8), i.e. from the evaluation of eqn. (9) with $\delta=-2$ if $\alpha=1$ and $\delta=2$ if $\alpha=2$ (the only terms that do not get canceled). Therefore, the knowledge of the truncated Laurent expansions (8) is all we need to evaluate all the coefficients $b_{i_1 i_2\cdots i_k}^{k}(\mathbf{y})$.

## 1.4. FINAL BOUNDARY INTEGRAL EQUATION

Collecting all the previous results the boundary integral equation (4), regardless of the order of singularity of the kernels involved, can be rewritten in terms of *regular* integrals and *bounded* free terms as

$$\sum_{\alpha=1,2}\int_{-1}^{1}\left(\tilde{\mathcal{W}}(\xi)J_{\alpha}(\xi)-\sum_{n=1}^{N+1}\left[\frac{\mathcal{W}_{\alpha}^{-n}(\mathbf{y})}{\delta(\xi)^{n}}\right]\right)\mathrm{d}\xi+\int_{\Gamma-\Gamma^{s}}\mathcal{W}(w;\mathbf{y},\mathbf{x})\,\mathrm{d}\ell$$

$$+\sum_{n=0}^{N}\left[a_{i_1 i_2\cdots i_n}^{n}(\mathbf{y})+b_{i_1 i_2\cdots i_n}^{n}(\mathbf{y})\right] w_{,i_1 i_2\cdots i_n}(\mathbf{y})=0. \quad (11)$$

Hence only the coefficients $\mathcal{W}_{\alpha}^{-n}$, $a_{i_1 i_2\cdots i_n}$ and $b_{i_1 i_2\cdots i_n}$ are required for the implementation of this approach. Moreover, the coefficients $b_{i_1 i_2\cdots i_n}$ directly flow from $\mathcal{W}_{\alpha}^{-n}$.

It is worth stressing that, as a general result, when $\Gamma$ is sufficiently smooth at $\mathbf{y}$, i.e. at least the first $N+1$ derivatives of the geometric coordinates w.r.t. $\xi$ (or $\delta$) are continuous, then $a_{i_1 i_2\cdots i_n}^{n}=0$, for $n<N$, and $a_{i_1 i_2\cdots i_N}^{N}w_{,i_1 i_2\cdots i_N}(\mathbf{y})=-1/2w_{,ab}$ $(=1/2w_{,abc})$ for the second order (third order, respectively) equation with $w^{\star}=W_{,ab}$ $(=W_{,abc})$.

## 2. NUMERICAL EXAMPLES

Some numerical tests are presented to check the proposed method for the direct evaluation of integrals with singularities two orders higher than the so-called hypersingular ones. Since the direct method operates in terms of the intrinsic coordinate $\xi$, it should be appreciated that the most difficult tests arise when curved or distorted elements (i.e., with non constant Jacobian) are employed.

We consider the following benchmark integral with the most singular kernels of the third order gradient equation

$$\mathcal{J}_{abc} = \lim_{\varepsilon \to 0} \left\{ \int_{\Gamma^s - e_\varepsilon + s_\varepsilon} Q_{,abc}(\mathbf{y}, \mathbf{x})\, \mathrm{d}\ell \right\} \tag{12}$$

where $\Gamma^s$ consists of one quadratic boundary element as shown in Figure 2. The kernels $Q_{,abc}(\mathbf{y}, \mathbf{x})$ behave like $r^{-4}$, when $r \to 0$, and are therefore two orders more singular than hypersingular.

A parameter $d$ controls the mesh distortion. The coordinates of the three nodes of the quadratic boundary element are (see Figure 2): $\mathbf{x}_1^s \equiv \mathbf{x}_{\text{in}} = (0,0)$, $\mathbf{x}_2^s = (1-d, d) \equiv \mathbf{y}$, $\mathbf{x}_3^s \equiv \mathbf{x}_{\text{fin}} = (2,0)$. The collocation point always coincides with the mid node $\mathbf{x}_2^s$.

It can be shown that integral (12) is equal to

$$\mathcal{J}_{abc} = \sum_{\alpha=1,2} \int_{-1}^{1} \left( Q_{,abc}(\mathbf{y}, \mathbf{x}(\xi)) J_\alpha(\xi) + \mathcal{L}[\Delta W_{,abcp}\, n_p J_\alpha] \right) \mathrm{d}\xi + b^0(\mathbf{y}). \tag{13}$$

and can be integrated analytically yielding

$$\mathcal{J}_{abc} = \frac{1}{\pi r^2} e_{ja} \left\langle 4r_{,j}\, r_{,b}\, r_{,c} - \delta_{jb} r_{,c} - \delta_{jc} r_{,b} - \delta_{bc} r_{,j} \right\rangle_{\mathbf{x}_{\text{in}}}^{\mathbf{x}_{\text{fin}}} \tag{14}$$

As shown in Figure 2, three different values for $d$ are tested: 0, 0.2 and 0.4. The relative errors (Euclidean norm of the relative error vector: $e =$

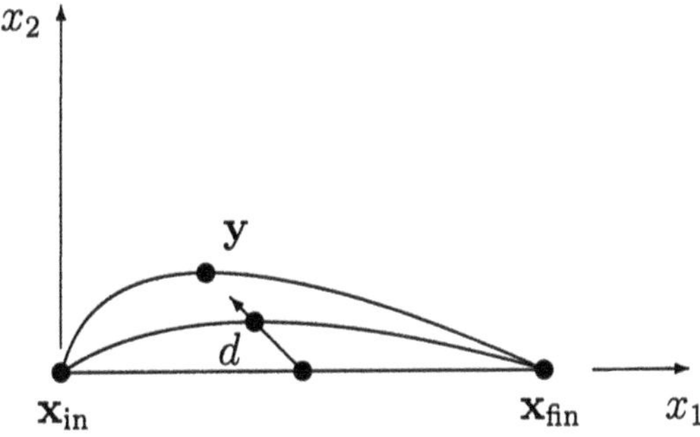


*Figure 2* Sequence of increasingly distorted quadratic elements.

$\sqrt{e_{abc}e_{abc}}$) for $\mathcal{J}_{abc}$ $(a,b,c=1,2)$ for the above values of the distortion parameter $d$ and number $N_g$ of Gauss points of the quadrature formula are collected in Table 1.

Exact results are as follows: $-\mathcal{J}_{111}=\mathcal{J}_{122}=\mathcal{J}_{212}=\mathcal{J}_{221}=A(d)$, $\mathcal{J}_{121}=\mathcal{J}_{112}=\mathcal{J}_{211}=-\mathcal{J}_{222}=B(d)$, with the following values $A(0)=0.$ and $B(0)=6.3661977367\times10^{-1}$, $A(0.2)=2.965870046\times10^{-1}$ and $B(0.2)=5.7666266700\times10^{-1}$, $A(0.4)=7.566462069\times10^{-1}$ and $B(0.4)=-9.377425426\times10^{-2}$.

If the element is undistorted the errors are negligible. With a mild distortion ($d=0.2$), the direct method provides extremely accurate results, even when dealing with $r^{-4}$ singular kernels (see Table 1). As expected, if the distortion of the element increases and reaches impractical values ($d=0.4$), the numerical results deteriorate, although the relative errors are as low as $10^{-6}$ with Gaussian formulas of order 10.

| $N_g$ | $d$=0 | $d$=0.2 | $d$=0.4 |
|---|---|---|---|
| 2 | 0. | $.48\times10^{-2}$ | $.41\times10^{1}$ |
| 4 | - | $.82\times10^{-5}$ | $.65\times10^{-1}$ |
| 6 | - | $.87\times10^{-8}$ | $.45\times10^{-3}$ |
| 8 | - | $.59\times10^{-11}$ | $.70\times10^{-5}$ |
| 10 | - | $.31\times10^{-11}$ | $.14\times10^{-6}$ |

*Table 1* Norm of the relative error for the third order gradient BIE.

## 3. CONCLUSIONS

BEM researchers and practitioners are now fairly familiar with boundary integral equations containing strongly singular and hypersingular kernels. However, kernels with higher orders of singularities may arise in some problems.

In this paper higher order gradient BIE's have been investigated. First it has been shown that all unbounded terms cancel each other, without the need of any kind of arbitrary interpretation. Then, the direct approach for the evaluation of singular integrals have been extended to deal with these new singularities. The final formulation is completely general, quite simple and does not involve any singular integration, since all the cancellations and singular integrals have been performed analytically. Very good accuracy is therefore expected.

## References


[1] FRANGI, A. & BONNET, M., A Galerkin symmetric and direct BIE method for Kirchhoff elastic plates: formulation and implementation, *Int. J. Num. Meth. Engng.*, 41 (1998), pp. 337–369.

[2] FRANGI, A. & GUIGGIANI, M., Boundary element analysis of Kirchhoff plates with direct evaluation of hypersingular integrals, *Int. J. Num. Meth. Engng.*, to appear, (1999).

[3] GUIGGIANI, M., KRISHNASAMY, G., RUDOLPHI, T.J. & RIZZO, F.J., A general algorithm for the numerical solution of hypersingular boundary integral equations, *ASME J. Appl. Mech.*, 59 (1992), pp. 604–614.

[4] GUIGGIANI, M., Formulation and numerical treatment of boundary integral equations with hypersingular kernels, in *Singular Integrals in Boundary Element Methods*, eds. SLÁDEK, V. & SLÁDEK, J., Computational Mechanics Publications, Southampton, Chap. 3, (1998) pp. 85–124.

[5] KARAMI, G. & DERAKHSHAN, D., An efficient method to evaluate hypersingular and supersingular integrals in boundary integral equations analysis, *Engng. Analysis with Boundary Elem.*, 23 (1999), pp. 317–326.

[6] KNÖPKE, B., The hypersingular integral equation for the bending moments $m_{xx}$ $m_{xy}$ and $m_{yy}$ of the Kirchhoff plate, *Computat. Mech.*, 15 (1994), pp. 19–30.

[7] WATSON, J.O., Singular boundary elements for the analysis of cracks in plane strain, *Int. J. Num. Meth. Engng.*, 38 (1995), pp. 2389–2411.

# BOUNDARY ELEMENT METHOD, HOMOGENIZATION AND HEAT CONDUCTION IN COMPOSITE MATERIALS

M. KAMINSKI
*Department of Civil Engineering, George Brown School of Engineering*
*Rice University, Houston, 6100 Main Street, 77005-1892 TX, USA*
*permanent address:*
*Division of Mechanics of Materials, Technical University of Lodz*
*Al. Politechniki 6, 93-590 Lodz, POLAND*
e-mail: marcin@kmm-lx.p.lodz.pl, http://kmm.p.lodz.pl/Marcin_Kaminski

Key words: Boundary Element Method, homogenization technique, heat conduction, composite materials, stochastic perturbation method

Abstract: The homogenization method for solution of the heat conduction problem is presented below in the context the Boundary Element Method. The approach is introduced for the fiber-reinforced composite with deterministically and randomly defined material properties. Computational implementation of the method can find the application for the composites with random interfaces where application of the Finite Element Method is very complicated. Numerical illustration shows deterministic and probabilistic sensitivity of effective composite heat conductivity with respect to fiber volume ratio and randomness level of composite constituents.

## 1. INTRODUCTION

Computational modeling of composite materials is a very complicated process due to the scale difference between the whole composite and its constituents, the randomness of reinforcement shape and location, local lacks of periodicity in composites as well as structural defects appearing in the macro-homogeneous components. One of the most powerful numerical tools intended for speeding up both the composite discretization process and the boundary value or boundary initial problems solution is the homogenization method [7,11]. The main idea is to find a globally

*T. Burczynski (ed.), IUTAM/IACM/IABEM Symposium on Advanced Mathematical and Computational Mechanics Aspects of the Boundary Element Method,* 145–160.

homogeneous medium equivalent to the original composite in the context of internal energy per unit volume, which is to be equal for both systems. Analyzing composites it should be mentioned that the homogenization method is only the intermediate tool used for efficient modeling of these materials – the next step is to solve original boundary value problem using the homogenized composite with effective parameters.

There are numerous methods of calculation of the effective characteristics for composites. In the case where the composite components volume fractions are given only, one can use the closed form algebraic equations on upper and lower bounds for the effective material tensors components [11]. On the other hand, for the composite with detailed information on the whole geometry together with material description the so-called cell problems are solved by corresponding Finite Element Method (FEM) numerical implementations [10,11], what makes it possible to compute directly the effective composite characteristics. However, one of the main disadvantages of the FEM approaches is difficulty in interface geometry randomness modeling in composites [15]; this difficulty can be omitted using Boundary Element Method (BEM) based approach proposed below, for instance.

Since that, mathematical description of the homogenization method for linear heat conduction problems in two-component composites in the context of BEM numerical algorithm and computational implementation is presented. Thanks to the method introduced, the effective thermal conductivity coefficient (general effective coefficient of linear potential field problem in fact) can be computed. Considering numerical aspects of the problem, homogenization procedure is based on the classical BEM algorithm (analogously to effective elasticity tensor modeling [10]) and may be implemented in any academic or commercial computer program. The effective heat conductivity tensor components are derived for the fiber-reinforced periodic structure but the method can be easily extended on n-component 1, 2 and 3D engineering composites in general configuration.

## 2. COMPOSITE MATERIAL MODEL

To introduce the homogenization method, let us consider a periodic two-component fiber-reinforced composite structure with a constant cross-section $Y \subset R^2$ along the $x_3$ axis. By the periodicity of the structure we understand the existence of such an element of composite, called the Representative Volume Element (RVE) $\Omega \subset Y$, that due to some homothety this element can be transformed onto structure Y. It is assumed, that the RVE has a rectangular cross-section and is constructed from two disjoint, coherent and perfectly bonded components - fiber ($\Omega_1$) and matrix ($\Omega_2$) (cf. Figs. 1,2) with the heat conductivity coefficients of the components defined as follows:

$$k(\mathbf{x}) = \chi_a(\mathbf{x})k_a\,;\ \mathbf{x} \in \Omega\,;\ a=1,2, \tag{1}$$

$$\chi_a = \begin{cases} 1;\ x \in \Omega_a \\ \text{elsewhere} \end{cases};\ \mathbf{x} \in \Omega. \tag{2}$$

Such a definition makes it possible to extend easily the proposed model to the general one for n-component composite. Finally, let us assume that the interface $\partial\Omega_{12}$ between fiber and matrix is continuous and sufficiently smooth contour. Finally, let us introduce scale parameter $\varepsilon>0$ relating both scales of a composite such that

$$\underset{\varepsilon\in\Re}{\exists}\, k^{\varepsilon}(\mathbf{y}) = k\left(\frac{\mathbf{y}}{\varepsilon}\right) = k(\mathbf{x}), \tag{3}$$

where: $\mathbf{x}$ - local coordinates connected with the RVE while $\mathbf{y}$ denote here composite global coordinates.

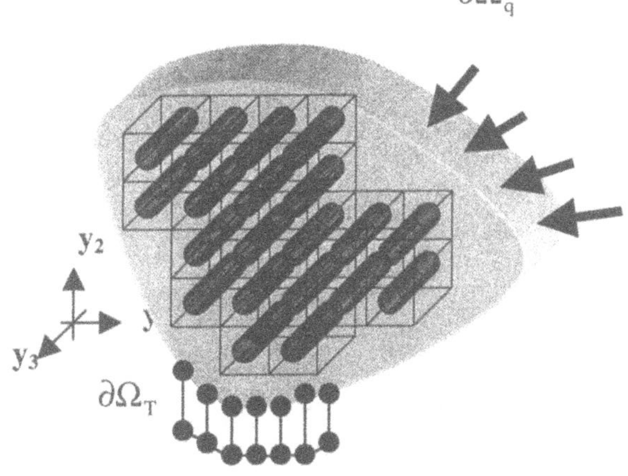


*Fig. 1. Periodic composite structure*

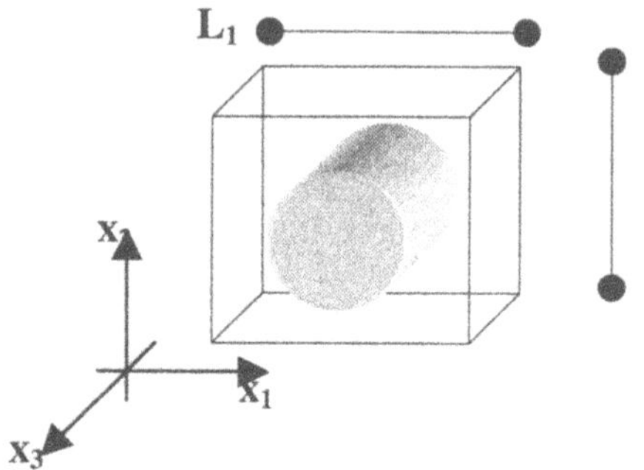


*Fig. 2. The Representative Volume Element (RVE)*

Considering above, the boundary value problem of the heat conduction is to find the temperature field $T^{\varepsilon}$ being a solution for the following set of partial differential equations:

$$k^{\varepsilon}(\mathbf{y})T^{\varepsilon}_{,i} - q^{\varepsilon}_{i} = 0, \tag{4}$$

$$q^{\varepsilon}_{i,i} + f = 0, \tag{5}$$

$$T^{\varepsilon} = \hat{T};\ \mathbf{y} \in \partial\Omega_{T},\ q^{\varepsilon} = \hat{q};\ \mathbf{y} \in \partial\Omega_{q}. \tag{6,7}$$

To propose the solution, the following Sobolev space is considered first:

$$H^{1}_{0}(Y) = \left\{ \xi \in L^{2}(Y) \middle| \frac{\partial \xi}{\partial x_{i}} \in L^{2}(Y), \xi = 0: x \in \partial Y \right\} \tag{8}$$

with the norm

$$\|\xi\| = \left[ \int_{Y} \left( \frac{\partial \xi}{\partial x_{i}} \right)^{2} dx \right]^{\frac{1}{2}}. \tag{9}$$

Further, a bilinear continuous form $a(T,\xi)$ is introduced on $H^{1}_{0}(Y)$

$$a(T,\xi) = \int_{Y} k^{\varepsilon}(x) \frac{\partial T}{\partial x_{i}} \frac{\delta \xi}{\partial x_{i}} dx \tag{10}$$

together with the linear one

$$L(\xi) = \int_Y f\xi dx. \tag{11}$$

Periodic solution $T^{\varepsilon} \in H_0^1(Y)$ is obtained for

$$\underset{\xi \in H_0^1(Y)}{\forall} \; a\left(T^{\varepsilon}, \xi\right) = L(\xi). \tag{12}$$

Thus, to arrive at the mathematical formulation for the homogenization problem, the periodic Sobolev space on $\Omega$ is considered

$$H_{per}^{1}(Y) = \left\{ \zeta \in L^2(\Omega) \middle| \frac{\partial \zeta}{\partial x_i} \in L^2(\Omega), \zeta \text{ is periodic} \right\} \tag{13}$$

with the norm

$$\|\zeta\| = \left[ \int_Y \zeta^2 dy + \int_Y \left( \frac{\partial \zeta}{\partial y_i} \right)^2 dy \right]^{\frac{1}{2}}, \tag{14}$$

and the following bilinear and linear forms $a_y(\chi,\zeta)$, $L_j(\zeta)$:

$$a_y(\chi, \xi) = \int_{\Omega} k(y) \frac{\partial \chi}{\partial y_i} \frac{\partial \zeta}{\partial y_i} dy, \; L_j(\zeta) = -\int_{\Omega} k(y) \frac{\partial \zeta}{\partial y_j} dy. \tag{15, 16}$$

Finally, the homogenization problem is to find the function $\chi \in H_{per}^{1}(Y)$ such that

$$\underset{\zeta \in H_{per}^1(Y)}{\forall} \; a_y(\chi, \zeta) = L(\zeta), \tag{17}$$

where the R.H.S. linear operator is equal to

$$L_j(\zeta) = -\int_{\partial \Omega_{12}} [k] \, \zeta n_j d\Omega. \tag{18}$$

In that case, the effective conductivity tensor is calculated as

$$k^{(\mathrm{eff})} = \langle k \rangle_\Omega + \langle k 1_i \Phi_{,i} \rangle_\Omega, \tag{19}$$

where the brackets indexed with $\Omega$ denotes spatial averaging of the internal function over this region, while $1_i$ is the unity tensor component.

## 3. NUMERICAL APPROACH

### 3.1. GENERAL LINEAR POTENTIAL FIELD PROBLEMS

The following equation is to be solved:

$$\Delta u \equiv \frac{\partial^2 u}{\partial x_1^2} + \frac{\partial^2 u}{\partial x_2^2} = f(\mathbf{x}) \tag{20}$$

with the fundamental solution

$$U^*(x,y) = -\frac{1}{2\pi} \ln\left(\frac{1}{r}\right); \ \mathbf{x} \in \Re^2 \tag{21}$$

and mixed boundary conditions for homogenization problem (17) as

$$\underset{\mathbf{x}\in\partial\Omega}{\forall} u(\mathbf{x}) = u^0, \quad \underset{\mathbf{x}\in\partial\Omega_{12}}{\forall} p(\mathbf{x}) = \frac{\partial u(\mathbf{x})}{\partial \mathbf{n}} = [k]. \tag{22, 23}$$

Then, the following BIEs pair is obtained [1,3]:

$$\underset{x\in\partial\Omega(\Gamma_1)}{\forall} : \quad \begin{aligned} &\tfrac{1}{2} u^0(\mathbf{x}) = \int_\Omega \mathbf{U}^*(x,y) f(y) d\Omega(y) + \\ &+ \int_{\partial\Omega} \left( P^*(x,y) u^0(y) - U^*(x,y) p(y) \right) d\Gamma(y) \end{aligned}, \tag{24}$$

$$\underset{x\in\partial\Omega_{12}(\Gamma_2)}{\forall} : \quad \begin{aligned} &\tfrac{1}{2} u(\mathbf{x}) = \int_\Omega \mathbf{U}^*(x,y) f(y) d\Omega(y) + \\ &+ \int_{\partial\Omega_{12}} \left( P^*(x,y) u(y) - U^*(x,y) p^0(y) \right) d\Gamma(y) \end{aligned} \tag{25}$$

It should be mentioned that extension of this model to the n-component fiber-like composite [11], where each next component is embedded into the previous one, results in n boundary integral equations, what makes it possible to solve the corresponding generalized homogenization problem.

### 3.2. NUMERICAL SOLUTION

Let us introduce a discretization of the boundary $\Gamma$ using K boundary elements $\Gamma^e$ and of the fields $\mathbf{u}(x)$ and $\mathbf{p}(x)$ in any element $\Gamma^e$ using local coordinates $\overline{\xi} = (\xi_i)$ as follows [1,3]:

$$\mathbf{u}(x(\xi)) \approx \mathbf{N}^w(\xi)\,(\mathbf{u})_e^w,\ \mathbf{p}(x(\xi)) \approx \overline{\mathbf{N}}^w(\xi)\,(\mathbf{p})_e^w; \tag{26}$$

$$\mathbf{N}^w(\xi) = \mathbf{I}\,\mathrm{N}_w(\xi),\ \overline{\mathbf{N}}^w(\xi) = \mathbf{I}\,\overline{\mathrm{N}}_w(\xi) \tag{27}$$

with $\mathrm{N}_w(\xi)$, $\overline{\mathrm{N}}_w(\xi)$ being the interpolation functions. We can rewrite the integral formulation for the boundary value problems as

$$\begin{aligned}
&\mathbf{c}(x)\,\mathbf{u}(x) + \sum_{e=1}^{K}\sum_{w=1}^{W_e}(\mathbf{u})_e^w \int_{\Gamma^e} \mathbf{P}(x, y(\xi))\,\mathbf{N}^w(\xi)\,\mathrm{J}(\xi)\,d\Gamma(\xi) = \\
&= \sum_{e=1}^{K}\sum_{w=1}^{W_e}(\mathbf{p})_e^w \int_{\Gamma^e} \mathrm{U}(x, y(\xi))\,\mathbf{N}^w(\xi)\,\mathrm{J}(\xi)\,d\Gamma(\xi) + \mathbf{B}(x)
\end{aligned} \tag{28}$$

$$\mathbf{B}(x) = \int_{\Omega} \mathbf{U}(x,z)\,\mathbf{b}(z)\,d\Omega(z),\ \mathbf{J}(\xi) = \frac{d\Gamma}{d\xi}. \tag{29}$$

Eqn (28) can be rewritten for global numbering of boundary nodes $x^\beta, \beta = 1,2,\ldots,W$ as

$$\mathbf{r}(x) = \sum_{\beta=1}^{W}\left[\mathbf{P}^\beta(x)\,\mathbf{u}_\beta - \mathbf{U}^\beta(x)\,\mathbf{p}_\beta\right] - \mathbf{B}(x) = \mathbf{0}, \tag{30}$$

$$\mathbf{P}^\beta(x) = \mathbf{c}(x) + \sum_m \int_{\Gamma^m} \mathbf{P}(x, y(\xi))\,\mathbf{N}^\beta(\xi)\,\mathrm{J}(\xi)\,d\Gamma(\xi), \tag{31}$$

$$\mathbf{U}^\beta(x) = \sum_m \int_{\Gamma^m} \mathrm{U}(x, y(\xi))\,\overline{\mathbf{N}}^\beta(\xi)\,\mathrm{J}(\xi)\,d\Gamma(\xi). \tag{32}$$

The unknown values of displacements $\mathbf{u}(x)$ and forces $\mathbf{p}(x)$ can be found by $\mathbf{r}(x)$ minimization using the following condition:

$$\int_{\Gamma^e} \mathbf{T}^{\alpha}(x)\,\mathbf{r}(x)\,d\Gamma(x) = 0, \tag{33}$$

where $\mathbf{T}^{\alpha}(x) = \mathbf{I}\,T^{\alpha}(x),\ x \in \Gamma^e$ is the weighted function matrix. In case of collocation with nodal points equivalent to the boundary nodes

$$\mathbf{T}^{\alpha}(x) = \mathbf{I}\,\delta\left(x - x^{\alpha}\right),\ \mathbf{r}(x^{\alpha}) = 0, \tag{34}$$

with $x^{\alpha}$ being the collocation nodes. As a result, it is obtained that

$$\begin{gathered} c_{\alpha\beta}u_{\beta} + H_{\alpha\beta}u_{\beta} = G_{\alpha\beta}p_{\beta} + B_{\alpha} \Leftrightarrow H_{\alpha\beta}u_{\beta} = G_{\alpha\beta}p_{\beta} + B_{\alpha} \Leftrightarrow \\ A_{\alpha\beta}X_{\beta} = F_{\alpha}, \end{gathered} \tag{35}$$

and

$$\mathbf{A} \equiv \begin{bmatrix} \mathbf{H}^{11} & -\mathbf{G}^{11} \\ \mathbf{H}^{21} & -\mathbf{G}^{21} \end{bmatrix} \text{ and } \mathbf{F} \equiv \begin{bmatrix} -\mathbf{H}^{11} & \mathbf{G}^{12} \\ -\mathbf{H}^{21} & \mathbf{G}^{22} \end{bmatrix} \begin{Bmatrix} \mathbf{u}^{1} \\ \mathbf{u}^{2} \end{Bmatrix} + \begin{Bmatrix} \mathbf{B}^{1} \\ \mathbf{B}^{2} \end{Bmatrix} \tag{36}$$

and

$$H_{\alpha e} = H_e\left(x^{\alpha}\right) = c\left(x^{\alpha}\right) - \int_{\Gamma_e} P^{*}\left(x^{\alpha}, y\right) d\Gamma(y), \tag{37}$$

$$G_{\alpha e} = G_e\left(x^{\alpha}\right) = -\int_{\Gamma_e} U^{*}\left(x^{\alpha}, y\right) d\Gamma(y), \tag{38}$$

$$B_{\alpha} = B\left(x^{\alpha}\right) = \int_{\Gamma_e} U^{*}\left(x^{\alpha}, y\right) f(y)\, d\Gamma(y). \tag{39}$$

Further, cf. eqns (17,18), the boundary value problem is solved by application of the periodicity conditions on the RVE and the difference of heat conductivity in the form of heat flux along the interface. Taking into account that components are coherent and disjoint, the problem is solved for matrix and fiber regions independently. Then, denoting solution of the homogenization problem as homogenization function $\chi$, the matrix description consists of the fiber problem

$$\begin{bmatrix} \mathbf{H}^1 & \mathbf{H}_i^1 \end{bmatrix} \begin{bmatrix} \boldsymbol{\chi}^1 \\ \boldsymbol{\chi}_{(i)}^1 \end{bmatrix} = \begin{bmatrix} \mathbf{G}^1 & \mathbf{G}_i^1 \end{bmatrix} \begin{bmatrix} \mathbf{p}^1 \\ \mathbf{p}_{(i)}^1 + \mathrm{k}^{(1)} \end{bmatrix} + \begin{Bmatrix} \mathbf{B}^1 \\ \mathbf{B}^2 \end{Bmatrix} \tag{40}$$

and the following boundary value problem in matrix:

$$\begin{bmatrix} \mathbf{H}^2 & \mathbf{H}_i^2 \end{bmatrix} \begin{bmatrix} \boldsymbol{\chi}^2 \\ \boldsymbol{\chi}_{(i)}^2 \end{bmatrix} = \begin{bmatrix} \mathbf{G}^2 & \mathbf{G}_i^2 \end{bmatrix} \begin{bmatrix} \mathbf{p}^2 \\ \mathbf{p}_{(i)}^2 + \mathrm{k}^{(2)} \end{bmatrix} + \begin{Bmatrix} \mathbf{B}^1 \\ \mathbf{B}^2 \end{Bmatrix}. \tag{41}$$

After application of interface boundary conditions, it is obtained that

$$\begin{bmatrix} \mathbf{H}^1 & \mathbf{H}_{(i)}^1 & -\mathbf{G}_{(i)}^1 & 0 \\ 0 & \mathbf{H}_{(i)}^2 & \mathbf{G}_{(i)}^2 & \mathbf{H}^1 \end{bmatrix} \begin{Bmatrix} \boldsymbol{\chi}^1 \\ \boldsymbol{\chi}_{(i)}^1 \\ [\mathrm{k}] \\ \boldsymbol{\chi}^2 \end{Bmatrix} = \begin{bmatrix} \mathbf{G}^1 & 0 \\ 0 & \mathbf{G}^2 \end{bmatrix} \begin{Bmatrix} \mathbf{p}^1 \\ \mathbf{p}^2 \end{Bmatrix} + \begin{Bmatrix} \mathbf{B}^1 \\ \mathbf{B}^2 \end{Bmatrix}. \tag{42}$$

If the n-component fiber-like composite is considered, the matrices **H**, **G**, **B** and vectors $\boldsymbol{\chi}$, **p** are decomposed into n submatrices **H**(n), **G**(n), **B**(n) as well as $\boldsymbol{\chi}$(n), **p**(n), cf. eqn (42). Next, the heat conductivity difference is applied at any interface [**k**(n-1,n)] into the global vector $\boldsymbol{\chi}$. It should be underlined that the main advantage of that problem is the general form of fundamental solution used for the boundary value problem being solved [4,17] typical for analysis of homogeneous media.

Finally, let us observe that in the case when the scale parameter of the composite cannot be treated as tending to 0, then closed form algebraic equation is used for effective parameter in the following form:

$$\mathrm{k}^{(\mathrm{eff})} = \mathrm{k}_2 \left( 1 + \mathrm{v}_\mathrm{f} \left( \frac{1 - \mathrm{v}_\mathrm{f}}{2} + \frac{\mathrm{k}_2}{\mathrm{k}_1 - \mathrm{k}_2} \right)^{-1} \right) \tag{43}$$

where $\mathrm{v}_\mathrm{f}$ is a fiber volume fraction while $\mathrm{k}_1 > \mathrm{k}_2$ are the heat conductivity coefficients for composite constituents. The RVE of composite is quadratic with round cross-section, while the methodology proposed before is valid for homogenization of composites with general configuration of the periodicity cell.

### 3.3. HOMOGENIZATION OF RANDOM COMPOSITES

Generally, we can solve the homogenization problem for the periodic two-component composite with randomly varying coefficients of heat conduction [14] by the use of Monte-Carlo simulation technique [9], stochastic spectral methods [6] or, alternatively, stochastic second order perturbation technique [13]. The last of these methods is applied here to derive the first two probabilistic moments of the effective heat conductivity coefficients and, for this purpose, the random variables of these coefficients are introduced using the expected values and variances definitions as

$$E\left[k(\mathbf{x};\omega)\right] = \chi_a(\mathbf{x})E\left[k^{(a)}(\omega)\right], \tag{44}$$

$$Var\left(k(\mathbf{x};\omega)\right) = \chi_a(\mathbf{x})Var\left(k^{(a)}(\omega)\right), \mathbf{x} \in \Omega_a. \tag{45}$$

As it is known [13], the stochastic second order second probabilistic method, contrary to the other methods [2], consists in the probabilistic extension of any state function $F(\mathbf{x};\omega)$

$$F(\mathbf{x};\omega) = F^0(\mathbf{x};\omega) + \theta F^{,r}(\mathbf{x};\omega)\Delta k^r + \tfrac{1}{2}\theta^2 F^{,rs}(\mathbf{x};\omega)\Delta k^r \Delta k^s. \tag{46}$$

The first and second order variations of $\Delta k^r$ about its expected value are equal to

$$\theta\Delta k^r = \delta k_r = \theta\left(k_r - k_r^0\right),$$
$$\theta^2 \Delta k^r \Delta k^s = \delta k_r \delta k_s = \theta^2\left(k_r - k_r^0\right)\left(k_s - k_s^0\right), \tag{47}$$

where symbols $(.)^0$, $(.)^{,r}$ and $(.)^{,rs}$ denote the partial derivatives orders with respect to input random variables. The probabilistic moments of effective material tensor are found using the solution of zeroth, first and second order equations [12] corresponding to eqn (35):

$$A^0_{\alpha\beta}\chi^0_\beta = F^0_\alpha, \tag{48}$$

$$A^0_{\alpha\beta}\chi^{,r}_\beta = F^{,r}_\alpha - A^{,r}_{\alpha\beta}\chi^0_\beta, \tag{49}$$

$$A^0_{\alpha\beta}\chi^{(2)}_\beta = [F^{,rs}_\alpha - 2A^{,r}_{\alpha\beta}\chi^{,s}_\beta - A^{,rs}_{\alpha\beta}\chi^0_\beta]S^{rs}. \tag{50}$$

The expected values and the covariance matrix for the discretized temperature homogenization function can be calculated as [11,13]

$$E[\chi] = \chi^0 + \tfrac{1}{2}\chi^{,rs}\mathrm{Cov}\left(k^r, k^s\right), \tag{51}$$

$$\mathrm{Cov}\left(\chi_{(\alpha)}, \chi_{(\beta)}\right) = \chi_{(\alpha)}^{;r}\chi_{(\beta)}^{;s}\,\mathrm{Cov}\left(k^r, k^s\right),\ \alpha,\beta=1,\dots,N. \tag{52}$$

As a result, we introduce the effective conductivity probabilistic characterization, what can be done using the expected values in the following form

$$E\left[k^{(eff)}\right] = \frac{1}{|\Omega|}\int_{\Omega} E[k(\mathbf{y})]d\Omega + \frac{1}{|\Omega|}\int_{\Omega} E\left[k(\mathbf{y})\chi_{,j}1_j\right]d\Omega. \tag{53}$$

Using the second order second probabilistic central moment approach it can be written that

$$\begin{aligned} E\left[k(\mathbf{y})\chi_{,j}\right] &= \int_{-\infty}^{+\infty} k^0(\mathbf{y})\chi_{,j}^0(\mathbf{y})p_R\left(k(\mathbf{y})\right)d\mathbf{k} + \\ &+ \int_{-\infty}^{+\infty} \Delta k^r k^{,r}(\mathbf{y})\Delta k^u \chi_{,j}^{,u}(\mathbf{y})p_R\left(k(\mathbf{y})\right)d\mathbf{k} + \\ &+ \tfrac{1}{2}\int_{-\infty}^{+\infty} k^0(\mathbf{y})\Delta k^u \Delta k^v \chi_{,j}^{,uv}(\mathbf{y})p_R\left(k(\mathbf{y})\right)d\mathbf{k} = \\ &= k^0(\mathbf{y})\chi_{,j}^0(\mathbf{y}) + \\ &+ \left\{k^{,r}(\mathbf{y})\chi_{,j}^{,u}(\mathbf{y}) + \tfrac{1}{2}k^0(\mathbf{y})\chi_{,j}^{,uv}(\mathbf{y})\right\}Cov\left(k^r, k^s\right) \end{aligned} \tag{54}$$

Further, using the definition of variance

$$\mathrm{Var}\left(k^{(eff)}\right) = \int_{-\infty}^{+\infty}\left(k^{(eff)} - E\left[k^{(eff)}\right]\right)^2 p(\mathbf{k})d\mathbf{k}, \tag{55}$$

it can be calculated that

$$\mathrm{Var}\left(k^{(eff)}\right) = \frac{1}{|\Omega|^2}\int_{\Omega}\mathrm{Var}(k(\mathbf{y}))d\Omega + \frac{1}{|\Omega|^2}\int_{\Omega}\mathrm{Var}\left(k(\mathbf{y})\chi_{,j}1_j\right)d\Omega, \tag{56}$$

where

$$\mathrm{Var}\left(k(\mathbf{y})\chi_{,j}\right) = \left(k^{,r}(\mathbf{y})\chi_{,j}^{;s}(\mathbf{y})\right)\mathrm{Cov}\left(k^r, k^s\right), \tag{57}$$

what gives the following result:

$$
\begin{aligned}
Var\left(k^{(eff)}\right) = & \frac{1}{(\Omega(^2} \int_\Omega Var(k(\mathbf{y}))\, d\Omega + \\
& + \frac{1}{(\Omega(^2} \int_\Omega \left(k^{,r} \chi_{,j}^{,s} 1_j\right) Cov\left(k^r, k^s\right) d\Omega
\end{aligned}
\qquad (58)
$$

Finally, it is observed that since eqns (53,58), the effective conductivity tensor expected values and variances can be computed for any configuration of the RVE, starting from statistical description of components heat conductivities obtained on the basis of laboratory experiments.

## 3.4. COMPUTATIONAL ILLUSTRATION

The essential purpose of the computational experiments performed is to make a visualization of the homogenization analysis proposed above for the fiber-reinforced composite with round fiber embedded into the rectangular RVE. The effective heat conductivity deterministic values, its sensitivity to fiber volume fraction as well as expetced values and coefficients of variation have been computed numerically using mathematical package for symbolic computations MAPLE, see Figs. 3-6 below: deterministic values of the effective heat conductivity (Fig. 3), its sensitivity to fiber volume fraction (Fig. 4), expected values of $k^{(eff)}$ (Fig. 5) as well as coefficient of variation of this variable (Fig. 6). The following input composite parameters have been adopted for numerical analysis: $E[k_2]=1.0$, $E[k_1]=[1.0,10.0]$, $\sigma(k_2)=\alpha^*E[k_2]$, $\sigma(k_1)=\alpha^*E[k_1]$, $\alpha=[0.0,0.5]$, $v_f=[0.0,0.5]$.

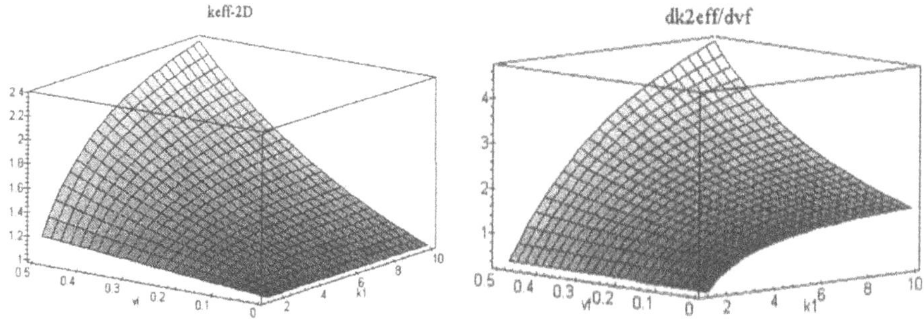


Fig. 3. Effective heat conductivity for fiber composite

Fig. 4. Sensitivity of $k^{(eff)}$ for fiber composite to $v_f$

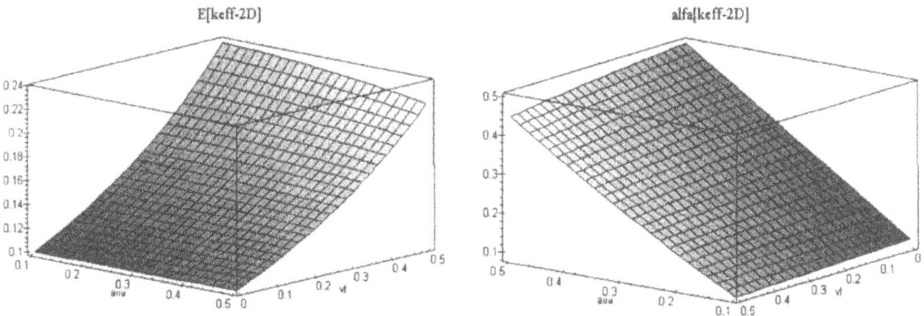


Fig. 5. Expected values of $k^{(eff)}$

Fig. 6. Coefficients of variation of $k^{(eff)}$

Analyzing the results shown in Figs. 3-6, the great sensitivity of $k^{(eff)}$ to the fiber heat conductivity coefficient $k_1$ with $v_f \rightarrow 0.5$ is observed with nonlinear character, while negligibly small is obtained for $v_f \rightarrow 0$. The character of sensitivity to the parameter $v_f$ is linear and, due to the values obtained, quite analogous - the visualization of the sensitivity gradients surface of $k^{(eff)}$ to this parameter is shown in Fig. 6 as a function of $v_f$ and $k_1$. Expected values of the effective heat conductivity shown in Fig. 5 is sensitive to the expected value of fiber heat conductivity, while shows small variability for various input coefficients of variation. The variability of the coefficient of variation of this variable have inverse character - great (almost linear) sensitivity with respect to coefficients of variation of input heat conductivities while extremely small to the fiber volume fraction. All these results can be used in computational both deterministic and stochastic modeling of the fiber-reinforced composites where instead of original composite, the homogenized medium is considered characterized by the effective heat conductivity (or its probabilistic moments) computed and shown above.

## 4. CONCLUDING REMARKS

1. The main value of the proposed BEM-based homogenization method is in the fact that, in the contrast to all BEM-based approaches to heterogeneous media, there is no need to modify the fundamental solution for numerical analysis of the composite [5]. Thanks to this method introduced it is possible to perform the numerical analysis the effective material tensors for the linear potential field problems with deterministic and random coefficients for fiber composites and to use this analysis results in BEM-based modeling of various composites. Taking into account the idea of the proposed approach in Figs. 3 and 4, it is easy

to extend the homogenization technique for n-component periodic composite in general configuration [10].

2. The second order perturbation based Stochastic Boundary Element Method (SBEM) proposed above can find its application in computational modeling of homogeneous as well as heterogeneous media in the context of reliability studies using FORM, SORM and especially W-SOTM (Weibull-Second Order Third Moment) approach [16]. Numerical Monte-Carlo simulation results obtained for homogenization analysis show that the effective heat conductivity coefficients have symmetric probability density functions (PDF), however usage of these data in further BEM or FEM computations leads to unsymmetric PDF where the W-SOTM is recognized as the most efficient method.
3. The proposed method can find its application in modeling of composites with random interfaces where usage of the FEM techniques is very complicated. Since the generality of the homogenization procedure, the methodology may be implemented in any BEM commercial or academic computer programs by the analogy to the research presented in [12]. Further, linking this approach with any of the sensitivity analysis techniques - Adjoint Variable Method (AVM) or Direct Differentiation Method (DDM) [7,13] it will be able to determine the crucial material parameters and their probabilistic moments influencing the effective material properties tensor how it is shown in computational experiment presented above.
4. Considering the fact that the homogenization problem is solved in each homogeneous subdomain independently, the computational parallelization strategy may be used in numerical analysis to decrease efficiently the total time of numerical experiments. Solution of any homogeneous boundary value problem is obtained separately together with corresponding spatial averages while effective parameters computations is performed at the same time. Analogous conclusion hold true in the case of stochastic second order perturbation approach where the solution character is the same while possible time savings significantly greater.

## ACKNOWLEDGMENTS

The final version of the paper has been completed during author's research stay at the Houston Advanced Research Center in Woodlands, Spring, Texas. The author would like to acknowledge the financial support of the Foundation for Polish Science for the postdoctoral studies at Rice University in Houston, Texas.

## 5. REFERENCES


[1] Brebbia, C.A., Dominguez, J.: *Boundary Elements. An Introductory Course*. Comp. Mech. Publ., 1996.

[2] Breitung, K., Casciati, F. and Faravelli, L.: A stochastic boundary element model for soil properties. *Proc. of IUTAM/IABEM BEM Symp.*, pp. 10-11, Cracow 1999.

[3] Burczynski, T.: *Boundary Element Method in Structural Mechanics* (in Polish). WNT, Warsaw, 1995.

[4] Duddeck F.M.E.: A boundary element method for general media via Parseval's theorem (in preparation).

[5] Furmanski, P.: Heat conduction in composites: homogenization and macroscopic behaviour, *Appl. Mech. Rev.* **50**(6) (1997), 327-355.

[6] Ghanem, R.G. and Spanos, P.D.: *Stochastic Finite Elements: A Spectral Approach*. Springer-Verlag, 1991.

[7] Hassani, B. and Hinton, E.: A review of homogenization and topology optimization: I– homogenization theory for media with periodic structure. *Comput. & Struct.* **69** (1998), 707-717; II – analytical and numerical solution of homogenization equations. *Comput. & Struct.* **69** (1998), 719-738; III – topology optimization using optimality criteria. *Comput. & Struct.* **69** (1998), 739-756.

[8] Haug, E.J., Choi, K.K. and Komkov, V.: *Design Sensitivity Analysis of Structural Systems*. Series Math. Sci. Engrg., Academic Press, 1986.

[9] Hurtado, J.E. and Barbat, A.H.: Monte-Carlo techniques in computational stochastic mechanics. *Arch. Comput. Meth. Engrg.* **5**(1) (1998), 3-30.

[10] Kaminski, M.: Boundary element method homogenization of linear elastic composites. *Engrg. Anal. Boundary Elem.* **23**(10): 815-823, 1999.

[11] Kaminski, M.: Homogenized properties of n-components composite materials. *Int. J. Engrg. Sci.* **38**(4): 405-427, 2000.

[12] Kaminski, M.: Stochastic second-order BEM perturbation formulation. *Engrg. Anal. Boundary Elem.* **23** (1999), 123-129.

[13] Kleiber, M. and Hien, T.D.: *The Stochastic Finite Element Method*. Wiley, 1992.

[14] Manolis, G.D. and Shaw, R.P.: Boundary integral formulation for 2D and 3D thermal problems exhibiting a linearly varying stochastic conductivity. *Comput. Mech.* **17**(1996), 406-417.

[15] Panagiotopoulos, P.D., Panagouli, O.K. and Koltsakis, E.K.: The BEM in place bodies with cracks and/or boundaries of fractal geometry. *Comput. Mech.* **15** (1995), 350-363.

[16] Peng, X.Q. et al.: A stochastic finite element method for fatigue reliability analysis of gear teeth subjected to bending. *Comput. Mech.* **21** (1998), 253-261

[17] Shaw, R.P. et al.: The 2D free space Green's function and BIE for a Poisson equation with linearly varying conductivity in two directions. In: Ertekin R.C. et al., eds., *Proc. of BETECH 11*, pp. 327-332, Comput. Mech. Publ., 1996.

# SOME MATHEMATICAL ASPECTS OF THE BEM IN ELASTIC SCATTERING OF ACOUSTIC WAVES

ANDRZEJ KARAFIAT
*The Cracow University of Technology,*
*Institute of Computer Modelling,*
*Section of Applied Mathematics,*
*ul. Warszawska 24, 31-155 Kraków, Poland*
akaraf@usk.pk.edu.pl

**Keywords:** Boundary Element Method, acoustic scattering, adaptive numerical integration, a-priori error estimate

**Abstract** The paper contains an a-priori error estimate for the hp- adaptive Boundary Element Method applied to solve the problem of acoustic wave scattering on an elastic body in the two-dimensional space. Furthermore a method of an hp- adaptive integration of functions arising in the boundary integral equations is considered.

## Introduction

The hp- adaptive Boundary Element Method (BEM) has been for a decade a subject of research. In many papers adaptation algorithms were presented, convergence was proved, especially the exponential convergence near singularities. Many numerical tests were performed. It is enough to quote [3],[7]-[9],[14]-[16].

However, the authors of these papers generally had not taken into consideration the numerical integration problem. The error caused by numerical quadratures is well investigated by the Finite Element Method (FEM), but integrands in FEM are bounded, in contrary to integrands in the BEM, especially in external problems. The classical error estimates based on Sobolev seminorms of integrands are not applicable here.

The purpose of this paper is to present a novel approach in which an hp- error estimate contains both errors, the one caused by the Galerkin method and the one generated by numerical integration. The extra error

*T. Burczynski (ed.), IUTAM/IACM/IABEM Symposium on Advanced Mathematical and Computational Mechanics Aspects of the Boundary Element Method,* 161–171.

included in this, arises from the approximation of a boundary. The latter result had been proven in [13]. The paper models the problem of acoustic waves' elastic scattering in the two-dimensional space and solves it by the Galerkin BEM, similarly to the analysis done by [6].

The paper is organized as follows: Section 2 - describes the elastic scattering problem. Section 3 - specifies general assumptions and computes interpolation estimates. Error estimates for internal integration are analyzed in Section 4 and finally, the outer integration error is considered in Section 5. This last section sums-up the two errors and presents their influence on the global error.

## 1. THE PROBLEM OF ACOUSTIC WAVE SCATTERING

The modeled problem is as follows. Let $\Omega \subset \mathbb{R}^2$ be a bounded domain, with a boundary $\Gamma$. Assuming, that $p^{inc}$ is a known pressure function of an incident wave and $p^s$ is an unknown pressure function of a scattered wave, we will try to find the following total pressure function

$$p(\boldsymbol{x}) = p^{inc}(\boldsymbol{x}) + p^s(\boldsymbol{x}), \ \boldsymbol{x} \in \Omega^e = \mathbb{R}^2 \backslash \Omega. \tag{1.1}$$

The above functions fulfil the three equations:

$$-\Delta p(\boldsymbol{x}) - k^2 p(\boldsymbol{x}) = 0, \ \boldsymbol{x} \in \Omega^e, \tag{1.2}$$

(Helmholtz differential equation),

$$\left| \frac{\partial p^s}{\partial r} - ikp^s \right| = o(r^{-\frac{1}{2}}), \text{ for } r = |\boldsymbol{x}| \to \infty, \tag{1.3}$$

(Sommerfeld radiation condition)

$$\frac{\partial p}{\partial n^x}(\boldsymbol{x}) = \varepsilon p(\boldsymbol{x}), \ \boldsymbol{x} \in \Gamma, \tag{1.4}$$

(a reduced boundary condition along $\Gamma$),
where $i$ is the imaginary unit, $k$ - the wave number and $\boldsymbol{\tau}^x, \boldsymbol{n}^x$ - tangent- and outward normal unit vectors at $\boldsymbol{x}$ (Fig.1).

The weak boundary formulation, for the boundary value problem (1.2) - (1.4) can be described as:

Find $p \in V = H^{\frac{1}{2}}(\Gamma)$, such that

$$a(p, q) = l(q), \ \forall q \in V, \tag{1.5}$$

which is equivalent to the above problem. This approach is known as a variational Burton - Miller formulation. The sesquilinear form $a$ and the semilinear form $l$ are given by the formulas:

$$a(p,q) = 0.5\alpha \int_\Gamma p(\boldsymbol{x})q(\boldsymbol{x})ds(\boldsymbol{x})$$

$$+0.5(1-\alpha)k^{-1}\varepsilon i \int_\Gamma p(\boldsymbol{x})q(\boldsymbol{x})ds(\boldsymbol{x})$$

$$+\alpha \int_\Gamma \int_\Gamma \left[\varepsilon\Phi(\boldsymbol{x},\boldsymbol{y})p(\boldsymbol{y})q(\boldsymbol{x}) - \frac{\partial\Phi}{\partial n^y}(\boldsymbol{x},\boldsymbol{y})p(\boldsymbol{y})q(\boldsymbol{x})\right] ds(\boldsymbol{y})ds(\boldsymbol{x})$$

$$+(1-\alpha)k^{-1}i \int_\Gamma \int_\Gamma \left[\Phi(\boldsymbol{x},\boldsymbol{y})\frac{\partial p}{\partial\tau^y}(\boldsymbol{y})\frac{\partial q}{\partial\tau^x}(\boldsymbol{x})\right.$$

$$\left. -k^2\Phi(\boldsymbol{x},\boldsymbol{y})p(\boldsymbol{y})q(\boldsymbol{x})\tau^x\tau^y + \varepsilon\frac{\partial\Phi}{\partial n^x}(\boldsymbol{x},\boldsymbol{y})p(\boldsymbol{y})q(\boldsymbol{x})\right] ds(\boldsymbol{y})ds(\boldsymbol{x}), \quad (1.6)$$

$$l(q) = \alpha \int_\Gamma p^{inc}(\boldsymbol{x})q(\boldsymbol{x})ds(\boldsymbol{x}) + (1-\alpha)k^{-1}i \int_\Gamma \frac{\partial p^{inc}}{\partial n^x}(\boldsymbol{x})q(\boldsymbol{x})ds(\boldsymbol{x}). \quad (1.7)$$

where

$$\Phi(\boldsymbol{x},\boldsymbol{y}) = \frac{i}{4}H_0^1(kr) \quad (1.8)$$

is the fundamental solution of the Helmholtz equation (1.2) and $H_0^1(x)$ is the Hankel function of the first kind. For details concerning its derivation see [6] or [12].

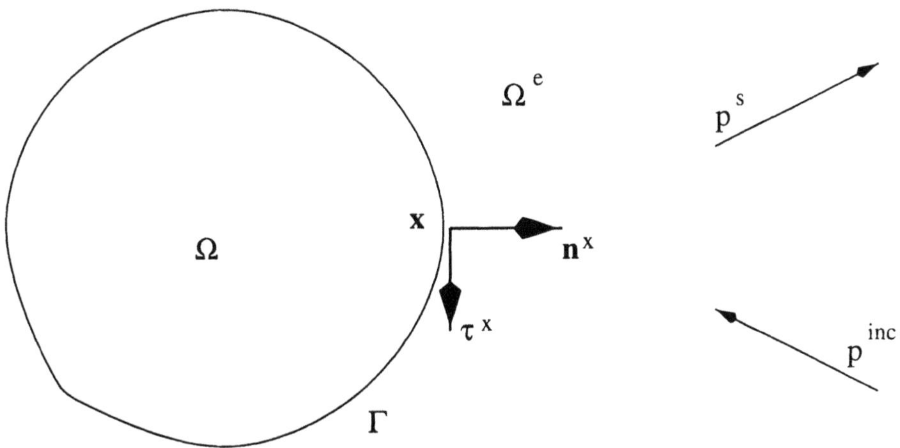


*Figure 1* The elastic scattering problem

## 2. INTERPOLATION AND APPROXIMATION ESTIMATES

The following notations will be used throughout the paper:
$G = [a,b] \subset \mathbb{R}$ - reference domain;

$\Gamma = \boldsymbol{X}(G)$ - parametrization of the boundary;
$\hat{T} = [0,1]$ - pattern element;
$G = \bigcup_{i=1}^{n} T_i$, $T_i = [t_{i-1} - t_i]$ - actual elements;
$h_i = diam(T_i)$, $h = \max_i h_i$;
$T_i$ - affine family of 1-D finite elements ([5]);
$P_d[A,B] = \{w(x) = \sum_{j=0}^{d} a_j t^j,\ t \in [A,B]\}$ - space of polynomials.
The space of interpolating functions is defined as:

$$I_d(G) = \{w_h : G \to \mathbb{R} : w_h \in C^0(G) : w_h|_{T_i} \in P_d(T_i)\ \forall i \in \{1,\dots,I\}\}. \tag{2.1}$$

We assume, that Sobolev spaces $H^m(G)$ with their norms $\|\cdot\|_{m,G}$ are defined as usual.

On each element $T_i$ a local shape function space $K_{i,d} \subset P_d(T_i)$ is defined. A global shape function space then becomes

$$V_h = \{v_h \in C^0(G),:\ v_h|_{T_i} \in K_{i,d}\ \forall i \in \{1,\dots,I\}\}. \tag{2.2}$$

We assume that $d$ is such that

$$\forall i \in \{1,\dots,I\}\ P_d(T_i) \subset K_{i,d} \ \Rightarrow I_d(G) \subset V_h. \tag{2.3}$$

In practice we often have to interpolate not only the solution $p$ and test functions $q$, but also the parametrization $\boldsymbol{X}$ of the boundary

$$\boldsymbol{X}_h = (X_h^1, X_h^2),\ X_h^j \in V_h,\ j = 1,2. \tag{2.4}$$

The approximate boundary we denote as $\Gamma_h = \boldsymbol{X}_h(G)$. To describe this approximation we use a Cartesian product of the Sobolev spaces

$$V^m(G) = H^m(G) \times H^m(G),\ \|\boldsymbol{X}\|_{m,G} = \left(\|X^1\|_{m,G}^2 + \|X^2\|_{m,G}^2\right)^{\frac{1}{2}}. \tag{2.5}$$

The known hp- interpolation theorem ([1],[2]) states that

**Theorem 2.1** *There exists a constant $C = C(m,d,G)$ such that for any function $v \in H^m(G)$, $m \geq 1$ and an integer $d \geq 1$ there exists a function $v_h \in V_h$ on each $T_i$ such that*

$$\sup\left\{\left|\frac{d^\alpha}{dt^\alpha}(v - v_h)(t)\right|, t \in G\right\} \leq C\frac{h^{\mu-\alpha-1}}{d^{m-\alpha-1}}\|v\|_{m,G}, \tag{2.6}$$

*for $\alpha = 0,1$, where*

$$\mu = min\{d+1, m\}. \tag{2.7}$$

In the Galerkin method we replace functions $p$ and $q$ of (1.5) by the functions $p_h, q_h \in V_h$, which are the solutions of the following approximate problem:

Find $p_h \in V_h$ such that

$$a(p_h, q_h) = l(q_h), \ \forall q_h \in V_h. \tag{2.8}$$

If the boundary $\Gamma$ is also approximated, integrals in formulas (1.6),(1.7) are evaluated over $\Gamma_h$. This transforms $a, l$ into $a_h, l_h$ and modifies the Galerkin's method into:

Find $p_h \in V_h$ such that

$$a_h(p_h, q_h) = l_h(q_h), \ \forall q_h \in V_h. \tag{2.9}$$

In the latter case also the given pressure functions $p^{inc}$ and $p^{inc}/\partial n^x$ are interpolated, by $p_h^{inc}$ and $p_h^{inc}/\partial n_h^x$ respectively, on $\Gamma_h$.

The following a-priori estimate was proved in [13]

**Theorem 2.2** *Let us assume that the parametrization $\boldsymbol{X}$ of the boundary $\Gamma$ belongs to $V^{m+1}(G)$, and that $p \in H^m(\Gamma)$ is the solution of the problem (1.5). Applying the approximation method mentioned before, gives $p_h \in V_h$ as the solution of (2.9). The obtained approximated error satisfies the inequality*

$$\|p - p_h\|_{\frac{1}{2},\Gamma} \leq C \left[ \|p^{inc} - p_h^{inc}\|_{-\frac{1}{2},\Gamma} + \left\|\frac{\partial p^{inc}}{\partial n^x} - \frac{\partial p_h^{inc}}{\partial n_h^x}\right\|_{-\frac{1}{2},\Gamma} \right.$$
$$\left. + \frac{h^{\mu-1}}{d^{m-1}} \left( \|\boldsymbol{X}\|_{m+1,G} \|p\|_{\frac{1}{2},\Gamma} + \|p\|_{m,\Gamma} \right) \right], \tag{2.10}$$

*where $C > 0$ is an appropriate constant independent of $h$ and $d$.*

There is, however, another reason, that the approximate forms $a_h, l_h$ differ from the original ones. It is the approximate integration. A method of this integration and its effect on the estimate (2.10) will be discussed in the next two sections.

## 3. INTERNAL INTEGRATION

In the integration process integrals usually have the form

$$\int_{T_1} \left[ \int_{T_2} \Phi(\boldsymbol{x}, \boldsymbol{y}) \varphi(\boldsymbol{y}) ds_y \right] \psi(\boldsymbol{x}) ds_x. \tag{3.1}$$

They can be singular, almost singular, or regular. The singularity of the integral (3.1) is a consequence of a coincidence $\boldsymbol{x} = \boldsymbol{y}$ in $X_h(T_1 \cap T_2)$. We call an integral form (3.1) singular, when $\boldsymbol{x} \in X_h(T_2)$; almost singular, when $\boldsymbol{x} \notin X_h(T_2)$ and $dist(\boldsymbol{x}, X_h(T_2)) < \varepsilon$; and regular in other cases.

The evaluation of a regular internal integral

$$I_i(\Phi\varphi_h, T_2) = \int_{T_2} \Phi(\boldsymbol{x}, \boldsymbol{y})\varphi(\boldsymbol{y})ds_y = \int_{t_1}^{t_2} \Phi(\boldsymbol{x}, \boldsymbol{y}(t))\varphi(\boldsymbol{y}(t)) \left|\frac{d\boldsymbol{y}}{dt}\right| dt \tag{3.2}$$

is performed using a Gaussian quadrature

$$Q_{N_i}(f, T) = \sum_{j=1}^{N_i} w_j^i f(a_j^i). \tag{3.3}$$

This quadrature is exact for each polynomial $w_h$ of order $2N_i - 1$

$$Q_{N_i}(w_h, T) = I_i(w_h, T). \tag{3.4}$$

We can subtract an interpolant $w_h$ of the integrand of (3.2). If $\boldsymbol{y} \in H^{m+1}(G)$, then

$$\begin{aligned} &|I_i(\Phi\varphi_h, T_2) - Q_{N_i}(\Phi\varphi_h, T_2)| \\ &= |I_i(\Phi\varphi_h - w_h, T_2) - Q_{N_i}(\Phi\varphi_h - w_h, T_2)| \\ &= \left| \int_{t_1}^{t_2} [\Phi(\boldsymbol{x}, \boldsymbol{y}(t))\varphi(\boldsymbol{y}(t)) - w_h(t)] \left|\frac{d\boldsymbol{y}}{dt}\right| dt \right. \\ &\quad \left. - Q_{N_i}(\Phi(\boldsymbol{x}, \boldsymbol{y}(t))\varphi(\boldsymbol{y}(t)) - w_h(t), T) \right| \\ &\le C \frac{h^{2N_i}}{(2N_i - 1)^{2N_i - 1}} \left\| \Phi\varphi_h \left|\frac{d\boldsymbol{y}}{dt}\right| \right\|_{2N_i, T_2}. \end{aligned} \tag{3.5}$$

Function $\Phi$ is regular and $\varphi_h$ is a polynomial. If $X_h$ is a polynomial on every $T_j$, then the norm (3.2) exists for any $N_i$.

The idea presented below is based on a division of the interval $G$ into primary finite elements $T_1^P, \dots T_N^P$, which are later subdivided into actual elements. Let $\boldsymbol{x} = \boldsymbol{X}_h(t_0) \in X_h(\operatorname{int} T_2)$ and $T_2 \subset T_j^P$. Now, we divide then $T_j^P = [t_{j-1}^P, t_j^P]$ into two subintervals $[t_{j-1}^P, t_0]$ and $[t_0, t_j^P]$ and each of them we consider as a range of the pattern element $\hat{T} = [0, 1]$ from the affine mapping $t = t(u)$, in such a way that $t_0 = t(0)$. After substituting

$$u = \tau^n, \ \tau \in [0, 1], \ n > 1, \tag{3.6}$$

the integral

$$\int_{t_0}^{t_j^P} \Phi(\boldsymbol{X}_h(t_0), \boldsymbol{X}_h(t))\varphi(\boldsymbol{X}_h(t)) \left|\frac{d\boldsymbol{X}_h}{dt}\right| dt \tag{3.7}$$

changes into:

$$\int_0^1 \Phi(\boldsymbol{X}_h(t_0), \boldsymbol{X}_h(t(\tau^n)))\varphi(\boldsymbol{X}_h(t(\tau^n))) \left|\frac{d\boldsymbol{X}_h}{dt}\right| \frac{dt}{du} n\tau^{n-1} d\tau, \tag{3.8}$$

i.e., into a regular integral. Definition (1.8) of function $\Phi$ implies that for any $\beta = 1, 2, \ldots$

$$\frac{d^\beta}{dr^\beta}\Phi(kr) \leq Cr^{-\beta}, \tag{3.9}$$

then for $r = \tau^n$, $n > \beta+1$ derivatives of the new integrand are bounded:

$$\frac{d^\beta}{d\tau^\beta}\left[\Phi(kr) \cdot n\tau^{n-1}\right] \leq C. \tag{3.10}$$

Inequality (3.10) implies that

$$\forall \varphi_h \in C^m[\tau_1, \tau_2],\ \boldsymbol{y} \in V_h,$$

$$\left\| \Phi\varphi \left|\frac{d\boldsymbol{y}_h}{du}\right| \tau^{n-1} \right\|_{m,T} \leq \left\| \Phi\varphi \left|\frac{d\boldsymbol{y}_h}{d\tau}\right| \right\|_{m,T} \leq C\|\boldsymbol{y}\|_{m+1,T} \cdot \|\varphi_h\|_{m,T}, \tag{3.11}$$

where $C$ depends on $m$ and $n$. Similarly, like in (3.5), we obtain

$$\begin{aligned} &|I_i(\Phi\varphi_h, T_j^P) - Q_{N_i}(\Phi\varphi_h, T_j^P)| \\ &\leq C\frac{h^{2N_i}}{(2N_i-1)^{2N_i-1}}\|\Phi\varphi\left|\frac{d\boldsymbol{y}_h}{du}\right| \tau^{n-1}\|_{2N_i,T_j^P} \\ &= C\frac{h^{2N_i}}{(2N_i-1)^{2N_i-1}}\|\Phi\varphi\left|\frac{d\boldsymbol{y}_h}{d\tau}\right|\|_{2N_i,T_j^P}. \end{aligned} \tag{3.12}$$

The case of almost singular integral, for which $\boldsymbol{x} \notin X_h(T_2)$, proceeds analogously.

## 4. EXTERNAL INTEGRATION

The following, different than before, Gaussian quadrature

$$Q_{N_o}(f, T) = \sum_{j=1}^{N_o} w_j^o f(a_j^o) \tag{4.1}$$

is applied to the outer integral

$$I_o(f, T) = \int_T f(\boldsymbol{x}(t)) \left|\frac{d\boldsymbol{x}}{dt}\right| dt. \tag{4.2}$$

We will use the Banerjee and Suri's theorem, ([4])

**Theorem 4.1** *Let the Gaussian quadrature with $N_o$ integration points be exact for polynomials of order $2N_o - 1$, all weights be positive and all nodes be internal*

$$w_j^o > 0,\ a_j^o \in int\hat{T},\ j = 1, \ldots, N_0. \tag{4.3}$$

*Then for each $m > 1$ there is a constant $C > 0$ such that for any $f \in H^m(\hat{T})$, $w_h \in P_{N_0}(\hat{T})$ the inequality*

$$\left| I_o(fw_h, \hat{T}) - Q_{N_o}(fw_h, \hat{T}) \right| \leq C(N_o - 1)^{\frac{1}{2}-m} \|f\|_{m,\hat{T}} \cdot \|w_h\|_{0,\hat{T}} \tag{4.4}$$

*holds.*

After evaluating the integral (3.1), we obtain

$$\begin{aligned} &I_o(I_i(\Phi(\boldsymbol{x}, \boldsymbol{y}), \varphi_h(\boldsymbol{y})), T_2) \cdot \psi_h(\boldsymbol{x}), T_1) \\ &\approx Q_{N_o}(Q_{N_i}(\Phi(\boldsymbol{x}, \boldsymbol{y}), \varphi_h(\boldsymbol{y})), T_2) \cdot \psi_h(\boldsymbol{x}), T_1). \end{aligned} \tag{4.5}$$

The error of this representation will be

$$\begin{aligned} &|I_o I_i(\Phi(\boldsymbol{x}, \boldsymbol{y})\varphi(\boldsymbol{y}), T_2) \cdot \psi(\boldsymbol{x}), T_1) \\ &-Q_{N_o} Q_{N_i}(\Phi(\boldsymbol{x}, \boldsymbol{y})\varphi(\boldsymbol{y}), T_2) \cdot \psi(\boldsymbol{x}), T_1)| \\ &\leq |(I_o - Q_{N_o})(I_i(\Phi(\boldsymbol{x}, \boldsymbol{y})\varphi(\boldsymbol{y}), T_2) \cdot \psi(\boldsymbol{x}), T_1)| \\ &+|Q_{N_o}(I_i - Q_{N_i})(\Phi(\boldsymbol{x}, \boldsymbol{y})\varphi(\boldsymbol{y}), T_2) \cdot \psi(\boldsymbol{x}), T_1)| \end{aligned} \tag{4.6}$$

Calculating the first term on the pattern interval using Theorem 4.1, we get the following

$$\begin{aligned} &|(I_o - Q_{N_o})(I_i(\Phi(\boldsymbol{x}, \boldsymbol{y})\varphi(\boldsymbol{y}), \hat{T})| \cdot \psi(\boldsymbol{x}), \hat{T})| \\ &\leq C(N_o - 1)^{\frac{1}{2}-m} \|I_i(\Phi \cdot \varphi_h, \hat{T}\|_{m,\hat{T}} \cdot \|\psi_h\|_{0,\hat{T}}. \end{aligned} \tag{4.7}$$

This estimate of the actual element $T_i$ includes in the standard way its diameter $h$

$$\begin{aligned} &|(I_o - Q_{N_o})(I_i(\Phi(\boldsymbol{x}, \boldsymbol{y})\varphi(\boldsymbol{y}), T_2) \cdot \psi(\boldsymbol{x}), T_1)| \\ &\leq Ch^{N_o}(N_o - 1)^{\frac{1}{2}-m} \|I_i(\Phi \cdot \varphi_h, T_2\|_{m,T_1} \|\psi \left| \frac{dx}{dt} \right| \|_{m,T_1}. \end{aligned} \tag{4.8}$$

Because the functions $\Phi$ and $\varphi$ are regular, the $m$ may be any positive integer. Using (3.5) or (3.12) we can estimate the second term, and obtain

$$\begin{aligned} &|Q_{N_o}(I_i - Q_{N_i})(\Phi(\boldsymbol{x}, \boldsymbol{y})\varphi(\boldsymbol{y}), T_2) \cdot \psi(\boldsymbol{x}), T_1)| \\ &\leq C \frac{h^{2N_i}}{(2N_i - 1)^{2N_i - 1}} \left\| \Phi\varphi_h \left| \frac{dy}{dt} \right| \right\|_{2N_i, T_2} \cdot \left\| \psi \left| \frac{dx}{dt} \right| \right\|_{m, T_1}. \end{aligned} \tag{4.9}$$

Estimates (4.8) and (4.9) use the $N_o$ and $N_i$ quadrature parameters. If we assume that

$$d = N_o - 1 = 2N_i - 1 \Rightarrow N_o = 2N_i \tag{4.10}$$

then we receive the following

$$\begin{aligned}&|I_o(I_i(\Phi(\boldsymbol{x},\boldsymbol{y}),\varphi_h(\boldsymbol{y})),T_2)\cdot\psi_h(\boldsymbol{x}),T_1)\\ &-Q_{N_o}(Q_{N_i}(\Phi(\boldsymbol{x},\boldsymbol{y}),\varphi_h(\boldsymbol{y})),T_2)\cdot\psi_h(\boldsymbol{x}),T_1)|\\ &\le C\frac{h^d}{d^d}\left\|\Phi\circ\varphi_h\left|\frac{d\boldsymbol{y}}{dt}\right|\right\|_{d+1,T_2}\cdot\left\|\psi_h\left|\frac{d\boldsymbol{x}}{dt}\right|\right\|_{0,T_1}.\end{aligned} \tag{4.11}$$

The Second Strang Lemma for this problem takes the form ([11], Th.4.3)

**Theorem 4.2** *If the above assumptions are satisfied, then there exists a positive constant $C$, such that*

$$\|p-p_h\| \le C \inf_{v_h\in V_h}\left[\|p-v_h\| + \sup_{0\ne q_h\in V_h},\frac{|a_h(v_h,q_h)-a(v_h,q_h)|}{\|q_h\|}\right]. \tag{4.12}$$

The final estimate concludes our analysis:

$$|a_h(v_h,q_h)-a(v_h,q_h)| \le C\frac{h^d}{d^d}\left\|\Phi\circ q_h\left|\frac{d\boldsymbol{y}}{dt}\right|\right\|_{d+1,G}\cdot\left\|v_h\left|\frac{d\boldsymbol{x}}{dt}\right|\right\|_{0,G}. \tag{4.13}$$

The norm in (4.12) is the norm in $V = H^{\frac{1}{2}}(\Gamma)$, i.e., the term arising from the approximate integration is bounded

$$C\frac{h^d}{d^d}\left\|\Phi\circ q_h\left|\frac{d\boldsymbol{y}}{dt}\right|\right\|_{d+1,G} \le C\frac{h^d}{d^d}\|v_h\|_{d+1,\Gamma} \le C\frac{h^d}{d^d}[\|p-v_h\|_{d+1,\Gamma}+\|p\|_{d+1,\Gamma}]. \tag{4.14}$$

If we include other terms and assume $d$ as being the same in Sections 2 and 4, from Theorem 4.2 we receive:

**Theorem 4.3** *With the condition, that corresponding assumptions, from Sections 3 and 5 are satisfied, if d is the order of shape functions' interpolation, internal quadratures are exact for polynomials of order d, outer quadratures are exact for polynomials of order 2d, the boundary is interpolated by polynomials of order d + 1, h is a maximal diameter of the element, the exact solution $p \in H^m(\Gamma)$, the approximate solution $p_h \in V_h$, $\boldsymbol{X} \in V^{m+1}(G)$, then for some $C > 0$ the following inequality*

$$\|p-p_h\|_{\frac{1}{2},\Gamma} \le C\left[\|p^{inc}-p_h^{inc}\|_{-\frac{1}{2},\Gamma} + \left\|\frac{\partial p^{inc}}{\partial n^x} - \frac{\partial p_h^{inc}}{\partial n_h^x}\right\|_{-\frac{1}{2},\Gamma}\right.$$

$$+ \frac{h^{\mu-1}}{d^{m-1}} \left( \|\boldsymbol{X}\|_{m+1,G} \|p\|_{\frac{1}{2},\Gamma} + \|p\|_{m,\Gamma} \right) \Bigg], \quad (4.15)$$

*holds, where $\mu$ is given by (2.7).*

**Corollary 4.1** *If $p^{inc}$ and $\partial p^{inc}/\partial n^x$ are interpolated using the same hp- method, then inequality (4.15) becomes*

$$\|p - p_h\|_{\frac{1}{2},\Gamma}$$

$$\leq C \frac{h^{\mu-1}}{d^{m-1}} \left[ \|p^{inc}\|_{m,\Gamma} + \|\frac{\partial p^{inc}}{\partial n^x}\|_{m,\Gamma} + \|\boldsymbol{X}\|_{m+1,G} \|p\|_{\frac{1}{2},\Gamma} + \|p\|_{m,\Gamma} \right] \quad (4.16)$$

## References


[1] Babuška, I., and Suri, M.: The h-p Version of the Finite Element Method with Quasiuniform Meshes, *RAIRO, Math.Mod.Numer.Anal.* **21** (1987), 199-238.

[2] Babuška, I., and Suri, M.: The Optimal Convergence Rate of the p-Version of the Finite Element Method, *SIAM J.Numer.Anal.* **24** (1987), 750-776.

[3] Babuška, I., Dorr, M.: Error estimates for the combined h and p version of the finite element method, *Numer.Math.* **37** (1981), 257-277.

[4] Banerjee, U., and Suri, M.: The Effect of Numerical Quadrature in the p- Version of the Finite Element Method *Math.Comp.* **59** (1992), 1-20.

[5] Ciarlet, P.G.: *The finite element method for elliptic problems*, North Holland, Amsterdam 1978.

[6] Demkowicz, L., Oden, J.T., Ainsworth, M., and Geng P.: Solution of elastic scattering problems in linear acoustics using $h - p$ boundary element methods, *J.Comput.Appl.Math.* **36** (1991), 29-63.

[7] Ervin, V.J., Heuer, N., and Stephan, E.P.: On the h-p version of the Boundary Element Method for Symm's integral equation on polygons, *Comput.Methods Appl.Mech.Engrg.* **110** (1993), 25-38.

[8] Heuer, N., and Stephan, E.P.: The hp- Version of the Boundary Element Method, *ZAMM* **74** (1994), T511-T513.

[9] Holm, H., Maischak, M., and Stephan, E.P.: The hp- version of the Boundary Element Method for Helmholtz screen problems, *Comput.* **57** (1996), 105-134.

[10] Karafiat, A.: *An Analysis of the Boundary Element Method for the Acoustic Scattering Problem*, Kraków, The Cracow University of Technology, Kraków 1996 (in Polish).

[11] Karafiat, A.: On hp- Error Estimation in the BEM for a Three - Dimensional Helmholtz Exterior Problem, *Comput.Methods Appl.Mech.Engrg.* **150** (1997), 199-214.

[12] Karafiat, A., Oden, J.T., and Geng, P.: Variational Formulations and *hp* Boundary Element Approximations for Hypersingular Integral Equations for Helmholtz Exterior Boundary Value Problems in Two Dimensions, *Int.J.Engrg.Sci.* **31** (1993), No. 4, 649-672.

[13] Karafiat, A.: On convergence of a hp- adaptive BEM for elastic scattering, *Computer Assisted Mechanics and Engineering Sciences* (submitted).

[14] Postell, F.V, and Stephan, E.P.: On the h, p, and h-p versions of the Boundary Element Method - Numerical Results, *Comput.Methods Appl. Mech.Engrg.* **83** (1990), 69-90.

[15] Rank, E.: Adaptive h- ,p-, and hp- versions for boundary integral element methods, *Internat.J.Numer.Methods Engrg.* **28** (1989), 1335-1349.

[16] Stephan, E.P.: The h-p Boundary Element method for solving 2- and 3-dimensional problems, *Comput.Methods Appl.Mech.Engrg.* **133** (1996), 183-208.

# NON-SINGULAR RECIPROCITY BASED BEM/FEM FORMULATIONS

Vladimír KOMPIŠ, Lenka JAKUBOVIČOVÁ, František KONKOĽ
*Faculty of Mechanical Engineering,*
*University of Žilina*
*Veľký diel, Sk-010 26 Žilina, Slovakia*
kompis@fstroj.utc.sk

**Keywords:** Non-singular BE, Trefftz polynomials, multi-domain, stress smoothing

**Abstract** In this paper, the Trefftz polynomials are used to obtain a non-singular boundary element formulation for linear elasticity problems. Tractions and boundary displacements are related by reciprocity principles over sub-domains (elements). The displacements between the sub-domains are compatible, the interdomain equilibrium is satisfied in a weak (integral) form only. If the stresses are computed using the Trefftz polynomial interpolants over some patches of nodal displacements, a smooth stress field is obtained over the whole domain with same rate of convergence as the displacements.

The convergence properties are studied on the fields defined by high order Trefftz fields of plane stress/strain problems.

## Introduction

The advantages of both FEM [1,2] and BEM [3-5] are well known. We can mention here some of them. The advantages of BEM over FEM are: higher accuracy with a similar degree of approximation, integration over the domain boundaries only (with exception of some nonlinear problems), simpler domain approximation. The advantages of the FEM over the BEM are the sparse, banded and symmetric structure of the resulting matrices, simpler numerical formulation and integration, simpler modeling of thin walled regions.

If polynomial or other non-singular (e.g., Kupradze's, harmonic, ... [6-11]) Trefftz (T-) functions, i.e., functions satisfying all governing equations inside the domain [12,13] are used in Betti's reciprocity theorem. Corresponding BIE does not contain any (weak, strong, quasi-, hyper-)

*T. Burczynski (ed.), IUTAM/IACM/IABEM Symposium on Advanced Mathematical and Computational Mechanics Aspects of the Boundary Element Method,* 173–188.

singularities. The integration is simpler. We need to find the fundamental solution; no free terms (no need to calculate intensities in the corner and edge points) and the evaluation of the field variables inside the domain is simpler.

In this paper, T-polynomials are used in reciprocity relations. If, however, the accuracy requires many polynomial terms (according to the number of d.o.f.), then high order polynomials must be used. This increases the complexity of the expressions. To overcome this drawback, the whole domain is divided into sub-domains, and a multi-domain variational formulation is introduced. The displacements are supposed to be compatible between the subdomain (elements), and the inter-element equilibrium is satisfied in a weak form, similarly as it is known from the hybrid-Trefftz-displacement formulations [14-16].

The stress recovery technique is used to obtain smooth stress fields between the elements [17-22] and to estimate the errors in the solution. As in our formulation, the only equations satisfied in a weak form are the inter-element equilibrium equations. We have shown [23,24] that reliable error estimators are the incompatibilities of the tractions between the elements. The smoothed stress fields obtained by Moving Least Square (MLS) interpolation using the T-polynomials with computed nodal displacements (and known boundary conditions in the boundary points and in the points near the boundaries) considerably improves the accuracy of the computed stress fields.

The examples in this paper and in our previous works on approximation of higher-order gradient fields show the advantages and accuracy of the proposed procedures.

## 1. GENERATION OF TREFFTZ POLYNOMIAL FORMULATIONS

T-polynomials can be found analytically (by symbolic algebra for more complicated problems, e.g., 3D T-functions) or numerically. The procedure is similar for both methods; only the form of the coefficients is different. Analytically we obtain the coefficients expressed by means of material constants, whereas in the second case the coefficients are given numerically, i.e., the whole procedure has to be repeated for a material to obtain the coefficients of the T-polynomials. The advantage of the second procedure is simpler expressions for the T-polynomials.

As an example we will describe here how to obtain the T-polynomials of the 4th order for 2D isotropic solids. Of course, all lower order terms will be used first in approximations, but this example serves as a presentation of the procedure.

The starting point are the Lame-Navier equations for 2D isotropic solid, disregarding body forces:

$$
\begin{aligned}
(\lambda + 2\mu)u_{1,11} + \mu u_{1,22} + (\lambda + \mu)u_{2,12} &= 0 \\
(\lambda + 2\mu)u_{2,22} + \mu u_{2,11} + (\lambda + \mu)u_{1,21} &= 0
\end{aligned} \tag{1}
$$

and 4th order polynomial terms for the approximation of displacements

$$
\left\{ \begin{array}{c} u1 \\ u2 \end{array} \right\} = \left[ \begin{array}{cccc} x_1^4 & x_2^4 & 0 & 0 \\ 0 & 0 & x_1^4 & x_2^4 \end{array} \right] \{ \boldsymbol{a} \} + \left[ \begin{array}{cccccc} x_1^3 x_2 & x_1^2 x_2^2 & x_1 x_2^3 & 0 & 0 & 0 \\ 0 & 0 & 0 & x_1^3 x_2 & x_1^2 x_2^2 & x_1 x_2^3 \end{array} \right] \{ \boldsymbol{b} \} \tag{2}
$$

$$
\begin{aligned}
\{\boldsymbol{a}\} &= \{ \begin{array}{cccc} a_1 & a_2 & a_3 & a_4 \end{array} \}^T \\
\{\boldsymbol{b}\} &= \{ \begin{array}{cccccc} b_1 & b_2 & b_3 & b_4 & b_5 & b_6 \end{array} \}^T
\end{aligned} \tag{3}
$$

where $u_i$ are displacement components, $\lambda$ and $\mu$ are the Lamé constants, and the subscript after the comma denotes a partial derivative in the corresponding coordinate direction.

Eq. (2) in matrix form is given as

$$
\{\boldsymbol{u}\} = [\boldsymbol{A}(x_i)]\{\boldsymbol{a}\} + [\boldsymbol{B}(x_i)]\{\boldsymbol{b}\} \tag{4}
$$

We express the second derivatives

$$
\left\{ \begin{array}{c} u_{1,11} \\ u_{1,12} \\ u_{1,22} \end{array} \right\} = \left[ \begin{array}{cc} 12x_1^2 & 0 \\ 0 & 0 \\ 0 & 12x_2^2 \end{array} \right] \left\{ \begin{array}{c} a_1 \\ a_2 \end{array} \right\} + \left[ \begin{array}{ccc} 6x_1x_2 & 2x_2^2 & 0 \\ 3x_1^2 & 4x_1x_2 & 3x_2^2 \\ 0 & 2x_1^2 & 6x_1x_2 \end{array} \right] \left\{ \begin{array}{c} b_1 \\ b_2 \\ b_3 \end{array} \right\} \tag{5}
$$

and similarly $\{u_{2,11} u_{2,12} u_{2,22}\}^T$.

We introduce (5) into the equilibrium equations (1) and obtain the relation between the subvectors of coefficients $b$ and $a$

$$
[\boldsymbol{M}(x_i)]\,\{\boldsymbol{b}\} + [\boldsymbol{N}(x_i)]\,\{\boldsymbol{a}\} = \{\boldsymbol{0}\} \tag{6}
$$

In order to satisfy the equilibrium conditions we express $b$ in terms of $a$ by evaluating (6) for three different points (j) which do not lie on a line. Thus, we get

$$\left[\boldsymbol{M}({}^{(j)}x_i)\right]\{\boldsymbol{b}\} = -\left[\boldsymbol{N}({}^{(j)}x_i)\right]\{\boldsymbol{a}\} \tag{7}$$

$$\{\boldsymbol{b}\} = -\left[\boldsymbol{M}^{-1}\right][\boldsymbol{N}]\{\boldsymbol{a}\} \tag{8}$$

Note that we have chosen as many coefficient terms in the sub-vector $b$ as polynomial terms contained in the polynomial expressions of the equilibrium equation (i.e. 2 x $[x^2, xy, y^2] = 2$ x 3 terms of 2nd order).

In this way we obtain T-polynomial displacements of the 4th order

$$\{\boldsymbol{U}\} = \left([\boldsymbol{A}(x_i)] - [\boldsymbol{B}(x_i)][\boldsymbol{M}]^{-1}[\boldsymbol{N}]\right)\{\boldsymbol{a}\} = [\boldsymbol{U}(x_i)]\ \{\boldsymbol{a}\} \tag{9}$$

Each column of $[\boldsymbol{U}]$ in (9) introduces a T- displacement function. Note that the number of T-functions which can be derived in this form is $(2n+1)$ for 2D problems and $(n+1)^2$ for 3D problems where $n$ is the polynomial order.

## 2. MULTI-DOMAIN BEM (RECIPROCITY BASED FEM) FORMULATION

From T-displacements, T-stresses are commonly obtained as

$$S_{ij} = \mu(U_{i,j} + U_{j,i} + \lambda\delta_{ij}U_{k,k} \tag{10}$$

and corresponding T-tractions on boundaries with outer normal $n_j$

$$T_i = S_{ij}n_j \tag{11}$$

For simplicity, consider the elasticity problem without body forces. Boundary displacements $u_i(x)$ and tractions $t_i(x)$ of each approximated subdomain (element) will be related by the Betti's reciprocity theorem

$$\int_{\Gamma_e} T_i(x)u_i(x)d\Gamma(x) = \int_{\Gamma_e} U_i(x)t_i(x)d\Gamma_e(x) \tag{12}$$

$x$ denotes a field variable, $U_i$ and $T_i$ are corresponding T-(displacement and traction) functions on the element boundary $\Gamma_e$.

The boundary displacements can be expressed by their nodal value $d^{(j)}$ and shape functions (the upper index corresponds to the nodal point) $N_u^{(j)}$

$$u_i(\xi) = N_u^{(j)}(\xi)d_i^{(j)} \quad \text{or} \quad u = N_u d^e \tag{13}$$

$\xi$ is the local co-ordinate of a point on the element boundary.

Tractions are given by their value $q^{(j)}$ at nodal points and by corresponding shape functions, $N_t^{(j)}$, as

$$t_i(\xi) = N_t^{(j)}(\xi) q_i^{(j)} \quad \text{or} \quad t = \boldsymbol{N}_t \boldsymbol{q}^e \tag{14}$$

which leads to the matrix form of Eq.(12)

$$\boldsymbol{T d}^e = \boldsymbol{U q}^e \tag{15}$$

The elements of the matrices $\boldsymbol{T}$ and $\boldsymbol{U}$ are given as

$$\begin{aligned} T_{kl} = \int_{\Gamma_e} T^{(k)}(x(\xi)) N_u^{(l)}(\xi) d\Gamma = \\ = \sum_j T^{(k)}(x(\xi^{(j)})) N_u^{(l)}(\xi^{(j)}) J(\xi^{(j)}) w^{(j)} \\ U_{kl} = \int_{\Gamma_e} U^{(k)}(x(\xi)) N_t^{(l)}(\xi) d\Gamma = \\ = \sum_j U^{(k)}(x(\xi^{(j)})) N_t^{(l)}(\xi^{(j)}) J(\xi^{(j)}) w^{(j)} \end{aligned} \tag{16}$$

$\xi^{(j)}$ and $w_{(j)}$ are co-ordinates and weights in the Gauss quadrature formulas and $J$ is the Jacobian.

Eq.(12) is an integral equation. Contrary to the use of fundamental solutions, this is a non-singular integral equation. The advantage of the non-singular formulation is both a simpler numerical evaluation of integral terms (compared to computing the strong, weak, hyper-, or quasi-singular integrals) and a simpler estimation of the required numerical integration order for its exact evaluation. The drawback of this formulation is a increasing number of polynomial terms with an increasing degree of approximation of the domain variables. This drawback can be overcome by using the multi-domain BEM-FEM formulation.

We will assume that the whole domain is decomposed into subdomains (elements) and the displacements between the subdomains are compatible. The tractions, however, are incompatible between the elements and so the interelement equilibrium and natural boundary conditions are satisfied only in a weak (integral) sense using the variational formulation

$$\int_{\Gamma_t} \delta u^T (t - \bar{t}) d\Gamma + \int_{\Gamma_i} \delta u^T (t^A - t^B) d\Gamma = \int_{\Gamma_e} \delta u^T t d\Gamma - \int_{\Gamma_t} \delta u^T \bar{t} d\Gamma = 0 \tag{17}$$

where $\Gamma_i$ and $\Gamma_t$ are the inter-element boundaries and the element boundaries with prescribed tractions, respectively. $A$ and $B$ denote the neighbouring elements. Prescribed values are denoted with a bar.

Eq.(17) can be expressed in discretized form:

$$\begin{aligned}\textstyle\sum_e \sum_j \sum_l N_u^{(k)}(\xi^{(j)}) N_t^{(l)}(\xi^{(j)}) J(\xi^{(j)}) w^{(j)} q^{(l)} = \\ \textstyle\sum_e \sum_i N_u^{(k)}(\xi^{(i)}) \bar{t}(\xi^{(i)}) J(\xi^{(i)}) w^{(i)}\end{aligned} \tag{18}$$

or in the equivalent matrix form

$$\sum_e \boldsymbol{M}^e \boldsymbol{q}^e = \sum_e \boldsymbol{p}^e \tag{19}$$

The summations in (18) are related to elements, Gauss integration nodes and nodal points, respectively.

From Eq.(15) we have

$$\boldsymbol{q}^e = \boldsymbol{U}^{-1} \boldsymbol{T} \boldsymbol{d}^e \tag{20}$$

Substituting this into Eq.(19) yields

$$\sum_e \boldsymbol{M}^e \boldsymbol{U}^{-1} \boldsymbol{T} \boldsymbol{d}^e = \sum_e \boldsymbol{p}^e \tag{21}$$

or

$$\boldsymbol{K} \boldsymbol{d} = \boldsymbol{p} \tag{22}$$

where $\boldsymbol{K}$ is the global stiffness matrix.

## 3. IMPORTANCE AND EFFICIENCY OF THE STRESS RECOVERY PHASE

Having obtained the nodal displacements from Eq.(22) the tractions of the nodal points of each element can be computed from Eq.(20). The stresses in the corner points of elements can be calculated directly from the tractions. Both tractions and stresses are incompatible between the elements. Smooth stress fields can be obtained in many ways. The simplest is by averaging the nodal values of the neighbouring elements.

Much more accurate continuous stress fields can be obtained using the MLS techniques from displacements and known tractions on the domain boundaries of the nodal points closest to the corresponding point of interest.

We assume the displacement field (at a field point $x$), $\boldsymbol{u}(x)$, given as

$$\boldsymbol{u}(x) = \boldsymbol{U}(x) \boldsymbol{a} \tag{23}$$

where $\boldsymbol{U}(x)$ is a matrix of T-displacement functions and $\boldsymbol{a}$ is the vector of unknown coefficients. The stress field is then given by

$$\boldsymbol{\sigma}(x) = \boldsymbol{S}(x) \boldsymbol{a} \tag{24}$$

Similarly, we can express T-tractions as

$$\boldsymbol{t}(x) = \boldsymbol{T}(x)\boldsymbol{a} = \boldsymbol{\sigma}(x)\boldsymbol{n}(x) \tag{25}$$

In this approximation we use the full T-polynomials of chosen order. The unknown coefficients $\boldsymbol{a}$ are computed by the Least Square (LS-) method from

$$\begin{aligned} &\sum_I [\boldsymbol{U}_{IJ}(x_I)\boldsymbol{a}_J - \boldsymbol{d}_I]^2 \to \min \\ &\sum_K [\boldsymbol{T}_{KJ}(x_K)\boldsymbol{a}_J - \boldsymbol{t}_K]^2 \to \min \end{aligned} \tag{26}$$

where $d_I$ and $t_K$ are the displacement and traction components of the nodal points, $U_{IJ}$ is the $J$-th T-displacement function for the $I$-th nodal component, and $T_{KJ}$ is the $J$-th T-traction for the K-th nodal component. The displacements are considered at the patch of nodal points in which the distance from the point of interest is smaller than a prescribed value.

Splitting the vector of the unknown coefficients $a$ and the matrices of T-functions $\boldsymbol{U}$ and $\boldsymbol{K}$ in Eq.(26) in a convenient way we solve the problem (26) in the form

$$\begin{aligned} \boldsymbol{U}_{11}^T\boldsymbol{U}_{11}\boldsymbol{a}_1 + \boldsymbol{U}_{11}^T\boldsymbol{U}_{12}\boldsymbol{a}_2 &= \boldsymbol{U}_{11}^T\boldsymbol{d} \\ \boldsymbol{T}_{22}^T\boldsymbol{T}_{21}\boldsymbol{a}_1 + \boldsymbol{T}_{22}^T\boldsymbol{T}_{22}\boldsymbol{a}_2 &= \boldsymbol{T}_{22}^T\boldsymbol{t} \end{aligned} \tag{27}$$

By Expressing the LS-problem in this form, we need not care about dimensioning in the first and second rows of Eq.(27).

# 4. ERRORS IN APPROXIMATION OF HIGHER ORDER GRADIENT FIELDS

The numerical experiments [25, 26] showed that the quadratic elements are a good choice both from the point of view of accuracy and algorithmic efficiency. The quadrilateral element, which is similar to the serendipity element in the displacement FEM formulation, is defined by four boundary elements over the subdomain. The definition of a subdomain of arbitrary polygonal form is a simple task.

The domain with boundary conditions corresponding to 6-th order T-polynomial tractions or displacements (test function) is a good demonstration of the proposed method. The undeformed and deformed mesh of 10 x 10 elements (sub-domains) is shown in Fig.1 with corresponding shear and von Mises stress fields given in Fig.2. The tractions are chosen so that lowest and largest gradients are at opposite corner points.

For this example, displacements, stresses and tractions are known at all points of the domain and along all domain and sub-domain boundaries, so errors can be evaluated. In this way, a valuable information is obtained concerning the performance of the proposed method.

Figures 3a and 3b show the distribution of the errors in the displacements of the plotted nodes only. Fig.3a depicts these values by given displacements along the whole boundary of the domain whereas Fig.3b gives the same problem by prescribed tractions. We can see that the errors in the displacements in the second case are much larger on the boundaries; however, the errors of the interior points are very similar in both cases. The relative error of the extension of the diagonal (i.e. the length between the corner nodes with the smallest and the largest gradients) is 2e-4.

It is also interesting to observe the interelement unbalance along the lines parallel to the element boundaries (see Fig.4). Differences in the value of the corner nodes and the midside nodes (they are often of opposite signs) as well as the zigzag character of the errors in the displacements (Fig.3) indicate that smoothing procedures can be a very effective tool to increase the accuracy of gradient (strains and stresses) fields. The maximum relative error of the nodal tractions (i.e., the error related to the maximum von Mises stress) is 6e-3.

The relative errors in the stresses were also related to the maximum von Mises stress in the domain. The errors of stresses in the corner nodes obtained from the element tractions (computed from the element displacements or those prescribed on the domain boundaries) are given in Fig.5. Their maximal relative value was 2.5e-3.

The recovered stresses computed by the smoothing procedure described in Section 4 were obtained, using the radius of the domain of influence equal to the length of the element diagonal and the 4th order Trefftz polynomials for the MLS interpolation the errors shown in the Figure 6. The maximum relative error of these values is 2e-4, one order smaller than that obtained by the simple averaging. Note that larger errors are obtained in the middle of the elements, where there are no nodes and thus, fewer nodes are included in the domain of influence of the corresponding point. Also, larger errors are obtained for the boundary points for two reasons; first, because of larger errors in the displacements (see Fig.3a); second, because of a smaller number of nodal points in corresponding domain of influence.

The traction discontinuities are the only discretization errors. Hence, they can be chosen as an error estimators in this formulation. In Figure 7, the integrals of the traction discontinuities are introduced for each element. Figure 8 contains the mean square error of displacements for each point of interest over its domain of influence.

Figure 9 shows the convergence characteristics of displacements and von Mises stresses.

# 5. CONCLUSIONS

In this paper, a non-singular reciprocity based BEM using T-polynomials is presented. Generation of the T-functions was shown first. The subdomain technique was then formulated. Compatible displacements and a weak form of inter-domain equilibrium between the sub-domain were assumed.

In this way, the stiffness matrix is obtainable by the non-singular BIE. Such a formulation also enables to connect both common finite elements and sub-domains defined by singular BIE's to the multi-domain defined by the present method. The sub-domain formulation can be understood as a reciprocity based FE. The form of the element, however, can be much more general. It can be a polygon with straight or curved sides, a multiply connected region, etc. A similar situation occurs for 3D problems.

Numerical experiments concerning approximation of fields with higher order gradients show that stress smoothing using the MLS method with T-polynomial interpolation from computed nodal displacements and known boundary conditions increases the accuracy of the gradient fields (strain and stress) by even more than one order.

Also, the formulation is very attractive for non-linear problems as well as other problems of computational mechanics.

## Acknowledgement:

The authors gratefully acknowledge the partial support of this research by the Slovak Grant Agency for Sience (Grant No. 1/6036/99)

# 6. REFERENCES

[1] Zienkiewicz, O.C. and Taylor, R.L.,*The Finite Element Method* 4th Edition , Vol I, McGraw Hill 1989, Vol II McGraw Hill (1991).

[2] K.-J. Bathe,*Finite Element Procedures*, Prentice Hall, Englewood Clifs, N.J., (1996).

[3] Brebbia, C.A.,*The boundary element method for engineers*, J. Wiley, (1978).

[4] Bausinger, R., Kuhn, G.,*The boundary element method* (in German), Expert Verlag, Germany, (1987).

[5] Balaš, J., Sládek, J. and Sládek, V.,*Stress Analysis by Boundary Element Methods*, Elsevier, (1989).

[6] Cheung, Y.K., Jin, W.G. and Zienkiewicz, O.C.,*Direct solution procedure for solution of harmonic problems using complete, non-singular, Trefftz function*, Commun. in Appl. Numer. Methods, **5**, 159-169 (1989).

[7] Cheung, Y.K., Jin, W.G. and Zienkiewicz, O.C.,*Solution of Helmholtz equation by Trefftz method*, Int. J. Numer. Meth. Engng., **32**, 63-78 (1991).

[8] Zieliński, A.P.,*On trial functions applied in the generalized Trefftz method*, Advancs in Eng. Software, **24**, 147-155 (1995).

[9] Kita, E., Kamiya, N. and Ikeda, Y.,*A new boundary-type scheme for sensitivity analysis using Trefftz formulation*, Finite Elements in Analysis and Design, **21**, 30-317 (1996).

[10] Kita, E., Kamiya, N. and Ikeda, Y.,*Application of the Trefftz Method to Sensitivity analysis of a three-dimensional potential problem*, Mech. Struct. & Mach,. **24**, 295-311 (1996).

[11] Karageorghis, A., Fairweather, G.,*The method of fundamental solutions for axisymmetric potential problems*, Int. J. Numer. Methods Engng., to appear.

[12] Treffez, E.,*Ein Gegenstück zum Ritzschen Verfahren*, Proc. 2nd Int. Congress of Apllied mechanics, Zürich, (1926).

[13] Kompiš, V.,*Finite element satisfyng all governing quations inside the element*, Computers & Structures, **4**, 273-278, (1994).

[14] Kompiš, V., Kaukič, M. and Žmindák, M.,*Modeling of local effects by hybrid-displacement FE*, J. Comput. Appl. Math., **63**, 265-269 (1995).

[15] Jirousek, J. and Wróblewski, A.,*T-elements: State of the art and future trends*, Achives of Comput. Mech., **3**, 323-434 (1996).

[16] Kompiš, V. and Fraštia, Ľ.,*Polynomial representation of hybrid finite elements*, Computer Assis. Mech. in Eng. Sci., **4**, 521-532 (1997).

[17] Hilton, E. and Campbell, J.,*Local and global smoothing of discontinuous finite element functions using a least square method potential problem*, Int. J. Numer. Meth. Eng. & Mach., **8**, 461-480 (1974).

[18] Zienkiewicz, O.C. and Zhu, J.Z.,*A simple error estimator and adaptive procedure for practical engineering analysis potential problem*, Int. J. Numer. Meth. Eng. & Mach., **24**, 337-357 (1987).

[19] Niu, Q. and Shephard, M.S.,*Superconvergent extraction tehniques for finite element analysis*, Int. J. Num. Meth. Eng., **36**, 811-836 (1993).

[20] Blacker, T. and Belytschko, T.,*Superconvergent patch recovery with equilibrium and conjoint interpolat enhancement*, Int. J. Num. Meth. Eng., **37**, 517-536 (1994).

[21] Ramsay, A.C.A and Maunder, E.A.W.,*Effective error estimation from continuous, boundary admissible estimated stress fields*, Comp. & Struct., **61**, 331-343 (1996).

[22] Wiberg, N.E.,*Superconvergent path recovery: A key to quality assessed FE solutions*, Advan. Eng. Software, **28**, 85-95 (1997).

[23] Kompiš, V. and Jakubovičová, L.,*Errors in modelling high order gradient fields using isoparametric and reciprocity based FEM.*

[24] Kompiš, V., Žmindák, M. and Jakubovičová, L.,*Error estimation in multy-domain BEM (Reciprocity based FEM)*, Proc. ECCM '99, European Conference on computional Mechanicseded, CD-ROM, München, Germany (1999).

[25] Kompiš, V., Oravec, J. and Búry, J.,*Reciprocity based FEM*, Mechanical Engineering, **50**, 187-202, No. 3., (1999).

[26] Kompiš, V., Fraštia, Ľ., Kaukič, M. and Novák, P., ed. Idelsohn S.R., Dvorkin E.F.N.,*Accuracy of direct Trefftz FE form*, Computational Mechanics, New Trends and Applications, CD-ROM, CIMNE Barcelona, (1998)

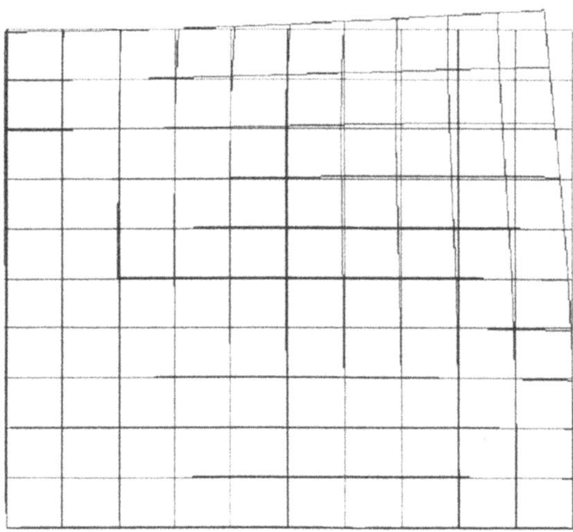

*Figure 1* Underformed and deformed mesh.

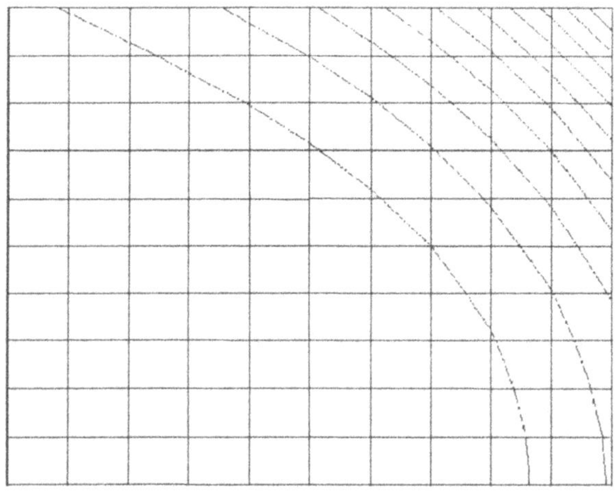

*Figure 2* Von Mises stress.

*Figure 3a* Displacement errors.

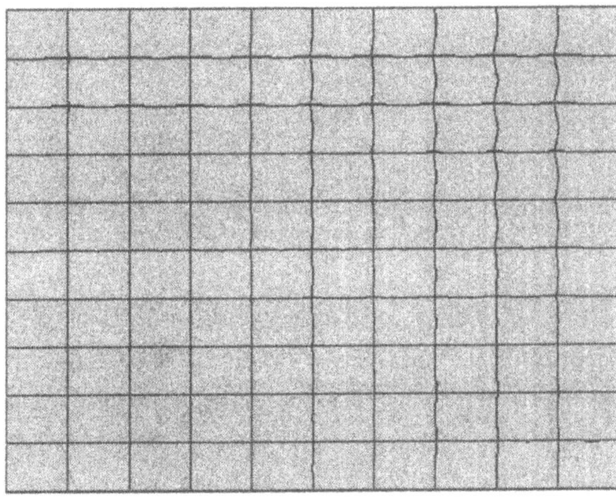

*Figure 3b* Displacement errors.

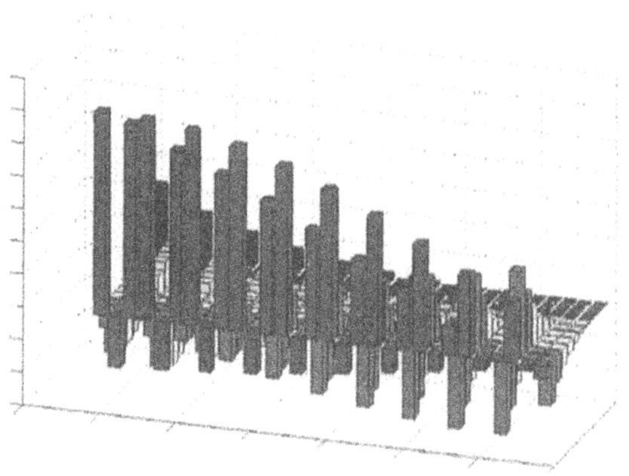

*Figure 4* **Errors in the tractions $t_x$ at nodal points of neighbouring elements.**

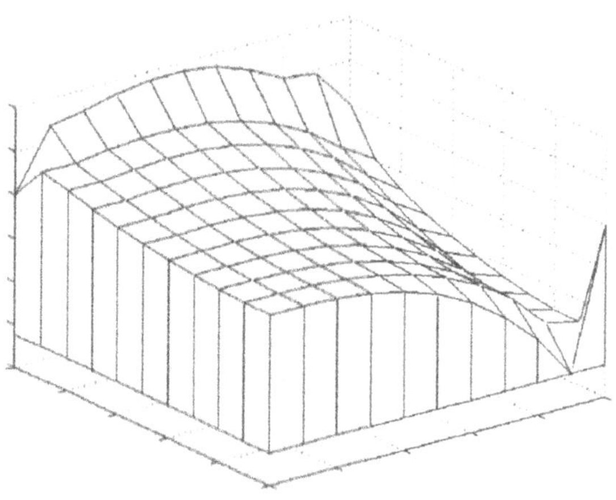

*Figure 5* **Errors in $\sigma_{11}$ computed from element tractions.**

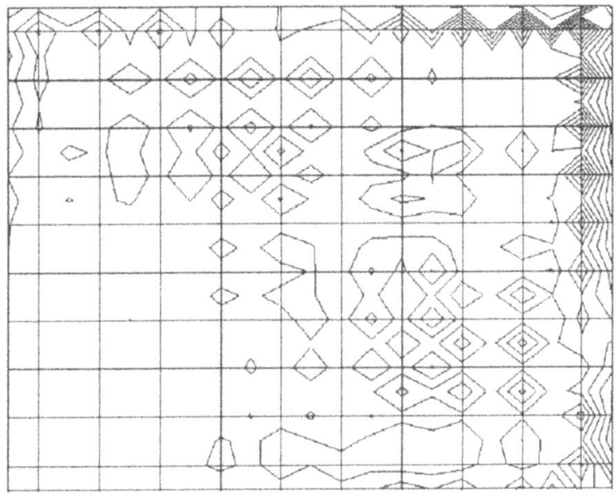

*Figure 6* Errors in the von Misses stress in smoothed fields.

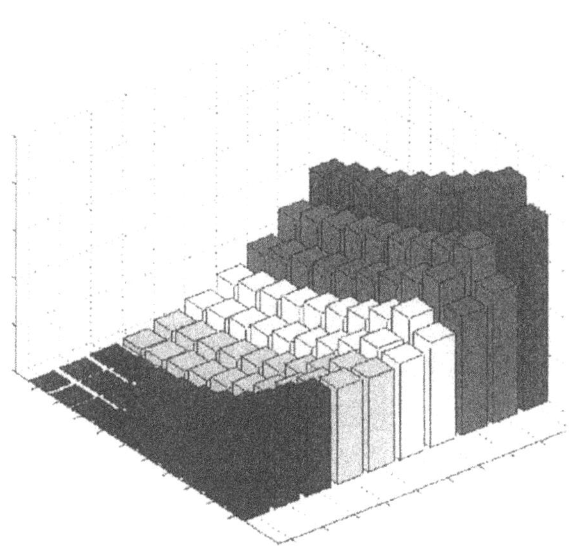

*Figure 7* Average element traction discontinuities.

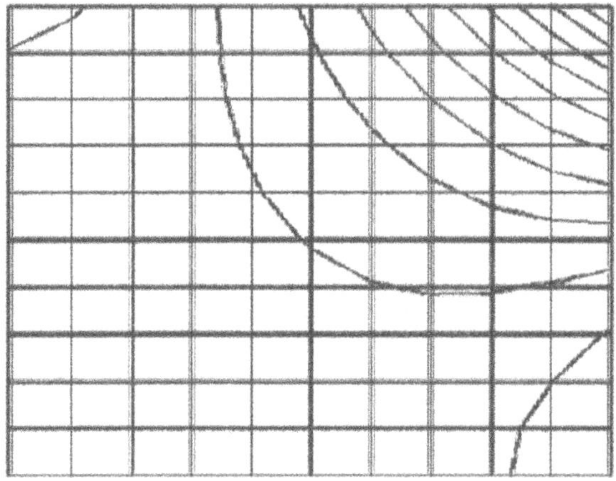

*Figure 8* Mean square error of displacements in points of interest.

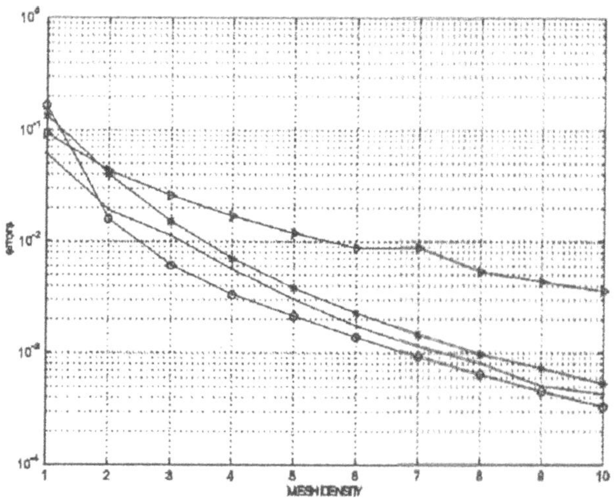


*Figure 9* Convergence of displacements and stresses. ▷▷▷ - $Err_{max}$ relative average element traction error, $***$ - $Err_u$ relative error in displacement norm, ∘∘∘ - $Err_{\sigma vM}$ relative error in Von misses stress.

# NUMERICAL ANALYSIS OF BIOLOGICAL TISSUE FREEZING PROCESS

E.MAJCHRZAK & M.DZIEWONSKI
*Department for Strength of Materials and Computational Mechanics*
*Silesian University of Technology*
*44-100 Gliwice, Konarskiego 18a, Poland*

Key words: thermal diffusion, tissue freezing, boundary element method, artificial heat source method

Abstract: The non-steady nonlinear thermal diffusion proceeding during the biological tissue freezing process is analyzed. The spherical cryoprobe is taken into account and its geometry determines the details of mathematical description of the process considered. In order to solve the problem discussed, the boundary element method and the approach called the artificial heat source method is used. The numerical experiments allow to determine the dependence between the depth of frozen region (the kinetics of ice ball generation) and the parameters of cryoprobe (the surface temperature and cryoprobe radius).

## 1. INTRODUCTION

Cryosurgery has a variety applications in medical treatment e.g. to causing a local necrosis of a tissue, the detachment a pathological tissue, destruction of cancer cells etc. The cryoprobe being in thermal contact with a biological tissue causes the freezing process to proceed in the domain considered. In this paper, the problem of tissue freezing under the spherical cryoprobe action is discussed – Figure 1.

On the basis of numerical experiments, the relation between the depth of frozen region, the cryoprobe temperature and its radius is found. From the mathematical point of view the bio-heat transfer proceeding in domain of

*T. Burczynski (ed.), IUTAM/IACM/IABEM Symposium on Advanced Mathematical and Computational Mechanics Aspects of the Boundary Element Method*, 189–200.

tissue is described by the nonlinear energy equation [1, 2], while in order to take into account the freezing process the one domain approach [2, 3, 4] is applied. The basic differential equation is supplemented by assumed boundary and initial conditions. The values of biological tissue thermophysical parameters are taken from [2]. The numerical solution is obtained on the basis of the BEM algorithm corresponding to non-steady diffusion problem described in the spherical co-ordinate system [5]. In order to take into account the nonlinearities resulting from temperature-dependent thermophysical parameters of biological tissue, the artificial heat source method [6, 7, 8] is applied. This approach requires the 'rebuilding' of the mathematical model of the process to the Kirchhoff convention. In the final part of the paper the results of numerical simulations are presented.

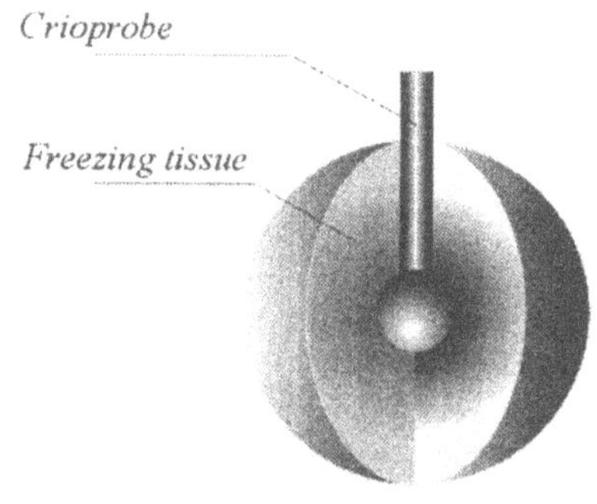


*Figure 1.* The domain considered

## 2. GOVERNING EQUATIONS

The following energy equation written in the spherical co-ordinate system is taken into account

$$c(T)\frac{\partial T(r,t)}{\partial t}=\frac{1}{r^2}\frac{\partial}{\partial r}\left[\lambda(T)r^2\frac{\partial T(r,t)}{\partial r}\right]+Q \tag{1}$$

where $c(T)$ is the volumetric specific heat, $\lambda(T)$ is the thermal conductivity, $Q$ is the source function connected with the latent heat $L$ evolution, while $T$, $r$, $t$ denote temperature, spatial co-ordinate and time.

The source function $Q(r, t)$ can be written in the form

$$Q(r,t) = L\frac{\partial f_S(r,t)}{\partial t} \tag{2}$$

where $f_S$ is the frozen state fraction at the neighbourhood of the point considered.
Let us assume that the solid state fraction is a known function of temperature $f_S = f_S(T)$ from the interval $[T_2, T_1]$ (the end and the beginning of the freezing process, respectively) and then

$$Q(r,t) = L\frac{\partial f_S(r,t)}{\partial t} = L\frac{\mathrm{d} f_S(T)}{\mathrm{d}T}\frac{\partial T(r,t)}{\partial t} \tag{3}$$

So, the equation (1) can be written in the form

$$C(T)\frac{\partial T(r,t)}{\partial t} = \frac{1}{r^2}\frac{\partial}{\partial r}\left[\lambda(T) r^2 \frac{\partial T(r,t)}{\partial r}\right] \tag{4}$$

where

$$C(T) = c(T) - L\frac{\mathrm{d} f_S(T)}{\mathrm{d}T} \tag{5}$$

is called the substitute thermal capacity of intermediate region. The energy equation (4) can be extended on the whole domain considered, because for $T > T_1$: $f_S(T) = 0$, while for $T < T_2$: $f_S(T) = 1$ and then

$$C(T) = \begin{cases} c_1(T), & T > T_1 \\ c_2(T) - L\dfrac{\mathrm{d} f_S(T)}{\mathrm{d}T}, & T_2 \le T \le T_1 \\ c_3(T), & T < T_2 \end{cases} \tag{6}$$

where $c_1$, $c_2$, $c_3$ are the specific heats for unfrozen tissue, intermediate region and frozen sub-domain. In this paper the functions $C(T)$ and $\lambda(T)$ have been constructed on the basis of experimental data quoted in [2].
Summing up, the equation (4) describes the heat transfer processes in the whole conventionally homogeneous volume and this approach is called the fixed domain method [2, 3, 4]. The problem is strongly non-linear - both the parameters $C(T)$ and $\lambda(T)$ are temperature-dependent.

Equation (4) is supplemented by the following boundary conditions

$$\begin{cases} r = R_1: & T(r,t) = T_c \\ r = R_2: & \dfrac{\partial T(r,t)}{\partial r} = 0 \end{cases} \tag{7}$$

and initial condition

$$t = 0: \quad T(r,t) = T_0 \tag{8}$$

where $R_1$ is the radius of cryoprobe, $R_2$ is the conventionally assumed external radius of domain, $T_c$ is the temperature of cryoprobe surface, $T_0$ is the initial temperature of tissue (as a rule $T_0$ =37°C).

## 3. ARTIFICIAL HEAT SOURCE METHOD

In order to use the boundary element method, the linearization of the problem discussed must be done. One of the possibilities which can be taken into account is the supplement of the basic BEM algorithm by the procedure called the artificial heat source method (AHSM) [6, 7, 8]. The AHSM requires the rebuilding of the energy equation and the boundary - initial conditions to the form in which in the place of temperature the Kirchhoff function appears. So, at first we define the physical enthalpy per unit of volume

$$H(T) = \int_{T_r}^{T} C(\mu)\, d\mu \tag{9}$$

and the Kirchhoff function

$$U(T) = \int_{T_r}^{T} \lambda(\mu)\, d\mu \tag{10}$$

where $T_r$ is an arbitrary assumed reference level.
Using these definitions, the energy equation (4) takes a form [4, 8]

$$\Phi(U)\frac{\partial U(r,t)}{\partial t} = \frac{1}{r^2}\frac{\partial}{\partial r}\left[r^2 \frac{\partial U(r,t)}{\partial r}\right] \tag{11}$$

where $\Phi(U)=\mathrm{d}H/\mathrm{d}U$. This function can be determined on the basis of numerical methods and the course of $\Phi(U)$ for the material considered is shown in Fig. 2.
The boundary and initial conditions are transformed immediately, namely

$$\begin{cases} r = R_1: & U(r,t) = U(T_c) \\ r = R_2: & \dfrac{\partial U(r,t)}{\partial r} = 0 \end{cases} \tag{12}$$

and

$$t = 0: \quad U(r,t) = U(T_0) \tag{13}$$

One can notice, that the right-hand side of equation (11) is a linear one. In order to use the boundary element method, which application requires also the constant value of function $\Phi(U)$, the following form of the equation discussed is considered

$$\Phi_0 \frac{\partial U(r,t)}{\partial t} = \frac{1}{r^2} \frac{\partial}{\partial r}\left[ r^2 \frac{\partial U(r,t)}{\partial r} \right] - \Delta\Phi(U) \frac{\partial U(r,t)}{\partial t} \tag{14}$$

where $\Delta\Phi(U) = \Phi(U) - \Phi_0$

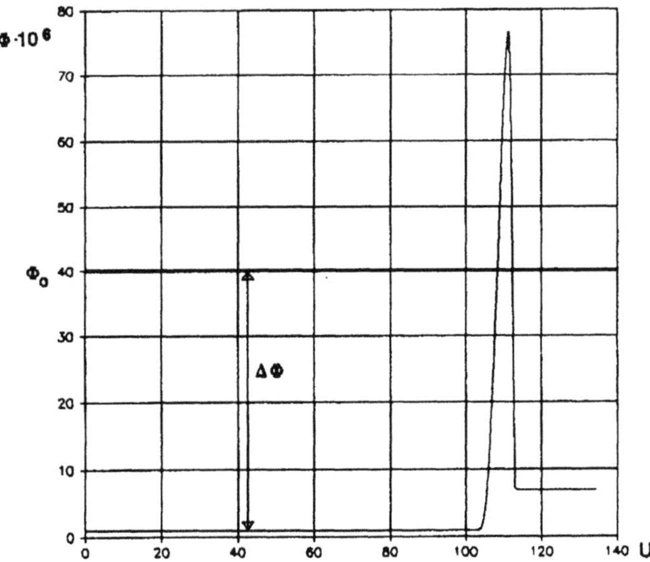


*Figure* 2. The course of function $\Phi(U)$

In other words

$$\Phi_0 \frac{\partial U(r,t)}{\partial t} = \frac{1}{r^2} \frac{\partial}{\partial r} \left[ r^2 \frac{\partial U(r,t)}{\partial r} \right] - Q(r,t) \tag{15}$$

where $Q(r, t)$ is called the artificial heat source. The last equation can be directly solved using the boundary element method, at the same time the local values of $Q$ can be found on the basis of the simple iterative procedure. The iteration process is convergent if $|\Delta\Phi(U)| < \Phi_0$.

# 4. THE BOUNDARY ELEMENT METHOD

In order to solve the problem discussed, the boundary element method is applied. At first, the time grid with step $\Delta t$ is introduced

$$0 = t^0 < t^1 < \ldots < t^{f-1} < t^f < \ldots < t^F < \infty \tag{16}$$

Using the weighted residual formulation to equation (15) one has [10, 11]

$$\int_{t^{f-1}}^{t^f} \int_{R_1}^{R_2} \left[ a \frac{\partial}{\partial r} \left( r^2 \frac{\partial U}{\partial r} \right) - r^2 \frac{\partial U}{\partial t} + a r^2 Q \right] U^*(\xi, r, t^f, t) \, \mathrm{d}r \mathrm{d}t = 0 \tag{17}$$

where $a = 1/\Phi_0$, $U^*$ is the fundamental solution and for the case considered it is a function of the form [5]

$$U^* = \frac{1}{2r\xi\sqrt{\pi a (t^f - t)}} \left[ \exp\left( -\frac{(r-\xi)^2}{4a(t^f - t)} \right) - \exp\left( -\frac{(r+\xi)^2}{4a(t^f - t)} \right) \right] \tag{18}$$

while $\xi$ is a point in which the concentrated heat source is applied.
After the mathematical manipulations the equation (17) takes a form

$$
\begin{aligned}
&U(\xi,t^f)+\left[ar^2\int_{t^{f-1}}^{t^f}U^*(\xi,r,t^f,t)q(r,t)\mathrm{d}t\right]_{r=R_1}^{r=R_2}=\\
&=\left[ar^2\int_{t^{f-1}}^{t^f}q^*(\xi,r,t^f,t)U(r,t)\mathrm{d}t\right]_{r=R_1}^{r=R_2}+\\
&+\int_{R_1}^{R_2}r^2U^*(\xi,r,t^f,t^{f-1})U(r,t^{f-1})\mathrm{d}r+\\
&+\int_{t^{f-1}}^{t^f}\int_{R_1}^{R_2}ar^2U^*(\xi,r,t^f,t)Q(r,t)\mathrm{d}r\mathrm{d}t
\end{aligned}
\tag{19}
$$

where

$$
\begin{aligned}
&q^*(\xi,r,t^f,t)=-\frac{\partial U^*(\xi,r,t^f,t)}{\partial r}\\
&q(r,t)=-\frac{\partial U(r,t)}{\partial r}
\end{aligned}
\tag{20}
$$

We consider the constant elements with respect to time, namely

$$
t\in[t^{f-1},t^f]:\quad U(r,t)=U(r,t^f),\quad q(r,t)=q(r,t^f)
\tag{21}
$$

and then the integration with respect to time is possible in exact way. So

$$
\begin{aligned}
&g(\xi,r)=ar^2\int_{t^{f-1}}^{t^f}U^*(\xi,r,t^f,t)\mathrm{d}t=\\
&=\frac{r}{\xi}\sqrt{\frac{a\Delta t}{\pi\lambda}}\cdot\left[\exp\left(-\frac{(r-\xi)^2}{4a\Delta t}\right)-\exp\left(-\frac{(r+\xi)^2}{4a\Delta t}\right)\right]+\\
&+\frac{r}{2\xi}\left[|r+\xi|\operatorname{erfc}\left(\frac{|r+\xi|}{2\sqrt{a\Delta t}}\right)-|r-\xi|\operatorname{erfc}\left(\frac{|r-\xi|}{2\sqrt{a\Delta t}}\right)\right]
\end{aligned}
\tag{22}
$$

and

$$
\begin{aligned}
h(\xi,r) &= ar^2 \int_{t^{f-1}}^{t^f} q^*(\xi,r,t^f,t)\,\mathrm{d}t = \\
&= \frac{1}{r}g(\xi,r) + \frac{r}{2\xi}\left[\operatorname{sgn}(r-\xi)\operatorname{erfc}\left(\frac{|r-\xi|}{2\sqrt{a\Delta t}}\right) - \operatorname{sgn}(r+\xi)\operatorname{erfc}\left(\frac{|r+\xi|}{2\sqrt{a\Delta t}}\right)\right]
\end{aligned}
\tag{23}
$$

where sgn(·) is the signum function, while erfc(·) = 1 – erf(·), erf(·) is the error function.
Finally, the equation (19) takes a form

$$
\begin{aligned}
& U(\xi,t^f) + g(\xi,R_2)q(R_2,t^f) - g(\xi,R_1)q(R_1,t^f) = \\
& = h(\xi,R_2)U(R_2,t^f) - h(\xi,R_1)U(R_1,t^f) + p(\xi) + z(\xi)
\end{aligned}
\tag{24}
$$

where

$$
p(\xi) = \int_{R_1}^{R_2} r^2 U^*(\xi,r,t^f,t^{f-1})U(r,t^{f-1})\,\mathrm{d}r \tag{25}
$$

and

$$
z(\xi) = \int_{R_1}^{R_2} Q(r,t^f)g(\xi,r)\,\mathrm{d}r \tag{26}
$$

For $\xi \to R_1^+$ and $\xi \to R_2^-$ the equation (24) leads to the system of algebraic equations in the form

$$
\begin{aligned}
& \begin{bmatrix} -g(R_1^+,R_1) & g(R_1^+,R_2) \\ -g(R_2^-,R_1) & g(R_2^-,R_2) \end{bmatrix} \begin{bmatrix} q(R_1,t^f) \\ q(R_2,t^f) \end{bmatrix} = \\
& = \begin{bmatrix} -h(R_1^+,R_1)-1 & h(R_1^+,R_2) \\ -h(R_2^-,R_1) & h(R_2^-,R_2)-1 \end{bmatrix} \begin{bmatrix} U(R_1,t^f) \\ U(R_2,t^f) \end{bmatrix} + \begin{bmatrix} p(R_1)+z(R_1) \\ p(R_2)+z(R_2) \end{bmatrix}
\end{aligned}
\tag{27}
$$

The first step of BEM algorithm consists in a determining of 'missing' boundary values (boundary heat flux for $r = R_1$ and boundary Kirchhoff's function for $r = R_2$). Next, the values of $U$ at internal nodes $\xi \in (R_1, R_2)$ are calculated from the formula

$$U(\xi,t^f) = g(\xi,R_1)q(R_1,t^f) - g(\xi,R_2)q(R_2,t^f) + \\ + h(\xi,R_2)U(R_2,t^f) - h(\xi,R_1)U(R_1,t^f) + p(\xi) + z(\xi) \tag{28}$$

In numerical realization the integrals $p(\xi)$ and $z(\xi)$ are substituted by a sum of integrals over the successive internal cells. In this connection, the local values of source function must be found and in this place the iterative procedure is introduced. The iterative process of source function $Q$ calculation is the following.

1. Transition from $t^0 = 0$ to $t^1 = t^0 + \Delta t$:
- it is assumed that $Q(r_i,t^1) = 0$, at the same time $r_i$ denotes a central point of internal cell distinguished in domain considered,
- for this assumption the Kirchhoff temperature field for whole domain is calculated,
- the local cooling rates $[U(r_i,t^1) - U(r_i,t^0)]/\Delta t$ are estimated,
- a local values of source function $Q(r_i,t^1)$ are corrected,
- the iterative process is stopped if required accuracy is obtained.

2. Transition from $t^{f-1}$ to $t^f$, $f = 2, 3, \ldots, F$:
- it is assumed that $Q(r_i,t^f)$ is equal to the last value of $Q$ found during the previous iterative process (at the point considered),
- for this assumption the Kirchhoff temperature field for whole domain is calculated,
- the local cooling rates $[U(r_i,t^f) - U(r_i,t^{f-1})]/\Delta t$ are estimated,
- a local values of source function $Q(r_i,t^f)$ are corrected,
- the iterative process is stopped if required accuracy is obtained.

As it was mentioned, the iteration process is convergent if $|\Delta\Phi(U)| < \Phi_0$.

## 5. RESULTS OF COMPUTATIONS

The cryoprobes of diameters $2R_1 = 3, 5, 10$[mm] have been considered. The surface temperatures were assumed as –90 °C, –145 °C and –190 °C. The external radius of domain: $R_2 = 25$[mm]. The following input data have been assumed $\Phi_0 = 4\cdot 10^7$, $T_1 = -1$ °C, $T_2 = -8$ °C. Initial temperature of tissue: $T_0 = 37$ °C. The domain of tissue was divided into 50 linear internal cells.
In Figure 3 the kinetics of ice ball generation for different cryoprobe radiuses and different surface temperatures are shown.
The numerical simulations show that dependence between the ice ball radius and time can be expressed in the form of formula

$$r(t) = R_1 + kt^n \tag{29}$$

where

$$k = 4.17188 \cdot 10^{-4} - 7.93231 \cdot 10^{-3} R_1 - 6.54342 \cdot 10^{-7} T_c \tag{30}$$

and

$$n = 0.321448 + 15.62205 R_1 - 2.284 \cdot 10^{-4} T_c \tag{31}$$

In equations (30), (31) $R_1$ [m] denotes the cryoprobe radius, $T_c$ [°C] is the cryoprobe surface temperature. The comparison between this simple mathematical model and results of successive simulations is shown in Figure 4.

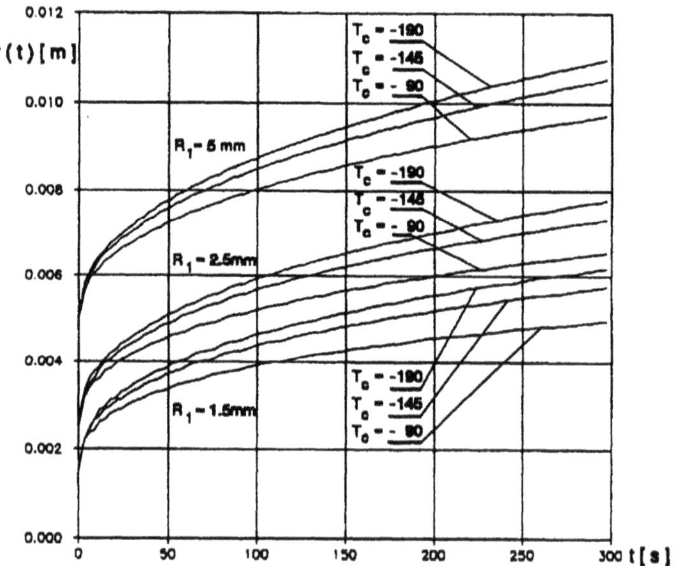


*Figure 3.* The kinetics of freezing process

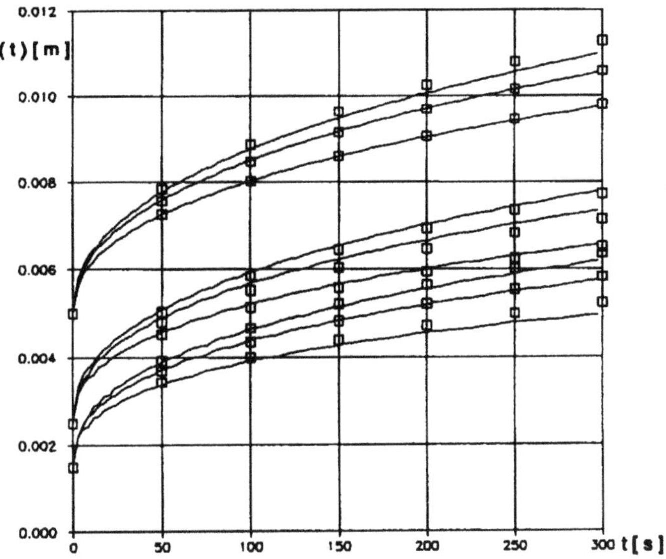


*Figure 4.* The comparison of simulations and formula (29)

So, it seems that the results presented in this paper can be used for prediction of freezing time assuring the required thickness of frozen region layer and also make possible the choice of optimum rate of the process.

## ACKNOWLEDGEMENT

This paper is sponsored by KBN (Grant No 8 T11F 004 19).

## REFERENCES

[1] Budman, H., Shitzer, A. and Dayan, J. (1995) Analysis of the inverse problem of freezing and thawing of a binary solution during cryosurgical processes, *Journal of Biomechanical Engineering*, **117**, 193-202.

[2] Comini, G. and Del Giudice, L. (1976) Thermal aspects of cryosurgery, *Journal of Heat Transfer, Transactions of the ASME*, 543-549.

[3] Idelsohn, S.R., Storti, M.A. and Crivelli, L.A. (1994) Numerical methods in phase change problems, *Archives of Computational Methods in Engineering*, **1**, 49-74.

[4] Mochnacki, B. and Suchy, J.S. (1995) *Numerical Methods in Computations of Foundry Processes*, PFTA, Cracow.

[5] Majchrzak, E. (1998) Solution of non-steady thermal diffusion problem in spherical domains using the boundary element method, *Publ. of the Sil. Techn. Univ.*, **7**, 1998.

[6] Mochnacki, B. (1996) Application of the BEM for numerical modelling of continuous casting, *Computational Mechanics*, **18**, 1, 62-71.

[7] Mochnacki, B. and Suchy, J.S. (1997) Numerical modelling of solidification: the concept of problem linearization, *AFS Transactions*, 96-11, 203-209.

[8] Majchrzak, E. and Mochnacki, B. (1996) The BEM application for numerical solution of non-steady and nonlinear thermal diffusion problems, *Computer Assisted Mechanics and Engineering Sciences*, **3**, 327-346.

[9] Brinck, H. and Werner, J. (1994) Estimation of the Thermal Effect of Blood Flow in a Branching Countercurrent Network Using a Three Dimensional Vascular Model, *Journal of Biomechanical Engineering, Transactions of the ASME,* **116**, 324-330.

[10] Brebbia, C.A., Telles J.C.F. and Wrobel, L.C. (1984) *Boundary Element Techniques*, Springer-Verlag, Berlin & New York.

[11] Majchrzak, E. and Mochnacki, B. (1995) Application of the BEM in the thermal theory of foundry, *Engineering Analysis with Boundary Elements*, **16**, 99-121.

# ANALYSIS OF BIO-HEAT TRANSFER IN THE SYSTEM BLOOD VESSEL - BIOLOGICAL TISSUE

E.MAJCHRZAK & B.MOCHNACKI
*Silesian University of Technology*
*44-100 Gliwice, Konarskiego 18a, Poland*
*Technical University of Czestochowa*
*Dabrowskiego 68, 42-200 Czestochowa, Poland*

Key words:

Abstract: In the paper the mathematical model of heat transfer processes in the system single blood vessel - biological tissue is presented. The mathematical model bases on the Pennes equation determining the temperature field in tissue sub-domain and the equation describing the change of blood temperature along the vessel. The coupling of these equations results from the heat flux continuity on the contact surface between vessel and tissue. In order to take into account the influence of local blood temperature on the capacity of internal heat source in the Pennes equation (the perfusion heat source) the simple iterative procedure is introduced. On the stage of numerical computations the combined algorithm basing on the boundary element method and control volume method is applied. In the final part of the paper the example of numerical simulation is shown.

## 1. MATHEMATICAL DESCRIPTION OF THE PROCESS

A single large blood vessel of radius $R_1$ embedded in a perfused tissue is considered - Fig. 1. The problem of heat transfer in the system is treated as the axially symmetrical one. The position of vessel corresponds to the axis of cylinder. The Pennes bio-heat transfer equation [1, 2] describing the steady temperature field in the tissue sub-domain is of the form

*T. Burczynski (ed.), IUTAM/IACM/IABEM Symposium on Advanced Mathematical and Computational Mechanics Aspects of the Boundary Element Method*, 201–211.

$$\frac{1}{r}\partial_r[r\lambda\partial_r T(r,z)]+\partial_z[\lambda\partial_z T(r,z)]+Q_{perf}+Q_{met}=0 \tag{1}$$

where

$$Q_{perf}=G\,c_b[T_b(z)-T(r,z)] \tag{2}$$

In equations (1) and (2) $\lambda$ is the thermal conductivity of tissue, $G$ [m$^3$ blood / m$^3$ tissue] is the perfusion coefficient, $c_b$ is the specific heat of blood per unit of volume, $T_b(z)$ is blood temperature, $Q_{met}$ [W/m$^3$] is the metabolic heat source. The following equation determines the temperature field in vessel domain [3, 4]

$$c_b w F\frac{\mathrm{d}T_b(z)}{\mathrm{d}z}+\alpha P[T_b(z)-T(R_1,z)]=0 \tag{3}$$

where F is the vessel lateral section, P its periphery, w is the blood velocity in the vessel, α is the heat transfer coefficient between blood and tissue.

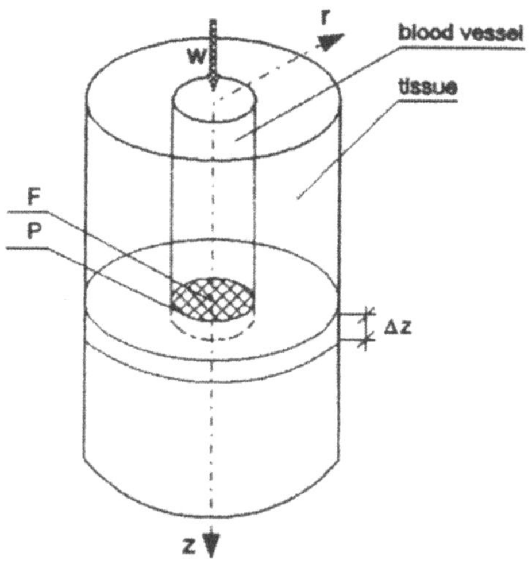


*Figure 1.* The system considered

The boundary conditions for the tissue sub-domain are the following

$$\begin{cases} r = R_1: & q(r,z) = \alpha[T(r,z) - T_b(z)] \\ r = R_2: & q(r,z) = 0 \\ z = 0: & q(r,z) = 0 \\ z = L: & q(r,z) = 0 \end{cases} \tag{4}$$

where $R_2$ is the conventionally assumed external radius of the system, $q(r, z)$ is the normal heat flux ($q = -\lambda \partial_n T$). The initial condition for equation (3) is determined by the initial blood temperature, namely for $z = 0$: $T_b(0) = T_{b0} = 37.5°C$.

## 2. THE METHOD OF NUMERICAL SOLUTION

Let us assume that the tissue thermal conductivity $\lambda$ is a constant value. Denoting $Q_{perf} + Q_{met} = Q(r, z)$ we obtain the following form of equation (1)

$$\lambda \partial_r [r \partial_r T(r,z)] + \lambda \partial_z [r \partial_z T(r,z)] + rQ(r,t) = 0 \tag{5}$$

The integral equation for the problem considered takes a form [5, 6, 7]

$$\begin{aligned} &B(\xi,\eta)T(\xi,\eta) + \int_\Gamma rT^*(\xi,\eta,r,z)q(r,z)\,\mathrm{d}\Gamma = \\ &= \int_\Gamma rq^*(\xi,\eta,r,z)T(r,z)\,\mathrm{d}\Gamma + \iint_\Omega rQ(r,z)T^*(\xi,\eta,r,z)\,\mathrm{d}\Omega \end{aligned} \tag{6}$$

where $(\xi, \eta) \in \Omega$ is the observation point, $T^*(\xi,\eta,r,z)$ is the fundamental solution, $q^*(\xi,\eta,r,z) = -\partial_n T^*(\xi,\eta,r,z)$, $B(\xi,\eta)$ is the coefficient from the scope (0, 1), at the same time [5, 6]

$$T^*(\xi,\eta,r,z) = \frac{1}{\lambda \pi \sqrt{(r+\xi)^2 + (z-\eta)^2}} \mathrm{K}(m) \tag{7}$$

where $\mathrm{K}(\cdot)$ is the elliptic integral of the first type [8], and

$$m = \frac{4r\xi}{(r+\xi)^2 + (z-\eta)^2} \tag{8}$$

The flux $q^*(\xi,\eta,r,z)$ is equal to

$$q^{*}(\xi,\eta,r,z)=\frac{1}{\pi\sqrt{(r+\xi)^{2}+(z-\eta)^{2}}}\cdot \cdot\left\{\frac{1}{2r}[\mathrm{K}(m)-m_{1}\mathrm{E}(m)]\cos\alpha_{r}+m_{2}\mathrm{E}(m)\cos\alpha_{z}\right\} \tag{9}$$

where

$$m_{1}=\frac{\xi^{2}-r^{2}+(z-\eta)^{2}}{(r-\xi)^{2}+(z-\eta)^{2}} \tag{10}$$

and

$$m_{2}=\frac{z-\eta}{(r-\xi)^{2}+(z-\eta)^{2}} \tag{11}$$

In equation (9) E(·) is the elliptic integral of the second type [8], $\cos\alpha_r$, $\cos\alpha_z$ are the directional cosines of normal versor $n$ at the boundary point $(r, z)$.
It should be pointed out that in numerical realization the values of functions K(·) and E(·) are calculated on the basis of approximate formulas presented among others in [8].
In order to solve the equation (6), the boundary $\Gamma$ is divided into $N$ linear boundary elements, while the interior $\Omega$ is divided into $L$ internal cells.
The discrete form of the boundary integral equation is the following

$$B(\xi_i,\eta_i)T(\xi_i,\eta_i)+\sum_{j=1}^{N}\int_{\Gamma_j} rT^{*}(\xi_i,\eta_i,r,z)q(r,z)\,\mathrm{d}\Gamma_j = =\sum_{j=1}^{N}\int_{\Gamma_j} rq^{*}(\xi_i,\eta_i,r,z)T(r,z)\,\mathrm{d}\Gamma_j+\sum_{l=1}^{L}P_{il} \tag{12}$$

where

$$P_{il}=\iint_{\Omega} rQ(r,z)T^{*}(\xi_i,\eta_i,r,z)\,\mathrm{d}\Omega_l \tag{13}$$

We introduce the linear boundary elements, this means

$$(r,z)\in\Gamma_j:\begin{cases}T(r,z)=N^I(u)T_j^I+N^{II}(u)T_j^{II}\\ q(r,z)=N^I(u)q_j^I+N^{II}(u)q_j^{II}\end{cases} \tag{14}$$

at the same time $N^I(u)=(1-u)/2,\ N^{II}(u)=(1+u)/2,\ u\in[-1,1]$ are the shape functions, $T_j^I=T(r_I,z_I),\ T_j^{II}=T(r_{II},z_{II})$ etc. are the temperatures and heat fluxes corresponding to the limits of the element considered.
So

$$\int_{\Gamma_j} r q^*(\xi_i,\eta_i,r,z)T(r,z)\,\mathrm{d}\Gamma_j=h_{ij}^I T_j^I+h_{ij}^{II}T_j^{II} \tag{15}$$

and

$$\int_{\Gamma_j} r T^*(\xi_i,\eta_i,r,z)q(r,z)\,\mathrm{d}\Gamma_j=g_{ij}^I q_j^I+g_{ij}^{II}q_j^{II} \tag{16}$$

where

$$h_{ij}^I=\frac{l_j}{2}\int_{-1}^{1}N^I(N^Ir_I+N^{II}r_{II})q^*(\xi_i,\eta_i,N^Ir_I+N^{II}r_{II},N^Iz_I+N^{II}z_{II})\,\mathrm{d}u \tag{17}$$

$$h_{ij}^{II}=\frac{l_j}{2}\int_{-1}^{1}N^{II}(N^Ir_I+N^{II}r_{II})q^*(\xi_i,\eta_i,N^Ir_I+N^{II}r_{II},N^Iz_I+N^{II}z_{II})\,\mathrm{d}u \tag{18}$$

$$g_{ij}^I=\frac{l_j}{2}\int_{-1}^{1}N^I(N^Ir_I+N^{II}r_{II})T^*(\xi_i,\eta_i,N^Ir_I+N^{II}r_{II},N^Iz_I+N^{II}z_{II})\,\mathrm{d}u \tag{19}$$

$$g_{ij}^{II}=\frac{l_j}{2}\int_{-1}^{1}N^{II}(N^Ir_I+N^{II}r_{II})T^*(\xi_i,\eta_i,N^Ir_I+N^{II}r_{II},N^Iz_I+N^{II}z_{II})\,\mathrm{d}u \tag{20}$$

In formulas (17)-(20) $l_j$ is the length of element $\Gamma_j$.
Assuming the constant internal cells, the integral (13) can be written as follows

$$P_{il} = Q(r_l, z_l) \iint_{\Omega_l} r T^*(\xi_i, \eta_i, r, z) \, \mathrm{d}\Omega_l \tag{21}$$

where

$$Q(r_l, z_l) = Q_{met} + c_b G_b [T_b(z_l) - T(r_l, z_l)] \tag{22}$$

As it was mentioned, this source function must be determined iteratively. The first iteration is done for $Q(r_l, z_l) = Q_{met}$, while the next iterations base on the values of temperatures $T_b(z_l)$ and $T(r_l, z_l)$ from the previous iteration.
It should be pointed out, that all integrals (boundary and internal) are calculated in numerical way using the Gauss method.
So, the equation (12) for the boundary node $(\xi_i, \eta_i)$ can be written in the form

$$B_i T_i + \sum_{k=1}^{M} g_{ik} q_k = \sum_{k=1}^{M} \hat{h}_{ik} T_k + \sum_{l=1}^{L} P_{il} \tag{23}$$

where $M$ is the number of boundary nodes (at the corners of domain considered the double nodes must be introduced). For the single boundary node $k$ we have

$$\begin{aligned} g_{ik} &= g_{ij}^{II} + g_{i,j+1}^{I} \\ \hat{h}_{ik} &= h_{ij}^{II} + h_{i,j+1}^{I} \end{aligned} \tag{24}$$

while for the double boundary node $k$, $k+1$

$$\begin{aligned} g_{ik} &= g_{ij}^{II}, \quad g_{i,k+1} = g_{i,j+1}^{I} \\ \hat{h}_{ik} &= h_{ij}^{II}, \quad \hat{h}_{k,k+1} = h_{i,j+1}^{I} \end{aligned} \tag{25}$$

One can write equation (23) for each '$i$' node, obtaining $M$ equations

$$\sum_{k=1}^{M} g_{ik} q_k = \sum_{k=1}^{M} h_{ik} T_k + \sum_{l=1}^{L} P_{il} \tag{26}$$

where

$$h_{ik}=\begin{cases}\hat{h}_{ik}, & i\neq k\\ -B_i=-\sum_{k\neq i}\hat{h}_{ik}, & i=k\end{cases} \tag{27}$$

Discretization of domain analyzed and its boundary is shown in Figure 2. For the needs of further considerations the equations (26) concerning the nodes along the boundary $\Gamma_1$ (the vessel wall) are considered separately.
So

$$\sum_{k=1}^{S} g_{ik}q_k+\sum_{k=S+1}^{M} g_{ik}q_k=\sum_{k=1}^{M} h_{ik}T_k+\sum_{l=1}^{L} P_{il} \tag{28}$$

where $S$ is the number of nodes corresponding to $\Gamma_1$.

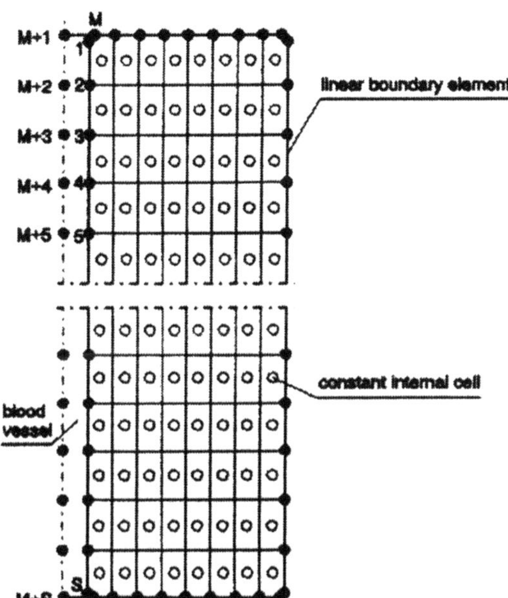


*Figure 2.* The mesh

Let us assume that $M$+1, $M$+2, ..., $M$+$S$ are the collocation points in domain of blood. The control volume method (or the FDM) leads to the following formula determining the values of temperatures at collocation points from the blood sub-domain (equation (3))

$$T_{M+1} = T_{b0}$$
$$T_{M+k} = T_{M+k-1} + A(T_{k-1} - T_{M+k-1}) \tag{29}$$

where $k$ =2, 3, ..., $S$ and $A = 2\alpha\Delta z / c_b w R_1$.
Finally

$$\begin{cases} \sum_{k=1}^{S} g_{ik}\alpha\left[T_k - T_{M+k}\right] + \sum_{k=S+1}^{M} g_{ik}q_k = \sum_{k=1}^{M} h_{ik}T_k + \sum_{l=1}^{L} P_{il}, i = 1,\ldots,M \\ T_{M+1} = T_{b0} \\ AT_{k-1} + (1-A)T_{M+k-1} - T_{M+k} = 0 \end{cases} \tag{30}$$

After the solving of above system of equations, all boundary temperatures and heat fluxes for tissue sub-domain are known. Next, one can find the internal temperatures at the points $(\xi_i, \eta_i)\in\Omega$ using the equation

$$T_i = \sum_{k=1}^{M} h_{ik}T_k - \sum_{k=1}^{M} g_{ik}q_k + \sum_{l=1}^{L} P_{il} \tag{31}$$

On the basis of the internal temperatures in $\Omega$ and the nodal blood temperatures $T_b(z)$, the values of perfusion heat source can be corrected. The iteration process is realized until the obtaining of required accuracy of the numerical solution.

## 3. THE RESULTS OF COMPUTATIONS

The blood vessel of radius $R_1$ = 0.0005 [m] is considered. The external radius of domain is assumed as $R_2 = 10R_1$, while $Z$ = 0.021 [m] (Figure 1). The following input data are take into account [2, 3]: $\lambda$ = 0.5 [W/mK], $Q_{met}$ = 24500 [W/m$^3$], $c_b = 4.134\times10^6$ [J/m$^3$K], $G_b$ = 0.01085[m$^3$/s/m$^3$ tissue], w = 0.08 [m/s], the Nusselt number Nu = $\alpha 2R_l/\lambda$ = 4 ($\alpha$ = 2000 [W/m$^2$ K]), $T_{b0}$ =37°C. The values $Q_{met}$ and $G_b$ correspond to the conditions of effort. In Figures 3, 4 and 5 the profiles of temperature for $z$ =1, 10 and 20 [mm] are shown, at the same time the results of successive iterations are also marked. The number of iterations assuring the good exactness of numerical solution was equal 5.

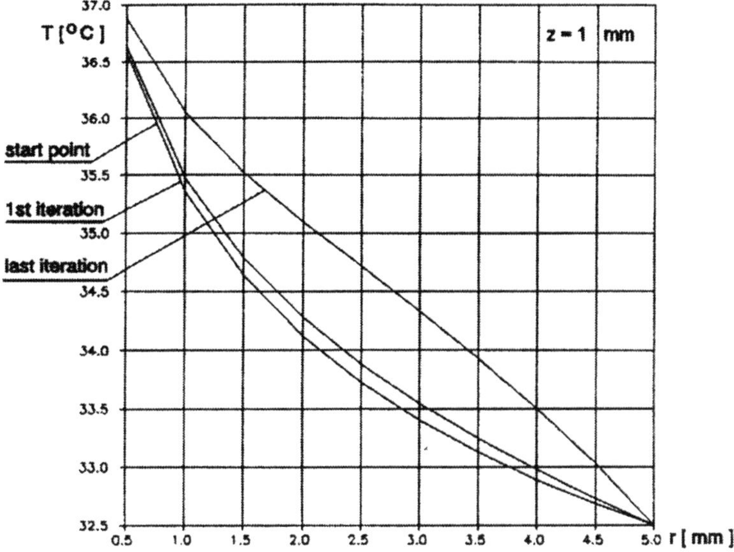


*Figures 3.* Temperature field in radial direction ($z = 0.001$ [m])

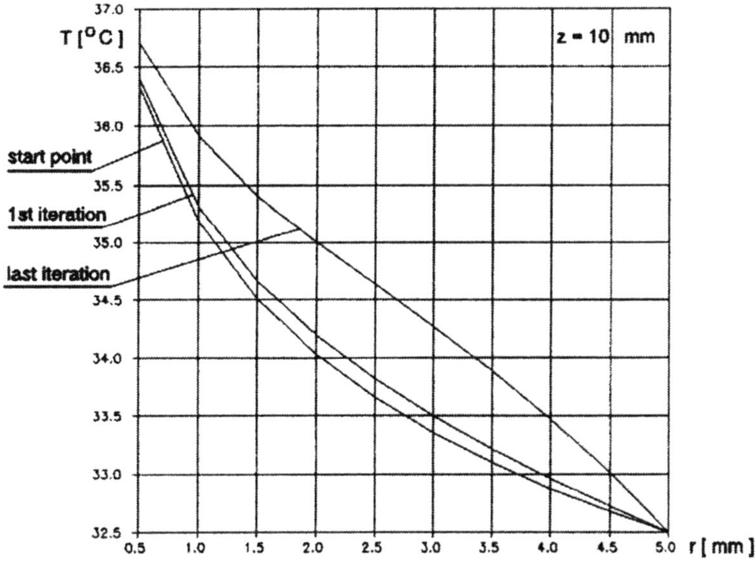


*Figures 4* Temperature field in radial direction ($z = 0.01$ [m])

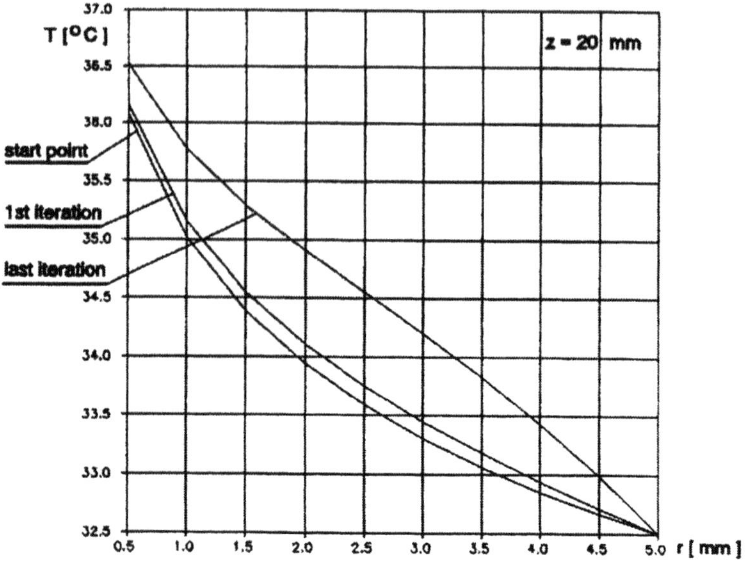


*Figures 4.* Temperature field in radial direction ($z = 0.02$ [m])

In Figure 6 a change of the blood temperature along z axis is shown.

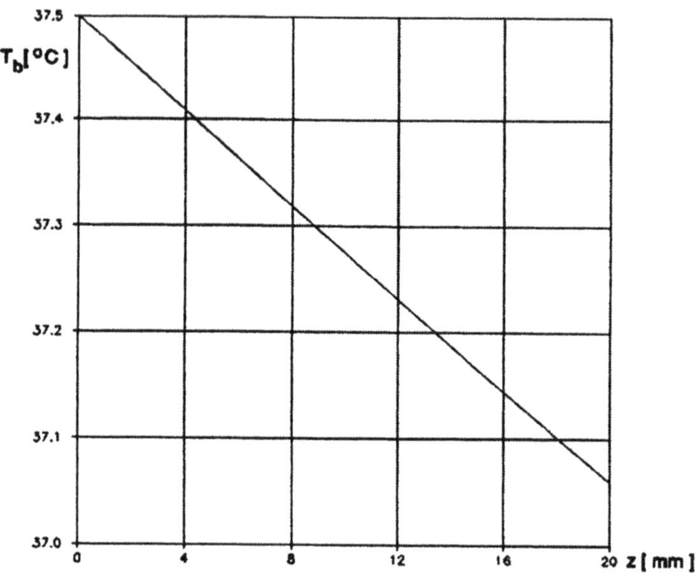


*Figure 6.* Cooling curve of blood in *z* direction

Summing up, the algorithm proposed seems to be quite effective. The composition of the BEM and CVM leads to the resolving system with the small number of unknown parameters (temperatures or heat fluxes). This fact and also the good approximation of boundary conditions resulting from the BEM application decide about the essential improvement of this algorithm over the others possible methods of the problem solution. Additionally, the iteration process is quickly convergent and allows to take into account the real form of the function describing the perfusion heat source.

From the practical view-point the algorithm presented in this paper can be used for numerical analysis of thermal processes proceeding in human body subjected to the extreme climatic conditions (for instance, very low temperatures). In this situation the effect of internal heat sources resulting from the presence of blood vessels has the essential importance for the course of heat transfer in the system biological tissue - environment.

## ACKNOWLEDGEMENT

This research work has been supported by KBN (Grant No 8 T11F 004 19).

## REFERENCES

[1] Brinck, H. and Werner, J. (1984) Estimation of the thermal effect of blood flow in a branching countercurrent network using a three-dimensional vascular model, Journal of Biomechanical Engineering, Transactions of the ASME, 116, pp. 324-330.

[2] Huang, H.W., Chan, L.C. and Roemer, L.B. (1994) Analytical solutions of Pennes bio-heat transfer equation with a blood vessel, Journal of Biomechanical Engineering, Transactions of the ASME, 116, pp. 208-212.

[3] Zhu, L. and Weinbaum, S.(1995) A model for heat transfer from embedded blood vessels in two-dimensional tissue preparations, Journal of Biomechanical Engineering, Transactions of the ASME, 116, pp. 64-73.

[4] Majchrzak, E. and Mochnacki, B. (1998) Application of the BEM for numerical modelling of heat transfer in the system single blood vessel - biological tissue, in: Numerical Methods and Computational Mechanics, University of Miscolc, Hungary, pp. 64-66.

[5] Brebbia, C.A. and Dominiquez, J. (1992) Boundary Elements - An Introductory Course. Comp. Mech. Publications and Mc Graw-Hill Book Co., Southampton and Boston.

[6] Brebbia, C.A., Telles, J.C.F. and Wrobel, L.C. (1984) Boundary Element Techniques. Springer-Verlag, Berlin, Heidelberg, New York, Tokyo.

[7] Majchrzak, E. (1991) Application of the BEM in the thermal theory of foundry, Publ. of the Silesian Technical University, Mechanics, 102, Gliwice.

[8] Abramowitz, M. and Stegun, I.A. (1985) Handbook of Mathematical Functions, Dover Publications, Inc., New York.

# Rigid inclusions in frictionless unilateral contact with the matrix: identification of maximum stiffness structures by means of BEM formulations and sensitivity analyses

V. Mallardo and C. Alessandri
*Department of Architecture, University of Ferrara - via Quartieri 8, 44100 Ferrara*

Key words: Optimisation, Boundary Element, Sensitivity Analysis, Inclusion.

Abstract: The Authors present a Boundary Element procedure for the solution of 2D optimisation problems in the presence of unilateral contact interfaces. The position which a rigid inclusion must occupy within the matrix in order to maximise the structural stiffness of the matrix-inclusion system under prescribed external loads is identified. The matrix is considered linear elastic. A minimisation problem is stated with design variables representing the size and the shape of the inclusion. The cost function is an error function which evaluates the strain energy accumulated by the matrix-inclusion system. The minimisation is performed by using a first-order nonlinear optimisation technique in which the cost function gradient is computed by implicit differentiation. Two numerical examples are presented and discussed. In the first example a comparison between the proposed procedure, the finite difference technique and the analytical strategy is performed, whereas the second example deals with the determination of the position of an internal circular inclusion such as to maximise the structural stiffness of a plane strain plate under a prescribed load.

## 1. INTRODUCTION

In practical engineering problems it might be necessary to make holes in a plate subjected to boundary loads on its mid-plane. Such cavities might be left empty or they might be filled with a different material. In the latter case,

*T. Burczynski (ed.), IUTAM/IACM/IABEM Symposium on Advanced Mathematical and Computational Mechanics Aspects of the Boundary Element Method*, 213–226.

during the deformation process, the initial contact surfaces between matrix and inclusion, as well as the position of the inclusion inside the hole, may change; that may affect considerably the mechanical response of the structure. Moreover, at the early stage of a design process, it might be important to know "a priori" position and size of the hole-inclusion system corresponding to the maximum structural stiffness.

On the basis of their recent experiences [1-3], the Authors intend to develop a two-dimensional BEM numerical procedure to optimise the mechanical response of structures with internal inclusions by maximising their structural stiffness.

In the present paper only one rigid inclusion with a prescribed area is considered; the material of the matrix is supposed linear elastic and the contact between matrix and inclusion is assumed as unilateral and frictionless. Existence and uniqueness of the elastic response of the system to applied loads are demonstrated in [4]. Fundamental results concerning the purely mathematical aspects of the problem and the use of the BEM for its numerical solution are reported in [5] and [6] respectively, whereas a numerical analysis of the contact problem by FEM is given in [7].

The problem is stated as an optimisation or inverse problem where the design variables identify the geometrical configuration of the hole-inclusion system. Moreover, for the presence of the unilateral boundary constraints, the problem is governed by variational inequalities with physical variables defined on the internal boundary only. A suitable cost function (the strain energy) is introduced into the iterative procedure of the inverse problem and minimised in order to provide the geometrical configuration corresponding to the maximum structural stiffness under prescribed loadings. The direct problem is discretised by means of the standard Boundary Element Method (BEM) based on collocations and solved as a Linear Complementarity Problem (LCP) with a nonsymmetric coefficient matrix [8]. The minimisation is performed by using a first-order nonlinear optimisation technique in which the error function gradient is computed by implicit differentiation and LCP. Because of the unilateral contact conditions, the cost function is nonconvex and nondifferentiable and its gradient can be computed along prescribed directions only.

Simple but meaningful examples will be presented and discussed in detail in order to show the applicability of the technique proposed.

## 2. THE IDENTIFICATION PROCEDURE

The "best" shape and position of the hole-inclusion system inside the solid is obtained by minimising the strain energy corresponding to applied loads (Figure 1); it can be defined by evaluating the elastic response of the solid. In order to obtain displacements and tractions on the whole boundary of the solid it is necessary to know the contact pressures and the position of the inclusion inside the hole at each step of the minimisation process.

The problem can be tackled [8] by finding first a stationary point of the following functional:

$$J(\mathbf{u};\lambda) = \frac{1}{2}\iint_{\Omega} C_{ijhk}\varepsilon_{ij}(\mathbf{u})\varepsilon_{hk}(\mathbf{u})\,d\Omega - \int_{\Gamma_t} \mathbf{t}\cdot\mathbf{u}\,d\Gamma + \\ - \int_{\Gamma_i} \lambda(\mathbf{u}\cdot\mathbf{n})\,d\Gamma \tag{1}$$

where $\mathbf{u}$ is an admissible displacement field, $C_{ijhk}$ is Hooke's tensor of elasticity, $\varepsilon_{ij} = 1/2(u_{i,j} + u_{j,i})$ and $\lambda \in H^{-1/2}(\Gamma_i)$. In mechanical terms, this means to consider the solution of the complete problem as the superposition of two partial solutions, the one corresponding to the body constrained at $\Gamma_u$ and loaded along $\Gamma_t$, the other to the body constrained at $\Gamma_u$ and loaded with the contact pressures $\lambda(\mathbf{x})\mathbf{n}(\mathbf{x})$ along $\Gamma_i$. Therefore, the direct problem is solved completely when the contact pressures are evaluated.

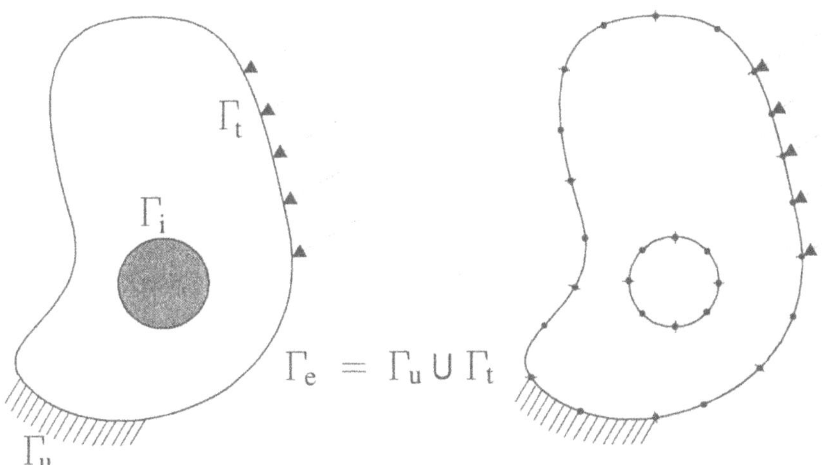


*Figure 1.* Solid with internal inclusion and its discretisation

After discretisation with quadratic boundary elements (Figure 1), the contact pressures between matrix and inclusion can be obtained by solving a LCP where the coefficients of the matrices involved are the BEM solutions of NC+1 elastostatic problems with suitable boundary conditions (NC is the number of nodes on the contact surface of the hole-inclusion system).

If $\mathbf{u}^q_{\Gamma_e}$ and $\mathbf{K}$ collect the nodal displacements at the outer boundary $\Gamma_e$ due, respectively, to the prescribed loads and to the unit normal pressures applied to the nodes of $\Gamma_i$, the final displacements can be evaluated by superposition as:

$$\mathbf{u}_{\Gamma_e} = \mathbf{u}^q_{\Gamma_e} + \mathbf{K}\,\mathbf{p} \tag{2}$$

where $\mathbf{p}$ collects the contact pressures between matrix and inclusion.

The strain energy can be therefore calculated by superposition and by applying Clapeyron's Theorem:

$$E = \frac{1}{2}\int_{\Gamma_e} \mathbf{t}\cdot\mathbf{u}_{\Gamma_e}\, d\Gamma + \frac{1}{2}\int_{\Gamma_i} p\, w\, d\Gamma \tag{3}$$

where $w = \mathbf{u}_{\Gamma_i} \cdot \mathbf{n}$ represents the normal displacement component in the matrix-inclusion contact zone due to the sum of the external load and of the contact pressures applied separately. The contact pressure vector $\mathbf{p}$ and the matrix $\mathbf{K}$ are computed by applying the BEM and the LCP.

The function (3) has to be minimised in the iteration process in order to obtain the position and shape of the hole-inclusion system corresponding to the maximum structural stiffness.

In this paper the internal inclusion, having a prescribed area, is modelled as circular; therefore, the design variables of the problem, collected in the vector $\mathbf{D}$, are the co-ordinates $(X_0, Y_0)$ of the centre of the inclusion. The vector $\mathbf{D}^* = (X_0, Y_0)$ corresponding to the minimum strain energy has to be computed. At every iteration step, i.e. at every position of the hole-inclusion system inside the solid, any first-order minimisation technique requires the evaluation of the strain energy and of its gradient.

Again it can be showed that the derivatives of displacement, traction and contact pressure with respect to the design variables can be found by solving a variational inequality or a minimisation problem [9-10]. By applying the BEM discretisation procedure discussed above for the evaluation of the strain energy, the boundary variational inequality gives a new LCP in which the involved terms are computed by implicit differentiation. The design sensitivity analysis procedure, based on such an implicit differentiation, uses the derivative form of the discretised boundary integral equations.

Design sensitivities $\mathbf{u}_{\Gamma_e,m}$ can be computed by differentiation of (2), i.e.:

$$\mathbf{u}_{\Gamma_e,m} = \mathbf{u}^q_{\Gamma_e,m} + \mathbf{K}\,\mathbf{p}_{,m} + \mathbf{K}_{,m}\,\mathbf{p} \tag{4}$$

where $,m$ denotes the partial derivative with respect to one of the design variables. $\mathbf{K}$ and $\mathbf{p}$ are known in advance, whereas $\mathbf{u}^q_{\Gamma_e,m}$ , $\mathbf{K}_{,m}$ and $\mathbf{p}_{,m}$ have to be evaluated; $\mathbf{u}^q_{\Gamma_e,m}$ and $\mathbf{K}_{,m}$ are computed by implicit differentiation technique, whereas $\mathbf{p}_{,m}$ is obtained by solving another LCP.

The gradient of the strain energy is computed by differentiation of (3):

$$(\nabla E)_i = \frac{\partial E}{\partial D_i} = \frac{1}{2}\int_{\Gamma_e}(\mathbf{t}_{,m}\cdot\mathbf{u} + \mathbf{u}_{,m}\cdot\mathbf{t})\,d\Gamma + \frac{1}{2}\int_{\Gamma_i}(p_{,m}\,w^p + p\,w^p_{,m})\,d\Gamma \tag{5}$$

It can be easily demonstrated that the last term of eq. (5) is always zero.

## 3. THE DIRECT PROBLEM

In this section the evaluation of the contact pressures between matrix and inclusion is discussed. The LCP is applied in order to compute the contact pressures, whereas the BEM is necessary both to evaluate the elements $\mathbf{K}$ and $\mathbf{u}^q_{\Gamma_e}$ of the eq. (2) and to build the vectors and matrices involved in the LCP. The external and internal boundaries of the solid are discretised with quadratic boundary elements; therefore all the involved terms are referred as nodal.

### 3.1 Linear Complementarity Problem

For each hole-inclusion position, the following system of relations must be satisfied:

$$\begin{bmatrix} Z_{11} & \cdots\cdots & Z_{1NC} \\ \vdots & & \vdots \\ Z_{NC1} & \cdots\cdots & Z_{NCNC} \end{bmatrix}\begin{pmatrix} p_1 \\ \vdots \\ p_{NC} \end{pmatrix} - \begin{bmatrix} n_1^1 & n_2^1 \\ \vdots & \vdots \\ n_1^{NC} & n_2^{NC} \end{bmatrix}\begin{pmatrix} u \\ v \end{pmatrix} + \begin{pmatrix} w_1^q \\ \vdots \\ w_{NC}^q \end{pmatrix} \geq 0 \tag{6a}$$

$$\begin{bmatrix} n_1^1 & \dots & n_1^{NC} \\ n_2^1 & \dots & n_2^{NC} \end{bmatrix} \begin{pmatrix} p_1 \\ \vdots \\ p_{NC} \end{pmatrix} = 0 \tag{6b}$$

or :

$$\begin{aligned} \overline{\mathbf{w}}^1 &= \mathbf{Z}\,\mathbf{p} - \mathbf{N}\,\mathbf{r} + \mathbf{w}^q \geq \mathbf{0} \\ \overline{\mathbf{w}}^2 &= \mathbf{N}^T\,\mathbf{p} = \mathbf{0} \end{aligned} \tag{7}$$

where the first NC relations express the unilaterality of the contact between inclusion and matrix, whereas the subsequent two equations impose the equilibrium conditions on the inclusion, i.e. the contact pressures applied on the inclusion must be self-equilibrated.

Matrix **Z**, according to [8], collects the influence coefficients $Z_{ij}$ due to the deformation of the matrix; matrix **N** contains the coefficients of the rigid body motion components of the inclusion inside the hole during the deformation process with $n_1^i, n_2^i$ as the components of the unit vector normal to the inclusion at the node i; **p** collects the contact pressures between inclusion and matrix, **r** collects the components of the rigid body motion in 2-D problems as horizontal and vertical displacements of the centre of the inclusion and $\mathbf{w}^q$ gives the normal displacements of the nodes of the hole due to the applied external loads.

The boundary element method is applied to calculate both $\mathbf{w}^q$ and **Z**.

In order to obtain a LCP, each rigid body motion component is split into a positive and negative part and the equilibrium equations are doubled.

The final LCP to be solved at every iteration step is:

$$\overline{\mathbf{w}} = \begin{pmatrix} \overline{\mathbf{w}}^1 \\ \overline{\mathbf{w}}^2 \end{pmatrix} = \begin{bmatrix} \mathbf{Z} & -\overline{\mathbf{N}} \\ \overline{\mathbf{N}}^T & \mathbf{0} \end{bmatrix} \begin{pmatrix} \mathbf{p} \\ \overline{\mathbf{r}} \end{pmatrix} + \begin{pmatrix} \mathbf{w}^q \\ \mathbf{0} \end{pmatrix} \geq \mathbf{0}$$

$$\begin{pmatrix} \mathbf{p} \\ \overline{\mathbf{r}} \end{pmatrix} \geq \mathbf{0} \qquad \overline{\mathbf{w}} \geq \mathbf{0} \qquad \begin{pmatrix} \mathbf{p} \\ \overline{\mathbf{r}} \end{pmatrix}^T \cdot \overline{\mathbf{w}} = 0 \tag{8}$$

## 3.2 Boundary Integral Equations

The displacement components of a boundary point are given by the following boundary integral equation:

$$c_{ij}(\xi)\, u_j(\xi) + \int_\Gamma T_{ij}(\xi,\mathbf{x})\, u_j(\mathbf{x}) d\Gamma(\mathbf{x}) = \int_\Gamma U_{ij}(\xi,\mathbf{x})\, t_j(\mathbf{x}) d\Gamma(\mathbf{x}) \tag{9}$$

where $\xi$ and $\mathbf{x}$ are the source point and the field point respectively, the first integral represents the Cauchy principle value, $c_{ij}(\xi)$ is a jump term depending on the boundary shape and $T_{ij}(\xi,\mathbf{x})$ and $U_{ij}(\xi,\mathbf{x})$ are the fundamental solutions in 2-D elastostatics [11].

The discretisation of the above equation allows to compute the vector $\mathbf{w}^q$ and the matrix $\mathbf{Z}$ by imposing suitable boundary conditions [8]. The Cauchy principle value integral requires only the continuity of the displacements at the boundary element nodes, and the quadratic elements used guarantee this continuity. The term of the first integral, when the element contains the source node, has a singularity in the form $1/r$ that cannot be computed by standard integration schemes. This term can be computed, together with the $c_{ij}$ term, by the rigid body condition, i.e. the application of a constant displacement to each node of the body, resulting in a traction free system: all the terms are known except the diagonal values including the free term and the unknown singular terms, which are simply computed as the negative sum of the off diagonal elements. The discretised form of the equation (9) is the following:

$$\begin{aligned} c_{ij}(\bar{\xi})u_j(\bar{\xi}) + \sum_{\gamma=1}^{NE}\sum_{n=1}^{3} u_j^n \int_{-1}^{1} T_{ij}(\bar{\xi}\mathbf{x}(\varsigma))\, \phi^n(\varsigma) J^\gamma(\varsigma) d\varsigma = \\ = \sum_{\gamma=1}^{NE}\sum_{n=1}^{3} t_j^n \int_{-1}^{1} U_{ij}(\bar{\xi}\mathbf{x}(\varsigma))\phi^n(\varsigma) J^\gamma(\varsigma)\, d\varsigma \end{aligned} \tag{10}$$

where $\phi^n(\varsigma)$ are the quadratic shape functions of the Jacobian $J(\varsigma)$, $\varsigma$ is a local dimensionless variable and NE is the number of elements on $\Gamma$.

The above equation can be used to set up the system of equations by collocating the source point in every node of the whole boundary. The final system can be written as:

$$\begin{bmatrix} \mathbf{H}_{\Gamma_e\Gamma_e} & \mathbf{H}_{\Gamma_e\Gamma_i} \\ \mathbf{H}_{\Gamma_i\Gamma_e} & \mathbf{H}_{\Gamma_i\Gamma_i} \end{bmatrix} \begin{pmatrix} \mathbf{u}_{\Gamma_e} \\ \mathbf{u}_{\Gamma_i} \end{pmatrix} = \begin{bmatrix} \mathbf{G}_{\Gamma_e\Gamma_e} & \mathbf{G}_{\Gamma_e\Gamma_i} \\ \mathbf{G}_{\Gamma_i\Gamma_e} & \mathbf{G}_{\Gamma_i\Gamma_i} \end{bmatrix} \begin{pmatrix} \mathbf{t}_{\Gamma_e} \\ \mathbf{t}_{\Gamma_i} \end{pmatrix} \tag{11}$$

where the first subscript of every sub-matrix indicates the position of the source node, the latter the position of the integration element. This subdivision of the matrices **H** and **G** in submatrices is useful in the iteration process because a change in the design variables involves a change only in some of the above submatrices.

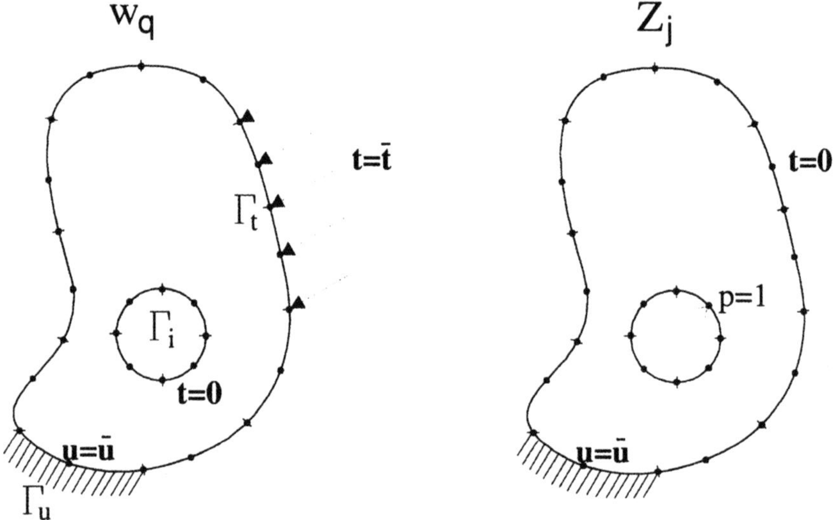


*Figure 2.* Boundary conditions.

After applying the boundary conditions, the system (11) becomes:

$$\mathbf{A}\,\mathbf{x} = \mathbf{b} \tag{12}$$

The boundary conditions for the evaluation of the vector $\mathbf{w}^q$ and of the generic column of the **Z** matrix are shown in Figure 2.

## 4. SENSITIVITY ANALYSIS

Design variables of the problem are the co-ordinates $\mathbf{D}=(X_0,Y_0)$ of the centre of the inclusion; therefore the vector $\mathbf{D}^*$ corresponding to the minimum strain energy has to be computed. At every iteration step, i.e. at every position of the hole-inclusion system inside the solid, any first-order minimisation technique requires the evaluation of the strain energy and of its gradient.

In this section the computation of the design sensitivities is carried out. In this work an implicit differentiation design sensitivity analysis is presented. The procedure uses the derivative form of the discretised integral equations (10) and the linear complementarity technique to obtain the gradient of the strain energy.

## 4.1 Linear Complementarity Problem

At each step of the minimisation process, the direct problem is solved first; therefore, displacements and tractions on the external boundary, contact pressures and relative normal displacements along the hole-inclusion contact line are known. Then, the design sensitivities must be computed.

Differentiation of (6) gives:

$$\begin{cases} \overline{\mathbf{w}}^{1}_{,m} = \mathbf{Z}_{,m}\,\mathbf{p} + \mathbf{Z}\,\mathbf{p}_{,m} - \mathbf{N}_{,m}\,\mathbf{r} - \mathbf{N}\,\mathbf{r}_{,m} + \mathbf{w}^{q}_{,m} = \\ \quad = \mathbf{Y}^{1}_{,m} + \mathbf{Z}\mathbf{p}_{,m} - \mathbf{N}\mathbf{r}_{,m} \\ \overline{\mathbf{w}}^{2}_{,m} = \mathbf{N}^{\mathrm{T}}_{,\mathrm{m}}\,\mathbf{p} + \mathbf{N}^{\mathrm{T}}\,\mathbf{p}_{,\mathrm{m}} = \\ \quad = \mathbf{Y}^{2}_{,m} + \mathbf{N}^{\mathrm{T}}\,\mathbf{p}_{,\mathrm{m}} = \mathbf{0} \end{cases} \tag{13}$$

By means of a static condensation of the zero and sign-free sensitivities of the contact pressures, the LCP can be formulated only in terms of the pressure sensitivities $\overline{\mathbf{p}}_{,m}$ referred to the nodes in contact but with zero contact pressures [9-10]. Besides, by splitting $\mathbf{r}_{,m}$ into its positive and negative parts , i.e. $\overline{\mathbf{r}}_{,m} = (\mathbf{r}^{+}_{,m}, \mathbf{r}^{-}_{,m})$, and by doubling the differentiated equilibrium equations, the new LCP can be written as:

$$\overline{\mathbf{w}}_{,m} = \begin{bmatrix} \mathbf{Z} & -\overline{\mathbf{N}} \\ \overline{\mathbf{N}}^{T} & \mathbf{0} \end{bmatrix} \begin{pmatrix} \overline{\mathbf{p}}_{,m} \\ \overline{\mathbf{r}}_{,m} \end{pmatrix} + \begin{pmatrix} \mathbf{Y}^{1}_{,m} \\ \mathbf{Y}^{2}_{,m} \end{pmatrix} \geq \mathbf{0}$$

$$\begin{pmatrix} \overline{\mathbf{p}}_{,m} \\ \overline{\mathbf{r}}_{,m} \end{pmatrix} \geq \mathbf{0} \qquad \overline{\mathbf{w}}_{,m} \geq \mathbf{0} \qquad \begin{pmatrix} \overline{\mathbf{p}}_{,m} \\ \overline{\mathbf{r}}_{,m} \end{pmatrix}^{T} \cdot \overline{\mathbf{w}}_{,m} = 0 \tag{14}$$

If the solution of the direct problem does not provide nodes in contact with zero contact pressure, the LCP becomes a linear system of equations with $p_{,m}$ of the contact nodes as unknowns.

Vector $\mathbf{w}^q_{,m}$ and matrix $\mathbf{Z}_{,m}$ are evaluated by performing NC+1 times the implicit differentiation of the BEM. The boundary conditions are provided by the differentiation of the boundary conditions of the direct problem, i.e.:

$$\begin{aligned} u_{i,m} &= 0 \quad \text{on } \Gamma_u \\ t_{i,m} &= 0 \quad \text{on } \Gamma_t \\ t_{i,m} &= 0 \quad \text{on } \Gamma_i \end{aligned} \tag{15}$$

The matrix $\mathbf{N}_{,m}$ collects the derivatives of the unit normal on the inclusion nodes and can be evaluated by performing the geometrical derivatives, i.e. by setting the shape of the inclusion.

## 4.2 Boundary Integral Equations

In order to evaluate $\mathbf{w}^q_{,m}$, $\mathbf{Z}_{,m}$ and $\mathbf{u}^q_{\Gamma_e,m}$ , $\mathbf{K}_{,m}$ it is necessary to differentiate the equation (10):

$$\begin{aligned} &\sum_{\gamma=1}^{NE}\sum_{n=1}^{3} u^n_{j,m}\int_{-1}^{1} T_{ij}(\bar{\boldsymbol{\xi}}\mathbf{x}(\varsigma))\phi^n(\varsigma)J^\gamma(\varsigma)d\varsigma + \\ &-\sum_{\gamma=1}^{NE}\sum_{n=1}^{3} t^n_{j,m}\int_{-1}^{1} U_{ij}(\bar{\boldsymbol{\xi}}\mathbf{x}(\varsigma))\phi^n(\varsigma)J^\gamma(\varsigma)d\varsigma = \\ &= -\sum_{\gamma=1}^{NE}\sum_{n=1}^{3} u^n_{j}\int_{-1}^{1} T_{ij,m}(\bar{\boldsymbol{\xi}}\mathbf{x}(\varsigma))\phi^n(\varsigma)J^\gamma(\varsigma)d\varsigma + \\ &+\sum_{\gamma=1}^{NE}\sum_{n=1}^{3} t^n_{j}\int_{-1}^{1} U_{ij,m}(\bar{\boldsymbol{\xi}}\mathbf{x}(\varsigma))\phi^n(\varsigma)J^\gamma(\varsigma)d\varsigma \end{aligned} \tag{16}$$

It must be pointed out that the derivative of the Jacobian and the derivative of the free term are both zero because the size and the shape of the inclusion do not change during the iteration process. Moreover, $U_{ij,m} = 0 = T_{ij,m}$ for every $\bar{\boldsymbol{\xi}},\mathbf{x} \in \Gamma_e$ and $\bar{\boldsymbol{\xi}},\mathbf{x} \in \Gamma_i$.

By collocating at every node of the whole boundary, the following system of equations is obtained:

$$\begin{bmatrix} \mathbf{H}_{\Gamma_e\Gamma_e} & \mathbf{H}_{\Gamma_e\Gamma_i} \\ \mathbf{H}_{\Gamma_i\Gamma_e} & \mathbf{H}_{\Gamma_i\Gamma_i} \end{bmatrix} \begin{pmatrix} \mathbf{u}_{\Gamma_e,m} \\ \mathbf{u}_{\Gamma_i,m} \end{pmatrix} - \begin{bmatrix} \mathbf{G}_{\Gamma_e\Gamma_e} & \mathbf{G}_{\Gamma_e\Gamma_i} \\ \mathbf{G}_{\Gamma_i\Gamma_e} & \mathbf{G}_{\Gamma_i\Gamma_i} \end{bmatrix} \begin{pmatrix} \mathbf{t}_{\Gamma_e,m} \\ \mathbf{t}_{\Gamma_i,m} \end{pmatrix} = \\ - \begin{bmatrix} \mathbf{0} & \mathbf{H}_{\Gamma_e\Gamma_i,m} \\ \mathbf{H}_{\Gamma_i\Gamma_e,m} & \mathbf{0} \end{bmatrix} \begin{pmatrix} \mathbf{u}_{\Gamma_e} \\ \mathbf{u}_{\Gamma_i} \end{pmatrix} + \begin{bmatrix} \mathbf{0} & \mathbf{G}_{\Gamma_e\Gamma_i,m} \\ \mathbf{G}_{\Gamma_i\Gamma_e,m} & \mathbf{0} \end{bmatrix} \begin{pmatrix} \mathbf{t}_{\Gamma_e} \\ \mathbf{t}_{\Gamma_i} \end{pmatrix} \tag{17}$$

where $\mathbf{u} = (\mathbf{u}_{\Gamma_e}, \mathbf{u}_{\Gamma_i})$ and $\mathbf{t} = (\mathbf{t}_{\Gamma_e}, \mathbf{t}_{\Gamma_i})$ collect the solution of the direct problem at each iteration step. After applying the derivatives (15) of the boundary conditions, the above system of equations can be re-written in a more compact form as:

$$\mathbf{A}\,\mathbf{x}_{,m} = -\mathbf{A}_{,m}\,\mathbf{x} + \mathbf{b}_{,m} \tag{18}$$

It is worth noting that the above equation has the same left-hand-side matrix **A** as equation (12). The LU decomposition of **A**, obtained for the solution of the direct problem, can be saved and re-used for the solution of system (18), with a considerable reduction of the computational effort.

The right hand side of system (17) does not have diagonal terms; therefore, no particular attention must be paid for the evaluation of the singular integrals as was done for the direct problem.

## 5. TEST EXAMPLE

In this section two numerical examples are presented. In both of them a plate with Young's modulus E = 1920000, Poisson's ratio $\nu$ = 0.2, all in compatible units with the geometrical dimensions, is considered.

In the first example the displacement and contact pressure derivatives, computed by using the procedure described previously, are compared with the results obtained by employing the forward finite difference formula and with the analytical values reported in [12]. The results are tabulated in Tables 1-2 where PP represents the proposed procedure (derivative form of the discretised boundary integral equation plus LCP) and the values 0.05, ...,0.0001 represent the step sizes for the finite difference approximations used to compute the derivatives.

The numerical accuracy of the finite difference formula depends on the step size used. In order to increase such an accuracy it is necessary to have a small step size; however, it is worth noting that as the step size decreases, the truncation error increases. That leads to relatively large errors in the approximation, unless the computation is performed with high computational precision.

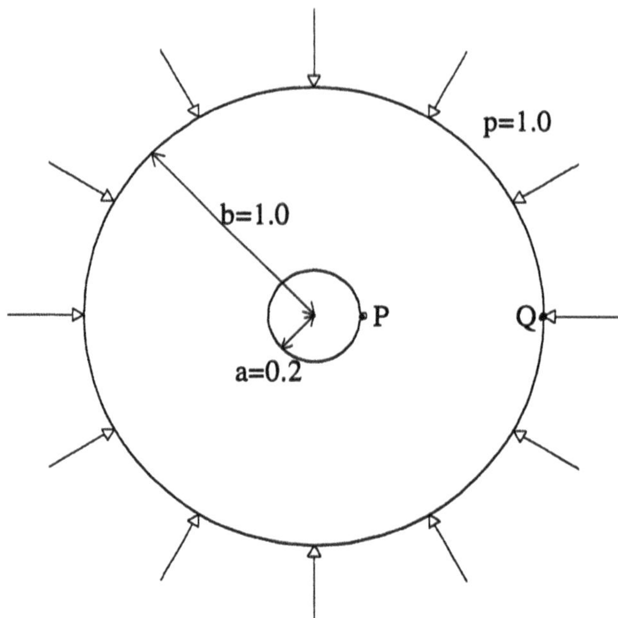


*Figure 3*. Elastic ring under external pressure.

The first example (Figure 3) deals with a plane stress elastic ring having an internal radius, $a$, an external radius, $b$, and subjected to unit external pressures.

Analytical expressions for radial displacement and contact pressure derivatives with respect to the internal radius can be obtained [12]; the former are evaluated at the two points P and Q, whereas the latter are constant on the whole contact line.

The results are collected in Table 1 and in Table 2 when the internal radius is 0.2.

*Table 1*. Derivatives of the horizontal displacement with respect to the hole radius normalised with respect to Young's modulus.

| Point | Analytical | PP | 0.05 | 0.01 | 0.001 | 0.0001 |
|---|---|---|---|---|---|---|
| P | -0.8681 | -0.8680 | -0.9990 | -0.8935 | -0.8705 | -0.8568 |
| Q | -2.2569 | -2.2568 | -2.3332 | -2.2708 | -2.2582 | -2.1980 |

*Table 2*. Derivatives of the contact pressure with respect to the radius of the hole.

| Analytical | PP | 0.05 | 0.01 | 0.001 | 0.0001 |
|---|---|---|---|---|---|
| -0.3662 | -0.3670 | -0.4067 | -0.3745 | -0.3690 | -0.3610 |

The second example deals with the determination of the position of an internal circular inclusion, with radius $R_{inc} = 0.1$, such as to maximise the structural stiffness of a plane strain plate with dimension 1x1, under a prescribed load.

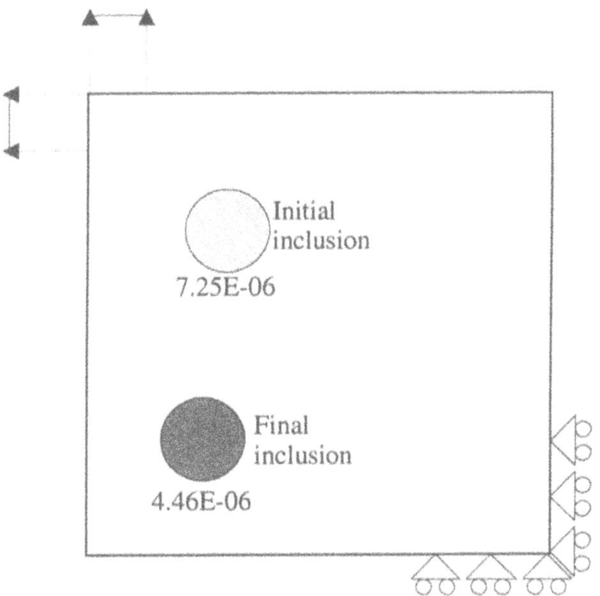


*Figure 4.* Test example

Figure 4 shows the guess position and the final position of the inclusion at the end of the iteration process; the values reported under each inclusion are the strain energies corresponding to those positions.

# 6. REFERENCES


1. Mallardo V., Aliabadi M.H. (1998) A BEM sensitivity and shape identification analysis for acoustic scattering in fluid-solid problems, Int. J. Num. Meth. Eng. **41**, 1527-1541.
2. Mallardo V., Aliabadi M.H. (1998) Boundary Element Method for acoustic scattering in fluid-fluidlike and fluid-solid problems, J. Sound and Vibrations **216** (3), 413-434.
3. Alessandri C., Mallardo V. (1999) Crack identification in two-dimensional unilateral contact mechanics with the boundary element method, in press on Comput. Mech..
4. Mikelic A., Shillor M., Tapiéro R. (1998) Homogenization of an elastic material with inclusions in frictionless contact, Mathl. Comput. Modelling **28**(4-8), 287-307.
5. Ekeland I., Temam R. (1974) *Analyse convexe et problèmes variationels*, Paris, Dunod.
6. Bui H.D. (1992) Mathematical Foundations of BEM Formulations for Static and Dynamic Analysis, in Advanced dynamic Analysis by BEM, 7, P.K. Banerjee, S. Kobayashi, Elsevier.

7. Kikuchi N., Oden J.T. (1988) *Contact problems in Elasticity: A study of Variational Inequalities and Finite Elements Methods*, SIAM, Philadelphia.
8. Alliney S., Tralli A., Alessandri C. (1990) Boundary variational formulations and numerical solution technique for unilateral contact problem, Comput. Mech. **6**, 247-257.
9. Alessandri C., Tralli A. (1995) Sensitivity analisys of fibre-reinforced composites with interphase unilateral constraints, IUTAM Symposium on "Anisotropy, Inhomogeneity and Nonlinearity in Solid Mechanics", Kluwer, 97-102.
10. Alessandri C., Tralli A. (1995) Sensitivity analysis for unilateral contact problems: Boundary integral formulations and B.E.M. discretisations, Comput. Mech. **15**, 287-300.
11. Brebbia C.A., Dominguez J. (1992) *Boundary elements: an introductory course*, C.M.P., Southampton, McGraw Hill, New York.
12. Timoshenko S.P., Goodier J.N. (1982) *Theory of Elasticity*, McGraw-Hill, Singapore.

# Numerical Model of Cast Composite Solidification

B. MOCHNACKI & R. SZOPA
*Technical University of Częstochowa*
*Dąbrowskiego 68, 42-200 Częstochowa, Poland*

Key words: cast composite, boundary element method, phase change

Abstract: In the paper the thermal interactions between the single leaden particle and its surrounding (Al-Si alloy) are analyzed. In the domain of cast composite the control volume containing the leaden particle (a central part of this volume) and spherical shell (a metal matrix) are distinguished. The external radius of volume considered results from the fraction of particles in domain of cast composite. In a stage of numerical realization the boundary element method using discretization in time is applied. The basic algorithm concerning only the heating and cooling processes in non-homogeneous spherical domain is supplemented by the procedures which allow (on the stage of numerical computations) to take into account the course of melting and solidification processes.

## 1. GOVERING EQUATIONS

At first, the following boundary-initial problem will be discussed. The spherical shell of dimension $R_0 < r < R_1$ and initial temperature $T_{10}$ is in thermal contact with the second spherical shell of dimension $R_1 < r < R_2$ and initial temperature $T_{20}$. On the contact surface the continuity of temperature and heat flux is assumed.
The transient temperature field of domain considered is described by the system of partial differential equations in the form

$$c_e \frac{\partial T_e(r,t)}{\partial t} = \frac{\lambda_e}{r^2} \frac{\partial}{\partial r}\left[ r^2 \frac{\partial T_e(r,t)}{\partial r} \right], \quad e = 1,2 \tag{1}$$

*T. Burczynski (ed.), IUTAM/IACM/IABEM Symposium on Advanced Mathematical and Computational Mechanics Aspects of the Boundary Element Method*, 227–236.

where $c_e$ is the volumetric specific heat, $\lambda_e$ is the thermal conductivity, $T$, $r$, $t$ denote temperature, spatial co-ordinate and time.
On the contact surface $r=R_1$ the continuity condition in the following form

$$r = R_1 : \begin{cases} -\lambda_1 \dfrac{\partial T_1(r,t)}{\partial r} = -\lambda_2 \dfrac{\partial T_2(r,t)}{\partial r} \\ T_1(r,t) = T_2(r,t) \end{cases} \tag{2}$$

is assumed, while for $r=R_2$:

$$r = R_2 : q(r,t) = -\lambda_2 \frac{\partial T_2(r,t)}{\partial r} = 0 \tag{3}$$

The same type of condition is assumed for $r=R_0$. The radius $R_0$ corresponds to the radius of a very small hole in the centre of leaden particle and it is introduced in order to avoid the singularities in mathematical model.
The initial condition

$$t = 0 : T_1(r,0) = T_{10} \;,\; T_2(r,0) = T_{20} \tag{4}$$

is also known.

## 2. THE BEM USING DISCRTIZATION IN TIME

We introduce the time grid defined as follows

$$0 = t^0 < t^1 < t^2 < \ldots < t^{f-1} < t^f < \ldots < t^F < \infty \tag{5}$$

and $\Delta t = t^f - t^{f-1}$ is a time step.
In the simplest version of the BEM using discretization in time the energy equations (1) for transition $t^{f-1} \rightarrow t^f$ are transformed to the form

$$\frac{T_e(r,t^f) - T_e(r,t^{f-1})}{\Delta t} = \frac{a_e}{r^2} \frac{\partial}{\partial r}\left[ r^2 \frac{\partial T_e(r,t^f)}{\partial r} \right], e = 1,2 \tag{6}$$

where $a_e = \lambda_e / c_e$. In other words

$$\frac{\partial}{\partial r}\left[ r^2 \frac{\partial T_e(r,t^f)}{\partial r} \right] - \frac{r^2}{a_e \Delta t}\left[ T_e(r,t^f) - T_e(r,t^{f-1}) \right] = 0 \tag{7}$$

The weighted residual method criterion [1, 2, 3] leads to the following formula

$$\int_{R_{r-1}}^{R_r} \left\{ \frac{\partial}{\partial r}\left[ r^2 \frac{\partial T_e(r,t^f)}{\partial r}\right] - \frac{r^2}{a_e \Delta t}\left[T_e(r,t^f) - T_e(r,t^{f-1})\right]\right\} \cdot T_e^*(\xi,r)\,\mathrm{d}r = 0 \tag{8}$$

where $\xi$ is a point in which a concentrated heat source is applied, $T_e^*(\xi, r)$ are the fundamental solutions for the problem considered and they are the functions of the form [4, 5]

$$T_e^*(\xi,r) = \frac{\sqrt{a_e \Delta t}}{2r\xi}\left[\exp\left[-\frac{|x-\xi|}{\sqrt{a_e\Delta t}}\right] - \exp\left[-\frac{|x+\xi|}{\sqrt{a_e \Delta t}}\right]\right] \tag{9}$$

It should be pointed out that the formulae (9) can be obtained by the integration of fundamental solution for 3D problem and objects oriented in rectangular co-ordinate system.

The fundamental solution fulfills the following equation

$$\frac{\partial}{\partial r}\left[r^2 \frac{\partial T_e^*(\xi,r)}{\partial r}\right] - \frac{r^2}{a_e \Delta t} T_e^*(\xi,r) = -\delta(\xi,r) \tag{10}$$

where $\delta(\xi, r)$ is the Dirac function.

Integrating twice by parts the first component of WRM equation and taking into account the properties of fundamental solutions one has

$$\begin{aligned} T_e(\xi,t^f) + \left[\frac{r^2}{\lambda_e} T_e^*(\xi,r) q_e(r,t^f)\right]_{R_{r-1}}^{R_r} = \\ = \left[\frac{r^2}{\lambda_e} q_e^*(\xi,r) T_e(\xi,t^f)\right]_{R_{r-1}}^{R_r} + \frac{1}{a_e \Delta t}\int_{R_{r-1}}^{R_r} r^2 T_e^*(\xi,r) T_e(\xi,t^{f-1})\,\mathrm{d}r \end{aligned} \tag{11}$$

where

$$q_e(r,t^f) = -\lambda_e \frac{\partial T_e(r,t^f)}{\partial r}, \quad q_e^*(r,t^f) = -\lambda_e \frac{\partial T_e^*(r,t^f)}{\partial r} \tag{12}$$

Above equations can be written as follows

$$\begin{aligned} T_e(\xi,t^f) + g_e(\xi,R_e) q_e(R_e,t^f) - g_e(\xi,R_{e-1}) q_e(R_{e-1},t^f) = \\ = h_e(\xi,R_e) T_e(R_e,t^f) - h_e(\xi,R_{e-1}) T_e(R_{e-1},t^f) + p_e(\xi) \end{aligned} \tag{13}$$

where

$$g_e(\xi,r)=\frac{r^2}{\lambda_e}T_e^*(\xi,r) \tag{14}$$

and

$$h_e(\xi,r)=\frac{r^2}{\lambda_e}q_e^*(\xi,r) \tag{15}$$

while

$$p_e(\xi)=\frac{1}{a_e\Delta t}\int_{R_{e-1}}^{R_e} r^2T_e^*(\xi,r)T_e(\xi,t^{f-1})\,\mathrm{d}r \tag{16}$$

For $\xi\to R_{e-1}^+$ and $\xi\to R_e^-$ one obtains the equations

$$\begin{bmatrix} g_e(R_{e-1}^+,R_{e-1}) & g_e(R_{e-1}^+,R_e)\\ g_e(R_e^-,R_{e-1}) & g_e(R_e^-,R_e)\end{bmatrix}\begin{bmatrix} q_e(R_{e-1},t^f)\\ q_e(R_e,t^f)\end{bmatrix}= \\ =\begin{bmatrix} h_e(R_{e-1}^+,R_{e-1})-1 & h_e(R_{e-1}^+,R_e)\\ h_e(R_e^-,R_{e-1}) & h_e(R_e^-,R_e)-1\end{bmatrix}\begin{bmatrix} T_e(R_{e-1},t^f)\\ T_e(R_e,t^f)\end{bmatrix}+\begin{bmatrix} p_e(R_{e-1})\\ p_e(R_e)\end{bmatrix} \tag{17}$$

This system of equations can be written in the form

$$\begin{bmatrix} g_{11}^e & g_{12}^e\\ g_{21}^e & g_{22}^e\end{bmatrix}\begin{bmatrix} q_e(R_{e-1},t^f)\\ q_e(R_e,t^f)\end{bmatrix}=\begin{bmatrix} h_{11}^e & h_{12}^e\\ h_{21}^e & h_{22}^e\end{bmatrix}\begin{bmatrix} T_e(R_{e-1},t^f)\\ T_e(R_e,t^f)\end{bmatrix}+\begin{bmatrix} p_1^e\\ p_2^e\end{bmatrix} \tag{18}$$

The final system for two-layers domain results from the coupling of these equations by the continuity conditions given for $r=R_1$.
The continuity condition for $r=R_1$ can be written in the form

$$r=R_1:\begin{cases} q_1(r,t^f)=q_2(r,t^f)=q(R_1,t^f)\\ T_1(r,t^f)=T_2(r,t^f)=T(R_1,t^f)\end{cases} \tag{19}$$

We put this equations to the resolving systems for $e=1,2$

$$\begin{bmatrix} g_{11}^1 & g_{12}^1\\ g_{21}^1 & g_{22}^1\end{bmatrix}\begin{bmatrix} q_1(R_0,t^f)\\ q(R_1,t^f)\end{bmatrix}=\begin{bmatrix} h_{11}^1 & h_{12}^1\\ h_{21}^1 & h_{22}^1\end{bmatrix}\begin{bmatrix} T_1(R_0,t^f)\\ T(R_1,t^f)\end{bmatrix}+\begin{bmatrix} p_1^1\\ p_2^1\end{bmatrix} \tag{20}$$

and

$$\begin{bmatrix} g_{11}^2 & g_{12}^2 \\ g_{21}^2 & g_{22}^2 \end{bmatrix} \begin{bmatrix} q(R_1,t^f) \\ q_2(R_2,t^f) \end{bmatrix} = \begin{bmatrix} h_{11}^2 & h_{12}^2 \\ h_{21}^2 & h_{22}^2 \end{bmatrix} \begin{bmatrix} T(R_1,t^f) \\ T_2(R_2,t^f) \end{bmatrix} + \begin{bmatrix} p_1^2 \\ p_2^2 \end{bmatrix} \tag{21}$$

The well-ordered systems can be written in the form

$$\begin{bmatrix} -h_{11}^1 & -h_{12}^1 & g_{12}^1 \\ -h_{21}^1 & -h_{22}^1 & g_{22}^1 \end{bmatrix} \begin{bmatrix} T_1(R_0,t^f) \\ T(R_1,t^f) \\ q(R_1,t^f) \end{bmatrix} = \begin{bmatrix} -g_{11}^1 q_1(R_0,t^f) + p_1^1 \\ -g_{21}^1 q_1(R_0,t^f) + p_2^1 \end{bmatrix} \tag{22}$$

and

$$\begin{bmatrix} -h_{11}^2 & g_{11}^2 & -h_{12}^2 \\ -h_{21}^2 & g_{21}^2 & -h_{22}^2 \end{bmatrix} \begin{bmatrix} T(R_1,t^f) \\ q(R_1,t^f) \\ T_2(R_2,t^f) \end{bmatrix} = \begin{bmatrix} -g_{12}^2 q_2(R_2,t^f) + p_1^2 \\ -g_{22}^2 q_2(R_2,t^f) + p_2^2 \end{bmatrix} \tag{23}$$

Finally we obtain the following system of equations (for $r = R_0$ and $r = R_2$ the adiabatic conditions are taken into account)

$$\begin{bmatrix} -h_{11}^1 & -h_{12}^1 & g_{12}^1 & 0 \\ -h_{21}^1 & -h_{22}^1 & g_{22}^1 & 0 \\ 0 & -h_{11}^2 & g_{11}^2 & -h_{12}^2 \\ 0 & -h_{21}^2 & g_{21}^2 & -h_{22}^2 \end{bmatrix} \begin{bmatrix} T_1(R_0,t^f) \\ T(R_1,t^f) \\ q(R_1,t^f) \\ T_2(R_2,t^f) \end{bmatrix} = \begin{bmatrix} p_1^1 \\ p_2^1 \\ p_1^2 \\ p_2^2 \end{bmatrix} \tag{24}$$

The knowledge of boundary values for $r = R_0$ , $r = R_1$ and $r = R_2$ allows to find the internal temperatures at time $t^f$ using the equation

$$\begin{aligned} T_e(\xi,t^f) = {} & g_e(\xi,R_{e-1}) q_e(R_{e-1},t^f) - g_e(\xi,R_e) q_e(R_e,t^f) + \\ & + h_e(\xi,R_e) T_e(R_e,t^f) - h_e(\xi,R_{e-1}) T_e(R_{e-1},t^f) + p_e(\xi) \end{aligned} \tag{25}$$

## 3. THE PARTICLE MELTING PROCESS

In order to simulate the melting process proceeding in the particle sub-domain the temperature recovery method [6] has been used. Let us assume that the thermophysical parameters of liquid and solidified part of the lead particle are constant and equal. The 'reserve' of temperature $\theta$ is defined as the quotient of the latent heat $L_1$ (related to an unit of volume) to the thermal capacity $c_1$, this means

$$\theta = \frac{L_1}{c_1} \tag{26}$$

If the node $r_i$ belongs to the particle sub-domain, then at the moment $t=0$ the temperature at this point corresponds to the initial temperature as well as the temperature reserve results from above equation.

On the basis of the optional numerical method we find a discrete temperature field at the set of points $r_i$ for successive levels of time. If during the interval $\Delta t = t^f - t^{f-1}$ the temperature $T_i^f$ at node $r_i$ rises above the solidification point, then it is assumed that the temperature at this node is equal to $T_{cr}$ and the reserve of temperature must be decreased, namely $\theta_i^f = \theta_i - \Delta\theta_i^f$, where $\Delta\theta_i^f = T_i^f - T_{cr}$. So, the temperature field obtained at time $t^f$ is corrected in following way:

*i.* For the nodes in which $T_i^f < T_{cr}$, the temperature reserve $\theta_i$ is untouched and equal to its initial value. The calculated temperature $T_i^f$ is, of course, accepted.

*ii.* For the nodes in which $T_i^{f-1} < T_{cr}$ and $T_i^f > T_{cr}$ it is assumed that $T_i^f = T_{cr}$ and the TRM procedure is initiated.

*iii.* For the nodes in which $T_i^{f-1} = T_{cr}$, $T_i^f > T_{cr}$ and $\theta_i^f > 0$ it is assumed that $T_i^f = T_{cr}$ and the temperature reserve is decreased according the formula: $\theta_i^f = \theta_i^{f-1} - (T_i^f - T_{cr})$.

*iv.* For the nodes in which $T_i^{f-1} \geq T_{cr}$ and $\theta_i^f \leq 0$ the obtained value of temperature is accepted.

Corrected in this way temperature field illustrates the thermal state in the particle sub-domain at the moment $t^f$, as well as this constitutes a pseudo-initial condition for the next step of computations.

# 4. THE SOLIDIFICATION OF METAL MATRIX

The concept of temperature correction resulting from temperature-dependent thermal capacity was presented by Szargut and Mochnacki [7], and next by Hong, Umeda and Kimura [8]. The most general variant of temperature field correction method was presented by Mochnacki [9]. The same results of temperature field correction assures the simpler (from the numerical point of view) procedure proposed by Szopa [4]. The procedure is very useful in the case of numerical modelling of alloys solidification because the thermal effect of phase change (which proceeds in the interval of temperature) can be taken into account by an introduction to the considerations a parameter called the substitute thermal capacity [10]. This approach is known as the one domain method [11].

The essence of the method consists in the correction of the local temperature values found for transition $t^{f-1} \rightarrow t^f$. So, the method can complete the algorithms for which the

time discretization is introduced.
The interval $[T^{\infty}, T_{10}]$ (ambient temperature, initial temperature of metal matrix) is divided into subintervals for which one assumes the constant value of thermal capacity. The limits of subinterval $[U_{m+1}, U_m]$ correspond to 'phase' $m$ - Fig. 1. Additionally the 'basic phase' (e.g. the hottest one) is introduced.

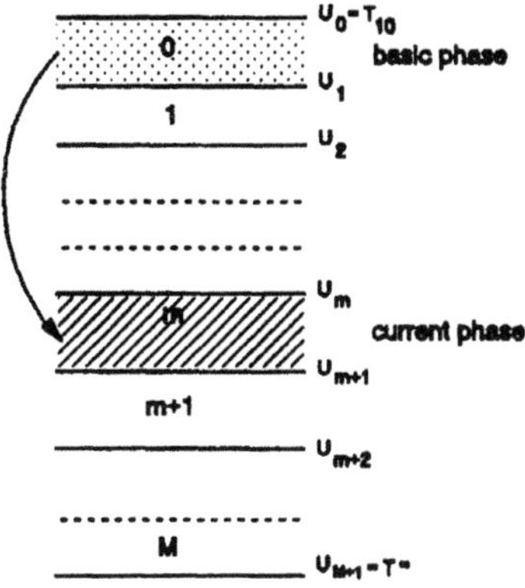


*Figure 1.* The phase corresponding to the interval $[T^{\infty}, T_{10}]$

The first stage of procedure reduces to the transformation of the results obtained for time $t^f$ to phase $m$ corresponding to temperature $T(r_i, t^{f-1})$. From the balance

$$c_B\left[T(r_i,t^{f-1})-T(r_i,t^f)\right]=c_m\left[T(r_i,t^{f-1})-\hat{T}(r_i,t^f)\right] \tag{27}$$

where $c_B$ is the basic thermal capacity, one obtains

$$\hat{T}(r_i,t^f)=T(r_i,t^f)-\frac{c_B}{c_m}\left[T(r_i,t^{f-1})-T(r_i,t^f)\right] \tag{28}$$

If the corrected temperature at point $r_i$ for time $t^f$ belongs to the limits of phase $m$ then the result is accepted, while for the case $U_{m+1}<T(r_i, t^f)<U_{m+2}$ one has

$$\check{T}(r_i,t^f)=U_{m+1}-\frac{c_m}{c_{m+1}}\left[U_{m+1}-\hat{T}(r_i,t^f)\right] \tag{29}$$

If the corrected temperature at point $r_i$ for time $t^f$ belongs to the interval $[U_{m+2}, U_{m+3}]$ then

$$\check{T}(r_i,t^f)=U_{m+2}-\frac{c_m}{c_{m+2}}\left[U_{m+1}-\hat{T}(r_i,t^f)\right]+\frac{c_{m+1}}{c_{m+2}}\left[U_{m+1}-U_{m+2}\right] \text{ etc.} \tag{30}$$

## 5. EXAMPLES OF NUMERICAL COMPUTATIONS

Particles of diameters from 1 [mm] to 2.5 [mm] are considered. The external radius of control volume (2.2 mm) results from the assumption concerning the volumetric fraction of particles and metal matrix in the domain of the cast composite. The various initial temperatures can be also taken into account.

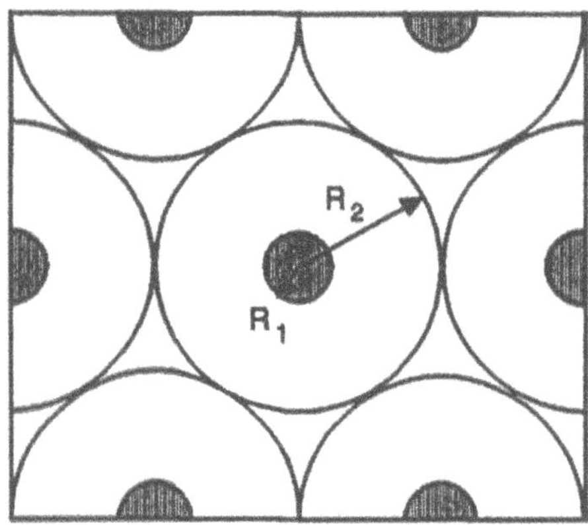


*Figure 2.* The control volumes

The control volume shown in Figure 2 is divided into 44 spherical internal cells, internal cells 1-19 belong to the particle sub-domain, while the cells 20-44 belong to the metal matrix sub-domain.

The following thermophysical parameters are assumed: thermal conductivity of lead $\lambda_1$=30 [W/mK], thermal conductivity of Al-Si alloy $\lambda_2$=90 [W/mK], latent heat of lead: $L_1=2.92\cdot 10^8$ [J/m$^3$], latent heat of Al-Si alloy $L_2=9.5\cdot 10^8$ [J/m$^3$]. Initial temperature of particle is assumed to be 20 °C, while the initial temperature of the Al-Si alloy is 650 ° C. The temperatures corresponding to the beginning and the end of solidification are equal 627 °C and 617 °C.

The part of the results obtained is presented in Figures 3 and 4. So, Figure 3 illustrates the course of heating curves at the central part of the particle. The successive heating curves correspond to the different diameters of leaden particle. Because the pure metal solidifies at the constant temperature therefore the typical 'halt' connected with the phase change is distinctly visible.

The next Figure shows the cooling curves near the contact surface in domain of metal matrix. One can notice that after a short time the temperature of this domain stabilizes (excepting the least particle) and the next stage of solidification process will be connected with the heat transfer between casting and mould.

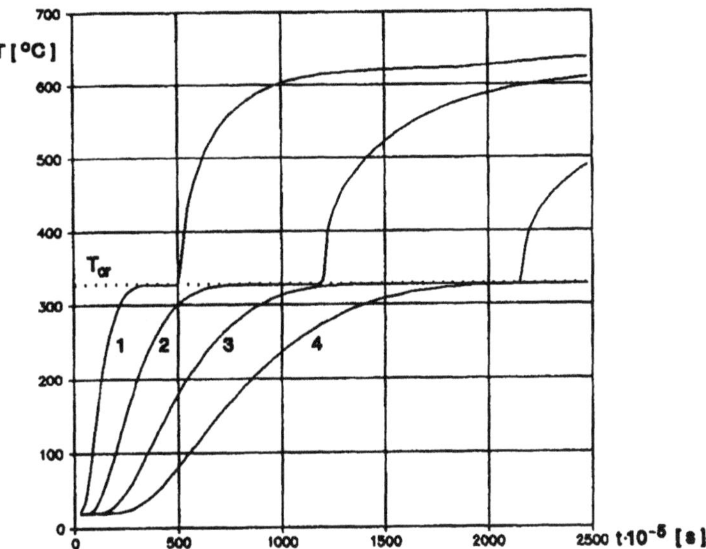


*Figure 3*. Heating curves at the center of particle (particle diameter: 1 – 1 mm, 2 – 1.5 mm, 3 – 2 mm, 4 – 2.5 mm)

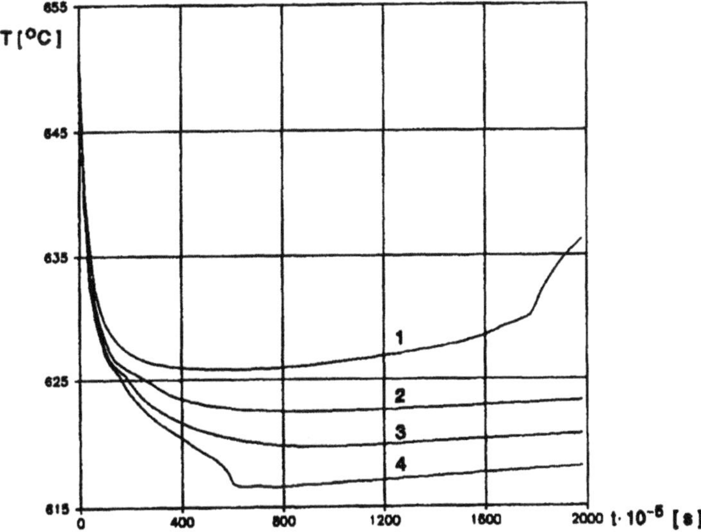


*Figure 4*. Cooling curves at point from metal matrix sub-domain near the contact surface (particle diameter: 1 – 1 mm, 2 – 1.5 mm, 3 – 2 mm, 4 – 2.5 mm)

Numerical simulations show that the process analyzed proceeds very quickly. The particle brings up to a temperature above the solidification point. Nevertheless, the particle acts to a certain degree as an internal chill because the temperature of the contact surface drops below the temperature corresponding to the beginning of Al-Si alloy solidification. The mean temperature of metal matrix becomes essentially less than the pouring temperature, and this fact is important from the technological point of view.

# ACKNOWLEDGEMENT

This paper is a part of the project No 7 T08B 008 18 sponsored by KBN.

# REFERENCES

[1] Brebbia, C.A., Telles, J.C.F. and Wrobel, L.C. (1984) *Boundary Element Techniques*, Springer-Verlag, Berlin & New York.

[2] Majchrzak, E. (1991) *Application of the BEM in the Thermal Theory of Foundry Processes*, Mechanics, 102, Publ. of the Sil. Techn. Univ., Gliwice.

[3] Mochnacki, B. and Suchy, J.S. (1995) *Numerical Methods in Computations of Foundry Processes*, PFTA, Cracow.

[4] Szopa, R. (1999) *Modelling of Crystallization and Solidification Using the Combined Variant of the BEM*, Metallurgy, 54, Publ. of the Sil. Techn. Univ., Gliwice.

[5] Szopa, R.(1998) Numerical modelling of heat diffusion in spherical domain using the combined variant of the BEM, in *Numerical Methods and Computational Mechanics,* University of Miscolc, pp. 110-111.

[6] Majchrzak, E. and Mochnacki, B. (1996) The BEM application for numerical solution of non-steady and nonlinear thermal diffusion problems, *Computer Assisted Mechanics and Engineering Sciences*, **3**, pp. 327-346.

[7] Szargut, J. and Mochnacki, B. (1971) Differential model of steel ingot solidification, *Arch. of Metall.*, **3**, pp. 270-289.

[8] Hong, C.P., Umeda, T. and Kimura, Y. (1984) Numerical models for casting solidification problems, *Metall. Trans. B*, **15B**, pp. 101-105.

[9] Mochnacki, B. (1996) Application of the BEM for numerical modelling of continuous casting, *Computational Mechanics*, **18**, **1**, pp. 62-71.

[10] Mochnacki, B. (1984) Substitute thermal capacity of metal solidifying in an interval of temperature, *Bull. of the Pol. Ac. of Sc., Techn. Sc.*, **3-4**, pp. 127-133.

[11] Idelsohn, S.R., Storti, M.A. and Crivelli, L.A. (1994) Numerical methods in phase change problems, *Archives of Computational Methods in Engineering*, **1**, pp. 49-74.

# RECENT DEVELOPMENTS ON A BOUNDARY ELEMENT METHOD IN AERODYNAMICS

L. Morino
*Università degli Studi Roma Tre*
*Dipartimento di Ingegneria Meccanica e Industriale*
L.Morino@uniroma3.it

G. Bernardini
*Università degli Studi Roma Tre*
*Dipartimento di Ingegneria Meccanica e Industriale*
nanni@gange.dimi.uniroma3.it

**Keywords:** Quasi–potential aerodynamics, trailing–edge conditions, high–order, boundary element method.

**Abstract** A high–order boundary–element method for quasi–potential flows, *i.e.*, flows that are potential everywhere except for the wake surface (vortex layer), is presented. This formulation is an extention of the high–order formulation introduced by the authors and their collaborators. The unknown is expressed through a bicubic Hermite interpolation, with derivatives at the nodes expressed in terms of suitable finite–difference approximations. The innovative aspects are related to the trailing–edge conditions; in particular, the issues related to the problem of three-dimensional multiply–connected domains are examined. An assessment of the formulation through numerical results is included.

## Introduction

The paper deals with recent developments on a boundary element method for quasi–potential flows (*i.e.*, flows that are potential everywhere with the exception of the wake points). In the past, the method, introduced by Morino (1974), has been implemented primarily as a zeroth–order (*i.e.*, piecewise constant) method (for a review, which is beyond the scope of this paper, see Morino, 1993). A piecewise–cubic

*T. Burczynski (ed.), IUTAM/IACM/IABEM Symposium on Advanced Mathematical and Computational Mechanics Aspects of the Boundary Element Method,* 237–247.

implementation was introduced by Gennaretti *et al.* (1998). In such a work, some issues (connected to the trailing–edge conditions) arise, which were later examined by Morino and Bernardini (1999), primarily from a theoretical point of view and are further analyzed here, primarily from a numerical point of view.

# 1. QUASI–POTENTIAL FLOW FORMULATION

A brief review of the formulation of quasi–potential flows is presented in this section (for details, see Morino 1993). Consider a body with a sharp trailing–edge (*e.g.*, wing or blade) moving in an inviscid incompressible fluid. Assume that at time $t = 0$, the wing and the fluid are at rest. Using Kelvin's theorem one obtains that the flow field remains irrotational at all times, for all points $\mathbf{x} \in \mathcal{V}_F$ (fluid volume) with the exception of those points, $\mathbf{x} \in \mathcal{S}_W$ (wake), that emanate from the trailing–edge. Thus, for $\mathbf{x} \in \mathcal{V}_F \backslash \mathcal{S}_W$, we have $\mathbf{v} = \nabla \varphi$ (quasi–potential flows, *i.e.*, flows potential everywhere except on $\mathcal{S}_W$). Combining with the continuity equation for incompressible flows, $\nabla \cdot \mathbf{v} = 0$, one obtains

$$\nabla^2 \varphi = 0. \tag{1}$$

Next, consider the boundary conditions for this equation. The surface of the body, $\mathcal{S}_B$, is assumed to be impermeable. This yields $(\mathbf{v} - \mathbf{v}_B) \cdot \mathbf{n} = 0$, *i.e.*,

$$\frac{\partial \varphi}{\partial n} = \mathbf{v}_B \cdot \mathbf{n} \qquad \text{for } \mathbf{x} \in \mathcal{S}_B, \tag{2}$$

where $\mathbf{v}_B$ is the velocity of a point $\mathbf{x} \in \mathcal{S}_B$, and $\mathbf{n}$ is the outward unit normal at $\mathbf{x}$. At infinity, in a frame of reference fixed with the unperturbed fluid, we have $\varphi = 0$. On the wake surface, $\mathcal{S}_W$, using the principles of conservation of mass and momentum across $\mathcal{S}_W$, one obtains that the wake surface is impermeable, and the pressure, $p$, is continuous across it. In terms of $\varphi$, the first condition yields

$$\Delta \left( \frac{\partial \varphi}{\partial n} \right) = 0 \tag{3}$$

(where $\Delta f = f_2 - f_1$ denotes discontinuity across $\mathcal{S}_W$), while the second one, using Bernoulli's theorem in the air frame of reference,

$$\frac{\partial \varphi}{\partial t} + \frac{1}{2} \|\mathbf{v}\|^2 + \frac{p}{\rho} = \frac{p_\infty}{\rho}, \tag{4}$$

yields

$$\frac{D_W}{Dt} \Delta \varphi := \left[ \frac{\partial}{\partial t} + \mathbf{v}_W \cdot \nabla \right] \Delta \varphi = 0, \tag{5}$$

where $\mathbf{v}_W = \frac{1}{2}(v_1 + v_2)$ denotes the velocity of a wake point (the average of the fluid velocity on the two sides of the wake). Equation 5 implies that the value of $\Delta\varphi$ remains constant in time following a wake point $\mathbf{x}_W$, and equals the value it had when $\mathbf{x}_W$ left the trailing–edge. This value is obtained by imposing the trailing–edge condition that, at the trailing–edge, $\Delta\varphi$ on the wake equals $\varphi_2 - \varphi_1$ on the body, where the subscripts 2 and 1 denote, respectively, the upper and lower sides of the body surface (for a detailed analysis of this issue, see Morino and Bernardini (1999); see also the end of Section 2).

The above problem for the velocity potential may be solved by a boundary integral formulation. Starting from the boundary integral representation for the Laplace equation with a surface that surrounds and is infinitesimally close to body and wake, and using the above boundary conditions, one obtains, in the limit as the surface approaches $\mathcal{S}_B \cup \mathcal{S}_W$,

$$\varphi(\mathbf{x}) = \int_{\mathcal{S}_B} \left( G\chi - \varphi \frac{\partial G}{\partial n} \right) d\mathcal{S}(\mathbf{y}) - \int_{\mathcal{S}_W} \Delta\varphi \frac{\partial G}{\partial n} d\mathcal{S}(\mathbf{y}), \tag{6}$$

with $\chi := \mathbf{v}_B \cdot \mathbf{n}$, whereas $G = -1/4\pi\|\mathbf{y} - \mathbf{x}\|$ denotes the fundamental solution for the Laplace equation in a three–dimensional space.

Note that Eq. 6, in the limit as $\mathbf{x}$ tends to $\mathcal{S}_B$, yields a compatibility condition between $\varphi$ and $\chi$, *i.e.*, a boundary integral equation for $\varphi$ on $\mathcal{S}_B$, since $\chi$ on $\mathcal{S}_B$ is known from the boundary condition, whereas by applying Eq. 5 and the trailing–edge condition, $\Delta\varphi$ on the wake may be expressed in terms of $\varphi$ over the body at preceding times. Once $\varphi$ on the body is known, $\varphi$ (and hence $\mathbf{v}$ and, by using Bernoulli's theorem, $p$) may be evaluated everywhere in the field.

## 2. NUMERICAL FORMULATION

In this section we present a high–order boundary–element formulation used for the analysis of potential incompressible flows around lifting bodies. First, we introduce the spatial discretization of the potential equation, in particular a piecewise cubic formulation. Finally, we analyze some issues that arise at the trailing–edge.

### 2.1. SPATIAL DISCRETIZATION FOR THE POTENTIAL EQUATION

Assuming an approximate representation for the velocity potential in terms of interpolation functions $M_k(\mathbf{x})$, one obtains:

$$\varphi(\mathbf{x}, t) = \sum_{k=1}^{N} \varphi_k(t) M_k(\mathbf{x}) \qquad \text{for } \mathbf{x} \in \mathcal{S} \tag{7}$$

where $\varphi_k(t) = \varphi(\mathbf{x}_k, t)$ and $N$ is the number of the nodes on the body surface. Similar expressions are used for the normalwash $\chi := \partial\varphi/\partial n$ (potential derivative in the direction normal to the body surface), and for the potential discontinuity on the wake, $\Delta\varphi$.

Combining Eqs. 7, and similar ones for $\chi$ and $\Delta\varphi$ with Eq. 6 and using the collocation method, yields the following system of $N$ equations in $N$ unknowns $\varphi_k$ ($\chi_i$ are known from the boundary conditions, $\Delta\varphi_j$ from the preceding time hystory; at the trailing edge, $\Delta\varphi = \varphi_2 - \varphi_1$),

$$\varphi_k = \sum_{i=1}^{N} B_{ki}\chi_i + \sum_{i=1}^{N} C_{ki}\varphi_i + \sum_{j=1}^{N_W} F_{kj}\Delta\varphi_j \quad (k = 1, .., N) \tag{8}$$

where, for instance

$$C_{ki} = \frac{1}{4\pi}\int_{S_B} \frac{\partial}{\partial n}\left(\frac{1}{\| \mathbf{x} - \mathbf{x}_k \|}\right) M_i(\mathbf{x}) dS. \tag{9}$$

Similar expressions hold for $B_{ki}$ and $F_{kj}$. These coefficients depend exclusively upon the surface geometry, the interpolation functions, and the collocation points $\mathbf{x}_k$.

The specific formulation used here is based upon a high-order formulation introduced by Gennaretti *et al.* (1998), to which the reader is referred for details. This consists of using bicubic quadrilateral elements, with Hermite interpolation in both directions, for both geometry and $\varphi$. This yields a bicubic representation in terms of the nodal values of $\varphi$, $\partial\varphi/\partial\xi$, $\partial\varphi/\partial\eta$, and $\partial^2\varphi/\partial\xi\partial\eta$; the nodal derivatives are then expressed in terms of the unknowns of the problem $\varphi_i$, through suitable finite-difference approximations, thereby leaving $\varphi_i$ as the only unknowns.

## 2.2. TRAILING–EDGE PROBLEM

Only steady flows are considered here; then, the wake geometry at the trailing–edge (typically tangent to one of the sides of the body surface at the trailing–edge) may be determined by the results of Mangler and Smith (1970); see also Morino and Bernardini (1999) for details, as well as Bassanini *et al.* (1999) for a review of a different approach.

Also, we note that the above formulation yields an issue connected to the trailing–edge: the existence of the trailing–edge potential discontinuity implies that at any trailing–edge node there exist two unknowns, but only one collocation point. In Gennaretti *et al.* (1998) this problem is circumvented by having, at the trailing–edge, two collocation points suitably located slightly ahead of the trailing edge (otherwise the collocation points coincide with the nodes). However, placing the collocation points at a very small distance from the trailing edge yields an ill–conditioned matrix. In order to avoid this, here we use the approach introduced by Morino and Bernardini (1999), who, at the trailing–edge, use (instead of two collocation points ahead of the the trailing–edge) only one control point but two different integral equations: the first is Eq. 6 with $\mathbf{x}$ at the trailing edge

$$\frac{1}{2}\left[\varphi_1(\mathbf{x}) + \left(1 - \frac{\gamma}{\pi}\right)\varphi_2(\mathbf{x})\right] = \int_{S_B}\left(G\frac{\partial\varphi}{\partial n} - \varphi\frac{\partial G}{\partial n}\right)dS(\mathbf{y}) - \int_{S_W}\Delta\varphi\frac{\partial G}{\partial n}dS(\mathbf{y}) \qquad (10)$$

where $\gamma$ denotes the dihedral angle (in Eq. 10, 1 denotes the side of the wing to which wake is tangent). The second is the derivative of Eq. 6 in the direction of $\mathbf{n}_{\mathbf{x}}$, where $\mathbf{n}_{\mathbf{x}}$ is the unit normal to the wake at the point $\mathbf{x}$ on trailing–edge,

$$\left(1 - \frac{\gamma}{2\pi}\right)\frac{\partial\varphi(\mathbf{x})}{\partial n_{\mathbf{x}}} = \frac{\partial}{\partial n_{\mathbf{x}}}\int_{S_B}\left(G\frac{\partial\varphi}{\partial n} - \varphi\frac{\partial G}{\partial n}\right)dS(\mathbf{y}) - \frac{\partial}{\partial n_{\mathbf{x}}}\int_{S_W}\Delta\varphi\frac{\partial G}{\partial n}dS(\mathbf{y}) \qquad (11)$$

with $\partial/\partial n_{\mathbf{x}} = \mathbf{n}_{\mathbf{x}} \cdot \nabla$. [The above approach is equivalent to taking the limit of the semi–sum and of the difference (divided by the distance) of the two equations used by Gennaretti *et al.* (1998), as the collocation points approach in a suitable manner the trailing–edge (for details, see Bernardini, 2000).]

In addition, note that a special approach is required for the evaluation of $\partial\varphi/\partial\xi^1$ at the trailing–edge (where the $\xi^1$-line is the grid line away from the trailing–edge). In Gennaretti *et al.* (1998), $\partial\varphi/\partial\xi^1$ at the trailing–edge is approximated by backward differences, on either side of the wing. Here, following again Morino and Bernardini (1999), to which the reader is referred for details, this requirement provides a very natural way to impose the following two additional conditions (stemming from the Joukowski hypothesis of smooth flow at the trailing edge): ($A$) the stagnation–point condition in the plane normal to the trailing–edge, which determines $\partial\varphi/\partial\xi^1$ on the inward dihedral line, and ($B$) Eq. 5, which determines $\partial\varphi/\partial\xi^1$ on the other side through the wake condition on $\Delta\varphi$ (Eq. 5 implies $\Delta p_{TE} = 0$, Kutta condition).

As pointed out by Morino and Bernardini (1999), conditions $A$ (stagnation) and $B$ (Kutta) appear to be satisfied by the numerical solution even when they are not imposed; this issue is further examined later in this paper.

## 3. NUMERICAL VALIDATION

The issue of the correct trailing–edge condition is important not only from a theoretical, but also from a numerical point of view. This issue has been raised in connection with the numerical results obtained in the past by the authors, where the condition $\Delta\varphi = \varphi_2 - \varphi_1$ appears sufficient to obtain a unique solution that satisfies the Kutta condition, $\Delta p_{TE} = 0$. This fact has been sometimes attributed to the low-order (zeroth order, *i.e.*, piecewise constant) numerical implementation, typically used in the past by the authors and their collaborators (in this case, the unknowns are evaluated at the centers of the elements and $\Delta\varphi$ at the trailing–edge is obtained from the values of $\varphi$ at the centers of the adjacent elements). It should be pointed out however, that numerical results with pressure discontinuities at the trailing–edge have been observed and remedies have been proposed (*e.g.*, including the values of $\Delta\varphi_{TE}$ among the unknowns and imposing the Kutta condition $\Delta p_{TE} = 0$), for instance by Kerwin (1987) and D'Alascio *et al.* (1997).

In order to gain some insight on this issue, we have explored the problem from a numerical point of view using a formulation that allows us to impose all the above conditions. We have compared our results with those presented by Roberts and Rundle (1972), which are considered to be highly accurate (Systma *et al.* 1979).

First, we present results for a RAE wing having aspect ratio 6, taper ratio 1/3, mid–chord sweep of 30°, angle of attack of 5° and symmetric NACA four–digit airfoil sections (Figure 1). Three different thickness ratios ($t/c = .02, .05, .15$) are considered in Fig. 2a–c, which shows the chordwise distribution of the pressure coefficient (at $2y/b = .549$), and in Fig. 3a–c which depicts the circulation $\Gamma = \Delta\varphi_{TE}$ as a function of $2y/b$. For all three cases, the results are obtained with a chordwise discretization $N_1 = 16$ and a spanwise discretization $N_2 = 12$, and are compared with those obtained by Roberts and Rundle (1972) with $N_1 = 39$ and $N_2 = 13$.

The agreement between the two formulations is good, except at the trailing–edge, where the pressure predicted in our case is higher (as it should, see below). For all the cases, the same level of agreement has been obtained; these results allow us to conclude that our formulation is able to describe the nature of the phenomenon even in the case in which the wings are relatively thin, a case particularly challenging because of the numerical evaluation of 'nearly singular' integrals (Hayami, 1992).

# 4. MULTIPLY–CONNECTED DOMAINS

As stated above, the results are nearly identical if conditions $A$ and $B$ above are removed. In order to explore this issue further, we have applied the formulation to multiconnected domains (multivalued potential), for which the solution is not unique. To see whether the trailing edge condition $\Delta\varphi = \varphi_2 - \varphi_1$ is sufficient, all the results presented in this section are obtained without imposing conditions ($A$) and ($B$).

Specifically, we have considered the axisymmetric flow around the annular nacelle shown in Fig. 4. In Fig. 5, for the above nacelle at an angle of attack (angle between the undisturbed flow and the nacelle axis) $\alpha = 0°$, we have compared our chordwise pressure distribution, obtained by a discretization $N_1 = 16$ and $N_2 = 8$, with that obtained by Roberts and Rundle (1972) using $N_1 = 55$ and $N_2 = 10$. Also in this case our results are in good agreement with those by Roberts and Rundle (1972) except at the trailing–edge where our formulation shows a higher pressure. Note that, this configuration is axisymmetric and, therefore there exists a stagnation point on both sides of the trailing–edge (see Morino and Bernardini, 1999). In contrast to Roberts and Rundle our method exactly predicts this situation.

In addition, we have analyzed the non–axisymmetric flow around the above mentioned nacelle, at an angle of attack $\alpha = 5°$. Figures 6a–c show the comparison between the chordwise distribution of the pressure coefficient, obtained with our formulation and that obtained by Roberts and Rundle (1972). In particular, the above figures correspond respectively to the sections characterized by azimutal coordinate $\theta = 9°, 81°, 171°$. Also in this case, the behavior of our results is in good agreement with those obtained by Roberts and Rundle (1972) for all three cases (except for the trailing–edge, as discussed above).

In order to further explore the issue of multiconnected domains and multivalued potential, consider again the case $\alpha = 0°$. It should be noted that in this case, the potential discontinuity is constant: $\Delta\varphi = \Gamma \neq 0$. Thus, there is no vorticity in the field and, akin to the steady two–dimensional case, we are dealing with a branch cut, not a wake. As mentioned above, it is surprising that the trailing–edge conditions ($A$) and ($B$), in particular the Kutta condition $\Delta p_{TE} = 0$, need not be imposed to obtain the correct solution, even though the potential is now multivalued and hence the solution not unique. In order to further analyze this, we have examined the transient solution (where the solution is unique, since the circulation is determined by the initial conditions and the Kelvin theorem); specifically, we have compared the asymptotic value of the transient solution with that obtained from the steady–

state formulation with the stagnation condition $A$ and the condition $B$ (Kutta) removed (*i.e.*, with the trailing–edge derivatives evaluated by backwards finite differences, akin to Gennaretti *et al.*, 1998; see Bernardini, 2000, for details). The results, shown in Fig. 7 (Case 1), suggest that somehow the numerics (in contrast to the theory) captures the correct value of the circulation, even without imposing $\Delta p_{TE} = 0$. This is somewhat baffling, and we report this result as an "open problem", for which we have no explanation.

To explore this case even further, we note that, in theory the solution ought to be independent of the branch–surface location, since the Kutta condition is not explicitly imposed at the trailing edge. Thus, we study the problem with a different branch cut: specifically, we use a branch surface parallel to the axis of symmetry, but emanating from the leading edge (which is not sharp). The results, shown in Fig. 8 (Case 2) disagree with those obtained with the branch surface emanating from the trailing edge (Fig. 8, Case 1); instead, they agree (within an appropriate change of sign) with those obtained through a transient response for an undisturbed flow in the opposite direction (Fig. 7, Case 2). This is also a baffling open problem. Further investigation of these issues is warranted and is now underway.

## References

[1] Bassanini, P., Casciola, C.M., Lancia, M.R., Piva, R., (1999) Edge singularities and Kutta condition in 3D aerodynamics, *Meccanica*, Vol. 34, pp. 199-229.

[2] Bernardini, G., (2000) Problematiche Aerodinamiche Relative alla Progettazione di Configurazioni Innovative, Ph.D. Thesis, Department of Aerospace Engineering, Politecnico di Milano.

[3] D'Alascio, A., Visingardi, A., Renzoni, P., (1997) Explicit Kutta condition correction for rotary wing flows, Proceedings of the 19th World Conference on the Boundary Element Method, Rome, Italy.

[4] Gennaretti, M., Calcagno, G., Zamboni, A., Morino, L., (1998) A high order boundary element formulation for potential incompressible aerodynamics, *The Aeronautical Journal*, Vol. 102, No. 1014, p. 211-219.

[5] Hayami, K., (1992) A Projection Transformation Method for Nearly Singular Surface Boundary Element Integrals, Springer Verlag, Heidelberg, Germany.

[6] Kerwin, J.E., Kinnas, S.A., Lee, J.T., Shih, W.Z., (1987) A Surface Panel Method for the Hydrodynamic Analysis of Ducted Propellers

*Trans. SNAME, 95.*

[7] Kutta, M.W., (1902) Auftriebskräfte in strömenden Flüssigkeiten, *Illustrierete Aeronautische Mitteilungen*, Vol. 6, pp. 133-135.

[8] Joukowski, N., (1905) On the adjunct vortices, *Obshchestvo liubitelei estestvoznaniia, antropologii i etnografii*, Trans. of the Phys. Section, Vol. XII.

[9] Mangler, K.W., Smith J.H.B., (1970) Behaviour of the vortex sheet at the trailing edge of a lifting wing, *The Aeronautical Journal*, Vol. 74, pp. 906-908.

[10] Morino, L., Bernardini, G., (2000) Singularities in discretized BIE's for Laplace's equation; trailing–edge conditions in aerodynamics, in M. Bonnet, A.–M., Sändig and W.L. Wendland (eds.), *Mathematical Aspects of Boundary Element Methods*, Chapman & Hall/CRC Press, London, UK.

[11] Morino, L., (1993) Boundary integral equation in aerodynamics, *Appl. Mech. Rev.*, Vol. 46, pp. 445-466.

[12] Morino, L., (1974) A General Theory of Unsteady Compressible Potential Aerodynamics, NASA CR-2464.

[13] Prandtl, L. (1924) Induced Drag of Multiplanes, NACA TN 182, pp. 1-22.

[14] Roberts, A., Rundle, K., (1972) Computation of incompressible flow about bodies and thick wings using the spline–mode system, BAC-(CAD) Rep. A. Ma 19.

[15] Sytsma, H.S., Hewitt, B.L., Rubbert, P.E., (1979) A comparison of panel methods for subsonic flow computation, AGARD AG 241.

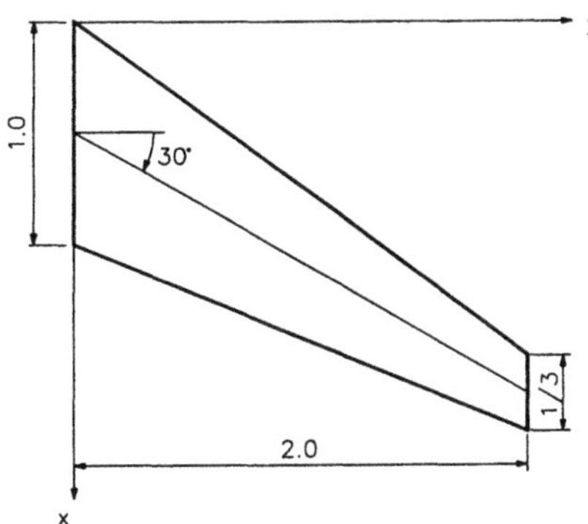

*Figure* 1. Planform of RAE wing.

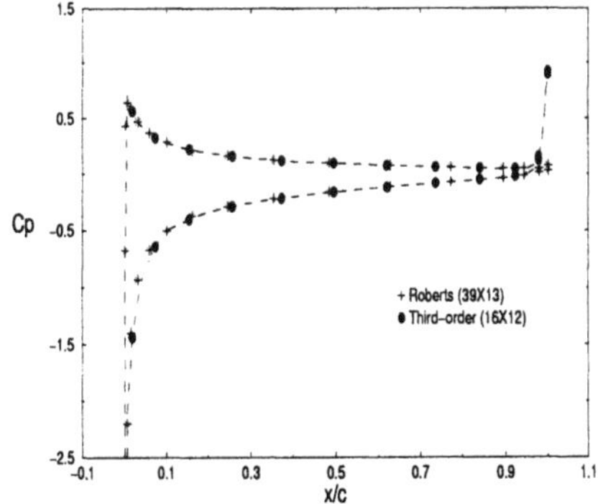

*Figure* 2a. Chordwise distribution of the $C_p$, $\alpha = 5°, t/c = 0.02, \eta = 0.549$.

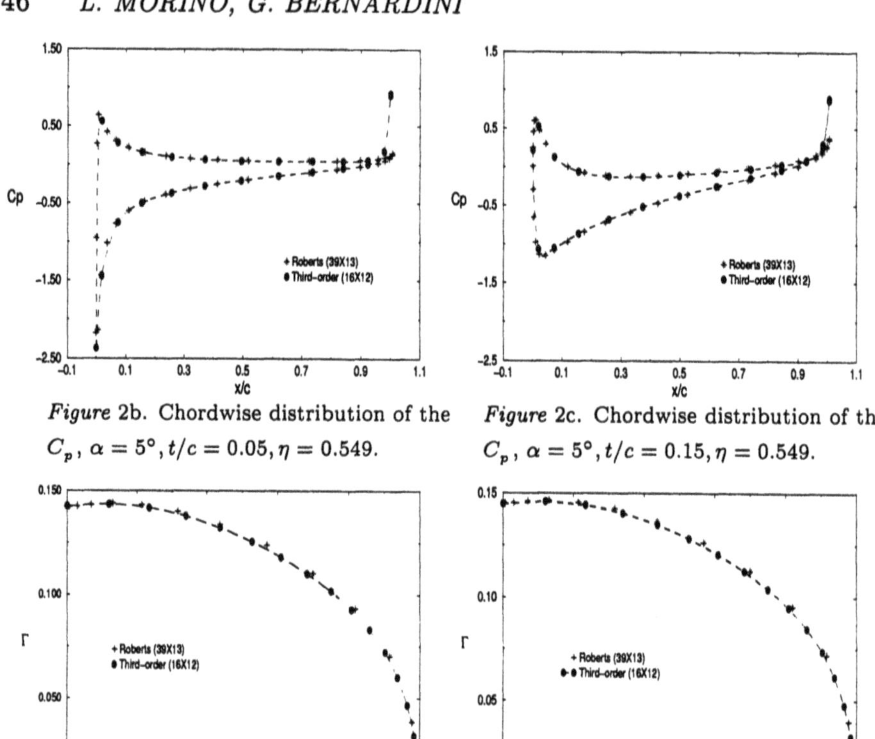


*Figure* 2b. Chordwise distribution of the $C_p$, $\alpha = 5°, t/c = 0.05, \eta = 0.549$.

*Figure* 2c. Chordwise distribution of the $C_p$, $\alpha = 5°, t/c = 0.15, \eta = 0.549$.

*Figure* 3a. Spanwise distribution of the circulation, $\alpha = 5°, t/c = 0.02$.

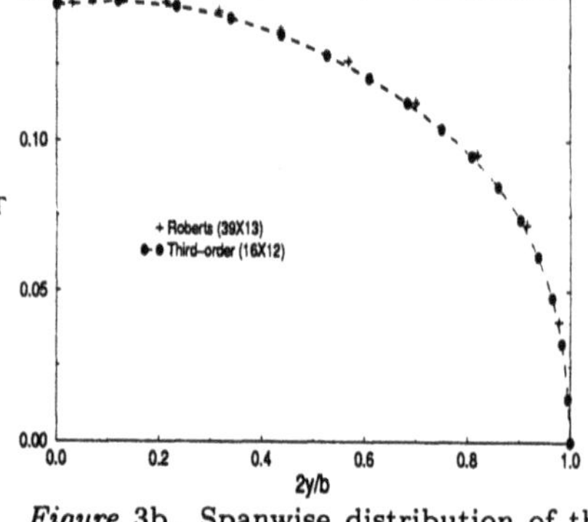


*Figure* 3b. Spanwise distribution of the circulation, $\alpha = 5°, t/c = 0.05$.

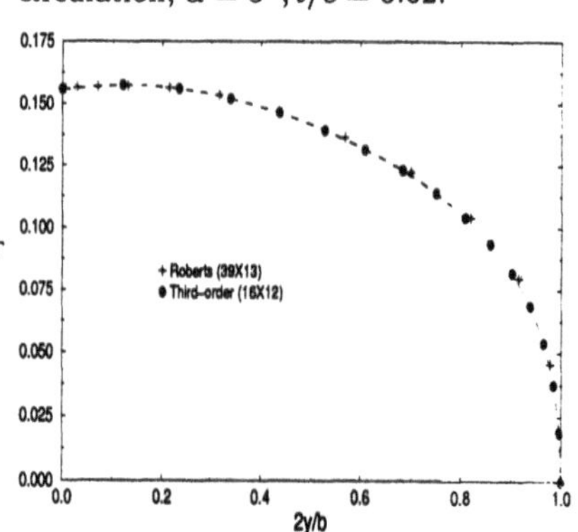


*Figure* 3c. Spanwise distribution of the circulation, $\alpha = 5°, t/c = 0.15$.

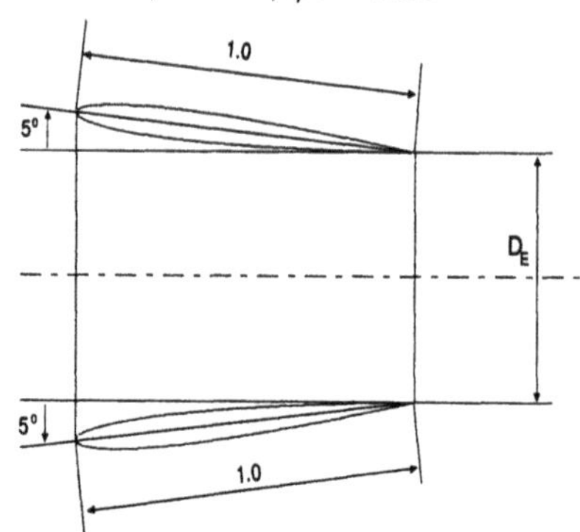


*Figure* 4. Cross section of nacelle.

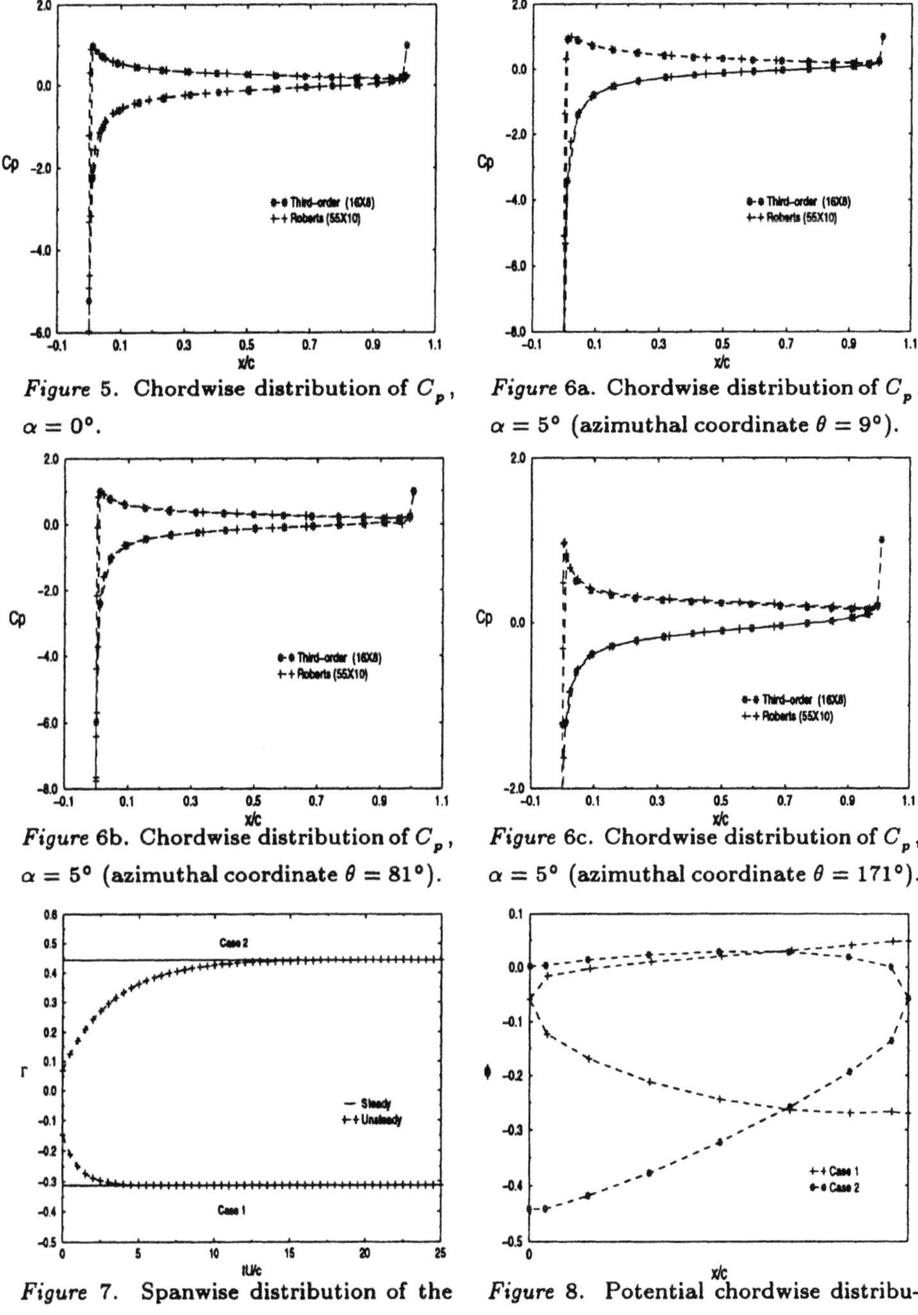


*Figure* 5. Chordwise distribution of $C_p$, $\alpha = 0°$.

*Figure* 6a. Chordwise distribution of $C_p$, $\alpha = 5°$ (azimuthal coordinate $\theta = 9°$).

*Figure* 6b. Chordwise distribution of $C_p$, $\alpha = 5°$ (azimuthal coordinate $\theta = 81°$).

*Figure* 6c. Chordwise distribution of $C_p$, $\alpha = 5°$ (azimuthal coordinate $\theta = 171°$).

*Figure* 7. Spanwise distribution of the circulation, $\alpha = 0°$.

*Figure* 8. Potential chordwise distribution, $\alpha = 0°$.

# APPLICATION OF A REGULARIZED SGBEM FORMULATION TO FEM-BEM COUPLING IN ELASTOSTATICS

Saida Mouhoubi[(a)], Laurent Ulmet[(a)], Marc Bonnet[(b)]

*(a) Laboraroire de Génie Civil d'Egletons, Université de Limoges, France*

*(b) Laboratoire de Mecanique des Solides (CNRS UMR 7649), Ecole Polytechnique, France*

mouhoubi@lgce0.unilim.fr, ulmet@unilim.fr, bonnet@lms.polytechnique.fr

**Abstract** This paper concerns a procedure, based on the symmetric Galerkin boundary element equations, for computing the stiffness matrix of an elastic region without body forces in terms of boundary nodal displacements. This domain can thus be treated as a macro-element in a FEM calculation. In the present case, the implementation has been done in the CASTEM 2000 FEM environment. Sample numerical examples are presented; they yield numerical results in good agreement with the corresponding exact solutions.

**Keywords:** Symmetric Galerkin BEM, FEM/BEM coupling, elastostatics

## 1. INTRODUCTION

The boundary element methods (BEMs) are often disadvantaged with respect to finite element methods (FEMs) due to the fully populated and unsymmetric character of the matrix resulting from the discretization of the integral equations. This fact has in particular hindered the development of coupled BEM-FEM solution procedures. Although the BEM-modelled region will in any case lead to fully populated matrices, it is possible to reduce both computer time and storage by devising symmetric coupling procedures.

In this paper, this goal is achieved by exploiting the fact that the Symmetric Galerkin BEM (SGBEM) equations, adequately combined, yield the strain energy corresponding to any given boundary displacement field, i.e. provide the (symmetric) stiffness matrix of the elastic region modelled by BEM. This technique thus treats the latter as a macro-element and is therefore very well suited for implementation into

*T. Burczynski (ed.), IUTAM/IACM/IABEM Symposium on Advanced Mathematical and Computational Mechanics Aspects of the Boundary Element Method,* 249–258.

a FEM code. It is currently being implemented by the present authors into the CASTEM 2000 [4] FEM environment. Some preliminary numerical results are presented in section 5

The choice of this approach, in which the SGBEM is introduced into FEM, can be justified by the historic advance of FEM, for which many powerful and popular packages are available, and whose environments offer consistent backgrounds for further developments (the CASTEM 2000 environment is, by design, very flexible in this respect).

## 2. SYMMETRIC GALERKIN BEM FORMULATION

Consider a three-dimensional elastic body $\Omega$, whose boundary $\Gamma$ is a piecewise closed surface: $\Gamma = \cup_\alpha \Gamma^\alpha$, $\alpha = 1, \ldots$ where each $\Gamma^\alpha$ is a closed surface that may be split into a displacement-prescribed part $\Gamma^\alpha_u$ and a traction-prescribed part $\Gamma^\alpha_t$ such that $\Gamma^\alpha = \Gamma^\alpha_u \cup \Gamma^\alpha_t$ and $\Gamma^\alpha_u \cap \Gamma^\alpha_t = \emptyset$. Let $(\boldsymbol{u}, \boldsymbol{t})$ denote any compatible pair of displacements and tractions on the boundary $\Gamma$, i.e. $\boldsymbol{u}, \boldsymbol{t}$ are the trace on the boundary of an elastostatic state $\boldsymbol{u}, \boldsymbol{\sigma}(\boldsymbol{u})$ in $\Omega$ (where $\boldsymbol{\sigma}(\boldsymbol{u}) \equiv \boldsymbol{C} : \boldsymbol{\nabla} \boldsymbol{u}$ is the tensor of elastic stresses, $\boldsymbol{C}$ denoting the fourth-order tensor of elastic moduli). In the absence of body forces, the elastic strain energy associated with $(\boldsymbol{u}, \boldsymbol{t})$ can be expressed in terms of boundary fields (Clapeyron formula):

$$W = \frac{1}{2} \int_\Omega \boldsymbol{\sigma}(\boldsymbol{x}) : \boldsymbol{\varepsilon}(\boldsymbol{x}) \, \mathrm{d}\Omega = \frac{1}{2} \int_\Gamma \boldsymbol{t}(\boldsymbol{x}) . \boldsymbol{u}(\boldsymbol{x}) \, \mathrm{d}\Gamma_x \tag{1}$$

Following Hsiao [5], we will refer to the second term in the above equation as *boundary energy*.

Since $(\boldsymbol{u}, \boldsymbol{t})$ are assumed to be compatible, the boundary energy is in fact a function of either $\boldsymbol{u}$ or $\boldsymbol{t}$ alone. In order to build a stiffness operator in the usual sense, the displacement $\boldsymbol{u}$ is taken as the primary variable and $W \equiv W(\boldsymbol{u})$ in Eq. (1).

The explicit form of the mathematical link between $\boldsymbol{u}$ and $\boldsymbol{t}$ on $\Gamma$ is either the displacement boundary integral equation (Somigliana identity) or, equivalently, the traction boundary integral equation. Here, both integral equations are invoked in Galerkin, i.e. weighted residual, form (see the survey paper [3] for more details and many references on the symmetric Galerkin boundary element methods).

First consider the traction $\boldsymbol{t}$ as being induced by given displacements $\boldsymbol{u}$ (i.e., $\Gamma^\alpha_u = \Gamma^\alpha$ and $\Gamma^\alpha_t = \emptyset$). The following governing Galerkin integral equation for $\boldsymbol{t}$ is obtained by weighting the displacement BIE with a traction-like test function $\tilde{\boldsymbol{t}}$ and following a regularization procedure:

$$\text{Find} \quad \boldsymbol{t} \in \mathcal{V}_T \qquad \mathcal{B}_{tt}(\boldsymbol{t}, \tilde{\boldsymbol{t}}) = \mathcal{B}_{ut}(\boldsymbol{u}, \tilde{\boldsymbol{t}}) + \mathcal{D}_{ut}(\boldsymbol{u}, \tilde{\boldsymbol{t}}) \quad (\forall \tilde{\boldsymbol{t}} \in \mathcal{V}_T) \tag{2}$$

where

$$\mathcal{B}_{tt}(\boldsymbol{t},\tilde{\boldsymbol{t}}) = \int_\Gamma\int_\Gamma t_k(\boldsymbol{x})\tilde{t}_i(\tilde{\boldsymbol{x}})U_i^k(\boldsymbol{x},\tilde{\boldsymbol{x}})\,\mathrm{d}\Gamma_{\tilde{x}}\,\mathrm{d}\Gamma_x \tag{3}$$

$$\begin{aligned}\mathcal{B}_{ut}(\boldsymbol{u},\tilde{\boldsymbol{t}}) &= \sum_\alpha \int_{\Gamma^\alpha}\int_{\Gamma^\alpha} \tilde{t}_k(\tilde{\boldsymbol{x}})[u_i(\boldsymbol{x}) - u_i(\tilde{\boldsymbol{x}})]T_i^k(\tilde{\boldsymbol{x}},\boldsymbol{x})\,\mathrm{d}\Gamma_{\tilde{x}}\,\mathrm{d}\Gamma_x \\ &\quad + \sum_{\tilde{\alpha}\neq\alpha} \int_{\Gamma^\alpha}\int_{\Gamma^{\tilde{\alpha}}} \tilde{t}_k(\tilde{\boldsymbol{x}})u_i(\boldsymbol{x})T_i^k(\tilde{\boldsymbol{x}},\boldsymbol{x})\,\mathrm{d}\Gamma_{\tilde{x}}\,\mathrm{d}\Gamma_x\end{aligned} \tag{4}$$

$$\mathcal{D}_{ut}(\boldsymbol{u},\tilde{\boldsymbol{t}}) = \sum_\alpha \kappa_\alpha \int_{\Gamma^\alpha} u_k(\boldsymbol{x})\tilde{t}_k(\boldsymbol{x})\,\mathrm{d}\Gamma_x \tag{5}$$

Then consider the displacement $\boldsymbol{u}$ as induced by given tractions $\boldsymbol{t}$ (i.e., $\Gamma_t^\alpha = \Gamma^\alpha$ and $\Gamma_u^\alpha = \emptyset$). The following governing Galerkin integral equation for $\boldsymbol{u}$ is obtained by weighing the traction BIE with a displacement-like test function $\tilde{\boldsymbol{u}}$ and following a regularization procedure [1] which is not shown here:

$$\text{Find} \quad \boldsymbol{u}\in\mathcal{V}_u \qquad \mathcal{B}_{uu}(\boldsymbol{u},\tilde{\boldsymbol{u}}) = \mathcal{B}_{tu}(\boldsymbol{t},\tilde{\boldsymbol{u}}) + \mathcal{D}_{tu}(\boldsymbol{t},\tilde{\boldsymbol{u}}) \quad (\forall\tilde{\boldsymbol{u}}\in\mathcal{V}_u) \tag{6}$$

where

$$\mathcal{B}_{uu}(\boldsymbol{u},\tilde{\boldsymbol{u}}) = \int_\Gamma\int_\Gamma (Ru)_{iq}(\boldsymbol{x})(R\tilde{u})_{ks}(\tilde{\boldsymbol{x}})B_{ikqs}(\boldsymbol{x},\tilde{\boldsymbol{x}})\,\mathrm{d}\Gamma_{\tilde{x}}\,\mathrm{d}\Gamma_x \tag{7}$$

$$\begin{aligned}\mathcal{B}_{tu}(\boldsymbol{t},\tilde{\boldsymbol{u}}) &= \sum_\alpha \int_{\Gamma^\alpha}\int_{\Gamma^\alpha} t_k(\boldsymbol{x})[\tilde{u}_i(\tilde{\boldsymbol{x}}) - \tilde{u}_i(\boldsymbol{x})]T_i^k(\boldsymbol{x},\tilde{\boldsymbol{x}})\,\mathrm{d}\Gamma_{\tilde{x}}\,\mathrm{d}\Gamma_x \\ &\quad + \sum_{\tilde{\alpha}\neq\alpha} \int_{\Gamma^\alpha}\int_{\Gamma^{\tilde{\alpha}}} t_k(\boldsymbol{x})\tilde{u}_i(\tilde{\boldsymbol{x}})T_i^k(\boldsymbol{x},\tilde{\boldsymbol{x}})\,\mathrm{d}\Gamma_{\tilde{x}}\,\mathrm{d}\Gamma_x\end{aligned} \tag{8}$$

$$\mathcal{D}_{tu}(\boldsymbol{t},\tilde{\boldsymbol{u}}) = \sum_\alpha (\kappa_\alpha - 1) \int_{\Gamma^\alpha} t_k(\boldsymbol{x})\tilde{u}_k(\boldsymbol{x})\,\mathrm{d}\Gamma_x \tag{9}$$

In Eqs. (2) and (6), $\mathcal{V}_u$ and $\mathcal{V}_T$ denote the spaces of admissible displacements and tractions on the boundary, and $\tilde{\boldsymbol{u}}$ and $\tilde{\boldsymbol{t}}$ are test functions. Also, $U_i^k(\boldsymbol{x},\tilde{\boldsymbol{x}})$ and $T_i^k(\boldsymbol{x},\tilde{\boldsymbol{x}})$ are the $i$-components of the Kelvin fundamental solution (displacement and traction generated at $\tilde{\boldsymbol{x}}$ by a unit point force applied at $\boldsymbol{x}$ along the $k$-direction). The unit normal $\boldsymbol{n}$ is exterior to $\Omega$. In Eqs. (5) and (9), one has either $\kappa_\alpha = 0$ (if $\boldsymbol{n}$ is exterior to $\Gamma_\alpha$, e.g. when $\Gamma_\alpha$ is the external boundary of a bounded body) or $\kappa_\alpha = 1$ (if $\boldsymbol{n}$ is interior to $\Gamma_\alpha$, e.g. when $\Gamma_\alpha$ is the boundary of an unbounded medium or of a cavity).

The SGBIE formulations (2)–(5) and (6)–(9) are in regularized form: the various double surface integrations consist of a weakly singular inner integral followed by a regular outer integral. This is achieved by means of a combined use of indirect regularization and Stokes theorem (see e.g. [1, 2]). In particular, the latter operation uses the surface curl operator, first introduced in [6] and defined (with $e_{jfq}$: components of the permutation tensor) by:

$$(Ru)_{iq} = e_{jfq} n_j u_{i,f} \tag{10}$$

and the fourth-order tensor $B_{ikqs}$, given by:

$$\frac{1}{\mu^2} B_{ikqs} = -4\delta_{qs} F_{,ik} + [4\delta_{ik}\delta_{qs} - 4\nu\delta_{is}\delta_{kq} - 2(1-\nu)\delta_{iq}\delta_{ks}] F_{,pp}$$

with

$$\begin{aligned} F(\tilde{\boldsymbol{x}} - \boldsymbol{x}) &= \frac{r}{16\pi\mu(1-\nu)} \\ F_{,pp}(\tilde{\boldsymbol{x}} - \boldsymbol{x}) &= \frac{1}{8\pi\mu(1-\nu)r} \\ F_{,ik}(\tilde{\boldsymbol{x}} - \boldsymbol{x}) &= \frac{1}{16\pi\mu(1-\nu)r}(\delta_{ik} - r_{,i}r_{,k}) \end{aligned}$$

# 3. CONSTRUCTION OF THE STIFFNESS MATRIX

Assume now that the SGBEM equations (2)–(5) and (6)–(9) have been discretized using boundary element interpolations of the surfaces $\Gamma^\alpha$ and the boundary fields $(\boldsymbol{u}, \boldsymbol{t})$ and associated test functions $(\tilde{\boldsymbol{u}}, \tilde{\boldsymbol{t}})$. Further, let the latter be chosen specifically as $\tilde{\boldsymbol{u}} = \boldsymbol{u}$ and $\tilde{\boldsymbol{t}} = \boldsymbol{t}$.

In that case, the discretized SGBEM equations (2) and (6) describing respectively the Dirichlet and Neumann problems yield the identities:

$$\{\boldsymbol{t}\}^T[\boldsymbol{B}_{tt}]\{\boldsymbol{t}\} = \{\boldsymbol{u}\}^T([\boldsymbol{B}_{ut}] + [\boldsymbol{D}_{ut}])\{\boldsymbol{t}\} \tag{11}$$

$$\{\boldsymbol{u}\}^T[\boldsymbol{B}_{uu}]\{\boldsymbol{u}\} = \{\boldsymbol{t}\}^T([\boldsymbol{B}_{tu}] + [\boldsymbol{D}_{tu}])\{\boldsymbol{u}\} \tag{12}$$

Then, subtracting Eq. (11) from Eq. (12), we obtain:

$$\{\boldsymbol{u}\}^T[\boldsymbol{B}_{uu}]\{\boldsymbol{u}\} - \{\boldsymbol{t}\}^T[\boldsymbol{B}_{tt}]\{\boldsymbol{t}\} = \{\boldsymbol{t}\}^T[\boldsymbol{D}_{tu}]\{\boldsymbol{u}\} - \{\boldsymbol{u}\}^T[\boldsymbol{D}_{ut}]\{\boldsymbol{t}\} \tag{13}$$

having used the symmetry property $[\boldsymbol{B}_{ut}]^T = [\boldsymbol{B}_{tu}]$. Besides, one can observe from Eqs. (5) and (9) that:

$$\begin{aligned} \mathcal{D}_{tu}(\boldsymbol{t}, \boldsymbol{u}) - \mathcal{D}_{tu}(\boldsymbol{u}, \boldsymbol{t}) &= [(\kappa - 1) - \kappa] \int_\Gamma \boldsymbol{t}(\boldsymbol{x}).\boldsymbol{u}(\boldsymbol{x}) \, \mathrm{d}\Gamma_x \\ &= -\int_\Gamma \boldsymbol{t}(\boldsymbol{x}).\boldsymbol{u}(\boldsymbol{x}) \, \mathrm{d}\Gamma_x \\ &= -2W(\boldsymbol{u}) \end{aligned}$$

Going back to the discretized problem, one then has in an approximate sense (see [8] for a detailed discussion of this issue):

$$\{\boldsymbol{t}\}^T[\boldsymbol{D}_{tu}]\{\boldsymbol{u}\} - \{\boldsymbol{u}\}^T[\boldsymbol{D}_{ut}]\{\boldsymbol{t}\} = -2W(\{\boldsymbol{u}\})$$

or, returning to Eq. (13:

$$2W(\{\boldsymbol{u}\}) = \{\boldsymbol{t}\}^T[\boldsymbol{B}_{tt}]\{\boldsymbol{t}\} - \{\boldsymbol{u}\}^T[\boldsymbol{B}_{uu}]\{\boldsymbol{u}\} \tag{14}$$

Finally, solving Eq. (11) for $\{\boldsymbol{t}\}$ (note that this system is always uniquely solvable) and substituting the result into Eq. (14) yields the result:

$$W = \frac{1}{2}\{\boldsymbol{u}\}^T[\boldsymbol{K}]\{\boldsymbol{u}\} \tag{15}$$

where the stiffness matrix $[\boldsymbol{K}]$ is given by:

$$[\boldsymbol{K}] = ([\boldsymbol{B}_{ut}] + [\boldsymbol{D}_{ut}])[\boldsymbol{B}_{tt}]^{-1}([\boldsymbol{B}_{tu}] + [\boldsymbol{D}_{ut}]^T) - [\boldsymbol{B}_{uu}] \tag{16}$$

Thus, the above equation provides the symmetric stiffness matrix for the subdomain $\Omega$, which can be added to the stiffness matrix for the complementary subdomain, obtained e.g. by the FEM approach. It is however important to point out that Eqs. (11) and (12) are not exactly satisfied by the same pair $\{\boldsymbol{u}\}, \{\boldsymbol{t}\}$ due to discretization error (in other words Eqs. (13) is true only in an approximate sense), although their continuous versions are exactly satisfied.

# 4. NUMERICAL IMPLEMENTATION

The evaluation of all integrals calls for the integration of kernels having at most a $1/r$ singularity, i.e. leading to simple surface integrations that are convergent in the usual sense; this is the main advantage of the analytical regularization. The present Galerkin procedure employs standard BEM modeling: the boundary $\Gamma$ of the subdomain $\Omega$ is divided into quadrilateral, eight-noded isoparametric elements. In order to evaluate numerically the entries of the matrices $[\boldsymbol{B}_{uu}]$ etc., the double surface integrals over all pairs of quadrilateral elements must be computed and three situations arise as follows:

(i) *Disjoint elements* : the two elements do not have any common edge or vertex, so the double integration is nonsingular and is evaluated by using the standard Gaussian quadrature product rule with a variable order. The severity index, as defined in [7], is computed, in order to adjust the number of Gauss points to the distance between both elements. When this distance is sufficiently large, the evaluation of the double regular integrals can be accurately achieved with (2x2) Gauss points on each element.

(ii) *Coincident elements* : the inner integral being weakly singular, its evaluation is based on a subdivision of the element into triangles for which the common shared vertex is a Gauss point for the (nonsingular) outer integration. This transformation introduces a Jacobian which cancels the weak singularity of the inner integral. The evaluation of the outer one is straightforward using Gaussian quadrature. One notice that in practice this integration procedure slightly alterates the overall symmetry due to the use of different quadrature schemes for inner and outer integrations. Another integration scheme, which preserves exact symmetry by treating the double integration as a whole and using relative intrinsic coordinates, is under implementation.

(iii) *Adjacent elements*: this case is specific to SGBEMs, and is like (ii), a case of singular integration, but the singularity is weaker and occurs when the outer point and the inner one are simultaneously located on the side or on the vertex shared by the two elements. This integration is, as of now, performed for simplicity using a nonsingular (Gaussian) integration scheme, and in particular the numerical examples presented in the next section rely on this imperfect treatment. However, another integration scheme, in which this weak singularity is correctly treated, is currently under implementation.

Once the stiffness matrix $[\boldsymbol{K}]$ is computed, coupling can directly be performed thanks to the standard high-level operators available in the CASTEM 2000 environment, which confers a great versatility to the method.

## 5. NUMERICAL EXAMPLES

Several example elastostatic problems for which analytical solutions are available are now presented in order to illustrate the accuracy and versatility of the three dimensional SGBEM-FEM coupling.

**Rectangular prism submitted to uniaxial tension.** An elastic bar (length: 3 units, $E = 1$, $\nu = 0.3$) with square section (Fig. 5) is divided into 3 subregions: the surface $\partial\Omega_2$ is discretized by quadrilateral boundary elements, while $\Omega_1$ and $\Omega_3$ are filled with standard parallelepipedic finite elements. Symmetry conditions (i.e. zero normal displacement and tangential tractions) are prescribed on the coordinate planes $x_i = 0$, $\forall i \in [1,2,3]$ (notice that displacements can be prescribed without difficulty along boundary elements thanks to standard CASTEM

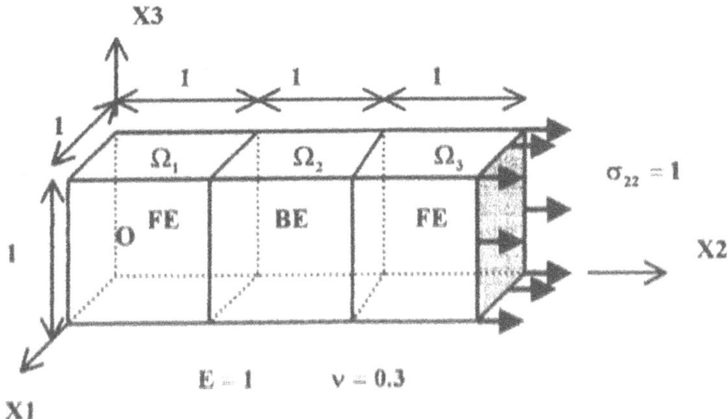


*Figure 1* Rectangular prism submitted to uniaxial tension : discretization of the 3 subregions

2000 operators). A uniform traction $\sigma_{22} = 1$ is imposed on the bar end defined by $x_2 = 3$.

Displacements in the BEM subregion are found to be in good agreement with the analytical solution. Table 1 contains the results for the displacements components when 24 quadrilateral boundary elements are used for $\partial\Omega_2$.

*Table 1* Nodal displacements along the edge $x_1 = x_3$ in the B.E. subregion.

| direction | $x_2$ | Numerical | Analytical | Relative error (%) |
|---|---|---|---|---|
| | 1.00 | 0.99722 | 1.00000 | -0.278 |
| | 1.25 | 1.24620 | 1.25000 | -0.380 |
| Longitudinal | 1.50 | 1.50000 | 1.50000 | +0.000 |
| | 1.75 | 1.75000 | 1.75000 | +0.000 |
| | 2.00 | 2.00107 | 2.00000 | +0.107 |
| | 1.00 | 0.29904 | 0.3 | -0.095 |
| | 1.25 | 0.30180 | 0.3 | +0.180 |
| Transversal | 1.50 | 0.29550 | 0.3 | -0.449 |
| | 1.75 | 0.30246 | 0.3 | +0.246 |
| | 2.00 | 0.30069 | 0.3 | +0.069 |

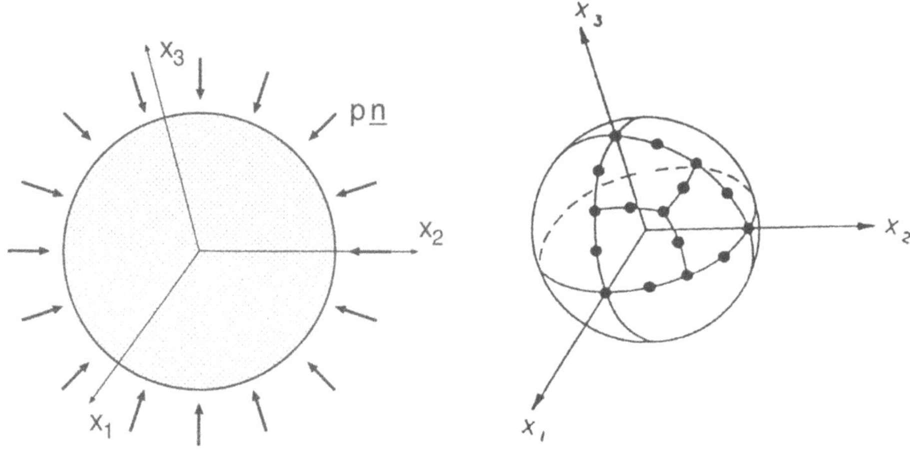


*Figure 2* Sphere submitted to external uniform pressure(a), three quadrilateral boundary elements (b).

One note that this problem is characterized by discontinuous tractions along the edges of the BE subdomain. This point, which is somewhat cumbersome to deal with in regular collocation methods (especially if one wants to equate the number of equations and unknowns), is taken into account without any special consideration, excepted the mention of these discontinuities in the data field. This point constitutes an additional useful feature of SGBIE.

**Sphere submitted to external uniform pressure.** The problem of a sphere of radius $r = 1$ submitted to external uniform pressure (Fig. 2a) is considered, again using $E = 1$ and $\nu = 0.3$. The FE-BE interface is made of 24 quadratic quadrilateral isoparametric elements (Fig. 3a) and 74 nodes. The load is applied on the external boundary of the FE subregion.

The relative error produced by the coupled FEM-SGBEM analysis, for the radial displacement $u_r$ in the BE subregion is found not to exceed 0.0325%, which is a quite good accuracy given the relatively coarse mesh.

**Spherical cavity in infinite body under uniform pressure.** Finally, the hollow spherical cavity in an infinite elastic body (with again $E = 1$ and $\nu = 0.3$), submitted to a uniform internal pressure, is studied. Figure 3b illustrates the discretization. The mesh of the BE-FE interface is again made of 24 elements. The spherical symmetry of the problem is, like in the previous example, not exploited. Here, the stiffness matrix

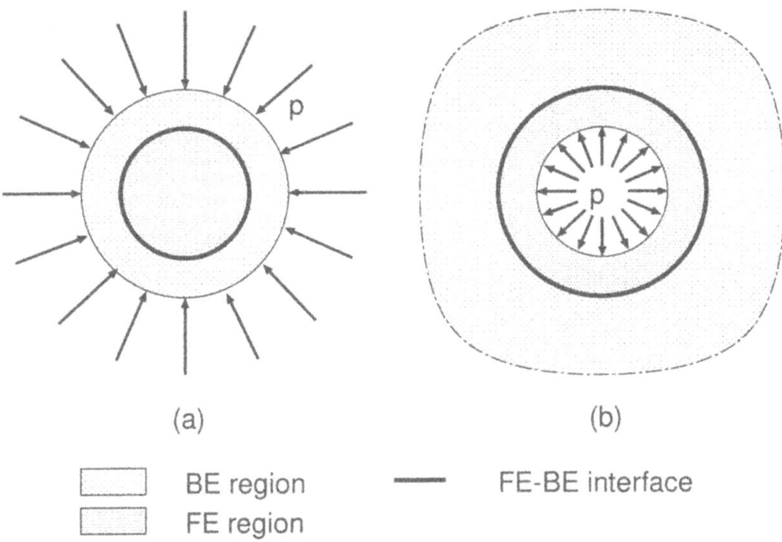


*Figure 3* BE and FE domains for (a) the sphere under uniform external pressure. (b) the spherical cavity in an infinite body under to uniform pressure

$[\boldsymbol{K}]$ evaluated with Eq. (16) accounts for the infinite elastic region, with implicitly enforced decay conditions at infinity.

The relative error produced by the coupled FEM-SGBEM analysis, for the radial displacement $u_r$ in the BE-FE interface is found not to exceed 0.23%, which is again good given the relatively coarse mesh. The higher relative error (the FE-BE mesh being the same as for the finite sphere problem) is probably explained by the fact that the BE-modelled region is much larger (recall that Eq. (16) relies on an equality which is only approximate for discretized problems).

## 6. CONCLUSION

A procedure for coupling the symmetric Galerkin boundary element method and the finite element method has been developed for three-dimensional problems. The procedure creates the stiffness matrix for the BEM-modelled subdomain, for which elastic equilibrium without body forces is assumed. This subdomain is thus treated as a macro-element. This procedure has been implemented into the CASTEM 2000 FEM environment.

Sample numerical examples (rectangular prism submitted to uniform tension, characterized by discontinuous tractions; sphere subjected to external pressure; spherical cavity in infinite body under uniform pressure) yield numerical results in good agreement with the corresponding exact solutions.

## References

[1] BONNET, M. Regularized direct and indirect symmetric variational BIE formulations for three-dimensional elasticity. *Engng. Anal. with Bound. Elem.*, **15**, 93–102 (1995).

[2] BONNET, M. *Boundary integral equations methods for Solids and Fluids*. John Wiley and sons (1999).

[3] BONNET, M., MAIER, G., POLIZZOTTO, C. On symmetric galerkin boundary element method. *Appl. Mech. Rev.*, **51**, 669–704 (1998).

[4] CASTEM 2000. see website `http://www.castem.org:8001` (1999). CASTEM 2000 is a research FEM environment; its development is sponsored by the French Atomic Energy Commission (*Commissariat à l'Energie Atomique - CEA*).

[5] HSIAO, G C. The coupling of boundary element and finite element methods. *Z. Angew. Math. Mech.*, **70**, 493–503 (1990).

[6] NEDELEC, J-C. Approximation des équations intégrales par éléments finis. Etude d'erreur. In J. L. Lions R. Dautray (ed.), *Analyse mathématique et calcul scientifique pour les sciences et les techniques (chap. 13)*, pp. 953–968. Masson, Paris (1985).

[7] REZAYAT, M., SHIPPY, D. J., RIZZO, F. J. On time-harmonic elastic wave analysis by the boundary element method for moderate to high frequencies. *Comp. Meth. in Appl. Mech. Engng.*, **55**, 349–367 (1986).

[8] SIRTORI, S, MICCOLI, S, KORACH, E. Symmetric coupling of finite elements and boundary elements. In J. H. Kane, G. Maier, N. Tosaka, S. N. Atluri (eds.), *Advances in Boundary Element Techniques*, pp. 407–427. Springer-Verlag (1993).

# ON A UNIFIED PROCEDURE FOR THE NUMERICAL EVALUATION OF SINGULAR AND QUASI-SINGULAR DOUBLE INTEGRALS

M. A. M. Noronha(1) and N. A. Dumont(2)

*(1) Civil Engineering Department, Escola Politécnica da Universidade de São Paulo – EPUSP 05424-970 São Paulo, Brazil (mnoronha@usp.br)*

*(2) Civil Engineering Department, Pontifícia Universidade Católica do Rio de Janeiro – PUC-Rio, 22453-900 Rio de Janeiro, Brazil (dumont@civ.puc-rio.br)*

Keywords: Boundary Element Methods, Numerical Integration

Abstract: A boundary integral implementation generally involves the evaluation of singular and quasi-singular integrals. Dealing with them requires elaborated codes, which quite frequently results in prohibitive computational effort as the price of adequate numerical accuracy. In the frame of a research line in progress at PUC-Rio, a simple and accurate scheme has been developed for the treatment of general singular linear integrals. It is now being implemented for double integrals. This technique makes use of fixed abscissas, as taken from Gauss-Legendre quadrature, with different weight sets calculated for different singularities – a procedure that demands low computational effort. This feature allows the development of robust codes, in which integration demands low computational effort and yields high precision. The core of the present contribution is the introduction of an adequate scheme for the global-to-local transformation of the coordinates of a given singularity pole. This transformation may be used in connection with polar coordinates, as singular integrals are usually dealt with. In the paper, however, it is an essential feature of the unified integration procedure the authors are proposing. The technique is straightforward to implement, independently of the degree of singularity or quasi-singularity.

*T. Burczynski (ed.), IUTAM/IACM/IABEM Symposium on Advanced Mathematical and Computational Mechanics Aspects of the Boundary Element Method, 259–268.*

# 1. INTRODUCTION

The effectiveness of a boundary integral equation formulation usually relies on a singularity function which is present in the required fundamental solution. Depending on the type of differential equation being dealt with in a three-dimensional problem, this singularity function may assume different shapes, such as $r^{-1}$, $r^{-2}$, $r^{-3}$, $r^{-5}$, and $r^{-7}$. A quasi-singularity is a more subtle issue than an actual singularity, particularly because its effects are more difficult to assess. Moreover, while an actual singularity may be dealt with indirectly, either by means of some spectral properties of the matrices involved in the formulation (use of rigid body motions or of constant stress states, for instance) or as a consequence of some re-formulation (non-singular boundary integral representations), there are no alternatives in case of quasi-singularities other than applying a suitable integration scheme. An inadequate numerical treatment of quasi-singular integrals may increase substantially the global computational effort in practical applications, and still not result in a satisfactory accuracy improvement (Dumont and Noronha, 1998). The present outline is an attempt to arrive at a formulation dealing adequately and in the same framework of a regular integration with both cases of singularity and quasi-singularity occurring in double integrals (Noronha 1998; Noronha and Dumont, 1999).

# 2. TYPES OF SINGULARITY POLES

Consider a generally curved boundary segment, expressed in terms of the integration variables $\xi$ and $\eta$ spanning from 0 to 1, according to Fig. 1. Depending on the position of a singularity source relative to the integration interval, three kinds of singularity may occur, as outlined in the following. The trivial case of the singularity source far from the integration interval (distance greater than the largest interval dimension) needs no consideration, as satisfactory accuracy may always be achieved with standard Gauss-Legendre integration.

If the pole is within the integration interval (point A in the figure) it is said to be a singularity pole. In case of Kelvin's fundamental solution, this singularity may be related to either a single-layer potential ($r^{-1}$), for which an improper integral arises, or to a double-layer potential ($r^{-2}$), with the corresponding singular integral being interpreted in terms of a Cauchy principal value. Higher singular kernels may arise in the frame of a hypersingular formulation. The proposed standardized procedure is applicable without restrictions to a $r^{-1}$ singularity, for general curved elements. For higher singularities, however, the technique has to assume that the integration interval is already normalized. This is still of no practical use, since the normalization of an integral, which has to be dealt with in terms of finite parts, is always related to a polar coordinates transformation (Dumont, 1995) – exactly what will be avoided in the present method (section 5).

A second case occurs if the singularity pole is outside the integration interval, but located on the ideal surface obtained as an extension of the integration interval (point B in the figure): it is a real quasi-singularity, which may be related to either $r^{-1}$ or $r^{-2}$, for Kelvin's fundamental solution.

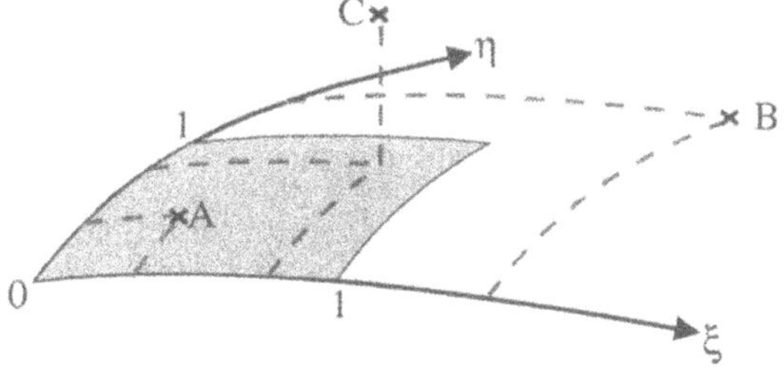


Figure 1. General singularity poles related to a curved boundary interval.

The third case occurs if the pole can only be expressed in terms of complex coordinates of $\xi$ and $\eta$ (point C in the figure), giving rise to a complex quasi-singularity (Dumont, 1994), as assessed in the next section.

## 3. EVALUATION OF A COMPLEX QUASI-SINGULARITY POLE IN LOCAL COORDINATES

Identifying complex poles is a very important issue for the correct interpretation of the singularity sources in an integral. Given a singularity pole $(x_0, y_0, z_0)$ in global coordinates, the first task is to obtain the corresponding coordinates $(\xi_0, \eta_0)$. In case of either a singularity or a real quasi-singularity, $\xi_0$ and $\eta_0$ are real numbers. In case of a complex quasi-singularity, however, both $\xi_0$ and $\eta_0$ are complex, reflecting the fact that the pole is outside the local $\xi\eta$ plane of the natural curvilinear coordinates. In either case, obtaining $(\xi_0, \eta_0)$ requires an iterative algorithm, for which convergence is a delicate issue, mainly because of the existence of multiple solutions.

The first equation for finding the local pole coordinates $(\xi_0, \eta_0)$ is

$$r(\xi_0,\eta_0)=\sqrt{[x(\xi_0,\eta_0)-x_0]^2+[y(\xi_0,\eta_0)-y_0]^2+[z(\xi_0,\eta_0)-z_0]^2}=0 \quad (1)$$

However, since one has a pair of unknowns, a second equation must be called upon. One may express $r(\xi, \eta)$ as a product of two functions:

$$r(\xi,\eta)=\rho(\xi,\eta)\,\bar{r}(\xi,\eta) \quad (2)$$

in which $\rho(\xi, \eta)$ is the distance from the singularity pole measured in the local coordinate system, such that $\rho(\xi_0, \eta_0) = 0$, and $\bar{r}(\xi,\eta)$ does not vanish at $(\xi_0, \eta_0)$. A criterion for the adequacy of eq. (2) is that $\bar{r}(\xi,\eta)$ be single-valued at ($\xi_0$, $\eta_0$), that is

$$\bar{r}(\xi_0,\eta_0)=\lim_{\substack{\xi\to\xi_0\\ \eta\to\eta_0}}\frac{r(\xi,\eta)}{\rho(\xi,\eta)} \tag{3}$$

independently of the path chosen for approaching ($\xi_0$, $\eta_0$). As it turns out, the solution is always conjugate complex, i. e., $\xi_0=a_\xi\pm b_\xi i$, $\eta_0=a_\eta\pm b_\eta i$, from which follows the expression for the local radius $\rho(\xi,\eta)$ as:

$$\begin{aligned}\rho(\xi,\eta)&=\sqrt{(\xi-a_\xi+b_\xi i)(\xi-a_\xi-b_\xi i)+(\eta-a_\eta+b_\eta i)(\eta-a_\eta-b_\eta i)}=\\&=\sqrt{(\xi-a_\xi)^2+(\eta-a_\eta)^2+b_\xi^2+b_\eta^2}\equiv\sqrt{(\xi-a_\xi)^2+(\eta-a_\eta)^2+d^2}\end{aligned} \tag{4}$$

Applying L'Hospital's rule to evaluate $\bar{r}(\xi_0,\eta_0)$ as $\xi\to\xi_0$ and $\eta\to\eta_0$ one obtains from eqs. (3) and (4)

$$\frac{\left.\dfrac{\partial r(\xi,\eta)}{\partial\xi}\right|_{\xi_0,\eta_0}}{\left.\dfrac{\partial\rho(\xi,\eta)}{\partial\xi}\right|_{\xi_0,\eta_0}}=\frac{\left.\dfrac{\partial r(\xi,\eta)}{\partial\eta}\right|_{\xi_0,\eta_0}}{\left.\dfrac{\partial\rho(\xi,\eta)}{\partial\eta}\right|_{\xi_0,\eta_0}}\Rightarrow\frac{\left.\dfrac{\partial r(\xi,\eta)}{\partial\xi}\right|_{\xi_0,\eta_0}}{b_\xi}=\frac{\left.\dfrac{\partial r(\xi,\eta)}{\partial\eta}\right|_{\xi_0,\eta_0}}{b_\eta} \tag{5}$$

since $\left(\left.\dfrac{\partial\rho(\xi,\eta)}{\partial\xi}\right|_{\xi_0,\eta_0}\right)\Big/\left(\left.\dfrac{\partial\rho(\xi,\eta)}{\partial\eta}\right|_{\xi_0,\eta_0}\right)\equiv {b_\xi}/{b_\eta}$.

Equations (1) and (5) are the necessary and sufficient conditions for evaluating the roots ($\xi_0$, $\eta_0$). Since they constitute a set of non-linear equations, ensuring convergence to the desired results is a key issue in the solution algorithm.

In eq. (4), $d$ is the distance of the complex singularity pole from the local $\xi\eta$ plane of the natural curvilinear coordinates. This concept of distance does not relate geometrically to the actual curved surface. As a matter of illustration, consider the curved 9-nodes Lagrangean element of Fig. (2). The edges form a unitary square in the $xy$ plane, the axes of which coincide with the local axes $\xi$ and $\eta$. The center node has an elevation of 0.2 units. Further, consider a singularity pole with coordinates (0.5, 0.5, 0.25), thus lying 0.05 units above the curved surface. Evaluating ($\xi_0$, $\eta_0$) for this example according to eqs. (1) and (5) yields $\xi_0=\eta_0=0.5\pm0.034041260i$, from which follows a value of $d=0.048141611$ which is different from the actual geometric distance of 0.05 units. The real parts of roots $\xi_0$ and $\eta_0$ coincide with the corresponding Cartesian values, as it should be the case because of symmetry reasons. In order to demonstrate how intriguing the results of the proposed coordinates transformation are, the singularity pole is evaluated again, referred to the middle line $z = 0.8\xi(1 - \xi)$, as a one-variable function obtained from the curved element for $\eta = 0.5$. In this case, one obtains for $r(\xi_0) \equiv \sqrt{[x(\xi_0) - x_0]^2 + [z(\xi_0) - z_0]^2} = 0$ that $\xi_0 = 0.5 \pm 0.048145601i$. On the other hand, considering $z = 0.8\xi(1-\xi)$ as the equation of a cylindrical surface, the distance $r(\xi_0, \eta_0)$ is expressed in terms of $(x, y, z)$ and, consequently, the solution of eqs. (1) and (5) is obtained as $\xi_0 = 0.5$ and $\eta_0 = 0.5 \pm 0.05i$, yielding a distance $d = 0.05$. Hence, each one of the three cases considered resulted in a different value of $d$ (15 digits precision used in the computation), although the Cartesian distance was always the same. A singularity pole outside the symmetry planes, say (0.6, 0.7, 0.205), may also be considered. In this case one obtains with eqs. (1) and (5) the result $\xi_0 = 0.59499357 \pm 0.038175635i$, $\eta_0 = 0.68876383 \mp 0.013073394i$.

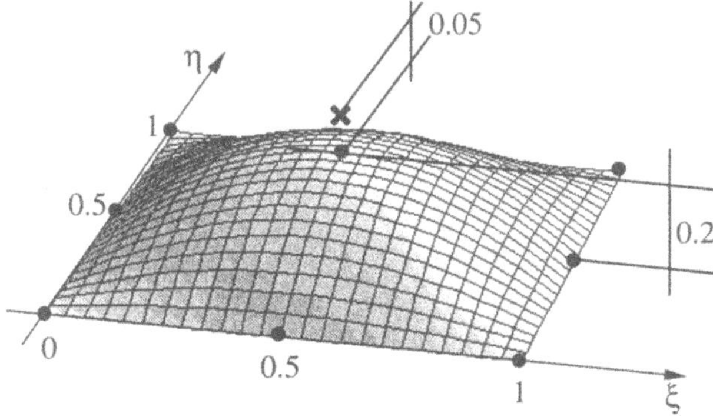


Figure 2: Curved Lagrangean element.

As a second example, consider the plane 4-nodes element of Fig. 3a together with a singularity pole given by global coordinates (0.5, 0.5, 0.5). The projection $(\xi_p, \eta_p)$ of this pole onto the local plane $\xi\eta$ is given by solving

$$\begin{cases} x(\xi_p, \eta_p) - x_0 = 0 \\ y(\xi_p, \eta_p) - y_0 = 0 \end{cases} \tag{6}$$

from which follows that $\xi_p = 0.28077641$, $\eta_p = 0.43844719$. On the other hand, solving eqs. (1) and (5) for this problem yields $\xi_0 = 0.31761122 \pm 0.55442933i$ and $\eta_0 = 0.32390947 \mp 0.36889734i$, which indicates that even for the case of a plane element $\xi_p \neq \Re(\xi_0)$ and $\eta_p \neq \Re(\eta_0)$.

These two illustrations show that complex singularity poles have to be dealt with as complex entities. The geometric projection of the pole, which makes use of a "tangent plane", as proposed by Hayami (1992), for instance, is only an approximation of the correct procedure.

Some applications might require the complex quasi-singularity pole expressed in a polar coordinate system, as illustrated in Fig. 3b with reference to the local coordinate system. In this case, one writes

$$\begin{cases} \xi = a_\xi + \rho_0 \cos(\theta) \\ \eta = a_\eta + \rho_0 \sin(\theta) \end{cases} \tag{7}$$

Then, it follows from eq. (4) that $\rho = \sqrt{\rho_0^2 + d^2}$ and, for $\rho = 0$,

$$\rho_0(\xi_0, \eta_0) = \pm d\, i \tag{8}$$

which yields the adequate interpretation of the distance *d*. Accordingly, eq. (2) may be expressed as

$$r(\xi(\rho_0, \theta), \eta(\rho_0, \theta)) = \sqrt{\rho_0^2 + d^2}\ \bar{r}(\xi(\rho_0, \theta), \eta(\rho_0, \theta)) \tag{9}$$

With the transformations introduced in this section one is able to treat the quasi-singularity occuring in a double integral in the framework of the procedure proposed by Dumont (1994) for linear integrals.

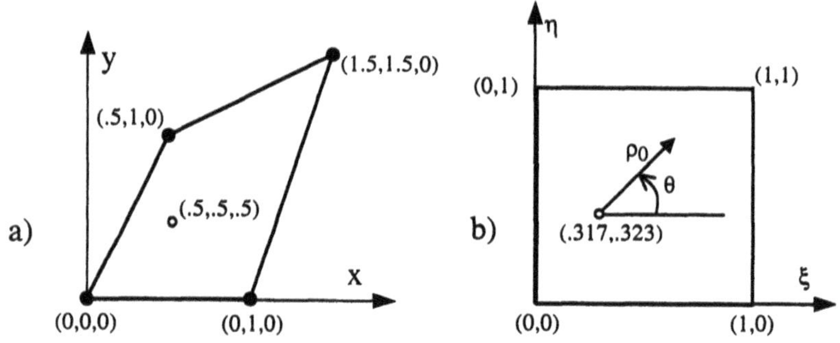


Figure 3: a) Plane 4-nodes element in global coordinates and quasi-singularity pole. b) Corresponding representation in local coordinates.

# 4. GRADIENTS AS POTENTIAL SOURCES OF QUASI-SINGULARITY

Consider, for illustration, the gradient of *r* in the direction x, expressed in terms of the coordinates $\xi$ and $\eta$:

$$\frac{\partial r}{\partial x} = \frac{x(\xi, \eta) - x_0}{r(\xi, \eta)} = \frac{x(\xi, \eta) - x_0}{\rho(\xi, \eta)\bar{r}(\xi, \eta)} \tag{10}$$

For both cases of singularity and real quasi-singularity, $(\xi_0, \eta_0) \equiv (\xi_p, \eta_p)$ is a real root of $r(\xi, \eta) = 0$. Then, according to eq. (6), $x(\xi_0, \eta_0) - x_0 = 0$ too, with the consequence that $\partial r(\xi_0, \eta_0)/\partial x$ is bounded.

In case of a complex quasi-singularity, however, $x(\xi_0,\eta_0)-x_0 \neq 0$ and $\partial r(\xi_0,\eta_0)/\partial x$ is unbounded. Then, use of Kelvin's fundamental solution results in the occurrence of higher-order quasi-singularities up to $r^{-7}$, as in case of evaluating stresses at interior points. Such singularities must be considered if one intends to avoid rough integration errors.

## 5. ON A UNIFIED INTEGRATION SCHEME

The outline of the preceding sections, especially of section 3, would suffice for the sake of a correct assessment of a quasi-singularity pole, if one was satisfied with the use of polar coordinates in an integration scheme by subdividing the integration interval into several triangles. However, this is a very cumbersome procedure, which requires lengthy and complicated coding.

The authors are proposing, as they successfully did in case of linear integrals, a unified procedure making use of fixed abscissas, as taken from the Gauss-Legendre quadrature, with different weight sets calculated for different singularities (Noronha, 1998).

Consider the general integration scheme

$$\int_0^1\int_0^1 \omega g \, d\xi \, d\eta \approx \sum_{i=1}^{m}\sum_{j=1}^{n} g(\xi_i,\eta_j)h_{ij} \tag{11}$$

in which $g \equiv g(\xi,\eta)$ is a regular function, in the sense that it may be well represented by a low-degree polynomial, and $\omega \equiv \omega(\xi,\eta)$ is a singular or quasi-singular function, that is, the part of the integrand that cannot be well represented by a polynomial. The integration is to be carried out over a normalized interval, say (0, 1), in Cartesian coordinates $(\xi,\eta)$. Note that depending on the singularity kernel $\omega(\xi,\eta)$ the integral represented at the left-hand side of eq. (11) may have to be interpreted in terms of finite parts (Dumont, 1995). The unified technique proposed in this section is not able to deal with such a normalization, which still has to be accomplished in the frame of a polar coordinates transformation.

In the proposed unified technique (Dumont, 1992, 1994; Dumont and Noronha, 1998), the integration scheme makes use of the same abscissas $(\xi_i,\eta_j)$ as the standard Gauss-Legendre quadrature. If the integrand were regular, that is, $\omega(\xi,\eta)=1$ in eq. (11), the weights $h_{ij}$ would correspond to those of the Gauss-Legendre quadrature, in which case $n$ integration points in a given coordinate direction would integrate exactly a polynomial of degree $(2n-1)$ representing the integrand $\omega(\xi,\eta)g(\xi,\eta)$ in this direction. In the present case, however, the weights $h_{ij}$ are to be evaluated for a general singularity kernel $\omega(\xi,\eta)$, but considering abscissas of a Legendre polynomial. As a consequence, with $n$ integration points in a given coordinate direction one can only achieve the exact value of the integral if the regular function $g(\xi,\eta)$ can be exactly represented by a polynomial of degree $(n-1)$ in this direction. Note that the $(2n-1)$ accuracy measurement of the Gauss-Legendre quadrature refers to the whole integrand $\omega(\xi,\eta)g(\xi,\eta)$, not only to $g(\xi,\eta)$. As a matter of fact, the regular part of the integrand in eq. (11) is usually related to low-degree polynomials, in the boundary element implementations, provided that the elements are not too distorted. Moreover, it is worth observing that the accuracy of the proposed scheme increases with the degree of singularity represented by $\omega(\xi,\eta)$ (Dumont, 1994).

For the evaluation of the weights $h_{ij}$, eq. (11) is applied to $g(\xi,\eta)$ taken as the Lagrange polynomial

$$\mathrm{L}_{ij}=\prod_{\substack{k=1\\k\neq i}}^{m}\frac{(\xi-\xi_k)}{(\xi_i-\xi_k)}\prod_{\substack{l=1\\l\neq j}}^{n}\frac{(\eta-\eta_l)}{(\eta_j-\eta_l)}\quad\text{for } i=1,\cdots,m, \text{ and } j=1,\cdots,n \tag{12}$$

of degrees ($m$-1) and ($n$-1) in the coordinates $\xi$ and $\eta$, respectively. As a consequence, one obtains

$$h_{ij}=\int_0^1\int_0^1\omega\mathrm{L}_{ij}d\xi d\eta\quad\text{for } i=1,\cdots,m, \text{ and } j=1,\cdots,n \tag{13}$$

provided that the double integral at the right-hand side of this equation can be obtained analytically, or at least by a sufficiently accurate numerical scheme. As a matter of fact, Noronha (1998) succeeded in arriving at the analytical solution of the general family of integrals

$$\int_0^1\int_0^1\frac{\xi^i\eta^j}{\rho^p}d\xi d\eta\quad\text{for } i=1,\cdots,m,\ j=1,\cdots,n, \text{ and } p=1,3,5,7\cdots \tag{14}$$

in which $\rho$ is defined by eq. (4). The analytical expressions of the different integrals comprised by eq. (14), as developed by Noronha (1998), are not found in mathematical handbooks. They shall be published opportunely.

## 6. SOME NUMERICAL EXAMPLES

As a matter of illustration, consider the numerical evaluation of the following integral, in which the numerator is a regular function (although the square roots involved precludes an adequate approximation by a polynomial of too low a degree):

$$\int_0^1 \int_0^1 \frac{\sqrt{\xi^2 - 3\xi + 2.5}\sqrt{\eta^2 - 3\eta + 2.5}}{\left(\sqrt{(\xi - \xi_0)^2 + (\eta - \eta_0)^2 + \zeta_0^2}\right)^p} d\xi\, d\eta \tag{16}$$

The following cases are investigated (Noronha and Dumont, 1999):

a) $p = 1$ and $(x_0, y_0, z_0)$ corresponding to points S and Q in Fig. 4, with integration errors shown in Figs. 5a and 5b, respectively. The first case is an improper integral, whereas the second is quasi-singular.

b) $p = 7$ and $(x_0, y_0, z_0)$ are out-of-plane singularity poles $\overline{S} = (0,0,.002)$ and $\overline{Q} = (-.002,-.002,.002)$ corresponding to points S and Q in Fig. 4, with integration errors shown in Figures 6a and 6b, respectively. Such type of integral is required for evaluating stresses at internal points.

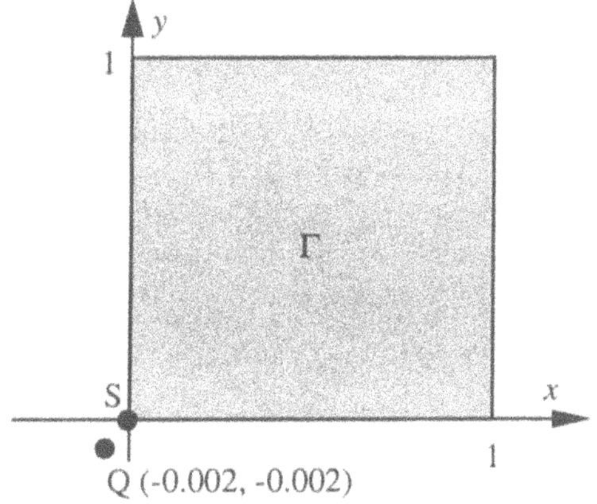


Figure 4: Square integration interval with a singularity pole S and a quasi-singularity pole Q.

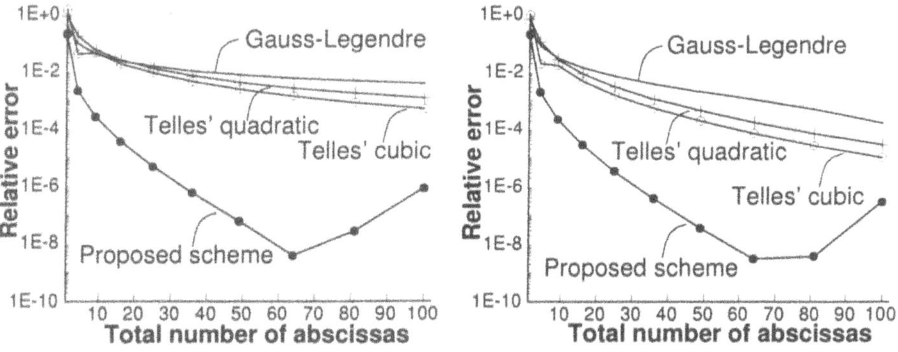


Figures 5a and 5b: Errors in the evaluation of the integral of eq. (16) for $p = 1$ (see text).

In all graphics, the results obtained by a standard Gauss-Legendre quadrature are also displayed. The relative errors are given in logarithmic scale as referred to the total number of integration points (a square mesh with the same number of abscissas in each direction). In both graphics of case $p = 1$ a comparison with the results given by the quadratic and cubic coordinates transformations proposed by Telles (1987) is performed. Transformation to polar coordinates before application of standard Gaussian quadrature would give good results only in case of $p = 1$ for the singularity pole S. In all other cases, this transformation is useless (Noronha, 1998).

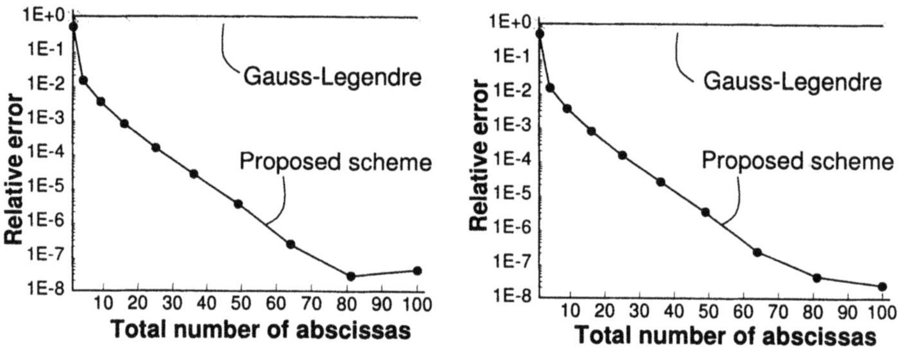


Figures 6a and 6b: Errors in the evaluation of the integral of eq. (16) for $p = 7$ (see text).

# REFERENCES

N. A. Dumont (1994), "On the Efficient Numerical Evaluation of Integrals with Complex Singularity Poles", Engineering Analysis with Boundary Elements 13, pp 155-168.

N. A. Dumont (1995), "A Procedure for the Semi-Analytical Evaluation of Generally Singular Integrals that Occur in the Three-Dimensional Boundary Element Analysis", Boundary Elements XVII, Eds.: C. A. Brebbia, S. Kim, T. A. Osswald, H. Power, Comput. Mech. Publ., Southampton, pp 83-90.

N. A. Dumont, and M. A. M. Noronha (1998), "A Simple, Accurate Scheme for the Numerical Evaluation of Integrals with Complex Singularity Poles", Computational Mechanics 22, pp 42-49.

N. A. Dumont and R. M. de Souza (1992), "A Simple Unified Technique for the Evaluation of Quasi-Singular, Singular and Strongly Singular Integrals", Boundary Elements XIV, Comput. Mech. Publ., Vol 1, pp 619-632.

K. Hayami (1992), "A Projection Transformation Method for Nearly Singular Surface Boundary Element Integrals", Lecture Notes in Engineering, Springer-Verlag, Heidelberg.

M. A. M. Noronha (1998), "Técnicas Avançadas de Integração Numérica e Programação Orientada a Objetos Aplicadas a Métodos de Elementos de Contorno", Ph.D. Thesis, PUC-Rio, Brazil.

M. A. M. Noronha, N. A. Dumont (1999), "A Numerical Procedure for the Evaluation of Singular and Quasi-Singular Integrals That Occur in the Three-Dimensional Boundary Element Analysis", PACAM'99, Pan-American Congress of Applied Mechanics, Vol. 6, pp 361-364, Rio de Janeiro, Brazil, 4-8 January 1999.

J. C. F. Telles (1987), "A Self-Adaptive Co-ordinate Transformation for Efficient Numerical Evaluation of General Boundary Element Integrals", Int. J. for Num. Meth. in Engng., Vol. 24, pp 959-973.

# SOLVING INVERSE BOUNDARY PROBLEMS IN CONTINUOUS CASTING UTILIZING SENSITIVITY COEFFICIENTS AND THE BOUNDARY ELEMENTS METHOD

Iwona NOWAK
*Institute of Mathematics, Technical University of Silesia,*
*44-101 Gliwice, Kaszubska 23, Poland*
iwonowak@polsl.gliwice.pl

Andrzej J. NOWAK
*Institute of Thermal Technology, Technical University of Silesia,*
*44-101 Gliwice, Konarskiego 22, Poland*
nowak@itc.ise.polsl.gliwice.pl

**Keywords:** inverse boundary problem, sensitivity coefficients, BEM solution of continuous casting

**Abstract** This paper deals with the determination of the unknown heat fluxes along the external boundary of the ingot during continuous casting process by making use of the sensitivity coefficients. As a result of the research, the complete temperature field was obtained and compared to the measurements. All calculations were made using the Boundary Element Method.

The main goal of the analysis was to assess the influence of the number of measurement sensors and their accuracy on estimated heat fluxes. The analysis shows that several factors are influencing the accuracy of the solution. It also shows that the search for the best sensor locations based only on sensitivity coefficients is a rather difficult task.

## 1. INTRODUCTION

Boundary problems with unknown boundary conditions which must be estimated by an optimization procedure are referred to as the Inverse

*T. Burczynski (ed.), IUTAM/IACM/IABEM Symposium on Advanced Mathematical and Computational Mechanics Aspects of the Boundary Element Method,* 269–280.

Boundary Problems. The solution of such a problems generally employs numerical methods based on space discretization.

The mathematical formulation of the inverse thermal problem is, in principle, similar to the formulation of the direct problem. The problem description requires the following components:

- a governing equation which involves a partial differential operator acting on the temperature $T$,
- boundary conditions defining heat transfer processes along the external boundary. It must be remembered, however, that for inverse boundary problems some of these conditions (or even all of them) are not known (or at least uncertain). Hence, quantities defining the boundary conditions are usually collected within the vector of design variables $\mathbf{Y}$ and then estimated,
- because some of the boundary conditions are unknown, the mathematical formulation needs to be supplemented with additional information resulting from measurements, *e.g.*, temperature measurements. However, measurements do not only carry useful information but also noise. This, together with the ill-posed nature of the problem, deteriorates the accuracy and can make the estimated heat fluxes very inaccurate.

In this paper a numerical heat transfer solution of the inverse boundary problem for continuous casting is discussed. The main goal at this stage of the analysis is to estimate heat fluxes along the external surface of the ingot. Verified information about these fluxes is a condition for a rational design of a mould and cooling system as well as generally of the whole caster.

Thermal inverse analysis discussed in the paper is based on temperature measurements collected by thermocouples immersed to the liquid metal, carried by cast and finally pulled out of the system by the solidified ingot. It is demonstrated that number and locations of sensors inside the cast are quite essential. Since it is usually cumbersome to carry out the measurements, the definition of the number of sensors which guarantees a reasonable inverse solution is crucial.

In this paper it is also shown how sensitivity analysis (based on the sensitivity coefficients) can be applied to simplify the solution procedure of discussed inverse heat transfer problem. Namely, sensitivity coefficients, calculated by the Boundary Element Method, are introduced into the objective function and, as a consequence, the unknown boundary conditions can be estimated.

## 2. FORMULATION OF THE INVERSE BOUNDARY PROBLEM

The inverse thermal problem in a continuous casting process, schematically shown in Fig.1, is governed by the following equations (see also the geometry of the solidified ingot shown in Fig.2):

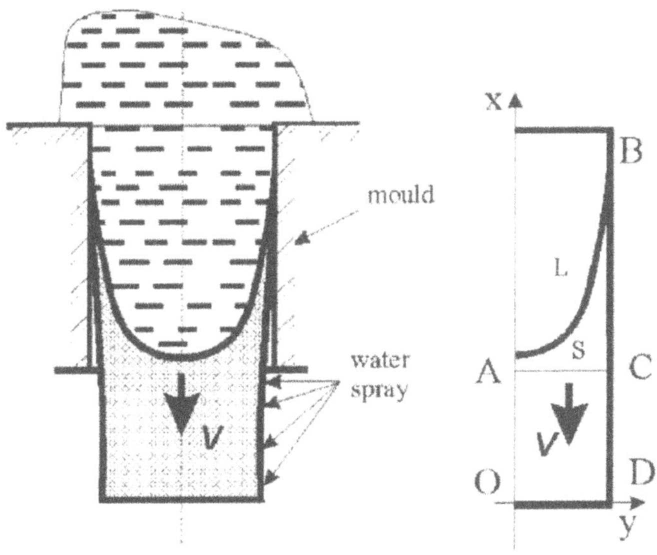


*Figure 1* Scheme of the continuous casting system and the domain under consideration.

- governing differential equation for steady-state diffusion-convection without internal heat sources [1]

$$\nabla^2 T(\mathbf{r}) - \frac{1}{a} v_x \frac{\partial T}{\partial x} = 0 \tag{1}$$

  where $T(\mathbf{r})$ stands for temperature at any point $\mathbf{r}$, $a$ is the thermal diffusivity and $v_x$ is the $x$-component of the velocity vector (equal to the casting velocity)

- two groups of boundary conditions

  1 known boundary conditions resulting from the phase change process, the known temperature of the ingot leaving the system (end temperature) and symmetry, respectively, *i.e.*,

$$T(\mathbf{r}) = T_m(\mathbf{r}), \qquad \mathbf{r} \in \Gamma_{AB} \tag{2}$$

$$T(\mathbf{r}) = T_s(\mathbf{r}), \qquad \mathbf{r} \in \Gamma_{DO} \tag{3}$$

$$q(\mathbf{r}) = -\lambda \frac{\partial T}{\partial n} = 0, \quad \mathbf{r} \in \Gamma_{OA} \tag{4}$$

where $T_m$ stands for the melting temperature, $T_s$ for the end temperature and $n$ for the outward normal to the boundary $\Gamma$

2 unknown boundary conditions (these conditions are being estimated) written here as the Neumann condition

$$q(\mathbf{r}) = -\lambda \frac{\partial T}{\partial n} = f(x), \quad \mathbf{r} \in \Gamma_{BCD} \tag{5}$$

where the function $f(x)$ defines the boundary heat flux along the side face $\Gamma_{BCD}$.

Because of the incomplete mathematical description of the problem it must be supplemented by temperature measurements

$$T(\mathbf{r}_i) = U_i, \quad i = 1, 2, \ldots, I \tag{6}$$

where $U_i$ stands for the temperature measured at point $\mathbf{r}_i$. During the analysis up to twenty sensors were selected inside the ingot at different locations (particularly 0.01m, 0.015m and 0.02m deep under the surface).

As mentioned already, the principal goal of the analysis was to estimate heat fluxes along the external surface (*i.e.*, to find the function $f(x)$ along $\Gamma_{BCD}$). From previous experience [2, 3], it was assumed that the heat flux varies linearly along the mould (face BC) and exponentially along the water spray (face CD), *i.e.*,

$$f(x) = q_{kk} - (q_{kk} - q_{gk}) \frac{1.22 - x}{0.22}, \quad \text{along } \Gamma_{BC} \tag{7}$$

$$f(x) = q_{gn} \exp\left[9.3\,(x - 1)\right], \quad \text{along } \Gamma_{CD} \tag{8}$$

where $q_{kk}, q_{gk}$ and $q_{gn}$ are heat fluxes at points B and C, respectively (see Fig.2). In this way only these three quantities need to be estimated.

## 3. SOLUTION PROCEDURE

The process of retrieving the unknown quantities is realised in two stages:

- unknown quantities are assumed and the temperature field $T_{cal}$ inside the body is calculated. It should be noted that this stage of analysis is equivalent to solving the direct problem

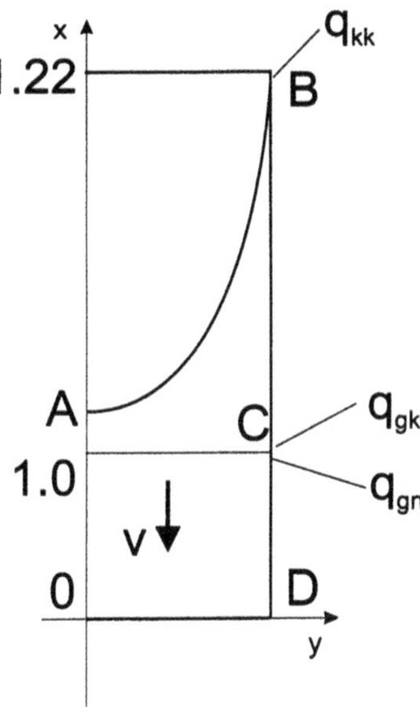


*Figure 2* Geometry and estimated boundary condition characteristics.

- the relevant objective function is minimized [2, 3, 4]. Using matrix notation this function reads

$$\begin{aligned} \Delta &= (\mathbf{T}_{cal} - \mathbf{U})^T \mathbf{W}^{-1} (\mathbf{T}_{cal} - \mathbf{U}) \\ &+ (\mathbf{Y} - \tilde{\mathbf{Y}})^T \mathbf{W}_Y^{-1} (\mathbf{Y} - \tilde{\mathbf{Y}}) \rightarrow \min \end{aligned} \tag{9}$$

  where $\mathbf{W}$ stands for the covariance matrix, the vector $\tilde{\mathbf{Y}}$ represents prior estimates of the sought quantities $\mathbf{Y}$ and $\mathbf{W}_Y$ is the covariance matrix of these estimates. The superscript $()^T$ indicates the transposed matrix.

The first step of the solution procedure is realised in this work by the boundary element method [5] resulting in the following system of equations

$$\mathbf{H}\,\mathbf{T} = \mathbf{G}\,\mathbf{q} \tag{10}$$

where the vector $\mathbf{q}$ contains heat fluxes whereas $\mathbf{H}$ and $\mathbf{G}$ stand for the BEM influence matrices. The coefficients of these matrices result from integration of the fundamental solution $u^*$ (and its normal derivative $q^*$) along the boundary elements only. Depending on the dimension of the problem, the fundamental solution is expressed by the following formulae

$$u^* = \begin{cases} \dfrac{1}{2\pi\lambda} \exp\left(-\dfrac{v_x r_x}{2a}\right) \mathrm{K}_0\left(\dfrac{|v_x| r}{2a}\right) & \text{2–D problems} \\ \dfrac{1}{4\pi r\lambda} \exp\left[\dfrac{v_x (r - r_x)}{2a}\right] & \text{3–D problems} \end{cases} \tag{11}$$

where $\mathrm{K}_0$ stands for the Bessel function of the second kind and zero order, $r$ is the distance between source and field point, with its component along the $x$-axis denoted by $r_x$.

It should be noted at this stage that in order to guarantee an adequate accuracy of the calculations, the phase change front ($\Gamma_{AB}$ in Fig.2) should be modelled by a smooth curve. This requires utilization of at least quadratic elements resulting in $\mathrm{C}^1$ continuity.

Considering a quadratic element (*cf.* Fig.3 and reference [1]) whose end nodes 1 and 3 have co-ordinates and slopes denoted as $x_1$, $y_1$, $s_1$, $x_3$, $y_3$ and $s_3$, respectively, one can obtain the location of the midnode 2 from the relationships

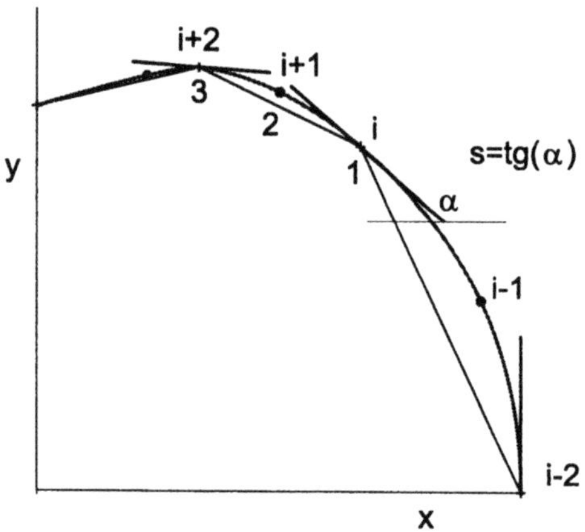


*Figure 3* Quadratic elements with $\mathrm{C}^1$ continuity.

$$\begin{cases} x_2 = \dfrac{1}{2(s_1 - s_3)} (a_1 + a_3 - b_1 s_1 - b_3 s_3) \\ y_2 = \dfrac{1}{2} (b_1 s_1 + 2 s_1 x_2 - a_1) \end{cases} \tag{12}$$

where

$$\begin{cases} a_1 = -1.5 y_1 - 0.5 y_3 & b_1 = -1.5 x_1 - 0.5 x_3 \\ a_3 = 0.5 y_1 + 1.5 y_3 & b_3 = 0.5 x_1 + 1.5 x_3 \end{cases}$$

It can easily be seen that a unique solution for the position of the midnode exists, provided that $s_1 \neq s_3$. If $s_1 = s_3$, a straight element with its midnode located at the centre should be used.

Formulae (12), valid for quadratic elements, can lead to erroneous results in the presence of an inflection point, because the outward normal changes its direction. Such a point appears when neither $s_1 < s_{13} < s_3$ nor $s_1 > s_{13} > s_3$ holds (where $s_{13}$ denotes the slope of the straight line connecting the end nodes of the element – *cf.* Fig.3). It is again recommended that in such a situation a straight element with the midnode located at the centre should be used.

The above-described procedure cannot be applied to the first or the last element along the interface because these elements have only one neighbouring element. Hence, it is proposed that the element starting on the axis of symmetry remains quadratic but its tangent line at the relevant limiting node remains horizontal. The element starting on the mould has to be degenerated into a linear one.

The process of modification of the assumed heat fluxes utilises the concepts of sensitivity analysis. Fundamentals of the sensitivity analysis are discussed, for instance, in [2, 3, 4].

The sensitivity coefficients $Z_{ij}$ are defined as derivatives of the measured quantity, *i.e.*, temperature $T_i$ at a certain location, with respect to an assumed and then identified design variable $Y_j$

$$Z_{ij} = \frac{\partial T_i}{\partial Y_j} \tag{13}$$

The most effective method of determination of the sensitivity coefficients is differentiation of the equations (1) ÷ (5) and (7) ÷ (8) with respect to the design variables. In problem discussed in the paper the sensitivity coefficients result in three systems of equations [2, 3, 4] (the number of equation systems has to be equal to the number of estimated heat fluxes). The known boundary conditions (2) ÷ (4) produce homogeneous adjoint boundary equations while non-zero quantities are generated only by estimated fluxes in conditions (7) and/or (8). The solutions of these systems are again obtained by making use of boundary element method

$$\mathbf{H}\,\mathbf{Z} = \mathbf{G}\,\mathbf{q}_Z \tag{14}$$

where the vector $\mathbf{q}_Z$ contains the normal derivatives of the sensitivity coefficients.

Assuming unknown heat fluxes $q^*_{kk}, q^*_{gk}$ and $q^*_{gn}$ at the points B and C on the external boundary one can expand temperature into Taylor series in the vicinity of temperature $T^*$ resulting from these trial values. Using matrix notation and truncating the Taylor series after the first term one obtains

$$\mathbf{T}_{cal} = \mathbf{T}^* + \mathbf{Z}\,(\mathbf{Y} - \mathbf{Y}^*) \tag{15}$$

where $\mathbf{Z}$ stands for the sensitivity coefficients matrix and $\mathbf{T}^*$ is a vector of temperatures at the sensor locations obtained by solving the direct problem with trial heat fluxes $\mathbf{Y}^*$.

Then, the sensitivity coefficients are introduced into the objective function (9) which is then minimized. This results in the following system of equations

$$\begin{aligned}\left(\mathbf{Z}^T\mathbf{W}^{-1}\mathbf{Z} + \mathbf{W}_Y^{-1}\right)\mathbf{Y} &= \mathbf{Z}^T\mathbf{W}^{-1}\,(\mathbf{U} - \mathbf{T}^*) \\ &+ \left(\mathbf{Z}^T\mathbf{W}^{-1}\mathbf{Z}\right)\mathbf{Y}^* + \mathbf{W}^{-1}\tilde{\mathbf{Y}}\end{aligned} \tag{16}$$

The solution of this set of equations results in the vector of sought design variables (*i.e.*, the vector $\mathbf{Y}$). It should also be borne in mind that the sensitivity analysis helps to determine which measurements are the most valuable. However, because of measurement errors the search for 'the best locations of sensors' unfortunately does not yield a unique solution.

## 4. NUMERICAL CALCULATIONS AND RESULTS

Continuous casting of copper was considered as a numerical example of the inverse BEM thermal analysis. The results obtained were compared to reference values produced by direct BEM modelling, *cf.* [1]. In these calculations the following heat fluxes $q_{kk} = -1.2 \cdot 10^6$ W/m$^2$, $q_{gk} = -10^5$ W/m$^2$, and $q_{gn} = -4.4 \cdot 10^5$ W/m$^2$ were assumed. All results were obtained for the melting temperature $T_m = 1083°$C. The end temperature $T_s$ equals 50°C.

The temperature measurements inside the ingot were simulated numerically by adding random errors to selected internal temperatures of reference field. The influence of measurement accuracy on the quality of heat fluxes estimation is discussed in the next subsections. The assumed measurement errors did not exceed 2% (namely, they were considered as 0.1%, 0.2%, 0.5%, 1% or 2%).

### 4.1. NUMBER AND LOCATION OF SENSORS

Because of the ill-posed nature of the inverse problem, it is recommended to make the problem overdetermined [2, 3, 4] which means that the number of measurements should to be larger than the number of the estimated quantities. However, it must be noted that more measurements do not necessarily improve results. This effect is caused by a 'noise' accompanying each measurement and is clearly visible on the left hand side of Fig.4. Presented there curves do not continuously tend to exact solutions when the number of temperature sensors is growing.

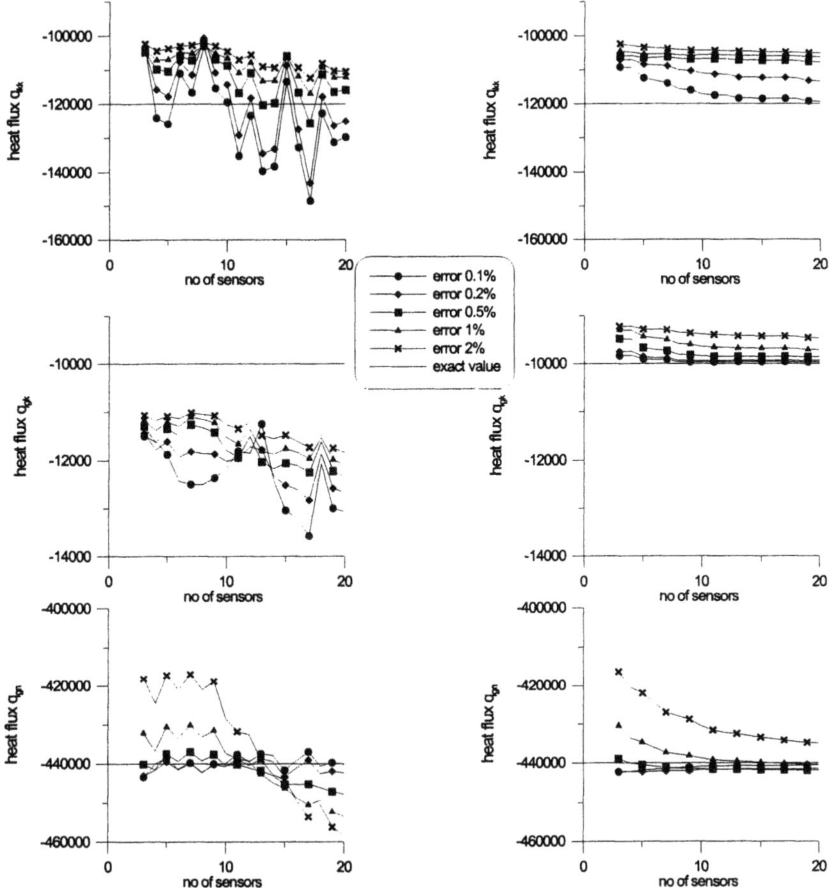


*Figure 4* Estimated heat fluxes $q_{kk}, q_{gk}$ and $q_{gn}$ against the number of sensors with and without random errors.

It was also checked how the inverse BEM solution behaves in the case the measurement errors are not random values but are known exactly. Nevertheless, it is still not known whether these errors are positive or negative. In other words, their sign was still lotted. Because an increasing number of sensors continuously improves the accuracy of the estimated heat fluxes (*cf.* right hand side column of Fig.4) the above conclusion is justified.

The next problem which was investigated is the location of sensors inside the ingot. The sensitivity coefficient is associated with the appropriate sensor and describes to what extent the temperature will change because of the imposed changes of the heat flux. Thus, generally the sensitivity coefficients should point out locations where their values are big and where temperature sensors should be placed. Unfortunately, a set of sensors which was not the best one from the viewpoint of sensitivity analysis (the sum of squares of the sensitivity coefficients was not a maximum) but gave better final solution, *i.e.*, more accurate estimates of the heat fluxes. This situation probably stems from measurement errors and will be further investigated in the future.

## 4.2. HEAT FLUX DISTRIBUTION ALONG THE BOUNDARY

As mentioned already, measurement errors have a significant influence on the accuracy of the estimated heat fluxes. A detailed analysis of these solutions helped to particularize the reasons for the few unsatisfactory results. One example is the worse accuracy of the estimation of the heat flux $q_{gn}$ (point C in Fig.2).

The heat fluxes vary linearly along the mould and exponentially along the face with the water spray (see eqs. (7) and (8)). It turned out that the discontinuity in the distribution of the heat flux at point C is the reason of the loss of the accuracy of the heat flux estimation. Making the heat flux continuous can improve the accuracy of the results considerably. Differences between these two options, (*i.e.*, discontinuous and continuous heat fluxes) are shown in Fig.5.

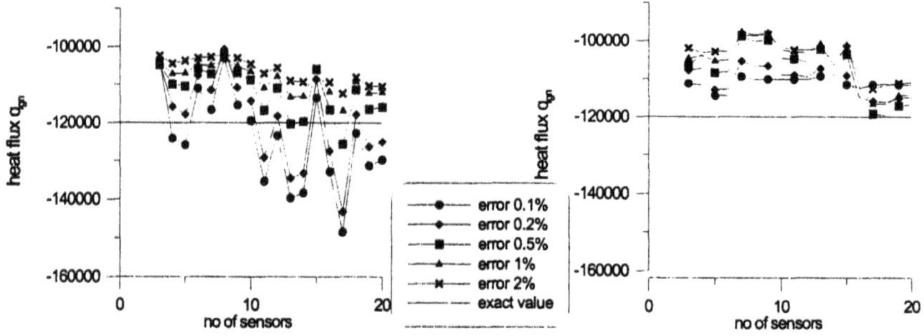


*Figure 5* Comparison of results for a discontinuous and continuous heat flux distribution on the mould surface.

The role of the prior estimates $\tilde{\mathbf{Y}}$ and their covariance matrix $\mathbf{W}_Y$ in the minimization process of the objective function (9) was also investigated. It was confirmed that the more accurate the prior estimates should be, the smaller elements of $\mathbf{W}_Y$ matrix must be.

Eventually, it was verified that the temperature field caused by the estimated heat fluxes is practically identical to the reference temperature field. Fig.6. shows this field in the vicinity of the phase change front (where the most rapid changes of temperatures occur).

*Figure 6* Fragment of the temperature field caused by the estimated heat fluxes in the vicinity of the phase change front.

## 5. CONCLUSIONS

The application of sensitivity analysis to the estimation of the boundary conditions in the inverse heat transfer problem in continuous casting was shown. The calculations made allow assessing the main influence factors on the accuracy of the results. It was found that the discontinuity in the heat flux distribution, the number of sensors and the measurement errors have the most significant influence on the accuracy of the results.

The best results were achieved after making the heat flux distribution along the boundary of the mould continuous. Although the number of sensors was increased considerably, the accuracy of the solution was not improved significantly. The reason for this is that more sensors offer valuable information, on the one hand, and disturbances spoiling the accuracy, on the other hand. Nevertheless, the final temperature field retrieved from the estimated boundary heat fluxes (the solution of the direct problem) is practically identical to the reference one.

The procedure described in the paper is currently extended for determination of the location of a phase change front.

## Acknowledgments

The financial assistance of the National Committee for Scientific Research, Poland, within the grants no 8 T10B 050 15 and SPUB-2/RIE-6/99 coordinated by European Union within the action COST P3 is gratefully acknowledged herewith. Thanks are also due to Prof. L.C. Wrobel from Brunel University, London, UK for many fruitful discussions.

## References

[1] Fic A., Nowak A.J., Białecki R.A.: Heat transfer analysis of the continuous casting problem by the front tracking method and the BEM, *Engineering Analysis with Boundary Elements*, in press.

[2] Kurpisz K., Nowak A.J.: *Inverse Thermal Problems*, Comp. Mech. Publications, Southampton 1995

[3] Nowak A.J.: BEM Approach to Inverse Thermal Problems, Chapter 10, in Ingham D.B., Wrobel L.C.: *Boundary Integral Formulations for Inverse Analysis*, Comp. Mech. Publications, Southampton 1997

[4] Beck J.V., Blackwell B., Inverse Problem, in Minkowycz W.J., Sparrow E.M., Schneider G.E. and Pletcher R.H., (eds.): *Handbook of Numerical Heat Transfer*, Wiley Intersc., New York 1988.

[5] Brebbia C.A., Telles J.C.F., Wrobel L.C., *Boundary Element Techniques - Theory and Applications in Engineering.* Springer-Verlag, 1984.

# BOUNDARY ELEMENT METHOD FOR TWO-DIMENSIONAL SHALLOW WATER ACOUSTIC WAVE PROPAGATION

J.A.F. Santiago * and L.C. Wrobel
*Brunel University*
*Department of Mechanical Engineering*
*Uxbridge, Middlesex UB8 3PH, UK.*

**Keywords:** underwater acoustics, boundary element method, subregions technique

**Abstract** This article presents a boundary element formulation for two-dimensional acoustic wave propagation in shallow water over irregular seabed, considering an isospeed channel with rigid bottom. It is assumed that the source of acoustic disturbance is time-harmonic and the medium in the absence of perturbations is quiescent.
For regions of constant depth, two fundamental solutions in the form of infinite series are employed in order to avoid the discretization of the free surface and bottom boundaries.

## 1. INTRODUCTION

Due to the advent of high-speed computers and the recent developments of numerical physics, sound propagation in the ocean can be studied and quantitatively described in greater detail with the more exact wave theory.

Increasing concern for coastal areas has, in recent years, focussed studies of ocean acoustic wave propagation on shallow water environments. The most common numerical techniques used to model underwater acoustic wave propagation are ray methods, normal mode methods, and parabolic equation methods, which are discussed by Jensen *et al.* [Jensen *et al.*, 1994].

The present paper proposes a novel boundary element formulation for the numerical modelling of shallow water acoustic propagation, in

*On leave from COPPE/Federal University of Rio de Janeiro, Brazil

*T. Burczynski (ed.), IUTAM/IACM/IABEM Symposium on Advanced Mathematical and Computational Mechanics Aspects of the Boundary Element Method,* 281–292.

the frequency domain, over irregular bottom topography. The model assumes a two-dimensional geometry, representative of coastal regions, which have little variation in the long shore direction. Application of the boundary element method (BEM) using a hybrid model which combines the BEM with eigenfunction expansions to solve this problem was presented by Grilli *et al.* [Grilli *et al.*, 1998]. The BEM model presented here makes use of two modified Green's functions, one of which only satisfies the free surface boundary condition, while the other directly satisfies the boundary conditions on the free surface and the horizontal part of the bottom. An alternative Green's function is employed to improve the convergence characteristics of the latter. Therefore, only bottom irregularities and vertical interfaces need to be discretized. Results for a region close to shore containing two sloping boundaries is included to assess the accuracy of the numerical solutions.

## 2. GOVERNING EQUATIONS OF THE PROBLEM

Consider the problem of acoustic wave propagation in a region $\Omega$ of infinite extent, shown in Figure 1. Assuming that this medium in the absence of perturbations is quiescent, the velocity of sound is constant and the source of acoustic disturbance is time-harmonic, the problem is governed by the Helmholtz equation [Kinsler *et al.*, 1982]:

$$\nabla^2 u + k^2 u = -\sum_{\alpha=1}^{Nes} B_\alpha \delta\left(\mathbf{E}_\alpha, \mathbf{S}\right) \qquad \text{in } \Omega \tag{1}$$

where $u$ is the velocity potential, $B_\alpha$ is the magnitude of the exciting source $\mathbf{E}_\alpha$ located at $(x_{E_\alpha}, y_{E_\alpha})$, $\mathbf{S}$ is the source point located at $(x_S, y_S)$, $Nes$ is the number of exciting sources, $\delta\left(\mathbf{E}_\alpha, \mathbf{S}\right)$ is the Dirac delta generalised function and $k = w/c$ is the acoustic wavenumber, with $w$ the angular frequency and $c$ the velocity of sound in water.

The problem is subject to the following boundary conditions:

a) Dirichlet condition

$$u\left(\mathbf{X}\right) = 0 \qquad \text{on } \Gamma_F \tag{2}$$

b) Neumann condition

$$\frac{\partial u}{\partial n}\left(\mathbf{X}\right) = 0 \qquad \text{on } \Gamma_B \tag{3}$$

c) Sommerfeld radiation condition at infinity

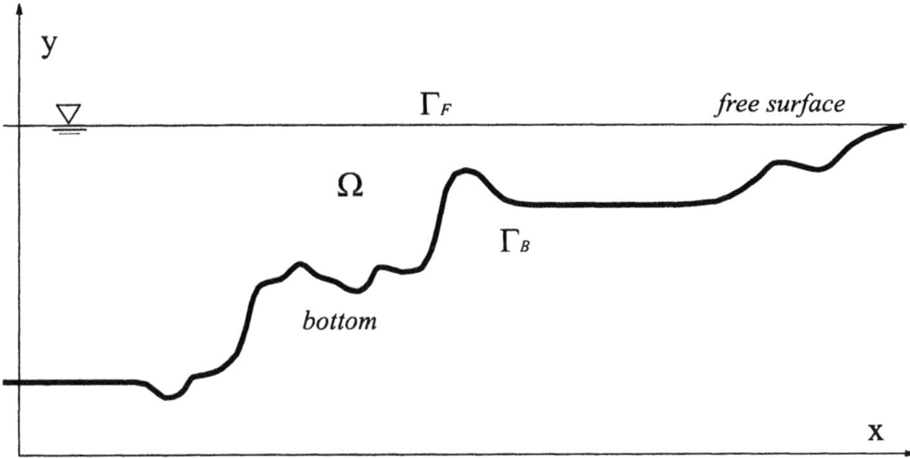


*Figure 1* General ocean section for 2D acoustic propagation problems in shallow water

$$\frac{\partial u}{\partial n}(\mathbf{X}) = iku(\mathbf{X}) \tag{4}$$

in which $\Gamma_F$ is the free surface, $\Gamma_B$ is the irregular bottom, $n$ is the outward normal vector and $i = \sqrt{-1}$.

According to Green's second identity, Equation (1) can be transformed into the following boundary integral equation [Chen and Zhou, 1992], [Lacerda, 1997]

$$c(\mathbf{S})\,u(\mathbf{S}) = \int_\Gamma G(\mathbf{S},\mathbf{X})\frac{\partial u}{\partial n}(\mathbf{X})\,d\Gamma - \int_\Gamma \frac{\partial G(\mathbf{S},\mathbf{X})}{\partial n}u(\mathbf{X})\,d\Gamma + \sum_{\alpha=1}^{Nes} B_\alpha G(\mathbf{E}_\alpha,\mathbf{S}) \tag{5}$$

where $\Gamma$ is equal to $\Gamma_F \cup \Gamma_B$, $\mathbf{X}$ is the field point located at $(x, y)$ and $G(\mathbf{S},\mathbf{X})$ is the free-space Green's function. The functions $u(\mathbf{X})$ and $\partial u/\partial n(\mathbf{X})$ represent the velocity potential and its normal derivative. The coefficient $c(\mathbf{S})$ depends on the boundary geometry at the source point $\mathbf{S}$. It is noted that the Green's function implicitly satisfies the Sommerfeld condition, therefore no discretization of the boundary at infinity is necessary.

# 3. NUMERICAL ANALYSIS

## 3.1. BOUNDARY ELEMENT METHOD

General propagation problems with irregular seabed topography can be dealt with by the subregions technique [Santiago *et al.*, 1999]. In this case the domain $\Omega$ is divided into several regions, as depicted in Figure 2. Two types of subregions can be seen in the figure, for which different fundamental solutions are employed. In the first the region has irregular bottom ($\Omega_2$ and $\Omega_4$) while the other has constant depth ($\Omega_1$ and $\Omega_3$).

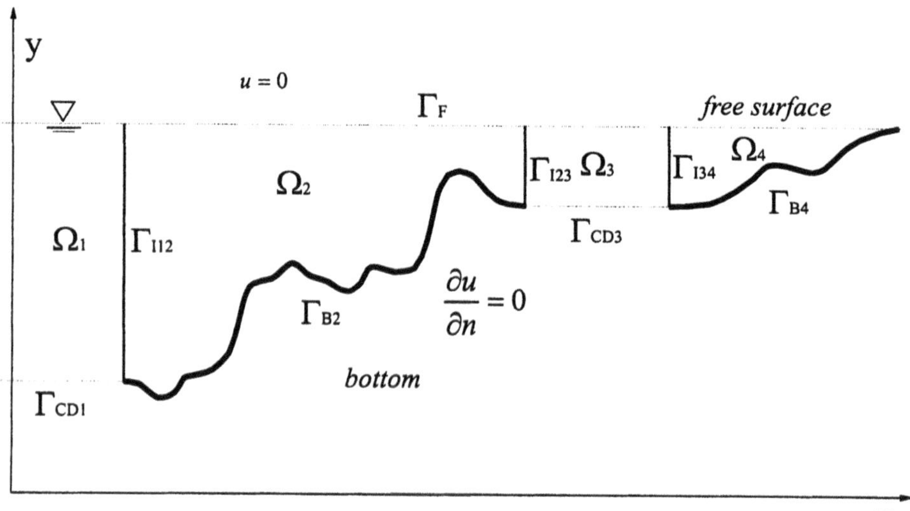


*Figure 2* Subregions dividing a typical shallow water section

Instead of using the fundamental solution of the Helmholtz equation for the free-space, functions which directly satisfy the boundary conditions either on $\Gamma_F$ or on $\Gamma_F$ and $\Gamma_{CD}$ are adopted. Therefore, only the irregular parts of the bottom boundaries ($\Gamma_{B2}$ and $\Gamma_{B4}$) and interfaces ($\Gamma_{I12}$, $\Gamma_{I23}$ and $\Gamma_{I34}$) need to be discretized.

Introducing the appropriate boundary conditions, given by (2) and (3), into Equation (5), using constant elements with linear geometry, and applying the collocation method to the integral equation gives, in terms of an intrinsic coordinate $\eta$, for region $\Omega_\chi$ with irregular bottom:

$$c\left(\mathbf{S}_p\right) u\left(\mathbf{S}_p\right) = \sum_{\alpha=1}^{Nes_\chi} B_\alpha G_f\left(\mathbf{E}_\alpha, \mathbf{S}_p\right) - \sum_{q=1}^{Neb_\chi} u_q \frac{L_q}{2} \int_{-1}^{+1} \frac{\partial G_f\left(\mathbf{S}_p, \mathbf{X}_q\right)}{\partial n} d\eta +$$
$$\sum_{q=1}^{Nei_\chi} \frac{L_q}{2} \left[ v_q \int_{-1}^{+1} G_f\left(\mathbf{S}_p, \mathbf{X}_q\right) d\eta - u_q \int_{-1}^{+1} \frac{\partial G_f\left(\mathbf{S}_p, \mathbf{X}_q\right)}{\partial n} d\eta \right] \tag{6}$$

with constant depth:

$$c\left(\mathbf{S}_p\right) u\left(\mathbf{S}_p\right) = \sum_{\alpha=1}^{Nes_\chi} B_\alpha G_D\left(\mathbf{E}_\alpha, \mathbf{S}_p\right) +$$
$$\sum_{q=1}^{Nei_\chi} \frac{L_q}{2} \left[ v_q \int_{-1}^{+1} G_D\left(\mathbf{S}_p, \mathbf{X}_q\right) d\eta - u_q \int_{-1}^{+1} \frac{\partial G_D\left(\mathbf{S}_p, \mathbf{X}_q\right)}{\partial n} d\eta \right] \tag{7}$$

where $p$ ranges from 1 to $Nf_\chi$, with $Nf_\chi$ the total number of functional nodes, $Nes_\chi$, $Neb_\chi$, $Nei_\chi$ are the number of exciting sources, external boundary elements and elements of all interfaces in the region $\Omega_\chi$, respectively; $\mathbf{S}_p$ are collocation points and $L_q$ is the length of element $\Gamma_q$. Finally, $u_q$ and $v_q$ are the velocity potential and its normal derivative, respectively, at the point $\mathbf{X}_q$.

In the above equations, the first modified Green's function $G_f(\mathbf{S}_p, \mathbf{X}_q)$ satisfies the boundary condition on the free surface $\Gamma_F$ while the second function $G_D(\mathbf{S}_p, \mathbf{X}_q)$ satisfies the boundary condition on the free surface $\Gamma_F$ and seabed $\Gamma_B$.

Equations (6) and (7) are computed for each specific region separately and are then coupled by imposing continuity of normal derivative and velocity potential over the interfaces. Hence, applying these equations to all functional nodal points yields

$$\mathbf{Ay} = \mathbf{b} \tag{8}$$

where matrix $\mathbf{A}$ contains the influence coefficients, vector $\mathbf{b}$ contains the contribution of the exciting sources and vector $\mathbf{y}$ contains unknown velocity potentials and normal derivatives.

## 3.2. FUNDAMENTAL SOLUTIONS

The fundamental solutions mentioned in the previous section were developed by two different means. The first was the image method using a single source point reflection and multiple source point reflections, while the second was the inhomogeneous Helmholtz equation for a line source in a plane geometry, using eigenfunctions (normal modes) of the depth-separated equation.

**Irregular bottom.** For the first subregion type (see Figure 2), the modified Green's function $G_f(\mathbf{S},\mathbf{X})$, obtained by means of the image method using a single source, is employed in the following form:

$$G_f(\mathbf{S},\mathbf{X}) = i\frac{1}{4}\left[H_0^{(1)}(kr) - H_0^{(1)}\left(kr^{(1F)}\right)\right] \tag{9}$$

where $H_0^{(1)}(\ )$ is the Hankel function of order 0 of the first kind, $r$ and $r^{(1F)}$ are the distances from the field point $\mathbf{X}$ to the source point $\mathbf{S}$ and its reflection with respect to the free surface, respectively.

**Constant depth.** For the second subregion type, the function $G_D(\mathbf{S},\mathbf{X})$ is used in the form of two infinite series, the modified Greens' function $G_F(\mathbf{S},\mathbf{X})$ or $G_B(\mathbf{S},\mathbf{X})$ and the Greens' function $G_m(\mathbf{S},\mathbf{X})$.

The first series, obtained through the image method using multiple source point reflections, can be written as [Santiago and Wrobel, 1999a], [Jensen *et al.*, 1994]:

$$G_F(\mathbf{S},\mathbf{X}) = i\frac{1}{4}\left\{\begin{array}{c}\left[H_0^{(1)}(kr) - H_0^{(1)}\left(kr^{(1F)}\right)\right] + \sum\limits_{m=1}^{\infty}(-1)^{m+1} \\ \sum\limits_{j=1}^{2}(-1)^{j+1}\left[H_0^{(1)}\left(kr_m^{(2jF)}\right) - H_0^{(1)}\left(kr_m^{((2j+1)F)}\right)\right]\end{array}\right\} \tag{10}$$

$$G_B(\mathbf{S},\mathbf{X}) = i\frac{1}{4}\left\{\begin{array}{c}\left[H_0^{(1)}(kr) + H_0^{(1)}\left(kr^{(1B)}\right)\right] \\ +\sum\limits_{m=1}^{\infty}(-1)^m\sum\limits_{j=2}^{5}H_0^{(1)}\left(kr_m^{(jB)}\right)\end{array}\right\} \tag{11}$$

where the superindices $(1F)$, $(2jF)$, $((2j+1)F)$, $(1B)$ and $(jB)$ identify the reflected source points.

The distances from the field point $\mathbf{X}$ to the reflections of the source point $\mathbf{S}$ with respect to free surface($F$) and bottom($B$) are denoted as $r_m^{(2jF)}$ and $r_m^{((2j+1)F)}$ for $G_F(\mathbf{S},\mathbf{X})$ and $r^{(1B)}$ and $r_m^{(2jB)}$ for $G_B(\mathbf{S},\mathbf{X})$ [Santiago and Wrobel, 1999b].

The modified Green's function $G_F(\mathbf{S},\mathbf{X})$ exactly satisfies the boundary condition on the free surface, but its normal derivative produces a small non-zero value at the bottom boundary. Alternatively, the modified Green's function $G_B(\mathbf{S},\mathbf{X})$ produces a small non-zero value at the free surface but its normal derivative exactly satisfies the boundary condition on the bottom.

The second series, in terms of normal modes, can be written as [Pedersen, 1996], [Jensen *et al.*, 1994]:

$$G_m = \frac{i}{H} \sum_{m=1}^{\infty} \sin\left[k_{ym}\left(Y_F - y_S\right)\right] \sin\left[k_{ym}\left(Y_F - y\right)\right] \frac{e^{ik_{xm}|x-x_S|}}{k_{xm}} \tag{12}$$

where $H$ and $Y_F$ are the depth and the coordinate $y$ of the free surface. The parameters $k_{xm}$ and $k_{ym}$ are horizontal and vertical wavenumbers, respectively:

$$k_{ym} = \left(m - \frac{1}{2}\right)\frac{\pi}{H} \qquad\qquad k_{xm} = \sqrt{k^2 - k_{ym}^2} \tag{13}$$

The calculation of the Green's function $G_m$ is much more efficient than the modified Green's functions $G_F$ and $G_B$, due to the exponential term in Equation (12), which makes the series decreases rapidly when $k_{xm}$ becomes an imaginary number, i.e. when $k_{ym}$ is greater than $k$.

Nevertheless, when coordinate $x$ of the source and field points is the same, the exponential term of the function $G_m$ is equal to one and convergence becomes slow.

## 4. APPLICATION

The idealised example shown in Figure 3 simulates the propagation of acoustic waves in infinite shallow water containing two different sloping bottoms close to shore. The depths of the deeper and shallower regions are equal to 1.0 $m$ and 0.5 $m$, respectively. The velocity of sound $c$ and the frequency $f$ are taken to be 1500 $m/s$ and 1000 $Hz$, respectively. The magnitude of the exciting source is equal to 1.0 $m^2/s$, located at $x_E = -2.0\ m$ and $y_E = 0.75\ m$.

The boundary conditions and element discretisation are depicted in Figure 4. A rigid bottom is adopted, therefore $\partial u/\partial n = 0$. Two different types of discretisation are employed for the second subregion: in the first, the bottom is fully discretised with 724 elements of graded length, concentrated close to node 1 on the interface and node 3 on the horizontal boundary, using the image method with a single reflection *(ms724)*. In the second, only the sloping boundaries are discretised with 360 graded elements, also concentrated close to node 1 on the interface, using infinite series for both subregions *(mm360)*.

In order to assess the accuracy of the results, a comparison was made with results obtained utilising an alternative BEM formulation in which

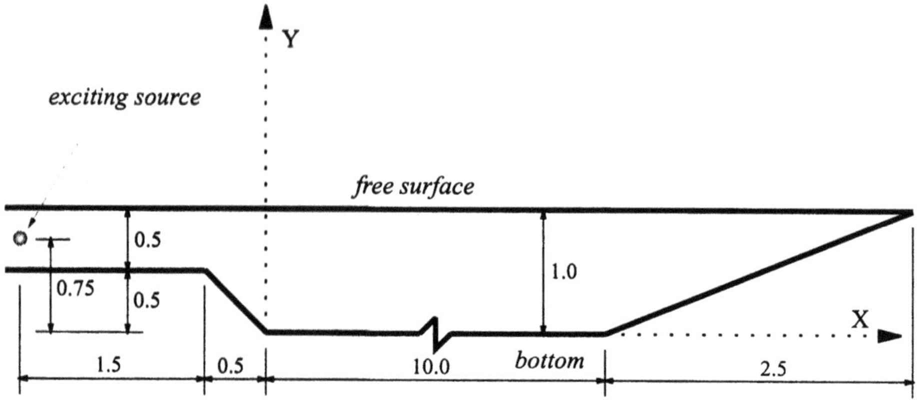


*Figure 3* Geometry for a region containing two sloping close to shore

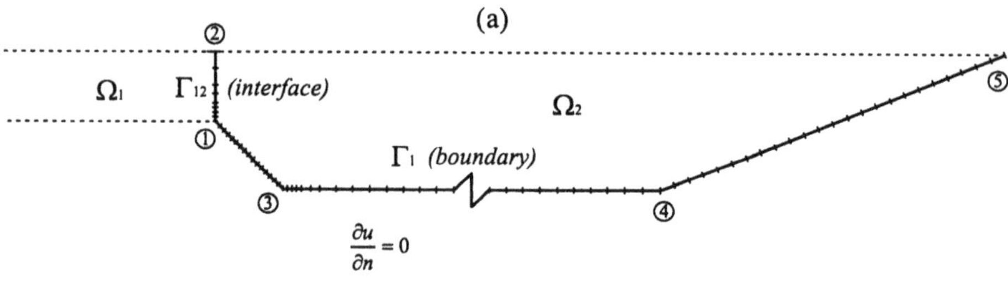


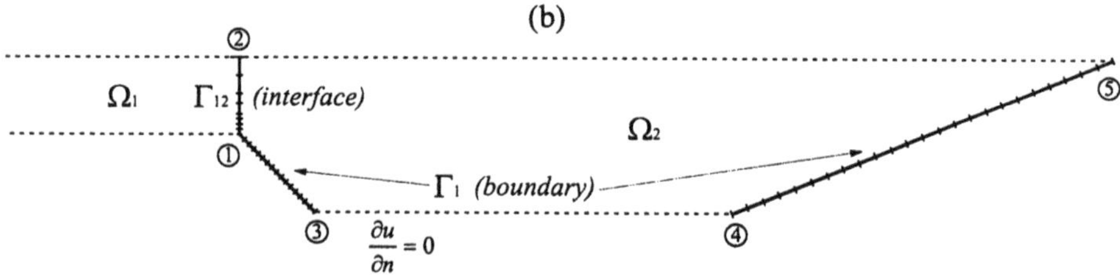


*Figure 4* Discretisation using graded elements and boundary condition of the problem. On the second subregion: (a) The bottom is fully discretised. (b) Only the sloping boundaries are discretised.

the bottom boundary is fully discretized using 8060 graded elements, but the free surface is eliminated using a single image source. The infinite bottom boundary was truncated at the distance $-1000.0\,m$ *(db8060)*.

Figure 5 presents the real part of the velocity potential at points along the right sloping bottom, where $s$ is the local coordinate along this boundary, i.e. $s = \sqrt{x^2 + y^2}$ .

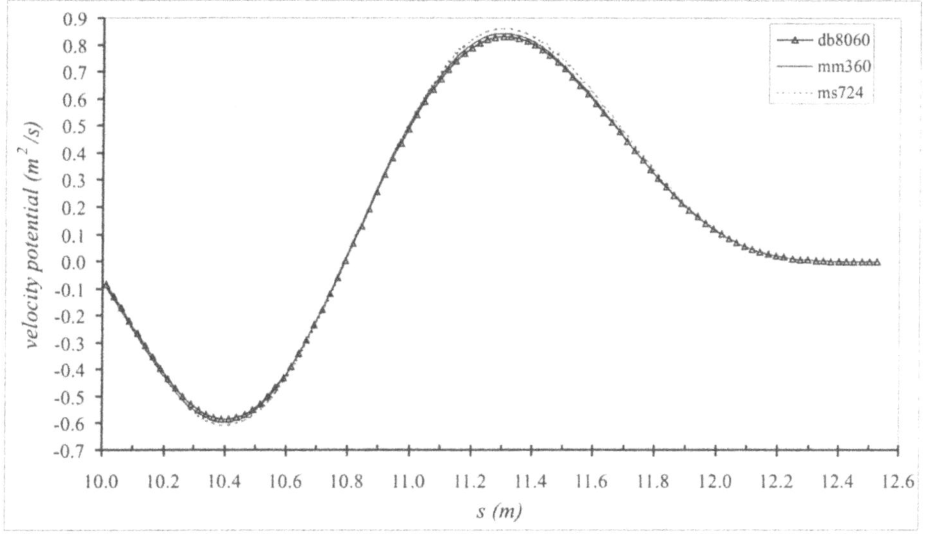


*Figure 5* Real part of the velocity potential at points along the right sloping bottom

Results for the velocity potential at internal points along three horizontal lines and two vertical lines are depicted in Figures 6 and 7, respectively. In the first figure the horizontal lines are located at $y = 0.75\,m$, from $x = -1.9\,m$ to $x = 11.7\,m$ ; $y = 0.50\,m$, from $x = -0.4\,m$ to $x = 11.1\,m$; and $y = 0.25\,m$, from $x = 0.0\,m$ to $x = 10.5\,m$. In the second figure the vertical lines are located at $x = 0.0\,m$ and $x = 10.0\,m$, from $y = 0.1\,m$ to $y = 0.9\,m$.

The real part of the velocity potential presents excellent agreement, for all discretizations. The imaginary part of the velocity potential is similar to the real part for all results presented here.

## 5. CONCLUSIONS

The subregions technique, validated here for a simple problem, is necessary to simulate complex problems of acoustic wave propagation in shallow water with irregular bottom.

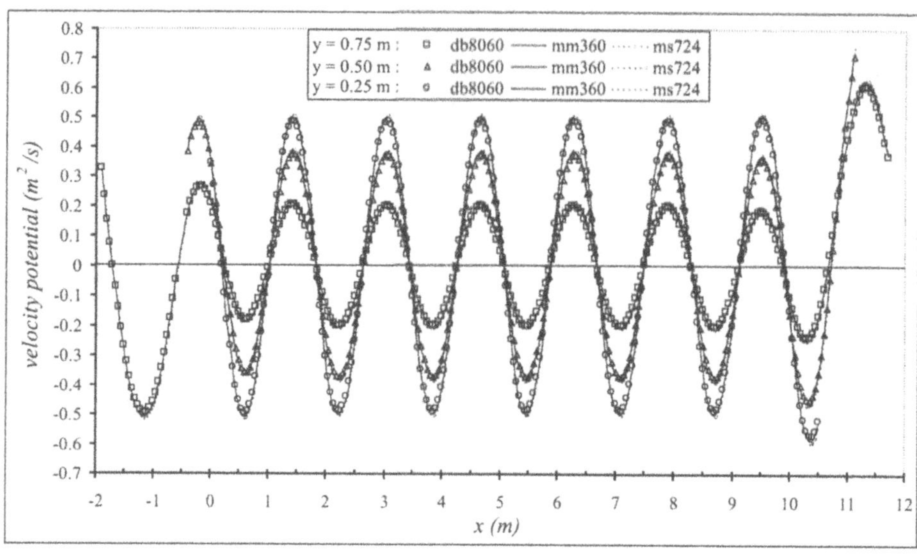


*Figure 6* Real part of velocity potential at internal points along three horizontal lines for different types of discretisation

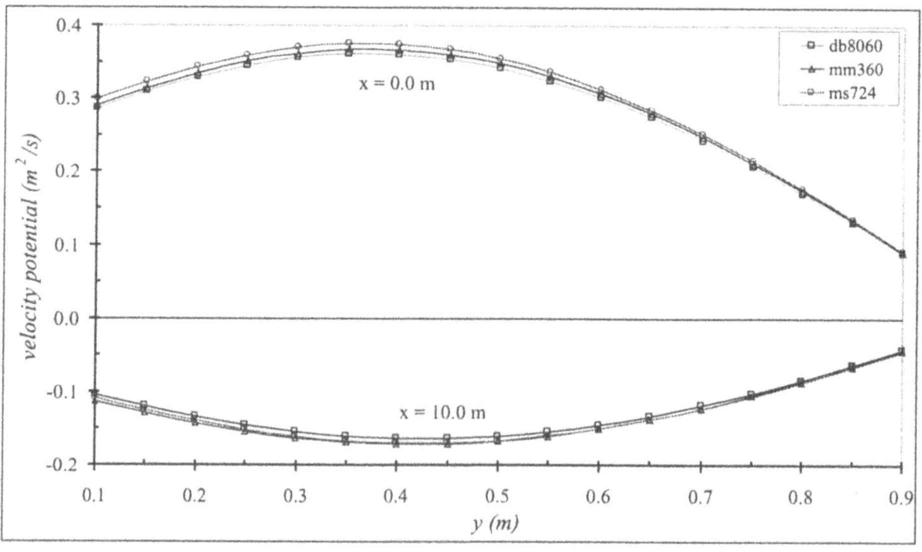


*Figure 7* Real part of velocity potential at internal points along two vertical lines for different types of discretisation

Although the series (12) in terms of normal modes dramatically improves the speed of convergence, it is still necessary to improve the convergence of the infinite series when the source and field points are located along the same vertical line.

## Acknowledgments

The first author would like to thank CNPq, the Brazilian Research Council, for providing financial support to this research.

## References

[Chen and Zhou, 1992] Chen, G. and Zhou, J. (1992) *Boundary Element Methods*, Academic Press, UK.

[Grilli *et al.*, 1998] Grilli, S., Pedersen, T. and Stepanishen, P.(1998) A Hybrid Boundary Element Method for Shallow Water Acoustic Propagation over an Irregular Bottom, *Engineering Analysis with Boundary Elements* **21**, 131-145.

[Jensen *et al.*, 1994] Jensen, F. B., Kuperman, W. A., Porter, M. B.and Schmidt, H. (1994) *Computational Ocean Acoustics*, American Institute of Physics, USA.

[Kinsler *et al.*, 1982] Kinsler, L.E., Frey, A.R., Coppens, A.B and Sanders, J.V. (1982) *Fundamentals of Acoustics*, Third edition, John Wiley.

[Lacerda, 1997] Lacerda, L.A. de (1997) *Boundary Element Formulations for Outdoor Sound Propagation*, D.Sc. Thesis, Department of Civil Engineering, COPPE / UFRJ, Brazil.

[Pedersen, 1996] Pedersen, T.K. (1996) *Modelling Shallow Water Acoustic Wave Propagation*, M.Sc. Thesis in Ocean Engineering, University of Rhode Island,USA.

[Santiago *et al.*, 1999] Santiago, J.A.F., Telles, J.C.F. and Valentim, V.A. (1999) An Optimized Block Matrix Manipulation for Boundary Elements with Subregions, *Advances in Engineering Software* **30**, 701-713.

[Santiago and Wrobel, 1999a] Santiago, J.A.F. and Wrobel,L.C. (1999a) Numerical Modelling of the Propagation of Underwater Acoustic Waves, *15th Brazilian Congress of Mechanical Engineering*, Águas de Lindóia, Brazil.

[Santiago and Wrobel, 1999b] Santiago, J.A.F. and Wrobel,L.C. (1999b) 2D Modelling of Shallow Water Acoustic Wave Propagation using Subregions Technique, *Boundary Element Techniques International Conference*, Queen Mary and Westfield College, University of London, pp. 415-424, UK.

# SOLUTION OF FLUID DYNAMICS PROBLEMS IN POROUS MEDIA BY THE DRBEM

Božidar Šarler
*Laboratory for Fluid Dynamics and Thermodynamics*
*Faculty of Mechanical Engineering, University of Ljubljana*
*Aškerčeva 6, SI–1 000 Ljubljana, Slovenia*
bozidar.sarler@fs.uni-lj.si

**Abstract** This paper describes the Dual Reciprocity Boundary Element Method (DRBEM) solution of steady natural convection problems in Darcy–Brinkman porous media. The solution of coupled mass, momentum, and energy equations in two–dimensions is structured by the fundamental solution of the Laplace equation, straight line geometry and discontinuous linear field boundary elements. The dual reciprocity method is based on inverse multiquadrics. Solutions are shown for natural convection in a rectangular cavity with differentially heated boundary, phase change, and internal heat generation.

**Keywords:** porous media, Darcy-Brinkman model, natural convection, phase change, boundary element method, dual reciprocity, multiquadrics

## 1. INTRODUCTION

The transport of multicomponent materials saturating porous media matrices [1] occurs extensively in nature and many engineering systems. The computational modelling of such systems has been extensively developed for the needs of safety assessments and site evaluation for nuclear waste repositories. The strategy for the groundwater flow and species transport modelling comprises nested models in the scale of region, site and canister. The boundary element method turns out to be particularly suitable for solving geometrically complicated transport phenomena where one or more coordinates extend to infinity. The use of the BEM in the context described has been proposed in [2]. The numerical implementation of the Darcy internal natural convection has been elaborated in [3] by using the DRBEM [4] based on augmented scaled thin

*T. Burczynski (ed.), IUTAM/IACM/IABEM Symposium on Advanced Mathematical and Computational Mechanics Aspects of the Boundary Element Method,* 293–302.

plate splines. This paper extends this implementation to the Darcy-Brinkman porous media which requires a complete modification of the solution procedure for the mass and momentum equations due to the involved viscous term.

## 2. GOVERNING EQUATIONS

This paper deals with the homogenous porous media with permeability $\mathcal{K}$, confined to domain $\Omega$ with the boundary $\Gamma$. The rigid porous matrix and the incompressible fluid with Darcy $\mu_{\mathcal{D}}$ and Brinkman viscosity $\mu_{\mathcal{B}}$ saturating the pores have the same constant density $\varrho$, effective thermal conductivity $k$ and the specific heat at constant pressure $c_p$. The mass conservation for the defined system is

$$\nabla \cdot \mathbf{v} = 0 \tag{1}$$

where $\mathbf{v}$ stands for the seepage velocity. The Darcy-Brinkman momentum conservation for the defined system is

$$0 = -\nabla P - \frac{\mu_{\mathcal{D}}}{\mathcal{K}} \mathbf{v} + \mu_{\mathcal{B}} \nabla^2 \mathbf{v} + \mathbf{f} \tag{2}$$

with $P$ standing for pressure and $\mathbf{f}$ for the body force

$$\mathbf{f} = \varrho \, \mathbf{a} \, (1 - \beta \, (T - T_{\text{ref}})) - c_{\mathcal{P}} \, (1 - f_{\mathcal{L}}(T)) \, \mathbf{v}, \tag{3}$$

which represents combined Boussinesq and solid-liquid phase change effect with $\mathbf{a}$ denoting the acceleration vector, $\beta$ the volumetric thermal expansion coefficient, $T_{\text{ref}}$ the reference temperature, $c_{\mathcal{P}}$ the phase change coefficient, and $f_{\mathcal{L}}$ the liquid volume fraction which is assumed to vary between the solidus $T_{\mathcal{S}}$ and the liquidus temperature $T_{\mathcal{L}}$. The energy conservation equation is

$$\varrho \, c_p \, \nabla \cdot (\mathbf{v} \, T) = k \, \nabla^2 T + q \tag{4}$$

with $q$ representing heat sources. The solution of the equation (2) is constructed by assuming the impermeable and no-slip velocity boundary conditions on the whole boundary $\Gamma$

$$\mathbf{v} \cdot \mathbf{n}_\Gamma = 0; \;\; \mathbf{p} \in \Gamma \qquad \mathbf{v} \cdot \mathbf{n}^\Gamma = 0; \;\; \mathbf{p} \in \Gamma \tag{5}$$

where $\mathbf{n}_\Gamma$ stands for the normal and $\mathbf{n}^\Gamma$ for the tangent on the boundary $\Gamma$, and $\mathbf{p}$ for the position vector. The solution of the equation (4) is constructed by assuming the division of the boundary $\Gamma$ into not necessarily connected parts $\Gamma^D$ and $\Gamma^N$ with the Dirichlet and Neumann thermal boundary conditions, respectively

$$T = T_\Gamma; \;\; \mathbf{p} \in \Gamma^D \qquad -k \, \frac{\partial T}{\partial n_\Gamma} = F_\Gamma; \;\; \mathbf{p} \in \Gamma^N \tag{6}$$

where $T_\Gamma$ and $F_\Gamma$ represent known functions. The solution of the natural convection problem posed represents the velocity, pressure, and temperature distribution over the domain $\Omega$ and the boundary $\Gamma$.

# 3. SOLUTION PROCEDURE

The construction of the solution is represented in two steps. The first step involves the conversion of the partial differential equations into integral equations as well the basic elements of the iterative procedure. The second step focuses on the discretization, setup and solution of the algebraic systems of equations. The momentum equation is coupled with the energy equation through the body force. The energy equation is coupled with the momentum equation through the velocity field. Consequently, the solution inherently involves iterations. The solution steps are explained in a continuous setting, where no reference need to be made regarding the discretization that is explained afterwards. Let us assume the velocity, pressure and temperature fields are all known at iteration level $m$. The discussion of iteration cycle that follows explains how the velocity, pressure and temperature fields are calculated at the next iteration level $m+1$. The solution of the momentum equation at the iteration level $m+1$ is obtained in the following way: The Pressure Poisson Equation (PPE) is constructed by taking the divergence of the momentum conservation (2)

$$\nabla^2 P^{m+1} = \nabla \cdot \mathbf{S}_P^m \tag{7}$$

with $\mathbf{S}_P^m$ standing for

$$\mathbf{S}_P^m = -\frac{\mu_\mathcal{D}}{\mathcal{K}}\,\mathbf{v}^m + \mu_\mathcal{B}\,\nabla^2\mathbf{v}^m + \mathbf{f}^m \tag{8}$$

The Neumann pressure boundary conditions can be defined on the whole boundary $\Gamma$ by taking the scalar product of the momentum equation with the normal on the boundary. This gives

$$\frac{\partial P}{\partial n_\Gamma}^{m+1} = \mathbf{S}_P^m \cdot \mathbf{n}_\Gamma; \;\; \mathbf{p} \in \Gamma \tag{9}$$

The PPE with the boundary conditions (9) is solved by weighting the equation (7) with the fundamental solution of the Laplace equation $T^*(\mathbf{p};\mathbf{s})$ (where parameter $\mathbf{s}$ stands for the source point position) over the domain $\Omega$. This gives the following integral equation after application of the Green's second identity for scalars

$$\int_\Gamma \frac{\partial P}{\partial n_\Gamma}^{m+1} T^*\, d\Gamma - \int_\Gamma P^{m+1} \frac{\partial T^*}{\partial n_\Gamma}\, d\Gamma - c_\mathbf{s}^*\, P_\mathbf{s}^{m+1} = \int_\Omega \nabla \cdot \mathbf{S}_P^m\, T^*\, d\Omega \tag{10}$$

Subscript $\mathbf{s}$ denotes evaluation of a quantity at the source point $\mathbf{s}$. $c_{\mathbf{s}}^{*}$ stands for the fundamental solution related coefficient. Present paper is limited to two-dimensional Cartesian system, e.g.

$$T^{*} = \frac{1}{2\pi} \log \frac{r_0}{r} \tag{11}$$

where $r_0$ represents the reference radius, and $r$ equals to

$$r^2 = \mathbf{r} \cdot \mathbf{r}; \quad r = r_x \mathbf{i}_x + r_y \mathbf{i}_y; \quad r_x = p_x - s_x, \quad r_y = p_y - s_y \tag{12}$$

$p_x$, $p_y$ denote Cartesian coordinates (base vectors $\mathbf{i}_x$, $\mathbf{i}_y$) of point $\mathbf{p}$, and $s_x$, $s_y$ Cartesian coordinates of fundamental solution source point $\mathbf{s}$ respectively. Equation (10) is first used for determining the pressure distribution on the boundary $\Gamma$ and subsequently explicitly in the domain $\Omega$. The pressure gradients on the boundary and in the domain can be explicitly calculated from the pressure gradient Poisson equation

$$\int_{\Gamma} \frac{\partial P}{\partial n_{\Gamma}}^{m+1} \nabla T^{*} \, d\Gamma - \int_{\Gamma} P^{m+1} \nabla \frac{\partial T^{*}}{\partial n_{\Gamma}} \, d\Gamma + \nabla (c_{\mathbf{s}}^{*} \, P_{\mathbf{s}}^{m+1}) = \int_{\Omega} \nabla \cdot \mathbf{S}_{P}^{m} \, \nabla T^{*} \, d\Omega \tag{13}$$

obtained by taking the gradient of the PPE acting on the fundamental solution source point. After calculating the pressure gradient field the velocity field at iteration level $m+1$ can be calculated from the momentum equation. Respectivelly, the momentum equation (2) is weighted by the fundamental solution of the Laplace equation over the domain $\Omega$. This gives the following integral equation after application of the Green's second identity for vectors

$$\int_{\Gamma} \frac{\partial \hat{\mathbf{v}}}{\partial n_{\Gamma}}^{m+1} T^{*} \, d\Gamma - \int_{\Gamma} \hat{\mathbf{v}}^{m+1} \frac{\partial T^{*}}{\partial n_{\Gamma}} \, d\Gamma - c_{\mathbf{s}}^{*} \, \hat{\mathbf{v}}_{\mathbf{s}}^{m+1} =$$

$$= \int_{\Omega} \frac{1}{\mu_{\mathcal{B}}} \left( \nabla P^{m+1} + \frac{\mu_{\mathcal{D}}}{\mathcal{K}} \, \hat{\mathbf{v}}^{m+1} - \mathbf{f}^{m} \right) T^{*} \, d\Omega \tag{14}$$

which is solved with respect to the Dirichlet representation

$$\mathbf{v} = \mathbf{0}; \quad \mathbf{p} \in \Gamma \tag{15}$$

of the impermeable and no-slip (5) velocity boundary conditions. The "hat" on the velocity denotes that the velocity field does not correspond to the mass conservation in general. Equation (14) is used for simultaneous determination of the velocity gradients on the boundary $\Gamma$ and velocity in the domain $\Omega$. The velocity gradients on the boundary and in the domain can be explicitly calculated from the velocity gradient Poisson equation

$$\int_{\Gamma} \frac{\partial \hat{\mathbf{v}}}{\partial n_{\Gamma}}^{m+1} \nabla T^{*} \, d\Gamma - \int_{\Gamma} \hat{\mathbf{v}}^{m+1} \nabla \frac{\partial T^{*}}{\partial n_{\Gamma}} \, d\Gamma + \nabla ( c_{\mathbf{s}}^{*} \, \hat{\mathbf{v}}_{\mathbf{s}}^{m+1} ) =$$

$$= \int_{\Omega} \frac{1}{\mu_{\mathcal{B}}} ( \nabla P^{m+1} + \frac{\mu_{\mathcal{D}}}{\mathcal{K}} \hat{\mathbf{v}}^{m+1} - \mathbf{f}^m ) \nabla T^* \, d\Omega \tag{16}$$

The derivatives $\frac{\partial^2}{\partial x^2} \hat{v}^{m+1}$ and $\frac{\partial^2}{\partial y^2} \hat{v}^{m+1}$ on the boundary that are needed in evaluation of the boundary condition (9) can be explicitly calculated from the integral representation

$$\int_{\Gamma} \frac{\partial \hat{\mathbf{v}}}{\partial n_{\Gamma}}^{m+1} \frac{\partial}{\partial p_{\xi}} \nabla T^* \, d\Gamma - \int_{\Gamma} \hat{\mathbf{v}}^{m+1} \frac{\partial}{\partial p_{\xi}} \nabla \frac{\partial T^*}{\partial n_{\Gamma}} \, d\Gamma + \frac{\partial}{\partial p_{\xi}} \nabla ( c^*_{\mathbf{s}} \, \hat{\mathbf{v}}^{m+1}_{\mathbf{s}} ) =$$

$$= \int_{\Omega} \frac{1}{\mu_{\mathcal{B}}} ( \nabla P^{m+1} + \frac{\mu_{\mathcal{D}}}{\mathcal{K}} \hat{\mathbf{v}}^{m+1} - \mathbf{f}^m ) \frac{\partial}{\partial p_{\xi}} \nabla T^* \, d\Omega \tag{17}$$

The incompressibility is in the case of Darcy–Brinkman law enforced through the pressure $\tilde{P}^{m+1}$ and velocity $\tilde{\mathbf{v}}^{m+1}$ correction which ensures

$$\nabla \cdot \mathbf{v}^{m+1} = \nabla \cdot ( \hat{\mathbf{v}}^{m+1} + \tilde{\mathbf{v}}^{m+1} ) = 0 \tag{18}$$

Consider that the velocity correction $\tilde{\mathbf{v}}^{m+1}$ occours exclusively due to action of the pressure correction $\tilde{P}^{m+1}$

$$\frac{\varrho}{\Delta t} \tilde{\mathbf{v}}^{m+1} = -\nabla \tilde{P}^{m+1} \tag{19}$$

with $\Delta t$ standing for heuristic velocity–pressure correction relaxation factor. The pressure correction can thus be calculated from the velocity field $\hat{\mathbf{v}}^{m+1}$ through the pressure correction Poisson equation

$$\nabla^2 \tilde{P}^{m+1} = \frac{\varrho}{\Delta t} \nabla \cdot \hat{\mathbf{v}}^{m+1} \tag{20}$$

deduced from equations (18) and (19). The boundary-domain integral equivalent for solving the pressure correction Poisson equation is

$$\int_{\Gamma} \frac{\partial \tilde{P}}{\partial n_{\Gamma}}^{m+1} T^* \, d\Gamma - \int_{\Gamma} \tilde{P}^{m+1} \frac{\partial T^*}{\partial n_{\Gamma}} \, d\Gamma - c^* \, \tilde{P}^{m+1}_s = \int_{\Omega} \frac{\varrho}{\Delta t} \nabla \cdot \hat{\mathbf{v}}^{m+1} \, T^* \, d\Omega \tag{21}$$

with the pressure correction Poisson equation Neumann boundary conditions

$$\frac{\partial \tilde{P}}{\partial n_{\Gamma}}^{m+1} = 0; \;\; \mathbf{p} \in \Gamma \tag{22}$$

The pressure correction gradient $\nabla \tilde{P}^{m+1}$ on the boundary and in the domain can be explicitly calculated from the integral representation

$$\int_{\Gamma} \frac{\partial \tilde{P}}{\partial n_{\Gamma}}^{m+1} \nabla T^* \, d\Gamma - \int_{\Gamma} \tilde{P}^{m+1} \nabla \frac{\partial T^*}{\partial n_{\Gamma}} \, d\Gamma + \nabla (c^* \, \tilde{P}^{m+1}_s) = \int_{\Omega} \frac{\varrho}{\Delta t} \nabla \cdot \hat{\mathbf{v}} \nabla T^* \, d\Omega \tag{23}$$

The pressure and pressure gradient fields are updated as

$$P^{m+1} = P^{m+1} + \tilde{P}^{m+1} \qquad \nabla P^{m+1} = \nabla P^{m+1} + \nabla \tilde{P}^{m+1} \tag{24}$$

The velocity field is updated through the pressure gradient corrections

$$\mathbf{v}^{m+1} = \hat{\mathbf{v}}^{m+1} - \frac{\Delta t}{\varrho} \nabla \tilde{P}^{m+1} \tag{25}$$

The iteration cycle is completed by calculating the temperature field at iteration level $m+1$ (and with this also $\mathbf{f}^{m+1}$). This is accomplished by weighting the energy conservation equation by the Fundamenal solution of the Laplace equation and by using the Green's second identity for scalars

$$\int_\Gamma \frac{\partial T}{\partial n_\Gamma}^{m+1} T^* \, d\Gamma - \int_\Gamma T^{m+1} \frac{\partial T^*}{\partial n_\Gamma} \, d\Gamma - c_\mathrm{s}^* \, T_\mathrm{s}^{m+1} =$$

$$= \int_\Omega \frac{1}{\alpha} \nabla \cdot (\, \mathbf{v}^{m+1} \, T^{m+1} \,) \, T^* \, d\Omega + \int_\Omega \frac{q}{k} T^* \, d\Omega \tag{26}$$

with $\alpha = k/(\varrho \, c_p)$ denoting thermal diffusivity. Equation (26), together with the thermal boundary conditions (6), is used to simultaneously solve the unknown temperature distribution in the Neumann part of the boundary, the unknown temperature derivative in the normal boundary direction in the Dirichlet part of the boundary and unknown temperatures in the domain. The temperature gradients on the boundary and in the domain can be explicitly calculated from the integral representation

$$\int_\Gamma \frac{\partial T}{\partial n_\Gamma}^{m+1} \nabla T^* \, d\Gamma - \int_\Gamma T^{m+1} \, \nabla \frac{\partial T^*}{\partial n_\Gamma} \, d\Gamma + \nabla (\, c_\mathrm{s}^* \, T_\mathrm{s}^{m+1}) =$$

$$= \int_\Omega \frac{1}{\alpha} \nabla \cdot (\, \mathbf{v}^{m+1} \, T^{m+1} \,) \, \nabla T^* \, d\Omega + \int_\Omega \frac{q}{k} \nabla T^* \, d\Omega \tag{27}$$

The iteration cycle is completed with calculation of the updated body force. The iterations are stopped when conditions

$$| \, |\mathbf{v}^{m+1}| - |\mathbf{v}^{m}| \, | < v_\epsilon \qquad | \, |T^{m+1}| - |T^{m}| \, | < T_\epsilon \tag{28}$$

are satisfied with $v_\epsilon$ and $T_\epsilon$ representing the velocity and temperature convergence criterions. In case iteration conditions (28) are not satisfied, new iteration cycle starts with the relaxed value of the body force

$$^{m+1}\mathbf{f} =^{m} \mathbf{f} + c_\mathrm{rel} \, (\, ^{m+1}\mathbf{f} -^{m} \mathbf{f} \,) \tag{29}$$

with $c_\mathrm{rel}$ representing the heuristic relexation factor.

The velocity, pressure, and temperature fields are all calculated in the same gridpoints $\mathbf{p}_n; n = 1, 2, \cdots, N; N = N_\Gamma + N_\Omega$. The first $N_\Gamma$ gridpoints are distributed on the boundary and the last $N_\Omega$ in the domain. Boundary geometry is approximated by $N^\Gamma$ straight line segments, and spatial variation of the fields on each of the boundary segments is represented by linear interpolation functions i.e., $N_\Gamma = 2\,N^\Gamma$. The domain integrals in equations (10,13,14,16,17,21,23,26,27) are transformed by considering the approximation of the spatial variation of the field $\mathcal{F}$ over $\Omega \cup \Gamma$ by the global interpolation functions $\psi_n$ and coefficients $\varsigma_n$

$$\mathcal{F}(\mathbf{p}) \approx \psi_n(\mathbf{p})\,\varsigma_n; \quad n = 1, 2, \cdots, N+1; \quad N = N_\Omega + N_\Gamma \tag{30}$$

The augmented inverse multiquadrics used in this work are for $n = 1, 2, \cdots, N$

$$\psi_n = (r_n^2 + r_0^2)^{-1/2} + (r_0^2/2)(r_n^2 + r_0^2)^{-3/2}, \qquad r_n^2 = (\mathbf{p} - \mathbf{p}_{\underline{n}})\cdot(\mathbf{p} - \mathbf{p}_{\underline{n}}) \tag{31}$$

and $\psi_{N+1} = 1$. These functions are (except the augmentation term) decaying as $r_n^{-1}$ for large $r_n$ and can be used also for semi–infinite problems. Their axisymmetric version can be expressed in a closed form.

The involved systems of linear equations that originate from solution of the integral equations are in present work solved by the standard subroutines from [5]. The regular $N \times N$ systems of linear equations that originate from integral equations (14,26) are solved by LU decomposition and backsubstitution by using subroutines LUDCMP and LUBKSB. The $N_\Gamma \times N_\Gamma$ systems that originate from integral equations (10,21) are singular since the pressure and pressure corrections are known only up to an additive constant. This two systems are thus solved by Householder reduction to bidiagonal form and QR diagonalization with shifts by using subroutines SVDCMP and SVBKSB. The temperature and velocity tolerances have been set to $10^{-4}$[m/s] and $10^{-4}$[K]. All involved boundary integrals are solved exactly [6]. The solution was assessed, similarly as for the Darcy model [3] by comparing the midplane velocity components, local Nusselt number, and streamfunction extreme with the standard fine-grid finite volume solution [7]. A detailed comparison study which shows excellent agreement between the two methods will appear elsewhere.

# 4. NUMERICAL EXAMPLES

Three numerical examples are shown in present text in order to illustrate the physical complexity for which the present method can be used. They are represented in a dimensional form because of the space constraints.

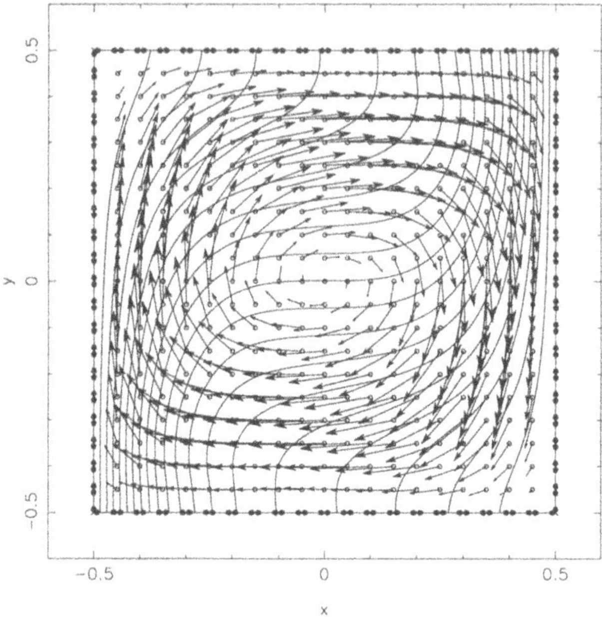


*Figure 1* Isotherms and velocity vectors of the Darcy–Brinkman model.

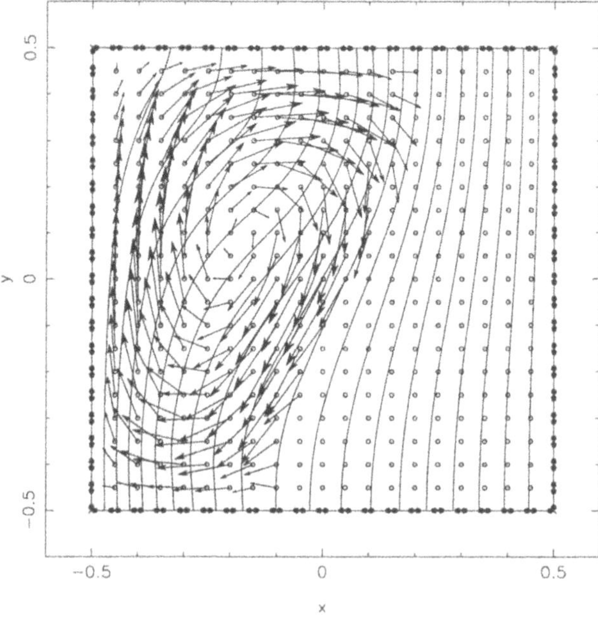


*Figure 2* Isotherms and velocity vectors of the Darcy–Brinkman model with phase change material saturating the pores. $T_S/[C] = 0.01$, $T_{\mathcal{L}}/[C] = 0.01$.

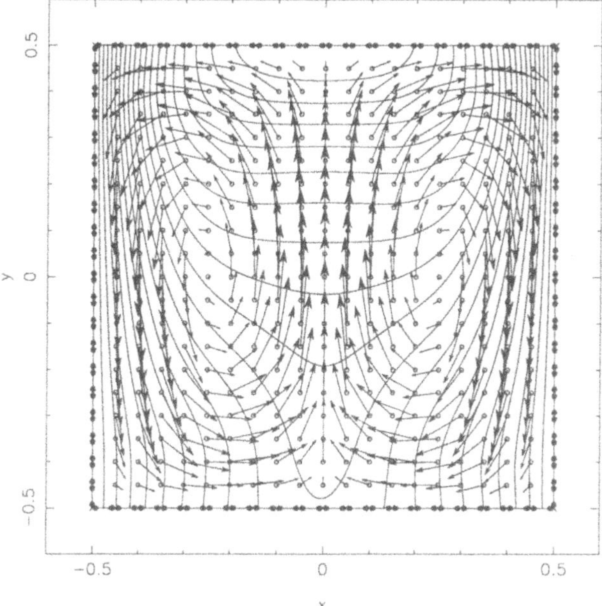


*Figure 3* Isotherms and velocity vectors of the Darcy–Brinkman model with internal heat generation $q/[\mathrm{W/m^3}] = 10$ and $T_- = T_+ = T_{\mathrm{ref}}$.

Geometry is a square with dimension 1 [m]. Upper and lower boundaries are insulated, the left boundary is subject to temperature $T_- = -0.5$ [C] and the right boundary is subject to temperature $T_+ = +0.5$ [C]. Reference temperature is $T_{\mathrm{ref}} = 0$ [C]. All material properties except $\beta = 1000/[\mathrm{C}]$ are set to unit values. $c^2$ was set to 0.05 (tipical mesh distance). The square is discretized into 80 geometrically straight line, linear discontinuous boundary elements and 361 domain nodes that give 521 meshpoints. Relaxation parameters has been set to $c_{\mathrm{rel}} = 0.1$, $\Delta t = 1.0$. The solution in Figures 1,2, and 3 is achieved in 50, 67, and 48 iterations, respectivelly.

# 5. CONCLUSIONS

This preliminary study gives basic framework for computational modelling of the natural convection in porous media as required in estimation of the radionuclides transport in geological systems. The structure of the solution procedure permits straighforward upgrading to transient situations and non-linear as well inhomogeneous response. Related complications of the energy equation are coped in [8, 9]. Treatment of the momentum equation convective term is elaborated in [10]. Realistic situations require upgrading of numerical implementation to axisymmetry and to three dimensions as well the use of the iterative methods [11] for solution of the involved large, fully populated systems of algebraic equations. The topics listed represent main subject of our current research.

## Acknowledgments

The author would like to acknowledge the Slovenian Nuclear Safety Administration and the European Community for financial support in the framework of the projects *Computational Modelling for Radionuclides Transport Estimation in Natural and Technological Systems* and *INCO-Copernicus ULTRAWAT*.

## References

[1] M. Sahimi. *Flow and Transport in Porous Media and Fractured Rock*, VCH Verlagsgesellschaft mbH, Weinheim, 1995.

[2] B. Šarler. Solid–liquid phase change in porous media: solution by Boundary Integral Method, *IABEM Symposium on Boundary Integral Methods for Nonlinear Problems*, (L. Morino and W.L. Wendland, eds.), Kluwer, Dordrecht, 185–190, 1997.

[3] B. Šarler, D. Gobin, B. Goyeau, J. Perko and H. Power. Natural convection in porous media - DRBEM solution of the Darcy model, *Int.J.Numer Methods Fluids*, 33:279-312, 2000.

[4] P.W. Partridge, C.A. Brebbia and L.C. Wrobel. *The Dual Reciprocity Boundary Element Method*, CMP, Southampton, 1992.

[5] W.H. Press, S.A. Teukolsky, W.T. Vetterling and B.P. Flannery. *Numerical Recipes in Fortran*, Cambridge University Press, Cambridge, 1992.

[6] Z. Rek and B. Šarler. Analytical integration of elliptic 2D fundamental solution and its derivatives for straight-line elements with constant interpolation, *E*ng.Anal., 23:515–525, 1999.

[7] B. Goyeau, J.-P. Songbe and D. Gobin. Numerical study of double-difussive natural convection in a porous cavity using the Darcy–Brinkman formulation, *I*nt.J.Heat Mass Transfer, 39:1363–1378, 1996.

[8] B. Šarler and G. Kuhn. Dual reciprocity boundary element method for convective-diffusive solid liquid phase change problems, Part I: Formulation, *E*ng.Anal, 21:53–63, 1998.

[9] B. Šarler and G. Kuhn. Dual reciprocity boundary element method for convective-diffusive solid liquid phase change problems, Part II: Numerical examples, *E*ng.Anal, 21:65–79, 1998.

[10] B. Šarler and G. Kuhn. Primitive variable dual reciprocity boundary element method solution of the Navier–Stokes equations, *E*ng.Anal., 23:443–455, 1999.

[11] V. Bulgakov, B. Šarler and G. Kuhn. Iterative solution of systems of equations in the dual reciprocity boundary element method for the diffusion equation, *I*nt.J.Numer.Methods.Eng., 43:713–732, 1998.

# DYNAMIC POROELASTICITY TREATED BY A TIME DOMAIN BOUNDARY ELEMENT METHOD

Martin Schanz

*Technical University Braunschweig, Institute of Applied Mechanics*

*P.O. Box 3329, D-38023 Braunschweig, Germany*

m.schanz@tu-bs.de

**Keywords:** Poroelasticity, Biot, time domain, Convolution Quadrature Method

**Abstract** The most powerful methodology to tackle semi-infinite poroelastic domains is the Boundary Element Method (BEM). The recently developed 'Convolution Quadrature Method', proposed by Lubich, utilizes the existing Laplace transform fundamental solution and performs the solution in the time domain. Hence, applying this quadrature formula to the time dependent poroelastic boundary integral equation a time stepping procedure is obtained based only on the Laplace domain fundamental solution and a linear multistep method. Finally, two examples show the accuracy of the proposed time stepping procedure.

## 1. INTRODUCTION

The dynamic responses of fluid-saturated semi-infinite porous continua subject to transient excitations such as seismic waves or ground vibrations are important in the design of soil-structure systems. For modeling wave propagation in such fluid-saturated media, Biot's theory may be applied. The most powerful numerical method to tackle semi-infinite poroelastic domains is the Boundary Element Method (BEM).

The efficiency of BEM in dealing with semi-infinite domain problems has long been recognized by researchers and engineers. There is a large number of applications (see e.g. Beskos [1, 2]). For modeling transient behavior, there exist two approaches: solving in the time domain by time-stepping, and solving in the integral transform domain via Laplace or Fourier transform with subsequent inverse transformation. The time stepping approach is often impossible due to the lack of a time dependent fundamental solution which is often only available in the trans-

*T. Burczynski (ed.), IUTAM/IACM/IABEM Symposium on Advanced Mathematical and Computational Mechanics Aspects of the Boundary Element Method*, 303–314.

formed domain. Therefore, the first poroelasto-dynamic BEM formulation based on Biot's theory was published in Laplace domain by Manolis and Beskos [15] expressed in terms of the solid and fluid displacements. However, it can be shown that only the solid displacements and one additional variable, the fluid pressure, are independent [5]. Based on these three unknowns in two-dimensions (2-d), formulations in frequency domain were published by Cheng et al. [9] and Domínguez [10]. The subsequent numerical inversion of the integral transform, to get time dependent results, can become unreliable when the transient behavior becomes complicated. Because of this a time dependent BEM formulation seems to be preferable. Such a formulation was developed by Wiebe and Antes [19], but with the restriction of vanishing damping between the solid skeleton and the fluid. Another time dependent formulation was proposed by Chen and Dargush based on analytical inverse transformation of the Laplace domain fundamental solutions [8]. However, as the authors have mentioned, this formulation is highly computer time demanding.

In this work, a time dependent formulation for poroelasto-dynamics is presented using the idea of the BEM formulation proposed for visco- and elastodynamics [17] based on the Convolution Quadrature Method developed by Lubich [14]. In this formulation, the convolution integral is numerically approximated by a quadrature formula whose weights are determined by the Laplace transform of the fundamental solutions and a linear multistep method. Hence, a time stepping procedure can be achieved using only the Laplace transformed fundamental solutions which are available in the literature [7].

In the following, the governing equations based on Biot's constitutive equations for poroelastic media are recalled, and subsequently a boundary integral formulation is deduced. After spatial discretization and application of the convolution quadrature for time discretization, a time stepping procedure for poroelasticity is presented. Two examples are presented to show the accuracy and efficiency (in view of [8]) of the proposed method.

## 2. GOVERNING EQUATIONS

For a wide range of fluid infiltrated materials, such as water saturated soils, oil impregnated rocks, or air filled foams, the elastic theory is a crude approximation for investigating wave propagation in such media. Due to their porosity, a different theory is necessary. A theory of porous materials containing a viscous fluid was presented by Biot [3]. The dynamic extension was done in [4]. Following Biot's approach to model the behavior of porous media, the constitutive equations can be expressed as

$$\sigma_{ij} = \mu\left(u_{i,j} + u_{j,i}\right) + \lambda\delta_{ij}u_{k,k} - \alpha\delta_{ij}p \quad \text{and} \quad \zeta = \alpha u_{k,k} + \frac{\phi^2}{R}p \tag{1}$$

in which $\sigma_{ij}$ denotes the total stress, $p$ the pore pressure, $\varepsilon_{ij}$ the strain of the solid frame, $\zeta$ the variation of fluid volume per unit reference volume, and $\delta_{ij}$ the Kronecker delta. The sign convention for stress and strain follows that of elasticity, namely, tensile stress and strain is denoted positive. The Latin indices take the values 1,2,3 or 1,2 in 3-d or 2-d cases, respectively, where summation convention is implied over repeated indices. The bulk material is defined by the material constants $\lambda$ and $\mu$, known from elasticity as the Lamé constants. The porosity $\phi$, Biot's effective stress coefficient $\alpha$, and $R$ complete the set of material parameters.

Now, the governing equations are completed by the dynamic equilibrium

$$\sigma_{ij,j} = \varrho\frac{\partial^2}{\partial t^2}u_i + \varrho_f\frac{\partial^2}{\partial t^2}v_i - F_i \tag{2}$$

and the continuity equation

$$\frac{\partial}{\partial t}\zeta + q_{i,i} = f\ , \tag{3}$$

where, with the densities of the solid $\varrho_s$ and the fluid $\varrho_f$, the bulk density $\varrho = \varrho_s(1-\phi) + \phi\varrho_f$ is used. The displacements of the solid are denoted by $u_i$ and the relative fluid to solid displacements by $v_i$. In Eqs. (2), (3), and in all subsequent equations, vanishing initial conditions for all variables are assumed. $F_i$ are the bulk body forces, and $q_i = \phi\frac{\partial v_i}{\partial t}$ denotes the specific flux of the fluid. Finally, the derivative with respect to the spatial variable $x_i$ is abbreviated by $()_{,i}$. The fluid is modeled with Darcy's law

$$q_i = -\kappa\left(p_{,i} + \varrho_f\frac{\partial^2}{\partial t^2}u_i + \frac{\varrho_a + \phi\varrho_f}{\phi}\frac{\partial^2}{\partial t^2}v_i\right) \tag{4}$$

where $\kappa$ denotes the permeability and $\varrho_a$ the apparent mass density. For simplicity, the apparent mass density $\varrho_a$ is assumed to be frequency independent as $\varrho_a = 0.66\phi\varrho_f$ [6].

As shown in [5], it is sufficient to use the solid displacements and the pore pressure as basic variables to describe a poroelastic continuum. Therefore, the above equations are reduced to these four unknowns. To do this, Eqs. (2), (3), and (4) are transformed to Laplace domain. Taking a linear strain-displacement relation $\varepsilon_{ij} = 1/2\left(u_{i,j} + u_{j,i}\right)$, i.e., assuming small deformation gradients, the final set of differential equations for the displacements $\hat{u}_i$ and the pore pressure $\hat{p}$ is achieved

$$\mu \hat{u}_{i,jj} + (\lambda + \mu)\,\hat{u}_{j,ij} - (\alpha - \beta)\,\hat{p}_{,i} - s^2\,(\varrho - \beta \varrho_f)\,\hat{u}_i = -\hat{F}_i \qquad \text{(5a)}$$

$$\frac{\beta}{s\varrho_f}\hat{p}_{,ii} - \frac{\phi^2 s}{R}\hat{p} - (\alpha - \beta)\,s\hat{u}_{i,i} = -\hat{f} \qquad \text{(5b)}$$

with the abbreviation $\beta = \frac{\phi^2 s\kappa\varrho_f}{\phi^2 + s\kappa(\varrho_a + \phi\varrho_f)}$ and $\hat{f}\,(s)$, denoting the Laplace transform of a function $f\,(t)$ with the complex Laplace variable $s$. With this set of equations, the behavior of a poroelastic continuum is completely given.

## 3. POROELASTIC BOUNDARY ELEMENT METHOD

The boundary integral equation for dynamic poroelasticity in Laplace domain can be obtained as usual using either the corresponding reciprocal work theorem [9] or the weighted residuals formulation. Here, the latter is used.

A short representation of the two coupled differential Eqs. (5) is given by

$$\mathbf{B}^* \begin{bmatrix} \hat{u}_i \\ \hat{p} \end{bmatrix} + \begin{bmatrix} \hat{F}_i \\ \hat{f} \end{bmatrix} = \mathbf{0}\,, \qquad (6)$$

with the not self-adjoint operator $\mathbf{B}^*$ and $\partial_i = ()_{,i}$.

The integral equation corresponding to (6) can be derived directly by equating the inner product of (6) and the matrix of the fundamental solutions $\mathbf{G}$ to a null vector, i.e.

$$\int\limits_{\Omega} \mathbf{G}^T \mathbf{B}^* \begin{bmatrix} \hat{u}_i \\ \hat{p} \end{bmatrix} \mathrm{d}\Omega = \mathbf{0} \qquad \text{with} \qquad \mathbf{G} = \begin{bmatrix} \hat{U}^s_{ij} & \hat{U}^f_i \\ \hat{P}^s_j & \hat{P}^f \end{bmatrix}, \qquad (7)$$

where the integration is performed over a domain $\Omega$ with boundary $\Gamma$, and vanishing body forces $F_i$ and sources $f$ are assumed. By this, essentially, the error in satisfying the governing differential Eqs. (5) is forced to be orthogonal to $\mathbf{G}$. Two integrations by part with respect to the spatial variable transfer all derivatives from the vector $[\hat{u}_i\ \hat{p}]^T$ to the matrix of the fundamental solutions $\mathbf{G}$. The result can be written as

$$\int\limits_{\Gamma} \begin{bmatrix} \hat{U}^s_{ij} & -\hat{P}^s_j \\ \hat{U}^f_i & -\hat{P}^f \end{bmatrix} \begin{bmatrix} \hat{t}_i \\ \hat{q} \end{bmatrix} \mathrm{d}\Gamma = \int\limits_{\Gamma} \begin{bmatrix} \hat{T}^s_{ij} & \hat{Q}^s_j \\ \hat{T}^f_i & \hat{Q}^f \end{bmatrix} \begin{bmatrix} \hat{u}_i \\ \hat{p} \end{bmatrix} \mathrm{d}\Gamma - \int\limits_{\Omega} (\mathbf{B}\mathbf{G})^T \begin{bmatrix} \hat{u}_i \\ \hat{p} \end{bmatrix} \mathrm{d}\Omega \qquad (8)$$

with $\mathbf{B}$ denoting the adjoint operator to $\mathbf{B}^*$. In (8), $\hat{t}_i = \hat{\sigma}_{ij} n_j$ is the traction vector and $\hat{q} = -\frac{\beta}{s\varrho_f}\left(\hat{p}_{,i} + \varrho_f s^2 \hat{u}_i\right) n_i$ the flux with the outward normal $\mathbf{n}$. Further the abbreviations

$$\hat{T}^s_{ij} = \left[\left(\lambda \hat{U}^s_{kj,k} + \alpha s \hat{P}^s_j\right)\delta_{i\ell} + \mu\left(\hat{U}^s_{ij,\ell} + \hat{U}^s_{\ell j,i}\right)\right] n_\ell \tag{9a}$$

$$\hat{Q}^s_j = \frac{\beta}{s\varrho_f}\left[\hat{P}^s_{j,i} - \varrho_f s \hat{U}^s_{ji}\right] n_i \tag{9b}$$

$$\hat{T}^f_i = \left[\left(\lambda \hat{U}^f_{k,k} + \alpha s \hat{P}^f\right)\delta_{i\ell} + \mu\left(\hat{U}^f_{i,\ell} + \hat{U}^f_{\ell,i}\right)\right] n_\ell \tag{9c}$$

$$\hat{Q}^f = \frac{\beta}{s\varrho_f}\left[\hat{P}^f_{,j} - \varrho_f s \hat{U}^f_j\right] n_j \tag{9d}$$

are used where (9a) and (9b) can be interpreted as being the adjoint term to the traction vector $\hat{t}_i$ and the flux $\hat{q}$, respectively.

The matrix of the fundamental solutions $\mathbf{G}$ is determined by

$$\mathbf{B}\begin{bmatrix} \hat{U}^s_{ij} & \hat{U}^f_i \\ \hat{P}^s_j & \hat{P}^f \end{bmatrix} + \begin{bmatrix} \delta(r)\mathbf{I} & \mathbf{0} \\ \mathbf{0}^T & \delta(r) \end{bmatrix} = \mathbf{0} \qquad r = |\mathbf{x} - \mathbf{y}| \tag{10}$$

using the adjoint operator $\mathbf{B}$ and the identity matrix $\mathbf{I}$. In Eq. (10), the load vector represents a point force $F_i = \delta(r) e_i$ in all three coordinate directions $e_i$ and a point source $f = \delta(r)$ at the point $\mathbf{y}$, with $\delta(r)$ denoting the Dirac distribution. The system of differential equations can be solved using the method of Hörmander [12]. Another possibility to find the fundamental solutions of the posed problem is using the analogy between thermoelasticity and poroelasticity in the transformed domain [5]. For the next step, it is necessary to know what happens with the fundamental solutions when $r$ approaches 0, i.e., the point $\mathbf{y}$ is put on the boundary close to the point $\mathbf{x}$. Six of the eight fundamental solutions, four in $\mathbf{G}$ and four calculated by Eqs. (9), are singular. The order of the singularity can be determined by the series representation of the fundamental solutions. This leads to: $\hat{P}^s_i, \hat{U}^f_i = O\left(r^0\right)$ regular, $\hat{U}^s_{ij}, \hat{P}^f, \hat{Q}^s_j, \hat{T}^f_i = O\left(r^{-1}\right)$ weakly singular, and $\hat{T}^s_{ij}, \hat{Q}^f = O\left(r^{-2}\right)$ strongly singular.

It can be shown that the singular parts of the displacement fundamental solution $\hat{U}^s_{ij}$ and of the traction fundamental solution $\hat{T}^s_{ij}$ are the same as in elastostatics, and those of the pressure $\hat{P}^f$ and of the flux $\hat{Q}^f$ are the same as in acoustics. Therefore, shifting in integral equation (8) the point $\mathbf{y}$ to the boundary $\Gamma$ results in the integral equation

$$\int_\Gamma \begin{bmatrix} \hat{U}^s_{ij} & -\hat{P}^s_j \\ \hat{U}^f_i & -\hat{P}^f \end{bmatrix}\begin{bmatrix} \hat{t}_i \\ \hat{q} \end{bmatrix} d\Gamma = \oint_\Gamma \begin{bmatrix} \hat{T}^s_{ij} & \hat{Q}^s_j \\ \hat{T}^f_i & \hat{Q}^f \end{bmatrix}\begin{bmatrix} \hat{u}_i \\ \hat{p} \end{bmatrix} d\Gamma + \begin{bmatrix} c_{ij} & 0 \\ 0 & c \end{bmatrix}\begin{bmatrix} \hat{u}_i \\ \hat{p} \end{bmatrix} \tag{11}$$

with the integral free terms $c_{ij}$ and $c$ known from elastostatics and acoustics, respectively. A transformation to time domain gives, finally, the time dependent integral equation for poroelasticity

$$\int_{\Gamma} \begin{bmatrix} U^s_{ij} & -P^s_j \\ U^f_i & -P^f \end{bmatrix} * \begin{bmatrix} t_i \\ q \end{bmatrix} \mathrm{d}\Gamma = \oint_{\Gamma} \begin{bmatrix} T^s_{ij} & Q^s_j \\ T^f_i & Q^f \end{bmatrix} * \begin{bmatrix} u_i \\ p \end{bmatrix} \mathrm{d}\Gamma + \begin{bmatrix} c_{ij} & 0 \\ 0 & c \end{bmatrix} \begin{bmatrix} u_i \\ p \end{bmatrix} \tag{12}$$

with the convolution integral $f * g = \int_0^t f(t-\tau)\, g(\tau)\, \mathrm{d}\tau$ and the Cauchy Principal value $\oint_{\Gamma}$.

According to the boundary element method, the boundary $\Gamma$ is discretized by $E$ iso-parametric elements $\Gamma_e$ where $F$ polynomial shape functions $N_e^f(\mathbf{x})$ are defined. Hence, with the time dependent nodal values $u_i^{ef}(t), t_i^{ef}(t)$, $p^{ef}(t)$, and $q^{ef}(t)$ the following ansatz functions are adapted

$$\left.\begin{matrix} u_i(\mathbf{x},t) \\ t_i(\mathbf{x},t) \end{matrix}\right\} = \sum_{e=1}^{E}\sum_{f=1}^{F} N_e^f(\mathbf{x}) \left\{\begin{matrix} u_i^{ef}(t) \\ t_i^{ef}(t) \end{matrix}\right. ; \quad \left.\begin{matrix} p(\mathbf{x},t) \\ q(\mathbf{x},t) \end{matrix}\right\} = \sum_{e=1}^{E}\sum_{f=1}^{F} N_e^f(\mathbf{x}) \left\{\begin{matrix} p^{ef}(t) \\ q^{ef}(t) \end{matrix}\right. . \tag{13}$$

When the period $t$ is discretized by $N$ equal time-increments $\Delta t$, the convolution integrals between the fundamental solutions and the nodal values in (12) can be approximated by the Convolution Quadrature Method proposed by Lubich [14]. This quadrature formula numerically approximates a convolution integral by the finite sum

$$f(t) * g(t) \approx \sum_{k=0}^{n} \omega_{n-k}\left(\hat{f}\right) g(k\Delta t) \ . \tag{14}$$

The integration weights $\omega_n$ are the coefficients of the power series for the function $\hat{f}\left(\frac{\gamma(z)}{\Delta t}\right)$ at the point $\frac{\gamma(z)}{\Delta t}$. Herein, $\gamma(z)$ is the quotient of the characteristic polynomials of a linear $A(\alpha)$-stable multistep method, e.g., $\gamma(z) = \frac{3}{2} - 2z + \frac{1}{2}z^2$ for the backward differential formula of second order (BDF 2). The coefficients of this power series are calculated by the integral

$$\omega_n\left(\hat{f}\right) = \frac{1}{2\pi i} \int_{|z|=\mathcal{R}} \hat{f}\left(\frac{\gamma(z)}{\Delta t}\right) z^{-n-1} \mathrm{d}z \approx \frac{1}{L}\sum_{\ell=0}^{L-1} \hat{f}\left(\frac{\gamma(\xi_\ell)}{\Delta t}\right) \xi_\ell^{-n}, \tag{15}$$

with $\xi_\ell = \mathcal{R}e^{i\ell\frac{2\pi}{L}}$, $\mathcal{R}$ being the radius of a circle in the domain of analyticity of $\hat{f}(z)$. After transformation to polar coordinates, the integral in equation (15) is approximated by a trapezoidal rule with $L$ equal steps

$\frac{2\pi}{L}$. Details are described in [17] as well as in the original work [14]. The quadrature rule (14) has the characteristic advantage that only the Laplace transformed function $\hat{f}$ is used.

Applying the quadrature formula (14) to the integral equation (12) gives the following boundary element time-stepping formulation for $n = 0, 1, \ldots, N$

$$\begin{bmatrix} c_{ij}u_i(n\Delta t) \\ cp(n\Delta t) \end{bmatrix} = \sum_{e=1}^{E}\sum_{f=1}^{F}\sum_{k=0}^{n}\left\{\begin{bmatrix} \omega_{n-k}^{ef}\left(\hat{U}_{ij}^{s}\right) & -\omega_{n-k}^{ef}\left(\hat{P}_{j}^{s}\right) \\ \omega_{n-k}^{ef}\left(\hat{U}_{i}^{f}\right) & -\omega_{n-k}^{ef}\left(\hat{P}^{f}\right) \end{bmatrix}\begin{bmatrix} t_i^{ef}(k\Delta t) \\ q^{ef}(k\Delta t) \end{bmatrix} - \begin{bmatrix} \omega_{n-k}^{ef}\left(\hat{T}_{ij}^{s}\right) & \omega_{n-k}^{ef}\left(\hat{Q}_{j}^{s}\right) \\ \omega_{n-k}^{ef}\left(\hat{T}_{i}^{f}\right) & \omega_{n-k}^{ef}\left(\hat{Q}^{f}\right) \end{bmatrix}\begin{bmatrix} u_i^{ef}(k\Delta t) \\ p^{ef}(k\Delta t) \end{bmatrix}\right\} \tag{16}$$

with the weights corresponding to (15), e.g.

$$\omega_n^{ef}\left(\hat{U}_{ij}^{s}\right) = \frac{1}{L}\sum_{\ell=0}^{L-1}\int_{\Gamma}\hat{U}_{ij}^{s}\left(\frac{\gamma(\xi_\ell)}{\Delta t}\right)N_e^f(\mathbf{x})\,\mathrm{d}\Gamma\,\xi_\ell^{-n}. \tag{17}$$

Note that the calculation of the integration weights is only based on the Laplace transformed fundamental solutions. Therefore, with the time stepping procedure (16), a boundary element formulation for poroelastodynamics is deduced without the fundamental solutions in time domain.

## 4. NUMERICAL EXAMPLES

In order to validate the proposed boundary element approach, two problems are investigated. First, the influence of the time step size is analyzed by comparing the approximated results achieved by the BEM to an analytical solution of a 1-d column, and, second, a half space under a vertical load is considered for studying wave propagation in different material modelings.

A one-dimensional (1-d) column of length $3m$, sketched in Fig. 1, is considered. It is assumed that the side walls and the bottom are rigid, frictionless, and impermeable. Hence, the displacements normal to the surface are blocked and the column is free to slide only parallel to the wall. At the top, the total stress vector $t_y = -1\frac{N}{m^2}H(t)$ and the pore pressure $p = 0\frac{N}{m^2}$ is prescribed, i.e., a normal pressure force of intensity one starts acting with $t > 0$ and fluid particles are assumed to be on a free fluid surface. Due to these restrictions, the 3-d continuum is reduced to a 1-d continuum with the only degree of freedom in $y$-direction. This 1-d problem has been analytically solved in [18] and its result is compared to the boundary element solution for a 3-d rod

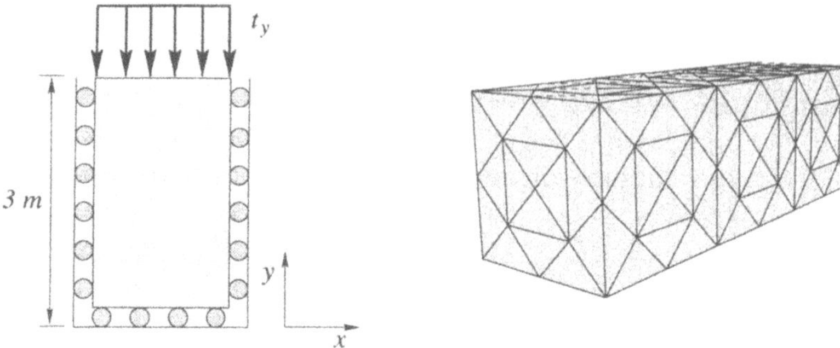


*Figure 1* One dimensional column under dynamic loading: model and discretization

($3m \times 1m \times 1m$). The boundary conditions are modeled as described above and the used mesh (224 linear triangular elements) is shown on the right side of Fig. 1. The material properties are those of Bera Sandstone [9] except that zero Poisson's ratio is set to model the 1-d behavior: $\mu = 7.2 \times 10^9 \frac{N}{m^2}, \lambda = 0 \frac{N}{m^2}, \varrho = 2458 \frac{kg}{m^3}, \phi = 0.19, \varrho_f = 1000 \frac{kg}{m^3}, \alpha = 0.777, R = 4.89 \times 10^8 \frac{N}{m^2}, \kappa = 1.9 \cdot 10^{-10}$. In Fig. 2 the vertical displacement on the top of the column is depicted versus time for different time step sizes of the numerical BEM model. Clearly, as in all BEM time stepping procedures, a critical time step size is observed below which the results become unstable. The graph for $\Delta t = 0.00001s$ is truncated for times $t > 0.012s$ because the oscillating is too strong. The stable results for $\Delta t > 0.00001s$ are sufficiently close to the analytical solution and show only a slight dependence on the time step size. This may be concluded since the results for $\Delta t = 0.00002s$ and $\Delta t = 0.00004s$ are very similar in spite of the doubled time step size.

The next example is a poroelastic half space loaded on area $A(1m^2)$ by a vertical total stress vector $t_z = -1000 \frac{N}{m^2} H(t)$ (shaded area in Fig. 3). The remaining surface is traction free. The pore pressure is assumed to be zero all over the surface, i.e., the surface is permeable. The used discretization with 180 linear triangles is also shown in Fig. 3. The material properties of the half space, i.e., the soil [13] are: $\mu = 1.45 \times 10^8 \frac{N}{m^2}, \lambda = 9.8 \times 10^7 \frac{N}{m^2}, \varrho = 1884 \frac{kg}{m^3}, \phi = 0.48, \varrho_f = 1000 \frac{kg}{m^3}, \alpha = 0.98, R = 1.2 \times 10^9 \frac{N}{m^2}, \kappa = 3.55 \cdot 10^{-9}$. The time history of the displacements at point P (see the mesh in Fig. 3) is compared for the poroelastic soil and for two other elastic soils with the same shear modulus as the poroelastic medium, but one with its undrained Poisson ratio ($\nu_u = 0.49$), and the other with its drained Poisson ratio ($\nu = 0.298$).

The results can better be understood when the wave speeds for the two elastic materials are known to identify the arrival time of the different

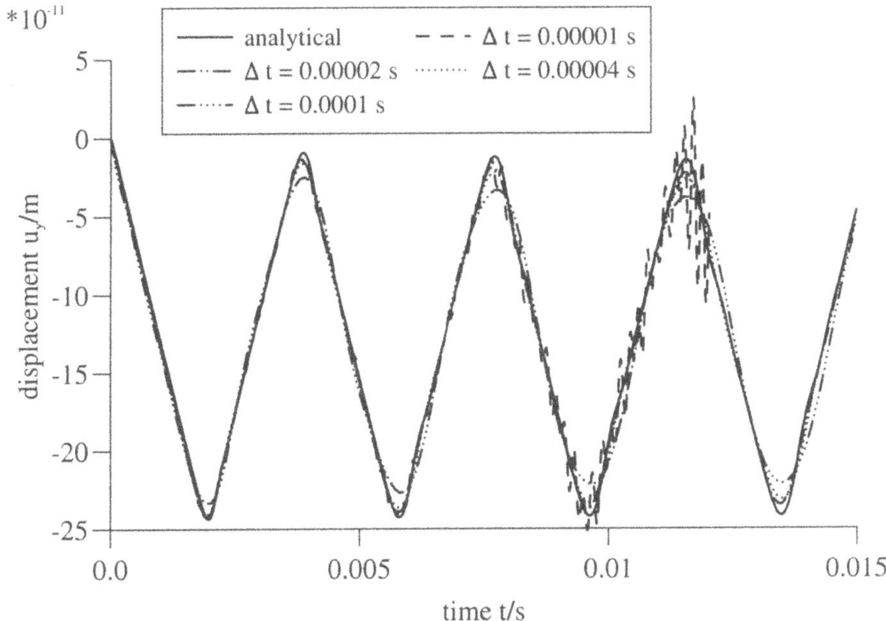


*Figure 2* Vertical displacement at top of column versus time: Influence of time step size

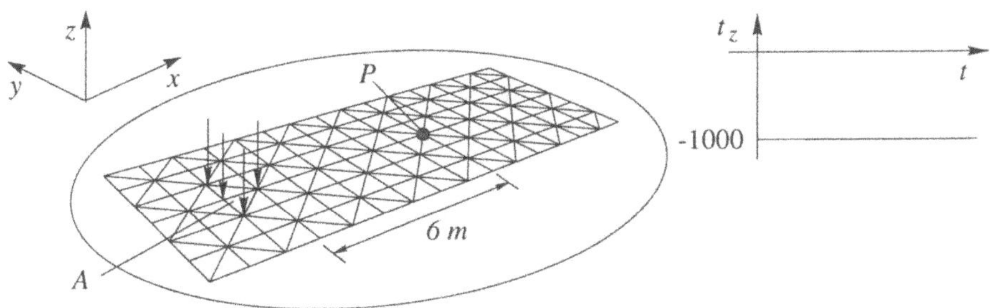


*Figure 3* Half space under vertical load: discretization and load history

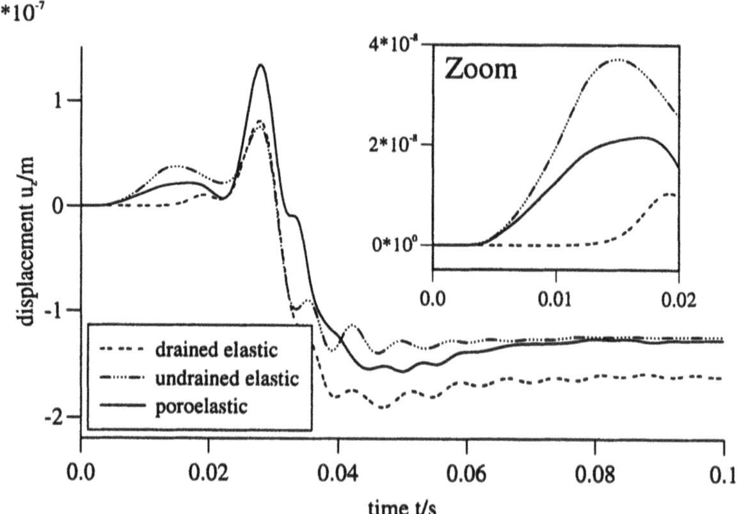


*Figure 4* Vertical displacements at point P versus time: different material modelings

waves at point P. In Tab. 1, the compression wave velocity $c_1$, the shear wave velocity $c_2$, and the Rayleigh wave velocity $c_R$ are given together with their arrival time, where the Rayleigh wave velocity is approximately determined by $c_R = \frac{0.87+1.12\nu}{1+\nu} c_2$ [11].

*Table 1* Wave velocities and corresponding arrival times for the soil

| | $c_1$ | $c_2$ | $c_R$ | $t_1$ | $t_2$ | $t_R$ |
|---|---|---|---|---|---|---|
| drained | 425 $\frac{m}{s}$ | 228 $\frac{m}{s}$ | 211 $\frac{m}{s}$ | 0.014 $s$ | 0.026 $s$ | 0.0284 $s$ |
| undrained | 1629 $\frac{m}{s}$ | 228 $\frac{m}{s}$ | 217 $\frac{m}{s}$ | 0.004 $s$ | 0.026 $s$ | 0.0276 $s$ |

In both elastic cases, the arrival of the compression wave agrees perfectly with the pre-calculated values (see Tab. 1). Since waves in poroelastic media are dispersive, i.e., the wave velocities are time-dependent, the waves are already damped when arriving at point P. This may be the reason for the modest increase of $u_z$ at the wave arrival. As Tab. 1 shows, the arrival time of the shear wave and of the Rayleigh wave is almost the same, and, hence, the exact wave fronts can not be distinguished. But, the Rayleigh wave, certainly, produces the large positive displacement since the analytical solution of Pekeris [16] shows even a singularity (pole) at the Rayleigh wave arrival. The decrease, on the other hand, is caused by the shear wave. For times after 0.04 s, the calculated displacements are only qualitatively correct, i.e., both elastic solutions tend to the correct static solution and enclose the poroelastic result. More quantitative correct results need an enlarged discretized area on the half space surface.

# 5. CONCLUSIONS

A novel poroelasto-dynamic boundary element formulation in time domain is presented for analyzing wave propagation in three-dimensional saturated porous continua described by Biot's theory for poroelasticity. Since no time dependent closed form fundamental solution exists, the "Convolution Quadrature Method"[14], a quadrature rule for convolution integrals, is applied whose integration weights are found from the Laplace transformed fundamental solution and a linear multistep method. This leads finally to a time-stepping boundary element formulation performing analysis in time domain although only the Laplace domain fundamental solution is known. By two examples, a 1-d poroelastic column and a poroelastic 3-d half space, the accuracy of the proposed formulation for wave propagation problems is demonstrated.

## References

[1] D. E. Beskos. Boundary element methods in dynamic analysis. *Applied Mechanics Review*, 40(1):1–23, 1987.

[2] D. E. Beskos. Boundary element methods in dynamic analysis: Part II (1986-1996). *Applied Mechanics Review*, 50(3):149–197, 1997.

[3] M.A. Biot. General theory of three-dimensional consolidation. *Journal of Applied Physics*, 12:155–164, 1941.

[4] M.A. Biot. Theory of propagation of elastic waves in a fluid-saturated porous solid. I. low-frequency range, II. higher frequency range. *Journal of the Acoustical Society of America*, 28(2):168–191, 1956.

[5] G. Bonnet. Basic singular solutions for a poroelastic medium in the dynamic range. *Journal of the Acoustical Society of America*, 82(5):1758–1762, 1987.

[6] G. Bonnet and J.-L. Auriault. Dynamics of saturated and deformable porous media: Homogenization theory and determination of the solid-liquid coupling coefficients. In N. Boccara and M. Daoud, editors, *Physics of Finely Divided Matter*, pages 306–316. Springer Verlag, Berlin, 1985.

[7] J. Chen. Time domain fundamental solution to Biot's complete equations of dynamic poroelasticity. part II: Three-dimensional solu-

tion. *International Journal of Solids and Structures*, 31(2):169–202, 1994.

[8] J. Chen and G.F. Dargush. Boundary element method for dynamic poroelastic and thermoelastic analysis. *International Journal of Solids and Structures*, 32(15):2257–2278, 1995.

[9] A. H.-D. Cheng, T. Badmus, and D.E. Beskos. Integral equations for dynamic poroelasticity in frequency domain with BEM solution. *Journal of Engineering Mechanics ASCE*, 117(5):1136–1157, 1991.

[10] J. Domínguez. Boundary element approach for dynamic poroelastic problems. *International Journal for Numerical Methods in Engineering*, 35(2):307–324, 1992.

[11] K. F. Graff. *Wave motion in elastic solids.* Oxford University Press, 1975.

[12] L. Hörmander. *Linear Partial Differential Operators.* Springer-Verlag, 1963.

[13] Y.K. Kim and H.B. Kingsbury. Dynamic characterization of poroelastic materials. *Experimental Mechanics*, 19:252–258, 1979.

[14] C. Lubich. Convolution quadrature and discretized operational calculus. I. *Numerische Mathematik*, 52:129–145, 1988.

[15] G.D. Manolis and D.E. Beskos. Integral formulation and fundamental solutions of dynamic poroelasticity and thermoelasticity. *Acta Mechanica*, 76:89–104, 1989. Also, Errata in Acta Mechanica, 83, 223-226, 1990.

[16] C.L. Pekeris. The seismic surface pulse. *Proc. of the National American Society*, 41:469–480, 1955.

[17] M. Schanz and H. Antes. A new visco- and elastodynamic time domain boundary element formulation. *Computational Mechanics*, 20(5):452–459, 1997.

[18] M. Schanz and A.H.-D. Cheng. Transient wave propagation in a one-dimensional poroelastic column. *Acta Mechanica*, page (accepted), 2000.

[19] Th. Wiebe and H. Antes. A time domain integral formulation of dynamic poroelasticity. *Acta Mechanica*, 90:125–137, 1991.

# ON ADAPTIVITY IN BOUNDARY ELEMENT METHODS

H. R. Schulz, O. Steinbach, W. L. Wendland
*Mathematisches Institut A, Universität Stuttgart,*
*Pfaffenwaldring 57, 70569 Stuttgart, Germany*

**Abstract** In this contribution we consider different global and local a posteriori error estimates for boundary integral equations arising in direct and indirect boundary element methods for regular boundary value problems of formally positive elliptic partial differential equations. Based on local error indicators, we design an adaptive mesh refinement strategy to improve the boundary element approximation.

**Keywords:** Boundary element, global and local a posteriori error estimators, adaptivity, discrete Sobolev norms

## 1. INTRODUCTION

Currently, there are two principally different philosophies underlying adaptivity: grid and degree adaptation in $h$–, $p$– and $hp$–finite or boundary element methods — or data adaption by appropriate compression as in wavelet methods on uniform hierarchical grids. In this contribution, we are concerned with grid adaption in $h$–methods. As a simple model problem, we consider the Dirichlet boundary value problem

$$Lu = 0 \quad \text{in } \Omega \subset \mathbb{R}^n \ (n = 2 \text{ or } 3) \quad \text{with } u = g \quad \text{on } \Gamma = \partial\Omega \qquad (1)$$

where $L$ is a second order formally positive elliptic partial differential system and $\Omega$ is a bounded domain with Lipschitz boundary $\Gamma$. Let $T$ be the associated conormal derivative and $U^*(x,y)$ a fundamental solution of $L$. Then the solution of (1) is given by the representation formula

$$u(\tilde{x}) = \int_\Gamma U^*(\tilde{x},y)t(y)ds_y - \int_\Gamma (T_y U^*(\tilde{x},y))^\top g(y)ds_y \quad \text{for } \tilde{x} \in \Omega. \qquad (2)$$

To find the yet unknown Neumann datum $t(x) = T_x u(x)$ on $\Gamma$ we may consider the boundary integral equation of the first kind resulting from

*T. Burczynski (ed.), IUTAM/IACM/IABEM Symposium on Advanced Mathematical and Computational Mechanics Aspects of the Boundary Element Method,* 315–325.

(2),

$$(Vt)(x) = (\frac{1}{2}I + K)g(x) =: f(x) \quad \text{for } x \in \Gamma \tag{3}$$

by using the single and double layer potentials, defined as

$$(Vw)(x) = \int\limits_\Gamma U^*(x,y)w(y)ds_y, \quad (Ku)(x) = \int\limits_\Gamma (T_y U^*(x,y))^\top u(y)ds_y\,,$$

respectively. Instead of solving (3) exactly one computes an approximation $t_h$ in some finite–dimensional trial space $Z_h$ satisfying a discrete system

$$V_h t_h = f_h \tag{4}$$

which results either from some Galerkin–Bubnov or Galerkin–Petrov method, from collocation or qualocation, based on finite or boundary elements, on wavelets or hybrid approximations, etc.. For the error $e_h := t_h - t$ we get the error equation

$$Ve_h = Vt_h - f =: r_h \tag{5}$$

with the computable *residue* $r_h$.

We are interested in finding error estimators for the global error $||e_h||$ in suitable norms and to derive local error indicators $\eta_k$ for driving an adaptive mesh refinement. A global error estimator $\eta$ is called

| | | | | | |
|---|---|---|---|---|---|
| *efficient* | $c_1 \cdot \eta$ | $\leq$ | $\|\|e_h\|\|$ | with | $c_1 > 0$; |
| *reliable* | $\|\|e_h\|\|$ | $\leq$ | $c_2 \cdot \eta$ | with | $c_2 > 0$; |
| *asymptotically exact* | $\|\|e_h\|\|/\eta$ | $\to$ | $1$ | for | $h \to 0$. |

Let $\{\eta_k\}_{k=1}^N$ denote local error indicators associated with boundary elements $\Gamma_k$, $k = 1, \ldots, N$. Then an adaptive strategy is to refine all boundary elements $\Gamma_k$ where

$$\eta_k \geq \theta \cdot \max_{\ell=1,\ldots,N} \eta_\ell \tag{6}$$

is satisfied with some parameter $\theta \in [0,1]$. In order to guarantee the convergence of $h$–methods, the refinement strategy must imply $h \to 0$, therefore $\theta \neq 1$. For $\theta = 0$ we have uniform refinement; practical choices are, for example, $\theta \in \{0.01, 0.05, 0.10\}$. For reliability, the refinement procedure must produce restricted grid families such as having the bounded local mesh ratio (BLMR) property or $\lambda$–regularity or stability of $L_2$–projections in other norms.

## 2. RESIDUAL BASED ERROR ESTIMATORS

The fundamental relation for localized a posteriori error estimates is the error equation (5). In general, for smooth boundaries, the operators arising in boundary element methods are *non-local* pseudodifferential operators. Such an operator is said to be of order $2\alpha$ if it is continuous:

$$A: \; H^s(\Gamma) \longrightarrow H^{s-2\alpha}(\Gamma) \quad \text{for all admissible } s \in \mathbb{R}.$$

For boundary integral equations of the first kind, one finds coercivity in the form of Gårding's inequality (strong ellipticity, see [6]),

$$\langle Av, v\rangle \geq \gamma_0 ||v||^2_{H^\alpha(\Gamma)} - \gamma_1 ||v||^2_{H^{\alpha-1}(\Gamma)}. \tag{7}$$

In our case of an elliptic second order operator $L$, the single layer potential operator $V$ is usually a continuous and bijective mapping from $H^s(\Gamma)$ onto $H^{s+1}(\Gamma)$. Due to the continuity and bijectivity of $V$, from the error equation (5) there follow the global estimates

$$||V||^{-1} \cdot ||r_h||_{H^{s+1}(\Gamma)} \leq ||e_h||_{H^s(\Gamma)} \leq ||V^{-1}|| \cdot ||r_h||_{H^{s+1}(\Gamma)} . \tag{8}$$

Hence, we obtain for the efficiency constant $c_1 = ||V||^{-1}$ and for the reliability constant $c_2 = ||V^{-1}||$. For Lipschitz boundaries, inequalities (8) remain true for $s \in [-1, 0]$, see [5].

There are two ways to compute local error indicators based on a posteriori error estimates:

1 One uses a global error estimator as e.g. the residue in (8) and localizes its norm. For boundary element methods see, e.g., [3, 4, 7, 20].

2 One proves *local* a posteriori error estimates, see, e.g., [12, 14, 15] and Section 3.

The basic properties for localization techniques in boundary element methods are the pseudolocality and the representation of the kernels of the boundary integral operators in terms of pseudohomogeneous asymptotic kernel expansions (see [16, 19] for details),

$$Aw(\tau) = \sum_{|\nu|\leq 2\alpha} \widetilde{c}_\nu(\tau) D^\nu_\tau w(\tau) + \text{p.f.} \int \Big\{ \sum_{j=0}^{L} |t|^{-2\alpha-(n-1)+j} \times \tag{9}$$

$$\left( f_{\beta-j}(\tau,\Theta) + \ln|t| Q_H^{(-2\alpha-(n-1)+j)}(\tau,\Theta)\Lambda\left(\frac{|t|}{r}\right)\right) \times$$

$$\left( w(\zeta) - \sum_{0\leq|\nu|\leq 2\alpha-j} \frac{1}{\nu!} t^\nu D^\nu_\tau w(\tau)\right) + K_{2\alpha-L+1}(\tau, t; r) w(\zeta) \Big\}_{|t=\zeta-\tau} d\zeta$$

with respect to a parametrization $\chi$ of the boundary: $y = \chi(\zeta)$, $x = \chi(\tau)$, $\Theta := \frac{t}{|t|}$. Here $Q_H^{(\ell)}(\tau, \theta)$ is a homogeneous polynomial of degree $\ell$ in $\theta$ and is set equal to zero for $\ell < 0$. (9) provides us with the decomposition:

$$A = A^{(0)} + C \tag{10}$$

where $A^{(0)}$ is the pseudolocal principal part and $C$ is a compact linear operator. Because of (7), the operator $A^{(0)}$ can always be modified such that it becomes invertible, and in view of (10) one can define

$$e_h^{(0)} := (A^{(0)})^{-1} r_h \tag{11}$$

as an approximation of the error which contains only the high frequencies. In the finite element method, estimators of this kind were introduced by Zienkiewicz & Zhu, see [23]. However, to obtain an efficient and reliable error estimator, one has also to include the global smooth, low frequency part $\widetilde{e}_h$ which is responsible for pollution. This is the case for all residual based error estimators. For finite element methods, such estimators were first introduced by Babuška & Rheinboldt [1, 2]. For boundary element methods we refer to [3, 7, 9, 20, 22], where also several localization techniques are presented. Note that all these estimates of Babuška–Rheinboldt type require that the BLMR property holds on the family of grids.

## 3. LOCAL ERROR ESTIMATORS

The aim of this section is to present residual–based **local** a posteriori error estimates. First results of this type were shown in [10, 11, 21] providing estimates only on fixed parts of the boundary, which is not appropriate for adaptive methods. Here we present some new results to overcome this disadvantage.

Let $\Gamma_\ell$ be a boundary element and let $\varepsilon_\ell > 0$ be given. We define an associated vicinity of $\Gamma_\ell$ by:

$$\widetilde{\Gamma}_\ell := \{x \in \Gamma : \ \text{dist}(x, \Gamma_\ell) \leq \varepsilon_\ell\} \ . \tag{12}$$

We call $\omega_\ell(x)$ an *admissible truncation function*, if

1 supp $\omega_\ell(x) \subset \widetilde{\Gamma}_\ell$,

2 $\omega_\ell(x) = 1$ on $\Gamma_\ell$, $0 \leq \omega_\ell(x) \leq 1$ for all $x \in \widetilde{\Gamma}_\ell$,

3 $|D^m \omega_\ell(x)| \leq c_D \, \varepsilon_\ell^{-|m|}$ for all $x \in \Gamma$, where $D^m$ denotes tangential derivatives of order $m$.

**Theorem 1** *Let $A$ be a bijective strongly elliptic pseudodifferential operator of order $2\alpha$. Furthermore, let $0 < \delta \leq 1$ be given and $\omega_\ell$ an admissible truncation function. In addition, we assume that the mesh family is $\lambda$–regular, i.e. there exist constants $\lambda \geq 1$ and $c_R > 0$ such that $h_{\min} \geq c_R h^\lambda$. The decay parameter $\varepsilon_\ell$ of $\omega_\ell$ is specified by the relations*

$$\varepsilon_\ell \geq c_0 \, h_\ell^\gamma \quad \textit{with } h_\ell := |\Gamma_\ell| \quad \textit{and} \quad 0 < \gamma \leq \frac{1-\delta}{\lambda(\ell+k+1)}, \tag{13}$$

*where either $\ell > n-1+|s-1|$ and $k > n-1+|s-2\alpha|$; or $\ell > n-1+|s-1-2\alpha|$ and $k > n-1+|s|$ are satisfied. Then, for the localized error one has*

$$\begin{aligned} c_1 ||\omega_\ell r_h||_{H^{s-2\alpha}(\Gamma)} &- c_3 h^{r-s+\delta} ||t||_{H^r(\Gamma)} \\ &\leq ||\omega_\ell e_h||_{H^s(\Gamma)} \leq c_2 ||\omega_\ell r_h||_{H^{s-2\alpha}(\Gamma)} + c_3 h^{r-s+\delta} ||t||_{H^r(\Gamma)} \end{aligned} \tag{14}$$

*provided $2\alpha - (d+1) \leq s \leq \alpha \leq t \leq d$, $\alpha < r + 1/2$ ($n = 2$) or $\alpha \leq r$ ($n = 3$). Here $d$ denotes the polynomial degree of the trial space.*

The proof can be found in [12, 14]. There, also a certain window technique was developed to replace the calculation of the residue norms occuring in (14). It turns out that the computation of $||\omega_\ell r_h||_{H^{-\alpha}(\Gamma)}$ is equivalent to the solution of *local Galerkin equations* with a globally defined commutator on the right hand side: *Find $\widetilde{t}_\ell \in \widetilde{H}^\alpha(\widetilde{\Gamma}_\ell)$ such that*

$$\langle A_0 \widetilde{t}_\ell, \widetilde{v} \rangle = \langle \omega_\ell f - (\omega_\ell A - A\omega_\ell) t_h - (A - A_0)\omega_\ell t_h, \widetilde{v} \rangle \text{ for all } \widetilde{v} \in H^\alpha(\widetilde{\Gamma}_\ell) .$$

These local equations can be solved, e.g., by using a local fine grid with mesh size $\widetilde{h}_\ell \leq c h_\ell^\varrho$ with some $\varrho \geq 1$. Then the *local error functions*

$$\widetilde{\eta}_\ell := \omega_\ell t_h - \widetilde{t}_\ell$$

satisfy the local error equation

$$\langle A_0 \widetilde{\eta}_\ell, \widetilde{v} \rangle = \langle \omega_\ell r_h, \widetilde{v} \rangle \qquad \text{for all } \widetilde{v} \in H^\alpha(\widetilde{\Gamma}_\ell) ,$$

which, in turn, implies the estimate [12, 14]:

$$\Big| ||\omega_\ell e_h||_{H^\alpha(\Gamma)} - ||\widetilde{\eta}_\ell||_{H^\alpha(\Gamma)} \Big| \leq c \left( h^{r-\alpha+\delta} + \widetilde{h}_\ell^{\,r - \frac{\gamma}{\varrho} r - \alpha} \right) ||t||_{H^r(\Gamma)} , \tag{15}$$

which shows that $||\widetilde{\eta}||_{H^\alpha(\Gamma)}$ gives us, indeed, asymptotically the localized error. However, the assumptions in Theorem 1 require that the adaptivity algorithm must produce $\lambda$–regular grid families.

## 4. AN EQUIVALENT ERROR EQUATION

Up to now both, global and local error estimators are based on the residue of the boundary integral equation (3). Alternatively, we may exploit an error equation to approximate the error itself. Details for this approach may be found in [13, 17].

The numerical approximation $t_h$ in the representation formula (2) provides us with an approximate solution of the boundary value problem (1),

$$u_h(\tilde{x}) := \int_\Gamma U^*(\tilde{x}, y) t_h(y) ds_y - \int_\Gamma (T_y U^*(\tilde{x}, y))^\top g(y) ds_y \quad \text{for } \tilde{x} \in \Omega \tag{16}$$

which satisfies the differential equation (1) exactly but admits only the approximate Cauchy data

$$\widetilde{g}(x) := u_h(x) \quad \text{and} \quad \widetilde{t}(x) := (T_x u_h)(x) \qquad \text{for } x \in \Gamma. \tag{17}$$

Then the error $e_h = t_h - t$ is a solution of the boundary integral equation of the second kind (see [13]),

$$(\frac{1}{2} I - K') e_h(x) = (\widetilde{t} - t_h)(x) \quad \text{for } x \in \Gamma, \tag{18}$$

where $K'$ is the adjoint double layer potential operator defined by

$$(K'w)(x) = \int_\Gamma T_x U^*(x, y) w(y) ds_y \quad \text{for } x \in \Gamma.$$

If we assume that the spectral radius $\varrho_K$ of $(\frac{1}{2}I + K')$ satisfies $\varrho_K < 1$ (see [18]), then the Neumann series for inverting (18) converges and the error $e_h$ can be computed by

$$e_h(x) = \sum_{\ell=0}^{\infty} (\frac{1}{2} I + K')^\ell (\widetilde{t} - t_h)(x) \quad \text{for } x \in \Gamma. \tag{19}$$

Hence, we may define an approximate error function by the truncation:

$$e_h^{(q)}(x) = \sum_{\ell=0}^{q} (\frac{1}{2} I + K')^\ell (\widetilde{t} - t_h)(x) \quad \text{for } x \in \Gamma. \tag{20}$$

If, in addition, one can show the estimate

$$||(\frac{1}{2} I + K') w||_V \leq c_K \cdot ||w||_V \quad \text{with } c_K < 1, \tag{21}$$

(see [18]), then (19) and (21) provide the equivalence inequalities

$$\frac{1}{1+c_K^{q+1}} \cdot ||e_h^{(q)}||_V \le ||e_h||_V \le \frac{1}{1-c_K^{q+1}} \cdot ||e_h^{(q)}||_V. \tag{22}$$

Therefore, an efficient and reliable global error estimator is given by

$$\eta^{(q)} = \sqrt{\langle V e_h^{(q)}, e_h^{(q)} \rangle_{L_2(\Gamma)}}. \tag{23}$$

Note that $\eta^{(q)}$ is *asymptotically exact* for $q \to \infty$.

To define *local error indicators* we may compute, correspondingly,

$$\eta_k^{(q)} = \sqrt{|\langle V e_h^{(q)}, e_h^{(q)} \rangle_{L_2(\Gamma_k)}|}. \tag{24}$$

From (22) we get

$$||e_h||_V \le \frac{1}{1-c_K^{q+1}} \cdot \sqrt{\sum_{k=1}^{N} [\eta_k^{(q)}]^2}. \tag{25}$$

In practical computations, in general, it is not possible to evaluate $e_h^{(q)}$ in (20) exactly. Hence, we need an additional approximation. To this end, let $Z_{\tilde{h}}$ be a trial space defined with respect to a refined triangulation $\Gamma_{\tilde{h}}$ with $\tilde{h} \le c_0 h$ and $c_0 > 0$ sufficiently small. Let $Z_h \subset Z_{\tilde{h}}$. The corresponding $L_2$–projection $Q_{\tilde{h}} : L_2(\Gamma) \to Z_{\tilde{h}}$ is defined by

$$\langle Q_{\tilde{h}} w, \tau_{\tilde{h}} \rangle_{L_2(\Gamma)} = \langle w, \tau_{\tilde{h}} \rangle_{L_2(\Gamma)} \quad \text{for all } \tau_{\tilde{h}} \in Z_{\tilde{h}}. \tag{26}$$

Instead of (20) we now compute an approximate error function as

$$\widetilde{e}_h^{(q)}(x) = \sum_{\ell=0}^{q} [Q_{\tilde{h}}(\frac{1}{2}I + K')]^\ell Q_{\tilde{h}}(\widetilde{t} - t_h)(x) \quad \text{for } x \in \Gamma \tag{27}$$

and replace $e_h^{(q)}$ by $\widetilde{e}_h^{(q)}$ to define both, the global error estimator (23) and the local error indicators (24). Note that the latter provide only the reliability inequality (25). In particular, we are not able to show efficiency yet. Instead of using the energy norm in (23) one may use any other equivalent Sobolev norm in $H^{-1/2}(\Gamma)$. Let $Z_J = Z_{\tilde{h}}$ be the trial space of the refinement level $J$ arising from all previous adaptive and some additional uniform refinements defining $Z_{\tilde{h}} \supset Z_h$, providing us with a hierarchy of trial spaces

$$Z_1 \subset Z_2 \subset \cdots \subset Z_J \subset H^{-1/2}(\Gamma)$$

where each space $Z_j$ is associated with a mesh size $h_j$. Let $Q_j$ denote the $L_2$–projection onto $Z_j$ for every $j = 1, \ldots, J$, and set $Q_0 = 0$.

For $s \in \mathbb{R}$ and $w \in Z_J$ we define the multilevel operator

$$A^{(s)}w = \sum_{j=1}^{J} h_j^{-2s} \cdot (Q_j - Q_{j-1})w \tag{28}$$

which satisfies the relation

$$A^{(s)} = A^{(s/2)} \cdot A^{(s/2)} . \tag{29}$$

Moreover, the energy norm induced by the operator $A^{(-1/2)}$ is an equivalent Sobolev norm in $H^{-1/2}(\Gamma)$, see [8]. Consequently, with (29) we see that

$$||A^{(-1/4)}w||^2_{L_2(\Gamma)} = \sum_{k=1}^{N_J} ||A^{(-1/4)}w||^2_{L_2(\Gamma_{J,k})} \tag{30}$$

is an equivalent Sobolev norm for $w \in Z_J \subset H^{-1/2}(\Gamma)$ and, by definition,

$$\eta = \sqrt{\sum_{k=1}^{N} \eta_k^2}, \qquad \eta_k = \sqrt{\sum_{\Gamma_{J,\ell} \subset \Gamma_k} ||A^{(-1/4)}w||^2_{L_2(\Gamma_{J,\ell})}} \tag{31}$$

is an efficient and reliable error estimator yielding local error indicators.

Here the refinement procedure must produce grids providing the $H^{-\frac{1}{2}}$– or $H^s$–stability for $Q_j$ as given in [17].

**Remarks.** As a byproduct of (27), $t^*(x) := t_h(x) - \widetilde{e}_h^{(q)}(x)$ defines a significantly improved approximate solution. Also note that the error function $\widetilde{e}_h^{(q)}$ can be used to compute almost any other norm of interest, for example local $L_2$ and $L_\infty$ norms.

## 5. NUMERICAL EXAMPLES

In addition to the two–dimensional examples in [12, 14, 15] for the local estimators of Section 3, we consider now for the approach of Section 4 the Laplace equation in 3D with Dirichlet boundary conditions, where $\Omega$ is the so–called Fichera cube defined by $\Omega = [0,2]^3 \backslash [1,2]^3$. The exact solution is chosen as

$$u(x) = \frac{1}{|x - x^*|} \qquad \text{with } x^* = (0.9, 0.9, 1.1)^\top .$$

The boundary element discretization is performed by the collocation method, and the discrete linear system is solved by a BiCGStab iteration with diagonal preconditioning. The initial triangulation with 48 boundary elements is shown in Figure 1. Then the adaptive refinement strategy leads to a final triangulation with 1748 boundary elements shown in Figure 2. The number of iterations for the discrete equation's relative error reduction of $\varepsilon = 10^{-8}$, the exact and estimated $L^2$–error are given in Table 1 for all mesh levels; ratio means the quotient $h_{\max}/h_{\min}$.

| N | M | Ratio | Iter | $\|\|t - t_h\|\|_{L^2(\Gamma)}$ | $e_h(t_h)$ |
|---|---|---|---|---|---|
| 48 | 26 | 1 | 12 | 0.9705 | 1.3831 |
| 100 | 52 | 4 | 16 | 0.8438 | 0.9479 |
| 146 | 75 | 16 | 19 | 0.6299 | 0.7446 |
| 182 | 93 | 64 | 25 | 0.4004 | 0.5914 |
| 266 | 135 | 256 | 36 | 0.2229 | 0.3533 |
| 414 | 209 | 1024 | 45 | 0.1619 | 0.2103 |
| 896 | 450 | 4096 | 61 | 0.1217 | 0.1327 |
| 1748 | 876 | 8192 | 69 | 0.0878 | 0.0954 |

*Table 1* Results for the Dirichlet problem

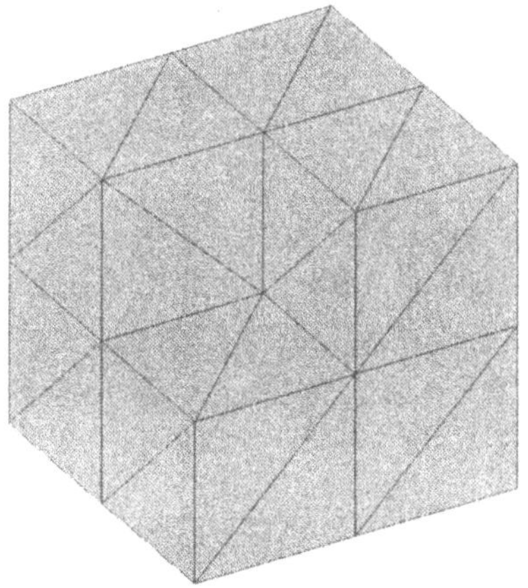

*Figure 1*: Initial grid

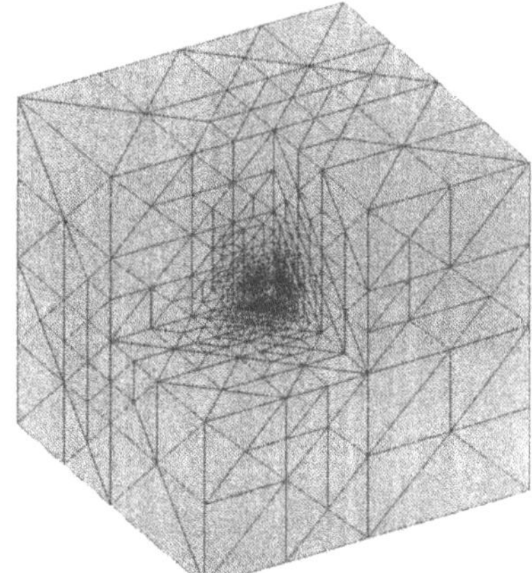

*Figure 2*: Adaptive refined grid

## References

[1] I. Babuška, W. C. Rheinboldt, Error estimates for adaptive finite element computations. SIAM J. Numer. Anal. 15 (1978) 736–754.

[2] I. Babuška, W. C. Rheinboldt, A posteriori error estimates for the finite element method. Int. J. Numer. Meth. Engrg. 12 (1978) 1597–1615.

[3] C. Carstensen, An a posteriori error estimator for a first kind integral equation. Math. Comp. 66 (1997) 139–156.

[4] C. Carstensen, E. P. Stephan, Adaptive boundary element methods for some first kind integral equations. SIAM J. Numer. Anal. 33 (1996) 2166–2183.

[5] M. Costabel, Boundary integral operators on Lipschitz domains — elementary results. SIAM J. Math. Anal. 19 (1987) 613–626.

[6] M. Costabel, W.L. Wendland, Strong ellipticity of boundary integral operators. Crelle's J. Reine u. Angew. Math. 372 (1986) 36–63.

[7] B. Faermann, Lokale a–posteriori–Fehlerschätzer bei der Diskretisierung von Randintegralgleichungen. Dissertation, Christian–Albrechts–Universität Kiel, 1993.

[8] P. Oswald, Multilevel norms for $H^{-1/2}$. Computing 61 (1998) 235–255.

[9] E. Rank, A–posteriori–Fehlerabschätzungen und adaptive Netzverfeinerung für Finite–Element– und Randintegralelement–Methoden. Dissertation, Mitteilungen aus dem Bauingenieurwesen I, Heft 16, TU München, 1985.

[10] J. Saranen, Local error estimates for some Petrov–Galerkin methods applied to strongly elliptic equations on curves. Math. Comp. 48 (1987) 485–502.

[11] J. Saranen, W. L. Wendland, Local residual–type error estimates for adaptive boundary element methods on closed curves. Appl. Anal. 48 (1993) 37–50.

[12] H. Schulz, Über lokale und globale Fehlerabschätzungen für adaptive Randelementmethoden. Dissertation, Universität Stuttgart, 1997.

[13] H. Schulz, O. Steinbach, A new a posteriori error estimator in adaptive direct boundary element methods. The Dirichlet problem. Submitted, 1999.

[14] H. Schulz, W. L. Wendland, Local, residual–based a posteriori error estimates forcing adaptive boundary element methods. In: Bound-

ary Element Topics (W. L. Wendland ed.), Springer–Verlag Berlin (1997), pp. 445–470.

[15] H. Schulz, W. L. Wendland, Local a posteriori error estimates for boundary element methods. In: Proceedings of the ENUMATH 97, Heidelberg 28.9.97–3.10.97 (H. G. Bock, G. Kanschat, R. Rannacher, F. Brezzi, R. Glowinski, Y. A. Kuznetsov, J. Périaux eds.), World Scientific Singapure - New Jersey - London - Hong Kong, (1998), pp. 564–571.

[16] C. Schwab, W. L. Wendland, Kernel properties and representations of boundary integral operators. Math. Nachr. 156 (1992) 187–218.

[17] O. Steinbach, Computational schemes for Sobolev norms in adaptive boundary element methods. Bericht 99/04, SFB 404, Universität Stuttgart, 1999.

[18] O. Steinbach, W. L. Wendland, On the spectral radius of double layer potentials. In preparation.

[19] W. L. Wendland, On boundary integral equations and applications, In: Tricomi's Ideas and Contemporary Applied Mathematics. Accad. Naz. dei Lincei, Rome Atti dei convegni lincei 147 (1998) 49–71.

[20] W. L. Wendland, De–Hao Yu, Adaptive boundary element methods for strongly elliptic integral equations. Numer. Math. 53 (1988) 539–558.

[21] W. L. Wendland, De–Hao Yu, Local error estimates of boundary element methods for pseudo–differential operators of non–negative order on closed curves. J. Comput. Math. 10 (1990) 273–289.

[22] De–Hao Yu, A posteriori error estimates and adaptive approaches for some boundary element methods. In: Boundary elements IX, Vol. 1 (Stuttgart, 1987), 241–256, Comput. Mech., Southampton, 1987.

[23] O. C. Zienkiewicz, J. Z. Zhu, The superconvergent patch recovery and a posteriori error estimates, Part 1 and Part 2. Int. J. Numer. Methods Engrg. 33 (1992) 1331–1382.

# ELECTROPHORETIC MOTION OF A CHARGED PARTICLE

A. Sellier
*Ladhyx*
*Ecole Polytechnique, 91128 Palaiseau Cedex*
*FRANCE*
sellier@buck.polytechnique.fr

**Keywords:** Electrophoresis, integral equations, ellipsoid

**Abstract** The electrophoretic motion of a charged particle is determined under the "thin" double-layer assumption by solving seven well-posed boundary integral equations. Special attention is paid to the case of ellipsoidal particles.

## 1. INTRODUCTION

Electrophoresis is the transport of charged particles by an external electric field. In several applications, such as particle separation in Chemical Industry, this phenomenon plays a key role.

Neglecting proximal boundaries, we restrict ourselves to a single particle $\mathcal{P}$ (see Figure 1) embedded in the electric field $\mathbf{E}_\infty$. The study also rests on the following physical and mechanical assumptions:

(1) The particle is solid. One thereafter describes its rigid body motion by translational and angular velocities $\mathbf{U}$ and $\boldsymbol{\omega}$. For convenience, $\mathbf{U}$ is the velocity of one point $O$ attached to the particle.

(2) The particle is freely suspended in a viscous and unbounded electrolyte with constant viscosity and permittivity respectively denoted by $\mu$ and $\epsilon$. The charge on the particle's surface $S$ is balanced by a diffuse cloud $\mathcal{C}$ of counter-ions surrounding $\mathcal{P}$. The double-layer $S \cup \mathcal{C}$, whose outer boundary is $S^+$, has a thickness of order of the Debye screening length $\kappa^{-1}$ [Anderson].

(3) The charge distribution within the double-layer is the same as that for $\mathbf{E}_\infty = \mathbf{0}$ (no polarization effects). For $\mathbf{E}_\infty \neq \mathbf{0}$, the electrically driven flow of the counter-ions in $\mathcal{C}$ induces particle and fluid motions.

*T. Burczynski (ed.), IUTAM/IACM/IABEM Symposium on Advanced Mathematical and Computational Mechanics Aspects of the Boundary Element Method*, 327–337.

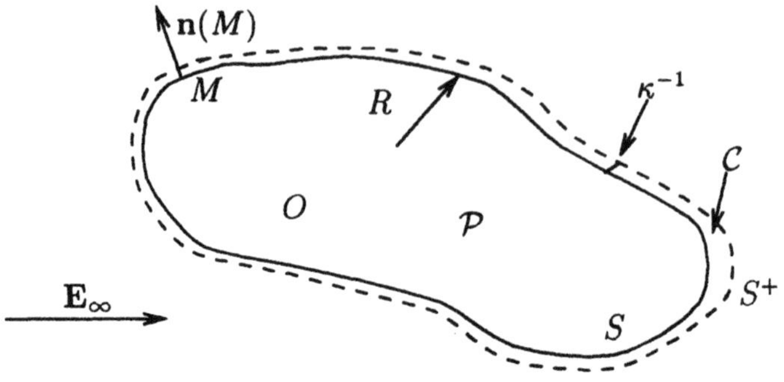

*Figure 1* Our charged particle embedded in the external field $\mathbf{E}_\infty$.

(4) The typical radius of curvature $R$ of the smooth surface $S$ obeys $\kappa R \gg 1$. Thus, $S^+ \sim S$. For this model of "thin" double-layer the fluid velocity $\mathbf{u}$ and the electric field $\mathbf{E} = \mathbf{E}_\infty - \nabla\phi$ (where $\phi$ denotes the harmonic perturbation potential) fulfill the boundary condition [Hunter]

$$\mathbf{u}(M) = \mathbf{u}_d(M) := \mathbf{U} + \boldsymbol{\omega} \wedge \mathbf{OM} - \frac{\epsilon\zeta(M)}{\mu}[\mathbf{E} - (\mathbf{E}.\mathbf{n})\mathbf{n}](M); \qquad \text{on } S, \tag{1.1}$$

where $\mathbf{n}$ is the outer unit normal on $S$ and $\zeta$ denotes the so-called zeta potential which can be related to the charge density on $S$ [Hiemenz & Rajagopalan].

(5) The particle surface is nonconducting, i. e. one imposes $\mathbf{E}.\mathbf{n} = 0$ on $S^+ \sim S$. This condition will be taken into account in enforcing (1.1).

Far from the particle, $\phi$ and $\mathbf{u}$ vanish. If $p$ denotes the fluid pressure, $(\mathbf{u}, p)$ obeys the *quasi-static* form of the creeping motion equations, and $\mathbf{U}$ and $\boldsymbol{\omega}$ are determined by requiring zero net hydrodynamic force and torque on $S$ [Anderson]. Under the assumptions (6)-(7) such that

(6) the external field $\mathbf{E}_\infty$ is uniform,

(7) the zeta potential $\zeta$ is uniform on $S$,

one obtains the Smoluchowski's result

$$\mathbf{U} = \frac{\epsilon\zeta}{\mu}\mathbf{E}_\infty, \qquad \boldsymbol{\omega} = \mathbf{0}. \tag{1.2}$$

The nice solution (1.2) has been obtained first for a sphere by Smoluchowski [Smoluchowski] and later extented to the case of arbitrary shapes (see [Morrison], [Teubner]).

For applications one is however eager to relax assumption(s) (6) or/and (7). In these circumstances (for arbitrary data $\zeta, \mathbf{E}_\infty$ and particle shape), only a numerical treatment is possible. As shown in this paper, such a task may be achieved by solving seven Fredholm boundary integral equations on $S$, six of the first kind and one of the second kind.

## 2. THE INTEGRAL FORMULATION

This section establishes a system of six linear algebraic equations that governs the unknown quantities $\mathbf{U}$ and $\boldsymbol{\omega}$. In addition, we exhibit seven well-[posed boundary integral equations for solving this key system.

### 2.1. THE GOVERNING SYSTEM

Henceforth, Cartesian coordinates $(O, x_1, x_2, x_3)$ and tensor summation convention are adopted. The open set $\Omega$ denotes the domain outside $S$. Under the assumptions listed in the previous section, the fluid flow $(\mathbf{u}, p)$ obeys

$$\mu\nabla^2\mathbf{u} = \nabla p; \quad \nabla.\mathbf{u} = 0 \quad \text{in } \Omega, \tag{1.3}$$

$$(\mathbf{u}, p) \to (\mathbf{0}, 0) \quad \text{as } r \to \infty \text{ and } \mathbf{u} = \mathbf{u}_d \text{ on } S, \tag{1.4}$$

where $r := OM$ and $\mathbf{u}_d$ is given by (1.1) with $\mathbf{E}.\mathbf{n} = 0$. The associated stress tensor $\boldsymbol{\sigma} = \boldsymbol{\sigma}(\mathbf{u}, p)$ reads $\boldsymbol{\sigma}(\mathbf{u}, p) = \sigma_{ij}\mathbf{e}_i \otimes \mathbf{e}_j$ with

$$\sigma_{ij} = \sigma_{ij}(\mathbf{u}, p) = -p\delta_{ij} + \mu(\partial u_i/\partial x_j + \partial u_j/\partial x_i) \tag{1.5}$$

where $\mathbf{u} = u_i\mathbf{e}_i$ and $\delta_{ij}$ designates the Kronecker delta. Hence, the requirement of zero net hydrodynamic force and torque on $S$ becomes

$$\int_S \mathbf{e}_i.\boldsymbol{\sigma}.\mathbf{n}dS = 0; \quad \int_S [\mathbf{e}_i \wedge \mathbf{OM}].\boldsymbol{\sigma}.\mathbf{n}dS = 0, \quad i \in \{1, 2, 3\}. \tag{1.6}$$

By virtue of the Lorentz reciprocal theorem [Kim & Karrila] any new flow $(\mathbf{u}', p', \boldsymbol{\sigma}')$ obeying (1.3)-(1.4) for prescribed velocity $\mathbf{u}'_d$ satisfies the basic relation

$$\int_S \mathbf{u}'_d.\boldsymbol{\sigma}.\mathbf{n}dS = \int_S \mathbf{u}_d.\boldsymbol{\sigma}'.\mathbf{n}dS. \tag{1.7}$$

Thus, if translational motions $(\mathbf{u}_T^{(i)}, p_T^{(i)}, \sigma_T^{(i)})$ and rotational motions $(\mathbf{u}_R^{(i)}, p_R^{(i)}, \sigma_R^{(i)})$ obey (1.3)-(1.4) with the boundary conditions

$$\mathbf{u}_T^{(i)} = \mathbf{e}_i; \quad \mathbf{u}_R^{(i)} = \mathbf{e}_i \wedge \mathbf{OM} \quad \text{on } S, \tag{1.8}$$

it is thereafter possible to cast the conditions (1.6) into the following system of six linear algebraic equations

$$\mathbf{K}.\mathbf{U} + \mathbf{C}.\boldsymbol{\omega} = \mathbf{T}; \ \mathbf{T}.\mathbf{e}_i = -\frac{\epsilon}{\mu^2}\int_S \zeta[\mathbf{E}_\infty - \nabla\phi].\boldsymbol{\sigma}_T^{(i)}.\mathbf{n}dS, \tag{1.9}$$

$$\mathbf{D}.\mathbf{U} + \boldsymbol{\Omega}.\boldsymbol{\omega} = \mathbf{M}; \ \mathbf{M}.\mathbf{e}_i = -\frac{\epsilon}{\mu^2}\int_S \zeta[\mathbf{E}_\infty - \nabla\phi].\boldsymbol{\sigma}_R^{(i)}.\mathbf{n}dS, \tag{1.10}$$

where the translation tensor $\mathbf{K}$, rotation tensor $\mathbf{\Omega}$ and coupling tensors $\mathbf{C}$ and $\mathbf{D}$ admit the following cartesian components

$$K_{ij} = -\frac{1}{\mu}\int_S \mathbf{e}_j.\boldsymbol{\sigma}_T^{(i)}.\mathbf{n}dS; \quad \Omega_{ij} = -\frac{1}{\mu}\int_S [\mathbf{e}_j \wedge \mathbf{OM}].\boldsymbol{\sigma}_R^{(i)}.\mathbf{n}dS, \quad (1.11)$$

$$C_{ij} = -\frac{1}{\mu}\int_S \mathbf{e}_i.\boldsymbol{\sigma}_R^{(j)}.\mathbf{n}dS; D_{ij} = -\frac{1}{\mu}\int_S [\mathbf{e}_i \wedge \mathbf{OM}].\boldsymbol{\sigma}_T^{(j)}.\mathbf{n}dS. \quad (1.12)$$

Tensors $\mathbf{K}$ and $\mathbf{\Omega}$ are symmetric whilst $\mathbf{C}$ and $\mathbf{D}$ are transposed and the system (1.9)-(1.10) admits a unique solution [Happel] whose determination requires to calculate only the surface forces $\mathbf{f}_A^{(i)} := \boldsymbol{\sigma}_A^{(i)}.\mathbf{n}$ (for $A \in \{T, R\}$) and the field $\nabla\phi$ on $S$.

## 2.2. THE INTEGRAL EQUATIONS

By invoking the material available in [Pozrikidis] the required surface forces $\mathbf{f}_A^{(i)} = \boldsymbol{\sigma}_A^{(i)}.\mathbf{n}$ are found to satisfy the Fredholm boundary integral equations of the first kind

$$[\mathbf{u}_A^{(i)}.\mathbf{e}_k](M) = -\int_S \{\frac{\delta_{jk}}{PM} + \frac{(\mathbf{PM}.\mathbf{e}_j)(\mathbf{PM}.\mathbf{e}_k)}{PM^3}\}[\frac{\mathbf{f}_A^{(i)}.\mathbf{e}_j}{8\pi\mu}](P)dS. \quad (1.13)$$

If $H^s(S)$ denotes the usual Sobolev space, observe that $\mathbf{u}_A^{(i)} \in (H^{1/2}(S))^3$ and $\int_S \mathbf{u}_A^{(i)}.\mathbf{n}dS = 0$. Accordingly [Ladyzhenskaya] each equation (1.13) is well-posed and admits in $(H^{-1/2}(S))^3$ a solution $\mathbf{f}_A^{(i)}$ unique up to any constant multiple of the normal $\mathbf{n}$. The knowledge of those solutions $\mathbf{f}_A^{(i)}$ authorizes us to calculate our previous tensors $\mathbf{K}, \mathbf{\Omega}, \mathbf{C}$ and $\mathbf{D}$. Since $\nabla\phi.\mathbf{n} = \mathbf{E}_\infty.\mathbf{n}$ is known on $S$, one only needs to evaluate the tangential derivatives of $\phi$ on $S$ in computing the right-hand sides of (1.11)-(1.12). By applying the third Green's identity [Kellogg] to the exterior Neumann boundary value problem

$$\Delta\phi = 0 \text{ in } \Omega; \quad \nabla\phi.\mathbf{n} = \mathbf{E}_\infty.\mathbf{n} \text{ on } S; \quad \nabla\phi \to \mathbf{0} \text{ as } r \to 0 \quad (1.14)$$

one arrives at the following Fredholm boundary integral equation of the second kind

$$-4\pi\phi(M) + \int_S [\phi(P) - \phi(M)]\frac{\mathbf{PM}.\mathbf{n}(P)}{PM^3}dS = \int_S \frac{[\mathbf{E}_\infty.\mathbf{n}](P)}{PM}dS. \quad (1.15)$$

For $\mathbf{E}_\infty.\mathbf{n} \in H^{-1/2}(S)$ equation (1.15) admits a unique solution in $H^{-1/2}(S)$. From this solution we thereafter numerically estimate the required tangential derivatives of $\phi$ on $S$. Note that one might also resort to a *single-layer* distribution on $S$ (i. e. use $\phi = \int_S q(P)dS/PM$ in

$\Omega \cup S$) whose density $q \in H^{-1/2}(S)$ is obtained by solving a well-known Fredholm integral equation of the second kind. In this approach, if $\mathbf{t}_1$ and $\mathbf{t}_2 := \mathbf{n} \wedge \mathbf{t}_1$ designate vectors tangential to $S$, one gets

$$[\nabla\phi.\mathbf{t}_l](M) = vp \int_S q(P)\mathbf{MP}.\mathbf{t}_l(M)dS/MP^3; \quad l \in \{1,2\} \qquad (1.16)$$

where $vp$ indicates the principal value of Cauchy [Kupradze]. Unfortunately, this occurrence of $vp$ makes the accurate computation of (1.16) so costly that we rather suggest to compute $\nabla\phi$ on $S$ by employing the numerical solution of (1.15).

# 3. APPLICATION TO THE ELLIPSOÏDAL PARTICLE

This Section presents numerical results for the ellipsoidal particle.

## 3.1. DECOUPLED EQUATIONS FOR U AND $\omega$

For our orthotropic body [Happel] tensors **C** and **D** vanish and the diagonal tensors **K** and **Ω** are respectively given by [Oberbeck] and [ Edwardes] (also corrected by [Perrin]). Moreover, if $S$ is described by the equation $x_1^2/a_1^2 + x_2^2/a_2^2 + x_3^2/a_1^3 = 1$, the required surface forces $\mathbf{f}_A^{(i)}$ read (see (1.8))

$$\mathbf{f}_A^{(i)}(M) = c_A^{(i)}[s\mathbf{u}_A^{(i)}](M); \quad s(M) := \{x_1^2/a_1^4 + x_2^2/a_2^4 + x_3^2/a_3^4\}^{-1/2} \quad (1.17)$$

with constants $c_A^{(i)}$ deduced from [Jeffery]. Hence, the system (1.9)-(1.10) becomes (no summation over $i$ in (1.19))

$$\frac{\epsilon}{\mu}U_i = \frac{1}{4\pi a_1 a_2 a_3}\int_S \zeta(\mathbf{E}_\infty - \nabla\phi).\mathbf{e}_i s(M)dS; \quad U_i := \mathbf{U}.\mathbf{e}_i, (1.18)$$

$$\frac{\epsilon}{\mu}\omega_i = \frac{3\int_S \zeta(\mathbf{E}_\infty - \nabla\phi).[\mathbf{e}_i \wedge \mathbf{OM}]s(M)dS}{4\pi a_1 a_2 a_3(a_1^2 + a_2^2 + a_3^2 - a_i^2)}; \quad \omega_i := \boldsymbol{\omega}.\mathbf{e}_i. (1.19)$$

## 3.2. NUMERICAL PROCEDURE

Since the numerical implementation of boundary integral techniques is treated in detail in many textbooks (among others see [Brebbia], [ Beskos], [Bonnet]) we briefly report the employed steps in numerically solving equation (1.15) and computing $(\mathbf{U}, \boldsymbol{\omega})$ through (1.18)-(1.19). The surface $S$ is discretized into $N_e$ quadratic triangular elements $\Delta_e(1 \leq e \leq N_e)$ mapped to the standard triangle $\Delta$ (see Figure 2A) that obeys,

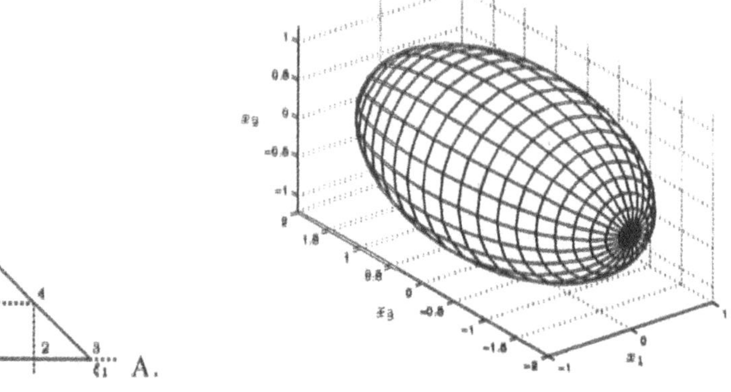


*Figure 2* A) Standard triangle $\Delta$. B) A 530 nodes mesh obtained for $(E, N_\varphi) = (1, 12), a_1 = 1, a_2 = 1.2$ and $a_3 = 2$.

in intrinsic coordinates $\xi = (\xi_1, \xi_2)$, the inequalities: $0 \leq \xi_1 \leq 1; 0 \leq \xi_1 \leq 1; \xi_1 + \xi_2 \leq 1$. More precisely, each boundary element $\Delta_e$ is described by six shape functions $M_q$ and six nodal points $\mathbf{y}(e, q)$ lying on $S$ (three corners and three midsides) such that, if $\xi_3 := 1 - \xi_1 - \xi_2$,

$$M_1(\xi) = (2\xi_3 - 1)\xi_3;\ M_2(\xi) = 4\xi_1\xi_3;\ M_3(\xi) = (2\xi_1 - 1)\xi_1, \quad (1.20)$$

$$M_4(\xi) = 4\xi_1\xi_2;\ M_5(\xi) = (2\xi_2 - 1)\xi_2;\ M_6(\xi) = 4\xi_2\xi_3. \quad (1.21)$$

Those shape functions are also employed to approximate $\phi$ over each element (isoparametric elements), i. e. for $P(\xi) \in \Delta_e$ the vector $\mathbf{y} = \mathbf{OP}$ and the potential $\phi(P)$ admit the following approximations

$$\mathbf{y}(\xi) = \sum_{q=1}^{6} M_q(\xi)\mathbf{y}(e, q); \quad \phi(P) = \phi(\xi) = \sum_{q=1}^{6} M_q(\xi)\phi(e, q). \quad (1.22)$$

Each point $M(x_1, x_2, x_3)$ of $S$ is located by ellipsoidal angles $\theta \in [0, 2\pi]$ and $\varphi \in [0, \pi]$ such that

$$x_1 = a_1 \sin\varphi\cos\theta; \quad x_2 = a_2 \sin\varphi\sin\theta; \quad x_3 = a_3\cos\varphi \quad (1.23)$$

and we choose as nodal points, for given integers $E \geq 0$ and $N_\varphi \geq 3$, the points $A(x_3 = a_3), B(x_3 = -a_3)$ and $M(\theta, \varphi)$ of $S$ given by

$$\theta = 2\pi(n_\theta - 1)/N_\theta; \quad \varphi = \pi n_\varphi/[2N_\varphi]; \quad N_\theta = 12 \times 2^E \quad (1.24)$$

with the following values of integers $n_\theta$ and $n_\varphi$ :

$$1 \leq n_\theta \leq N_\theta \quad \text{if} \quad 2 \leq n_\varphi \leq 2(N_\varphi - 1), \quad (1.25)$$

$$n_\theta = 2k \quad \text{with} \quad k \in \{0, ..., N_\theta/2 - 1\} \quad \text{if} \quad n_\varphi \in \{1, 2N_\varphi - 1\}. \quad (1.26)$$

Thus, the knowledge of $(E, N_\varphi)$ determines a mesh of $N$ nodal points and $N_e$ boundary elements with:

$$N = 2[1 + N_\theta(N_\varphi - 1)]; \quad N_e = N_\theta(N_\varphi - 1); \quad N_\theta = 12 \times 2^E. \quad (1.27)$$

For example, the Figure 2B shows a 530 nodes mesh for $(E, N_\varphi) = (1, 12)$ (note that one has to discard half of the drawn nodal points lying near ends $A(x_3 = a_3)$ and $B(x_3 = -a_3)$). If $J(\xi)$ denotes the Jacobian of the mapping of Cartesian coordinates $\mathbf{y}$ to triangular coordinates $\xi$ and the function $F(P)$ admits on each element $\Delta_e$ an approximation similar to (1.22), one obtains

$$\int_S F(P)K(P, M)dS = \sum_{e=1}^{N_e}\sum_{q=1}^{6} I_e^q, \quad (1.28)$$

$$I_e^q = \int_{\Delta_e} F(e, q)K[\xi, M]J(\xi)d\xi \quad (1.29)$$

where $K$ designates a kernel function. Equation (1.15) is discretized by applying (1.28) for each nodal point $M \in S$. The arising integrals $I_e^q(M)$ are numerically evaluated as explained in [Rezayat *et al.*] (by using the "elastic severity' if $M$ is not on the element $\Delta_e$ and, for $M \in \Delta_e$, analytical removal of inverse powers of $PM$ through the introduction of polar coordinates (with pole $M$) in the space of intrinsic coordinates $\xi$). Moreover, Gaussian integration formulas (both for the standard square and triangle (see [Lyness and Jespersen]) are employed in calculating each encountered (regular) integral. Finally, (1.15) becomes the matrix system $AX = Y$ of fully populated and unsymmetric $N \times N$ influence matrix $A$ and $N \times 1$ vectors $Y$ and unknown $X$ with ${}^TX = (\phi(1), ..., \phi(N))$. This equation $AX = Y$ is solved by applying a $LU$ factorization algorithm (subroutines DGETRF and DGETRS of the LAPACK Library). From $X$ one thereafter computes the tangential derivatives of $\phi$ and the vector $\nabla\phi$ on $S$ (use $\nabla\phi = \mathbf{E}_\infty.\mathbf{n}$). The integrals arising in (1.18)-(1.19) are evaluated with seven Gauss points on each boundary element $\Delta_e$. The accuracy of the numerical procedure is illustrated in Table 1. This Table reports, versus the couple $(E, N_\varphi)$, the average value $< f >$ (use nodal points $M$ such that $x_i \neq 0$) of functions $f = \phi/x_i$ and $f = |\phi/x_i - < \phi/x_i > |$ for a spheroid embedded in the uniform electric field $\mathbf{E}_\infty = -\mathbf{e}_i$. In such circumstances, the exact solution reads [Lamb]

$$[\frac{\phi}{x_i}](M) = \frac{\alpha_i}{2 - \alpha_i}; \quad \frac{\alpha_i}{a_1 a_2 a_3} = \int_0^\infty \frac{(a_i^2 + t)^{-1/2}dt}{\{(a_1^2 + t)(a_2^2 + t)(a_3^2 + t)\}^{1/2}} \quad (1.30)$$

*Table 1* Numerical comparisons for a spheroid ($a_1 = a_2 = 1, a_3 = 2$).

| E | $N_\varphi$ | $N$ | $\mathbf{e}_i = -\mathbf{E}_\infty$ | $< \phi/x_i >$ | $< \lvert\phi/x_i - < \phi/x_i >\rvert >$ | $\frac{\alpha_i}{2-\alpha_i}$ |
|---|---|---|---|---|---|---|
| 0 | 8 | 170 | $\mathbf{e}_2$ | 0.700499 | 2.4291E-3 | 0.704210 |
| 1 | 12 | 530 | $\mathbf{e}_2$ | 0.703838 | 6.1146E-4 | 0.704210 |
| 2 | 24 | 2210 | $\mathbf{e}_2$ | 0.704181 | 1.4097E-4 | 0.704210 |
| 0 | 8 | 170 | $\mathbf{e}_3$ | 0.209590 | 1.6859E-4 | 0.210015 |
| 1 | 12 | 530 | $\mathbf{e}_3$ | 0.209985 | 6.6689E-5 | 0.210015 |
| 2 | 24 | 2210 | $\mathbf{e}_3$ | 0.210012 | 1.9932E-5 | 0.210015 |

*Table 2* Computed integrals $I_A^{(i)}$ versus $(\phi_\infty, E, N_\varphi)$. Only cases $|I_A^{(i)}| \geq 1.E-6$ are reported and $i$ is indicated in parentheses ($a_1 = 1, a_2 = 1.2, a_3 = 2$).

| $-\phi_\infty$ | $(E, N_\varphi)$ | $-I_T^{(i)}$ | $-I_R^{(i)}$ |
|---|---|---|---|
| $x_1$ | $(0, 8)$ | 1.32E-3(1) | −3.70E-5(1) |
| $x_1$ | $(1, 12)$ | 1.18E-4(1) | 1.43E-5(1) |
| $x_1$ | $(2, 24)$ | 5.32E-6(1) | 2.57E-6(2) |
| $x_1 x_2$ | $(1, 12)$ | −2.92E-5(3) | −3.83E-4(3) |
| $x_1 x_2$ | $(2, 24)$ | −2.82E-6(3) | −2.80E-5(3) |

with available analytical value of $\alpha_i$. As Table 1 shows, the results are in excellent agreement with the exact value of $\alpha_i/[2-\alpha_i]$.

## 3.3. NUMERICAL RESULTS

Thus Subsection gives the velocities $\mathbf{U}$ and $\boldsymbol{\omega}$ for different cases. First we consider a uniformly charged ellipsoid (for instance $\zeta = 1$) embedded in different electrostatic potential $\phi_\infty$ (with $\mathbf{E}_\infty = -\nabla\phi_\infty$). As recently established ([Sellier]), the contributions of $\nabla\phi$ to formulas (1.18)-(1.19) vanish; more precisely (see (1.8))

$$I_A^{(i)} = \int_S \nabla\phi.\mathbf{u}_A^{(i)} s(M) dS = 0; \quad \text{for} \quad A \in \{T, R\}. \qquad (1.31)$$

Our numerical results (see Table 2) agree with (1.31) and show that for $(E, N_\varphi) = (1, 12)$ the errors for $\epsilon U_i/\mu$ or $\epsilon\omega_i/\mu$ are about $1.E-4$. Keeping the values $(E, N_\varphi) = (1, 12), a_1 = 1, a_2 = 1.2$ and $a_3 = 2$ we computed formulas (1.18)-(1.19) for the one-parameter ($\alpha \geq 0$) zeta

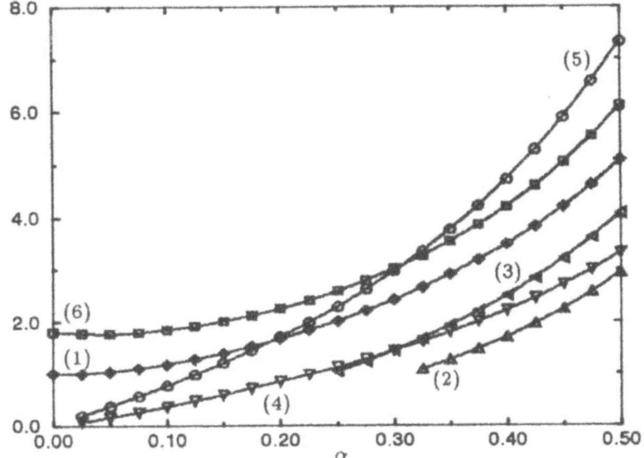


*Figure 3* Velocity components. (1): $\epsilon U_1/\mu$ for $\phi_\infty = -x_1$. (2): $-10^3\epsilon U_2/\mu$ for $\phi_\infty = -x_1$. (3): $10^3\epsilon\omega_1/\mu$ for $\phi_\infty = -x_1$. (4): $\epsilon\omega_2/\mu$ for $\phi_\infty = -x_1$. (5): $-10\epsilon\omega_3/\mu$ for $\phi_\infty = -x_1x_2$. (6): $-10\epsilon\omega_3/\mu$ for $\phi_\infty = -x_1x_2x_3$.

potential:

$$\zeta_\alpha(M) = A_\alpha[2 - (1 - x_3/a_3)^{2\alpha}]; \int_S \zeta_\alpha dS = \int_S dS. \tag{1.32}$$

Thus, each function $\zeta_\alpha$ admits a unit average value on S and this condition settles the numerical value of $A_\alpha$. Remark that Fair and Anderson [Fair & Anderson] handled the case $\alpha = 1, a_1 = a_2$. Only velocity components $\epsilon U_i/\mu$ or $\epsilon\omega_i/\mu$, at least of 1.E-3 magnitude, are plotted in Figure 3 as a function of $\alpha$ and for several potential $\phi_\infty$. Clearly, the results deeply depend on $\alpha$, i. e. on the zeta potential pointwise distribution. For instance, if $\phi_\infty = -x_1$ the solution $\epsilon U_1/\mu$ quickly differs from the Smoluchowski's result $\epsilon U_1/\mu = 1$ (see (1.2)).

## 4. CONCLUDING REMARKS

The theoretical result (1.2) of Smoluchowski holds for a particle of uniform zeta potential embedded in a uniform external electric field $\mathbf{E}_\infty$. In many applications this assumption of a uniform field $\mathbf{E}_\infty$ is valid but the encountered particles are likely to present non-uniform zeta potential distributions. It has been found in this work that the rigid-body motion of an ellipsoid $(\mathbf{U}, \boldsymbol{\omega})$ is very sensitive to the zeta potential. Other zeta potential distributions have been addressed but, for brevity, the results are not reproduced here. The proposed method allows us to deal with other particle's shapes (not ellipsoids). In such cases, the numerical treatment of equations (1.13) rests on the material previously mentioned for equation (1.15). Additional troubles are not encountered. Finally, it is worth noting that the advocated method is likely to apply to the case of an assemblage of particles. Such a work is in current investigation.

## References


Anderson, J. L. (1989). Colloid transport by interfacial forces *Ann. Rev. Fluid Mech.*, 21: 61–99.

Beskos, D. E. (1987). *Introduction to Boundary Element Methods.* In D. E. Beskos, editor, *Computational Methods in Mechanics*, pages 23–106. Elsevier Science Publishers.

Bonnet, M. (1999). *Boundary Integral Equation Methods for Solids and Fluids.* John Wiley & Sons Ltd.

Brebbia, C. A., Telles, J. C. L. and Wrobel, L. C. (1984). *Boundary Element Techniques.* Springer-Verlag, Theory and Applications in Engineering, Berlin Heidelberg New York Tokyo.

Edwardes, B. A. (1892). Steady motion of a viscous liquid in which an ellipsoid is constrained to rotate about a principal axis. *Quart. J. Math.*, 26: 70–78.

Fair, M. C. and Anderson, J. L. (1989). Electrophoresis of nonuniformly charged ellipsoidal particles. *J. Colloid. Interface Sci.*, 127: 388–400.

Happel, J. and Brenner, H. (1973). *Low Reynolds number hydrodynamics.* Martinus Nijhoff.

Hiemenz, P. C. and Rajagopalan, R. (1986). *Principles of Colloid and Surface Chemistry.* , Marcel Dekker, New York.

Hunter, R. J. (1981). *Zeta Potential in Colloid Science.* Academic Press, New York, 1981.

Jeffery, G. B. (1922). The motion of ellipsoidal particles immersed in a viscous fluid. *Proc. Roy. Soc. Lond. A.*, 102: 161–179.

Kellogg, O. D. (1967). *Foundations of potential theory.* Springer-Verlag, Berlin.

Kim, S. and Karrila, S. J. (1991). *Microhydrodynamics: Principles and Selected Applications.* Butterworth.

Kupradze, V. D. (1963). Dynamical problems in elasticity. In *Progress in solid mechanics*, North-Holland, New York.

Ladyzhenskaya, O. A. (1969). *The Mathematical Theory of Vicous Incompressible Flow.* Gordon & Breach.

Lamb, H. (1932). *Hydrodynamics.* 6th edn, Cambridge University Press.
Lyness, J N. and Jespersen, D. (1975). Moderate Degree Symmetric Quadrature Rules for the Triangle. *J. Inst. Maths Applics*, 15: 19–32.
Morrison, F. A. (1970). Electrophoresis of a particle of arbitrary shape. *J. Colloid. Interface Sci.*, 34: 210–214.
Oberbeck, A. (1876). Uber stationare Flussigkeitsbewegungen mit Berucksichtigung der inneren Reibung. *J. Reine. Angew. Math.*, 81: 62–80.
Perrin, F. (1934). Mouvement Brownien d'un ellipsoide (I). Dispersion Diélectrique pour des molécules ellipsoidales. *J. Phys. Radium.*, 5: 497–519.
Pozrikidis, C. (1992). *Boundary integral and singularity methods for linearized viscous flow.* Cambridge University Press.
Rezayat, M., Shippy, D. J. and Rizzo, F. J. (1986). On time-harmonic elactic-wave alaysis by the Boundary Element Method for moderate to high frequencies. *Comp. Meth. in Appl. Mech. Engng.*, 55: 349–367.
Sellier, A. (2000). A note on the Electrophoresis of a uniformly charged particle. *submitted to Quart. J. Math.*
Smoluchowski, M. V. (1921). In *Handbuch der Elektrizität und des Magnetismus.* Ed. L. Graetz. Leipzig: J. A. Barth.
Teubner, M. (1982). The motion of charged colloidal particles in electric fields *J. Chem. Phys.*, 76(11): 5564–5573.

# ON INCREMENTAL BOUNDARY ELEMENT PROCEDURES FOR FRICTIONALLY CONSTRAINED INTERFACES

A.P.S.SELVADURAI
*Department of Civil Engineering and Applied Mechanics,*
*McGill University,*
*Montreal, QC, Canada*

Key words: incremental plasticity, non-linear interfaces, embedded disc inclusion, cylindrical inclusion, penny-shaped cracks, stress intensity factors

Abstract: The paper presents the application of an incremental boundary element technique to the study of material interfaces, which are frictionally constrained. The effectiveness of the method is illustrated by appeal to examples which deal with penny-shaped edge-cracks at the extremities of a frictionally constrained cylindrical inclusion and the in plane translation of a disc inclusion with frictional surfaces which is located at a pre-compressed elastic interface.

## 1. INTRODUCTION

Problems dealing with inclusions embedded in elastic media have applications in the modelling related to materials science, mechanics of composites and geomechanics (Willis, 1981; Mura, 1988; Bilby et al., 1985; Selvadurai, 1994). In the classical analyses involving such inclusion problems, it is invariably assumed that the contact between the inclusion and the surrounding elastic solid exhibits complete continuity. This is at variance with observations where delamination and debonding can be introduced at the interface as a result of thermal loads, impact loads and environmental effects such as moisture migration by diffusive processes and accumulation at the interface. Further influences such as chemical incompatibility between the inclusion and the matrix, as is the case of alkali-aggregate reactions in

*T. Burczynski (ed.), IUTAM/IACM/IABEM Symposium on Advanced Mathematical and Computational Mechanics Aspects of the Boundary Element Method*, 339–349.

concrete, can include delaminations at the inclusion matrix interface. The debonding of the contact does not necessarily result in the inability of the inclusion to contribute to the composite action in the material. The debonded interfaces usually possess an irregular affine structure consisting of asperities and frictional properties that tend to support the load transfer mechanism. Since these interactions are non-linear but restricted to pre-defined interfaces, it becomes convenient to examine the stress analysis of such problems by appeal to a boundary element or boundary integral equation approach. The objective of this paper is to document a boundary approach to the study of inclusion problems, which exhibit non-linear interface phenomena. Since non-linear processes are involved, the boundary element formulation must of necessity adopt an incremental approach. The paper presents the basic equations associated with the incremental
formulation of the non-linear interface problem and the incremental constitutive responses associated with the inclusion-matrix interface. To illustrate the influence of the non-linearity of the interface on the behaviour of the system, we examine two specific problems. The first deals with the behaviour of penny shaped cracks, which occur at the extremities of a cylindrical elastic inclusion, the interface of which exhibits Coulomb friction phenomena (Figure 1). The second deals with the problem of the in-plane loading of a circular rigid disc of finite thickness, which is embedded in frictional contact between two precompressed elastic halfspace regions (Figure 2). In both problems, the influence of interface non-linearity on the response of primary interest is
presented. The first problem deals with the influence of fibre-matrix frictional contact on the stress intensity factor at the crack tip and the second problem establishes the non-linear load-displacement behaviour of the embedded rigid disc inclusion.

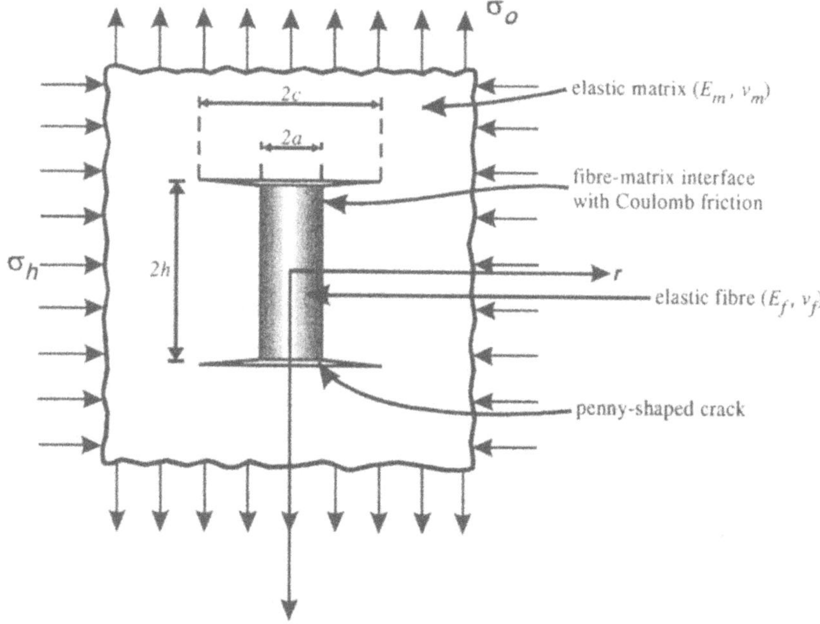


Figure 1.Penny-shaped edge cracks at a cylindrical elastic inclusion

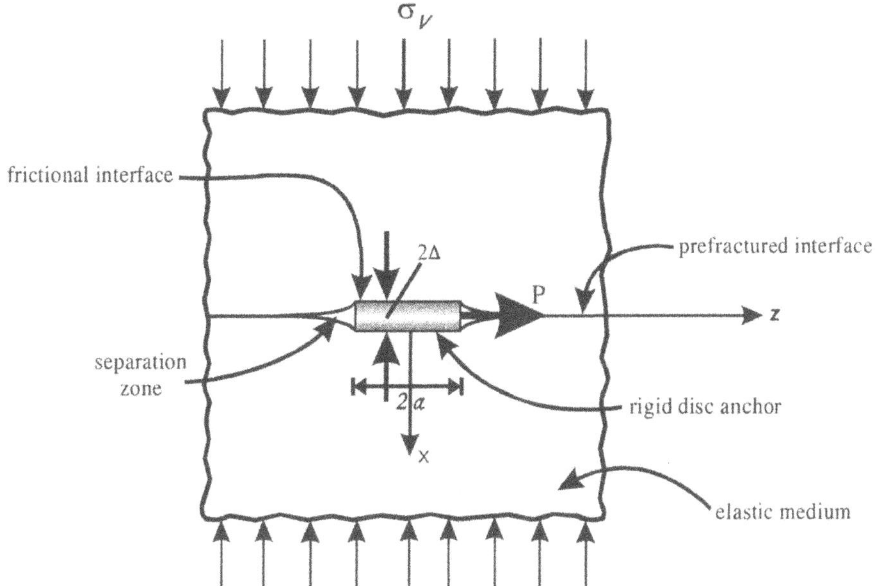


Figure 2. Rigid disc inclusion embedded at a pre-fractured plane

## 2. THE BOUNDARY ELEMENT METHOD

The formulation of the boundary element method for elastostatic problems is given by Brebbia et al. (1984) and in several recent texts on the subject. A brief outline of the incremental formulation applicable to, say, axisymmetric problems are presented here for completeness. For an isotropic bi-material region, the axisymmetric form of the boundary element equation takes the form

$$c_{lk}\dot{u}_k^{(\alpha)} + \int_{\Gamma_{(\alpha)}} \left\{ \overset{*\,(\alpha)}{T_{lk}}\; \dot{u}_k^{(\alpha)} - \overset{*}{u}{}_{lk}^{(\alpha)}\, \dot{T}_k^{(\alpha)} \right\} \frac{r}{r_0} d\Gamma = 0 \tag{1}$$

where $\overset{*(\alpha)}{T_{lk}}$ and $\overset{*(\alpha)}{u_{lk}}$ are, respectively, the tractions and displacement fundamental solutions which are given in the literature (see, e.g. Brebbia et al., 1984), $\dot{u}_k^{(\alpha)}$ are the incremental components of the displacement vector; $\dot{T}_k^{(\alpha)}$ are the incremental components of the traction vector; $c_{lk}$ are constants which are zero if the field point is outside the body; $c_{lk}=\delta_{lk}$, the Kronecker delta if the field point is within the body; $c_{lk}=\delta_{lk}/2$ if the point is located at a smooth boundary and $\Gamma_{(\alpha)}$ are the boundaries associated with the regions. These boundaries can be discretized into boundary elements and the integral equation (1) can be replaced by its discretized equivalent.

For an isoparametric boundary element, the geometric displacement and traction variations can be represented in the form

$$[x_i;u_i;T_i]=\sum_{\gamma=1}^{S}N^{\gamma}(\xi)[x_i^{\gamma};u_i^{\gamma};T_i^{\gamma}]=N(\xi)[\{x_i\};\{u_i\};\{T_i\}] \tag{2}$$

For a quadratic element $S$=3 and the shape functions take the forms

$$N^{(1)}(\xi)=\frac{\xi(\xi+1)}{2};N^{(2)}(\xi)=(\xi^2-1);N^{(3)}(\xi)=\frac{\xi(\xi+1)}{2} \tag{3}$$

with $-1\leq\xi\leq1$. The discretized version of (1) can be written as

$$\begin{aligned}&c_{lk}\dot{u}_k^{(\alpha)}+\sum_e\int_{-1}^{1}T_{lk}^{(\alpha)}[N(\xi)]J|\frac{r}{r_0}d\xi\{\dot{u}_k^{(\alpha)}\}^e\\&=\sum_e\int_{-1}^{1}\overset{*(\alpha)}{u_{lk}}[N(\xi)]J|\frac{r}{r_0}d\xi\{\dot{T}_k^{(\alpha)}\}^e\end{aligned} \tag{4}$$

where $e$ is the element number and $|J|$ is the boundary Jacobian matrix. For an axisymmetric problem, this takes the form

$$|J|=\left[\left(\frac{\partial r}{\partial\xi}\right)^2+\left(\frac{\partial z}{\partial\xi}\right)^2\right]^{1/2} \tag{5}$$

In conventional formulations, the inclusion-matrix interface region can be subjected to traction and displacement boundary conditions. With non-classical or non-linear interfaces, the inclusion-matrix interface can be subjected to incremental constraints of the form

$$\dot{T}_i=\dot{R}_i+\tilde{K}_{ij}\dot{\tilde{u}}_j \tag{6}$$

where $\dot{\tilde{u}}_j$ are the incremental relative displacement components; $\dot{R}_i$ are the incremental residual or initial tractions and $\tilde{K}_{ij}$ are the stiffness coefficients which are derived by considering the incremental constitutive relationships for the interface. From (4) we can formulate the boundary element matrix equation that can be written as

$$[\mathbf{H}]\{\mathbf{u}\}=[\mathbf{G}]\{\mathbf{T}\} \tag{7}$$

where [**H**] and [**G**] are the boundary element influence coefficients matrices which can be obtained by an integration of the fundamental solutions $\overset{*(\alpha)}{u_{ij}}$ and $\overset{*(\alpha)}{T_{ij}}$ applicable to the inclusion and matrix regions separately.

## 3. NON-LINEAR INTERFACES

The mechanics of interfaces generally involve complex micromechanical phenomena. In the mechanistic treatment of such interfaces, it is desirable to idealize the constitutive responses by considering in a macroscopic sense, processes associated with friction, slip, separation, yield, dilatancy, asperity degradation, etc. (Bowden and Tabor, 1951,1964; Johnson, 1985; Selvadurai and Voyiadjis, 1986; Selvadurai and Boulon, 1995). We can assume that the incremental relative displacements $\tilde{\dot{u}}_i$ at a non-linear interface are composed of elastic and plastic components $\tilde{\dot{u}}_i^{(e)}$ and $\tilde{\dot{u}}_i^{(p)}$ respectively, i.e.

$$\tilde{\dot{u}}_i = \tilde{\dot{u}}_i^{(e)} + \tilde{\dot{u}}_i^{(p)} \tag{8}$$

The linear elastic interface constitutive relations relate the elastic component of the incremental relativedisplacement to the incremental interface tractions

$$\dot{T}_i = K_{ij}^{(e)} \tilde{\dot{u}}_j^{(e)} \tag{9}$$

where $K_{ij}^{(e)}$ are the linear elastic stiffness coefficients of the interface. In order to establish the irreversible components of (8) we need to define the stress level at which plastic phenomena manifest at the interface. For this we require a yield function. There are, of course, as many yield functions, as there are material interfaces. It is convenient, however, to first examine a basic interface response which can be idealized by Coulomb friction. The yield function $F$ for the Coulomb model is given by

$$F = (T_t T_t)^{1/2} - \mu T_n \tag{10}$$

where $\mu$ is the coefficient of friction at the interface and $T_n$ and $T_t$ are, respectively, the components of the total traction at the interface in the *normal* and *tangential* directions at the Euclidean plane. When the interface tractions satisfy (10) failure can take place. We now need to postulate the manner in which irreversible displacements follow after failure. For this we utilize a flow/slip rule very similar to that used in the classical theory of plasticity, i.e.,

$$\tilde{\dot{u}}_i^{(p)} = \dot{\lambda} \frac{\partial \Phi}{\partial T_i} \tag{11}$$

where $\lambda$ is a proportionality factor defined as the plastic/slip multiplier and $\Phi$ is the plastic/slip potential. For an interface which exhibits Coulomb friction

$$\Phi(T_i) = T_t \tag{12}$$

Considering (9) and (12) we can write

$$\dot{T}_i = K_{ij}^{(e)}\left[\tilde{\dot{u}}_j - \dot{\lambda}\frac{\partial \Phi}{\partial T_j}\right] \tag{13}$$

where for problems involving axial symmetry $i,j$ = $r,z$. When slip occurs at the interface between the inclusion and the matrix

$$dF = \frac{\partial F}{\partial T_i} dT_i = 0 \tag{14}$$

and

$$\dot{\lambda} = \frac{1}{\Omega}\frac{\partial F}{\partial T_i} K_{ij}^{(e)}\tilde{\dot{u}}_j ; (i, j = r, z) \tag{15}$$

$$\Omega = \frac{\partial F}{\partial T_l} K_{lm}^{(e)} \frac{\partial \Omega}{\partial T_m} ; (l, m = r, z) \tag{16}$$

Using (13) and (15) we can write the elastic-plastic interface constitutive relationship as

$$\dot{T}_i = K_{ij}^{(ep)}\tilde{\dot{u}}_j \tag{17}$$

where the elastic-plastic stiffness matrix is given by

$$K_{ij}^{(ep)} = K_{ij}^{(e)} - \frac{1}{\Omega}\frac{\partial \Omega}{\partial T_l} K_{il}^{(e)} K_{mj}^{(e)} \frac{\partial F}{\partial T_m} \tag{18}$$

In essence, we have utilized the conventional elasto-plastic constitutive formulations approach used in finite element modelling of continua to model the incremental constitutive relationship applicable to an interface. These can be incorporated at the respective interface regions, which are modelled by either isoparametric elements, or special elements, which reflect the nature of the delamination.

## 4. FRICTIONALLY CONSTRAINED CYLINDRICAL INCLUSION WITH PENNY-SHAPED EDGE CRACKS

We consider the axisymmetric problem of a cylindrical elastic inclusion (diameter $2a$ and length $2h$), which is embedded, in an isotropic elastic solid of infinite extent. The cylindrical boundary of the inclusion exhibits Coulomb frictional behaviour and the plane ends of the inclusion contain symmetrically placed penny-shaped cracks of radius $c$ (Figure 1). The region is subjected to a radial compressive stress $\sigma_h$. The region is now subjected to an axial stress $\sigma_0$, which induces stress intensity factors at the tip of the penny-shaped crack. The objective of the modelling is to assess the influence of the interface friction on the stress intensity factor at the crack tip. For obtaining the stress intensity factors at the crack tip, special crack tip singularity elements need to be incorporated. The special singular traction quarter-point boundary elements where the displacement and traction fields are represented in the forms

$$u_i^{(\alpha)} = \sum_{n=0}^{2} b_n r^{n/2} \tag{19}$$

$$T_i^{(\alpha)} = \sum_{n=0}^{2} c_n r^{(n-1)/2} \tag{20}$$

where $b_n$ and $c_n$ are constants, are utilized. The accuracy of these elements is well established (Blandford et al., 1981; Smith and Mason, 1982; Selvadurai and Au, 1987, 1988; Selvadurai, 1996). The incremental value of the Mode I stress intensity factor is obtained by a displacement correlation technique which utilizes the radial displacement components at selected locations around the crack tip, i.e.

$$\dot{K}_I^{(\alpha)} = \frac{G_\alpha}{(k_\alpha + 1)} \sqrt{\frac{2\pi}{l_0}} \left[4\left\{\dot{u}_z^{(\alpha)}(B) - \dot{u}_z^{(\alpha)}(D) - \dot{u}_z^{(\alpha)}(E) - \dot{u}_z^{(\alpha)}(A)\right\}\right] \tag{21}$$

where $l_0$ is the length of the crack tip element and the points $A,B,C$ and $E$ are located along the boundary of the crack, and $k=(3-4\nu_\alpha)$.

For the purposes of the numerical computations, we have selected the following specific values for the parameters in the problem: $h/a=1.0$; $c/a=1.5$; $\nu_f=\nu_m=0.2$; $\sigma_h/E_m=0.0001$ and 0.001; $\mu \in (0.1,0.5)$. The incremental boundary element procedure described previously is applied to determine the influence of the fibre-matrix modular ratio $(E_f/E_m)$ and the interface friction between the fibre and the matrix, on the Mode I stress intensity factor at the tip of the penny-shaped crack located at the plane end of the cylindrical inclusion. Figure 3 illustrates typical results derived from the non-linear boundary element analysis. As can be observed, the frictional response has a significant influence on the amplification of the stress intensity factor at the crack tip.

## 5. IN-PLANE DISPLACEMENT OF A FRICTIONALLY EMBEDDED RIGID DISC.

We consider the problem of a rigid circular disc inclusion, which is embedded, in bonded contact at a pre-fractured interface (Figure 2). The thickness of the rigid circular disc inclusion (thickness 2D; diameter $2a$) induces a precompression stress at the inclusion-elastic medium interface. This problem can be visualized as the computational analogue of a plate anchor region that is created by using expansive cementaceous material. The shrinkage of the cementaceous material can induce delamination at the interface, such that the anchoring capacity is derived from the frictional effect generated by the precompression. The asymmetric loading by the horizontal movement of the anchor results in a three-dimensional state of stress in the elastic solid. It also results in a non-symmetric separation region at the prefractured interface. For the purposes of this study, we assume that the remote boundary where contact is present can be approximated by a circular boundary. For the purposes of the computational modelling we also assume the following specific values of the parameters involved: $\Delta G/a\sigma_v = 0.5, 1.0$ ; $K_n = 10^4\, kPa/mm$ ; $K_t = 10^3\, kPa/mm$ . The incremental boundary element technique is used to examine the influence of the frictional behaviour of the interface on the normalized load $\left(P/\sigma_v a^2\right)$ vs. normalized displacement $(\delta/a)$ response. Figure 4 illustrates typical results derived from the computational scheme.

## 6. CONCLUSIONS

The paper presents a summary of the incremental boundary element procedures that can be applied to examine the mechanical response of inclusions with frictional/elastic-plastic interfaces. It is shown that the incremental boundary element procedure can be conveniently adopted to obtain results of interest to engineering applications. The presence of the interface constraint at a specified boundary makes it convenient to apply the boundary scheme where the iterative scheme is used primarily to locate regions in which interface yield and slip has occurred. The methodology can be adapted to a wider class of interface responses, including interface regions that can experience degradation and damage during repeated or quasistatic cyclic loading.

## REFERENCES

Bilby, B.A.,Miller, K.J. and Willis, J.R. (Eds.)(1985) *Fundamentals of Deformation and Fracture*, Proc. IUTAM Symposium, Cambridge University Press, Cambridge.

Blandford, G.E., Ingraffea, A.R. and Liggett, J.A. (1981) Two-dimensional stress intensity factor computations using the boundary element method, *Int. J.Num. Meth. Engng.*, Vol. 17, 387-404.

Bowden, F.P. and Tabor, D. (1951) *Friction and Lubrication of Solids Vol.II*, Oxford University Press, Oxford.

Bowden, F.P. and Tabor, D. (1964) *Friction and Lubrication of Solids Vol.II*, Oxford University Press, Oxford.

Brebbia, C.A., Telles, J.C.F. and Wrobel, L.C. (1984) *Boundary Element Techniques*, Springer Verlag, Berlin.

Mura, T. (1988) *Micromechanics of Defects in Solids*, Sijthoff and Noordhoff, Alphen aan den Rijn, The Netherlands,

Johnson, K.L. (1985) *Contact Mechanics*, Cambridge University Press, Cambridge.

Selvadurai, A.P.S. (1994) Analytical methods for flat anchor problems in geomechanics, *Proc .IACMAG'94* (H.J.Siriwardane and M.M.Zaman, Eds.), Vol.1, pp. 305-321.

Selvadurai, A.P.S. (1996) On integral equation approaches to the mechanics of fibre-reinforced crack interaction, *Engineering Analysis with Boundary Elements*, Vol. 17, 287-294.

Selvadurai, A.P.S. and Au, M.C. (1987) Cracks with frictional closure: A boundary element approach, *Proc. 10th Int. Conf. Boundary Element Methods in Engng.* Southampton, (C.A.Brebbia, Ed.),pp.211-230.

Selvadurai, A.P.S. and Voyiadjis, G.Z. (Eds)(1986) *Mechanics of Material Interfaces, Studies in Applied Mechanics*, Vol.11, Elsevier Scientific Publ. Co., The Netherlands

Selvadurai, A.P.S. and Boulon, M.J. (Eds.)(1995) *Mechanics of Geomaterial Interfaces, Studies in Applied Mechanics*, Vol.42, Elsevier Scientific Publ. Co., The Netherlands.

Smith, R.N.L. and Mason, J.C. (1982) A boundary element method for curved crack problems in two dimensions, *Boundary Element Methods in Engineering*, (C.A.Brebbia, Ed.) Springer Verlag, Berlin.

Willis, J.R. (1981) Variational and related methods for the overall properties of composites, *Advances in Applied Mechanics*(C. -S. Yih, Ed.) Academic Press, New York

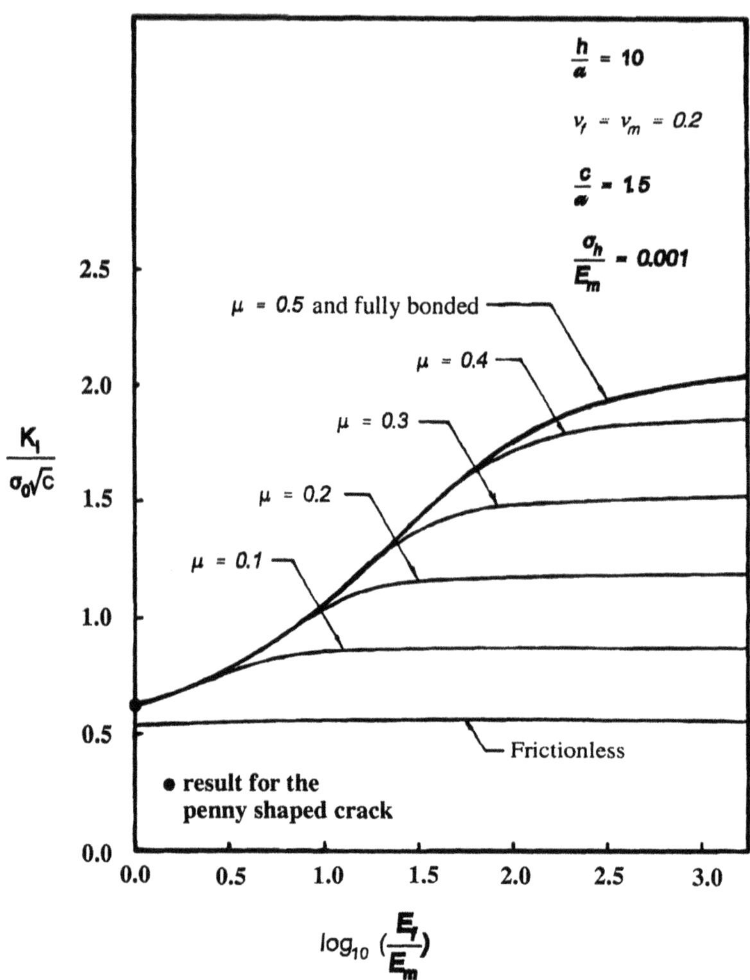


Figure 3. Mode I stress intensity factor at the tip of the penny-shaped crack

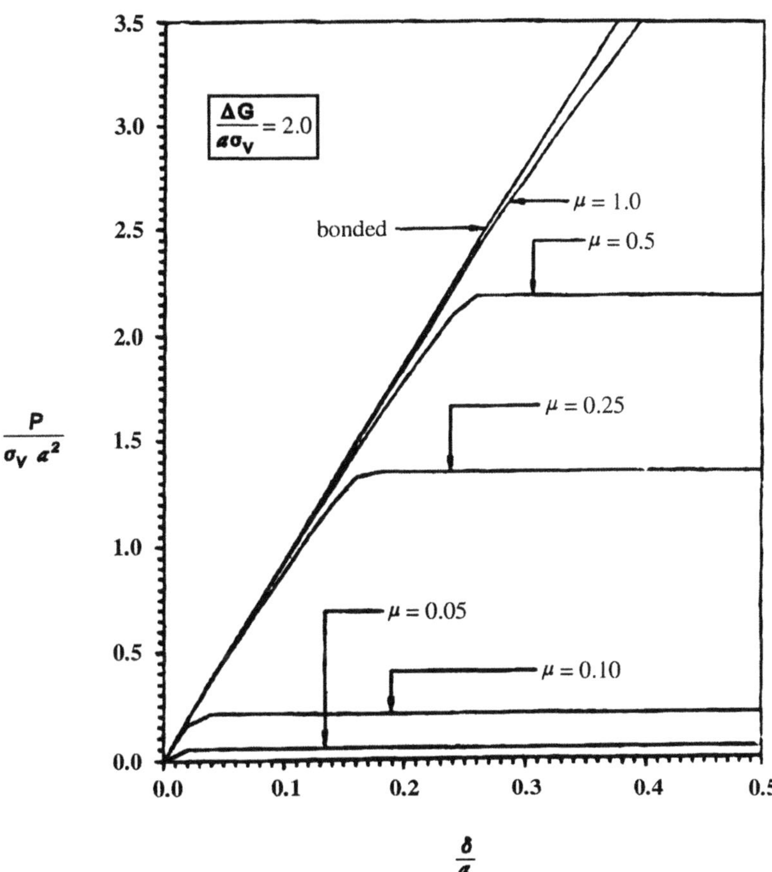


Figure 4. Load-displacement response for the disc inclusion with frictional interface

# THEORETICAL AND COMPUTATIONAL ASPECTS OF THE FUZZY BOUNDARY ELEMENT METHODS

J. Skrzypczyk and T. Burczyński
*Department of Theoretical Mechanics, Silesian University of Technology, Krzywoustego 7, 44-100 Gliwice, Poland*
*Department for Strength of Materials and Computational Mechanics, Silesian University of Technology, Konarskiego 18a, 44-100 Gliwice, Poland*

Key words: Fuzzy Boundary Element, Fuzzy Solutions, Fuzzy Boundary Conditions, Fuzzy Singular Integration

Abstract: In the paper basic concepts of a new methodology of the fuzzy boundary element method are presented. This article deals with fuzzy-set-valued mappings which are solutions of the fuzzy boundary integral equations. Exact fuzzy solutions of fuzzy boundary integral equations are defined as well as conditional solutions. Computational fuzzy problems and applications are considered in details for boundary potential problems with fuzzy Dirichlet and Neumann type boundary conditions and fuzzy density source functions in a fuzzy domain.

## 1. INTRODUCTION

When a physical problem is transformed into the deterministic boundary problem, we usually cannot be sure that this modelling is perfect. The nature of this uncertainty can be discussed generally under three headings: human based uncertainty, system uncertainty and random uncertainty. The prediction of these three types of uncertainty is difficult and present methods, embodied in reliability theory, tend to concentrate on random uncertainty. There is however, a fundamental difference between the nature

*T. Burczynski (ed.), IUTAM/IACM/IABEM Symposium on Advanced Mathematical and Computational Mechanics Aspects of the Boundary Element Method*, 351–364.

of random uncertainty and that of human and system uncertainty. To analyse this type of uncertainty a mathematics which is distinct from randomness is required and this is the potential role of fuzzy sets.

The boundary problem may not be known exactly and some functions i.e. the shape of a structure, material properties, boundary conditions, external or internal excitations, solutions etc. may contain unknown parameters. Many different interpretations are possible for terminology of uncertain aspects of the *Boundary Element Method* (BEM). We focus our attention on fuzzy-set-theoretic description of uncertain phenomena in BEM, and will refer to these approaches as *Fuzzy Boundary Element Method* (FBEM).

Applications of the FBEM appear to have been initiated in the 1995. The earliest application used the fuzzy boundary integral equation to solve fuzzy boundary value potential problem with uncertain boundary conditions and internal sources (cf. Burczyński & Skrzypczyk [13-15]). Then FBEM has been used for elastostatic problems [33,34]. Modelling uncertainties as fuzzy variables or fuzzy processes suggests the use of fuzzy-set-theoretical methods, which are closely related to convex modelling of uncertainties. Only linear static problems are studied and applications to non-linear or dynamic problems are left for a future study. Using BEM to solve boundary value problem in some domain with prescribed boundary conditions on the boundary, one can obtain the *Boundary Integral Equation*. From now we assume that values of some of boundary conditions, material properties, internal prescribed fields and the shape of a boundary are uncertain and we'll model this uncertainty using fuzzy variables. We obtain the *Fuzzy Boundary Integral Equation* where all operations are in the fuzzy sense.

Singular integrals are understood in the sense of fuzzy principal values. New results in singular integration and singular integration over a fuzzy domain are presented cf. [31,32,35,36].

Different types of fuzzy solutions are discussed as well as their existence and properties. Integral expressions can be discretized to find the system of *Fuzzy Algebraic Equations* from which the fuzzy boundary values can be found [16,17,33-36,38].

Illustrative examples from the potential theory are given to comment different aspects of the presented theory. Interval and trapezoid - type fuzzy boundary conditions are considered. To complete the presentation the potential problem in a fuzzy domain is discussed.

Presented methods give the complete methodology how to obtain good approximations of solutions of uncertain boundary problems with use of fuzzy analysis.

## 2. BASIC DEFINITIONS AND NOTION

In the paper we use the following notion. $R^n$ denotes the set of n-dimensional reals, $(R^n,|\cdot|)$ - n-dimensional Euclidean space with the metric $|\cdot|$, $\Gamma$ is reserved for k-dimensional (k<n) manifold in Euclidean space $R^n$.

Let further I(R) (similarly $I(R^n)$) denote the set of all closed, bounded intervals $\bar{z}=[z^-,z^+]$ on real line R ($R^n$ respectively), where $z^- i z^+$ denote their end points. We call further elements of sets I(R) ($I(R^n)$) interval numbers ( interval vectors respectively) cf. [1,3,25,28].

Let $F(R^n)$ be the class of fuzzy sets in $R^n$, i.e. the set of maps [19,27]

$$F(R^n):=\{\mu:R^n\to[0,1]\}.$$

We call a fuzzy number the set $\tilde{a}\in F(R^n)$ defined by the so called membership function $\mu(x;\tilde{a}), x\in R^n$ and satisfying some additional conditions [18,19,20,22,23]. Let further

$$\tilde{a}_\lambda:=\{x\in R^n:\mu(x;\tilde{a})\geq\lambda\}, 0<\lambda\leq 1 \tag{1}$$

$$\tilde{a}_0:=\overline{\{x\in R^n:\mu(x;\tilde{a})>0\}}, \tag{2}$$

denote α-level set of $\tilde{a}$. We can generalise the definition of fuzzy sets to any metric space H and F(H).

Let $\tilde{a},\tilde{b}\in F^*(R^n)$. Then $\tilde{a}$ is said to be more accurate than $\tilde{b}$, denoted by $\tilde{a}\subset\tilde{b}$, iff $\mu(x;\tilde{a})\leq\mu(x;\tilde{b}), \forall x\in R^n$.

If $f:R^n\times R^n\to R^n$ is a usual real function, then according to Zadeh extension principle we can extend f to fuzzy function $\tilde{f}:F^*(R^n)\times F^*(R^n)\to F^*(R^n)$ in the following way [18,19,20,20,23,27]

$$\mu(z;\tilde{f}(\tilde{u},\tilde{v}))=\sup_{z=f(x,y)}\mu(x;\tilde{u})\wedge\mu(y;\tilde{v}) \tag{3}$$

where ∧, ∨ denote max and min respectively. It's well known result, that

$$\tilde{f}_\lambda(\tilde{u},\tilde{v})=\tilde{f}(\tilde{u}_\lambda,\tilde{v}_\lambda), \forall\tilde{u},\tilde{v}\in F^*(R^n), 0\leq\lambda\leq 1 \tag{4}$$

if f is a continuous function [29]. Notice, that $I(R^n)\subset F^*(R^n)$ [21-23].

## 3. FUZZY SINGULAR INTEGRATION

Notice, that a fuzzy singular integral in a non-fuzzy domain, in the fuzzy principal value of Cauchy type (FPV) has been defined first in [31] and investigated further in papers [12-17,32-37]. Fuzzy singular integration over a fuzzy domain was defined first in [32,35].

Let $\mathcal{M}$ denote further a set of manifolds in $R^n$. By a fuzzy domain $\tilde{\Gamma}$ we understand a fuzzy subset of manifolds from $\mathcal{M}$, i.e. a map $\mathcal{M}$ into [0,1]. With each map $\Gamma$ we consider its membership function $\mu(\Gamma;\tilde{\Gamma}), \Gamma \in \mathcal{M}$. Let $F(\mathcal{M})$ denote a class of fuzzy subsets of $\mathcal{M}$.

One can define a fuzzy domain in another way:

- a classical map $\Gamma_u$ defined over a set of parameters $U \subset R^p$ with values in $\mathcal{M}$;
- a classical map defined over a fuzzy set $\tilde{A} \in F^*(R^p)$ with values in $\tilde{\Gamma}$, i.e. such that $\mu(\cdot;\tilde{A}) \leq \mu(\cdot;\tilde{\Gamma}) \circ \Gamma$.

We know, that the membership function of a fuzzy singular integral defined over some non-fuzzy domain - $\Gamma$ is as follows

$$\mu_\Gamma\left(w; \int_\Gamma \tilde{h}(y) d\Gamma(y)\right), \Gamma \in M \quad, w \in R$$

and can be considered as the conditional membership function.

Let further $\tilde{h}: \Gamma \to F^*(R^n)$ be a measurable fuzzy function [2,22-23,26], integrable in the FPV-sense over each subset $\Gamma \subset M$ . Then a singular integral [24] of the fuzzy function $\tilde{h}$ in the FPV-sense, over the fuzzy set $\tilde{\Gamma} \in F(M\ )$, will be denoted further as $(FPV)\int_{\tilde{\Gamma}} \tilde{h}(y) d\Gamma(y)$ and its membership function is defined in the following way:

$$\begin{aligned} &\mu\left(w; (FPV)\int_{\tilde{\Gamma}} \tilde{h}(y) d\Gamma(y)\right) = \\ &:= \sup_{\Gamma \in M} \left(\mu(\Gamma;\tilde{\Gamma}) \wedge \mu_\Gamma\left(w; (FPV)\int_\Gamma \tilde{h}(y) d\Gamma(y)\right)\right), w \in R \end{aligned} \tag{5}$$

Such defined integral will be called the generalized fuzzy singular integral of the fuzzy function $\tilde{h}$ over the fuzzy domain $\tilde{\Gamma}(x)$, in the fuzzy principal value of Cauchy-type (FPV) cf. details [31,35-36].

# 4. FUZZY BOUNDARY INTEGRAL EQUATIONS

Using BEM to solve potential boundary value problem in a domain $\Omega$ with prescribed boundary conditions on the $\Gamma$ boundary of $\Omega$: Dirichlet (essential) conditions of the type $u(x)=u_0(x)$, for $x \in \Gamma_1$ and Neumann (natural) conditions such as $q(x)=\partial u(x)/\partial n=q_0(x)$, for $x \in \Gamma_2$, $\Gamma=\Gamma_1 \cup \Gamma_2$, one obtains

$$\begin{aligned} &c(x)u(x) + \int_\Gamma Q(x,y)u(y)d\Gamma(y) + \int_\Omega U(x,y)\xi(y)d\Omega(y) = \\ &= \int_\Gamma U(x,y)q(y)d\Gamma(y), x \in \Gamma \end{aligned} \tag{6}$$

where $\xi(x)$, $x\in\Omega$ is a known source density function and U is a fundamental solution of the Laplace equation ( $Q=\partial U/\partial n$ ), see [4,5,12].

We now assume that values of some of boundary quantities, a source density function and a contour $\Gamma$ are uncertain and we shall model this uncertainty using fuzzy variables. Let $\tilde{u}_0, \tilde{q}_0, \tilde{\xi}$ and $\tilde{\Gamma}$ be fuzzy functions. Define

$$U_\lambda(x|\Gamma) := \left\{ \begin{array}{l} u : c(x)u(x) + \int_\Gamma Q(x,y)u(y)d\Gamma(y) + \int_\Omega U(x,y)\xi(y)d\Omega(y) = \\ = \int_\Gamma U(x,y)q(y)d\Gamma(y), \\ u_0(z) \in \tilde{u}_{0\lambda}(z)\big|_{z\in\Gamma_1}, q_0(z) \in \tilde{q}_{0\lambda}(z)\big|_{z\in\Gamma_2}, \xi(z) \in \tilde{\xi}_\lambda(z), z \in \Omega, \Gamma \in M \end{array} \right\} \tag{7}$$

The family $U_\lambda(x|\Gamma)$, $0<\lambda\le 1$ represents the $\lambda$-level set of a fuzzy number-valued function iff [22,23]

A1) $U_\lambda(x|\Gamma)$ is a non-empty compact convex subset of $R^n$ for $\forall\, 0<\lambda\le 1$;

A2) $U_\lambda(x|\Gamma) \subseteq U_\mu(x|\Gamma)\ \forall\, 0\le\mu\le\lambda\le 1$;

A3) if $\{\lambda_k\}$ is a non-decreasing sequence converging to $\lambda$ then

$$U_\lambda(x|\Gamma) = \bigcap_{k\ge 1} U_{\lambda_k}(x|\Gamma)$$

We assume further that the family $U_\lambda(x|\Gamma)$, $0<\lambda\le 1$ generates the fuzzy function, which we call ***conditional united fuzzy set-valued solution (CUFSVS)*** $\tilde{u}_1(x|\Gamma), x\in\Gamma\in M$ . The membership function of the fuzzy solution $\tilde{u}_1(x), x\in\tilde{\Gamma}$ is thus defined according to the following relation

$$\mu_\Gamma(y;\tilde{u}_1(x|\Gamma)) := \sup\{\lambda : y\in U_\lambda(x|\Gamma)\},\ x\in\Gamma\in M\ ,\ y\in R^1 \tag{8}$$

and the ***united fuzzy set-valued solution (UFSVS)*** $\tilde{u}_1(x), x\in\tilde{\Gamma}$ is thus defined according to min-max composition of fuzzy relations

$$\mu(y;\tilde{u}_1(x)) := \sup_{\Gamma\in M}\{\mu(\Gamma;\tilde{\Gamma}) \wedge \mu_\Gamma(y;\tilde{u}_1(x|\Gamma))\},\ y\in R^2 \tag{9}$$

Formula (9) describes the membership function of the first-type fuzzy solution of boundary potential problem defined over a fuzzy domain.
Now we can write formally the fuzzy version of eq.(6) in the form

$$\begin{aligned} &\tilde{c}(x)\tilde{u}(x) + \int_{\tilde{\Gamma}} Q(x,y)\tilde{u}(y)d\Gamma(y) + \int_{\tilde{\Omega}} U(x,y)\tilde{\xi}(y)d\Omega(y) = \\ &= \int_{\tilde{\Gamma}} U(x,y)\tilde{q}(y)d\Gamma(y), x\in\tilde{\Gamma} \end{aligned} \tag{10}$$

Let all operations be considered in the fuzzy sense. Fuzzy integrals are understood in the sense of fuzzy principal value. Fuzzy function $\tilde{u}_2(x), x\in\tilde{\Gamma}\in F(M)$ which satisfies formally eq.(10) will be called the

***conditional fuzzy algebraic solution (CFAS)***. The membership function of the fuzzy solution $\tilde{u}_2(x|\Gamma)$ is denoted by $\mu_\Gamma(y;\tilde{u}_2(x|\Gamma))$, $y \in R$. The ***fuzzy algebraic solution (FAS)*** $\tilde{u}_2(x), x \in \tilde{\Gamma}$ is thus defined according to min-max composition of fuzzy relations

$$\mu(y;\tilde{u}_2(x)) := \sup_{\Gamma \in M} \left(\mu(\Gamma;\tilde{\Gamma}) \wedge \mu_\Gamma(y;\tilde{u}_2(x|\Gamma))\right), \; y \in R \tag{11}$$

Alternatively let substitute $\tilde{u}_0, \tilde{q}_0, \tilde{\xi}$ and $\tilde{\Gamma}$ for $u_0$, $q_0$, $\xi$ and $\Gamma$ respectively and let all operations be considered in the fuzzy sense. Thus it is a difficult problem to solve eq. (10) in such a way, so we are looking rather for approximate methods cf. [6-11].

Now we discuss how to solve eq. (10) for the conditional fuzzy function $\tilde{u}(x), x \in \Gamma \in M$ . Let

$$\tilde{u}_{0\lambda}(x) = \left[u_{0\lambda}^{-}(x), u_{0\lambda}^{+}(x)\right], x \in \Gamma_1 \tag{12}$$

$$\tilde{q}_{0\lambda}(x) = \left[q_{0\lambda}^{-}(x), q_{0\lambda}^{+}(x)\right], x \in \Gamma_2 \tag{13}$$

$$\tilde{\xi}_{\lambda}(x) = \left[\xi_{\lambda}^{-}(x), \xi_{\lambda}^{+}(x)\right], x \in \Omega \tag{14}$$

Assume, that we are looking for the interval - type solution

$$\tilde{u}_{\lambda}(x) = \left[u_{\lambda}^{-}(x), u_{\lambda}^{+}(x)\right], x \in \Gamma \in M \tag{15}$$

where $0 \le \lambda \le 1$. Taking $\lambda$-cuts, $\forall 0 \le \lambda \le 1$ of the fuzzy eq. (10) we obtain formally the infinite set of interval boundary integral equations as follows

$$c(x)\left[u_{\lambda}^{-}(x), u_{\lambda}^{+}(x)\right] = \int_\Gamma Q(x,y)\left[u_{\lambda}^{-}(y), u_{\lambda}^{+}(y)\right]d\Gamma(y) +$$
$$+ \int_\Omega U(x,y)\left[\xi_{\lambda}^{-}(y), \xi_{\lambda}^{+}(y)\right]d\Omega(y) = \int_\Gamma U(x,y)\left[q_{\lambda}^{-}(y), q_{\lambda}^{+}(y)\right]d\Gamma(y), x \in \Gamma \in M \tag{16}$$

We solve eq. (16) for the interval values $u_{\lambda}^{-}(x)$ and $u_{\lambda}^{+}(x)$ producing the family of interval functions $\tilde{u}_{3\lambda}, 0 \le \lambda \le 1$. Following we define the fuzzy solution of the second type $\tilde{u}_3$ by the relation

$$\mu(y;\tilde{u}_3(x|\Gamma)) := \sup\{\lambda : y \in \tilde{u}_{3\lambda}(x|\Gamma)\}, x \in \Gamma \in M \;, y \in R^1 \tag{17}$$

Naturally we are now interested in the relationship between solutions $\tilde{u}_1$, $\tilde{u}_2$ and $\tilde{u}_3$. Following [13-17,32-37] and the results [6-11], we have different relations between solutions $\tilde{u}_1$, $\tilde{u}_2$ and $\tilde{u}_3$ depending on the understanding of the solution $\tilde{u}_3$. For more details look part 5[th].

# 5. FUZZY BOUNDARY ELEMENT METHOD - COMPUTATIONAL METHODOLOGY

Let us now consider how expression (16) can be discretized to find the system of fuzzy algebraic equations from which the boundary values can be found. Assume for simplicity that the body is two-dimensional and its boundary is divided into N elements. Let $M \supset \Gamma \cong \bigcup_{j=1}^{N} \Gamma_j$, where $\Gamma_j$ is the boundary of the j-th element. The fuzzy (interval) values of $\tilde{u}_\lambda$ and $\tilde{q}_\lambda$ are assumed to be fuzzy constant/linear/cubic etc. over each element.
After discretizing the boundary into a series of N elements eq.(16) can be written as

$$c_i \tilde{u}_\lambda(x_i) = \sum_{j=1}^{N} \int_{\Gamma_j} Q(x_i, y)\tilde{u}_\lambda(y)d\Gamma(y) + \int_{\Omega} U(x_i, y)\tilde{\xi}_\lambda(y)d\Omega(y) = \sum_{j=1}^{N} \int_{\Gamma_j} U(x_i, y)\tilde{q}_\lambda(y)d\Gamma(y), x_i \in \bigcup_{j=1}^{N} \Gamma_j \tag{18}$$

$c_i = \theta/2\pi$, where $\theta$ is the non-fuzzy internal angle of the corner in radians.
The values of $\tilde{u}_\lambda$ and $\tilde{q}_\lambda$ at any point on the element can be defined in terms of their nodal fuzzy values and some interpolating functions, similarly as in the non-fuzzy case [13-17, 31-37].
If we now assume that the position of i-th point can vary from 1 to N one obtains a system of N fuzzy algebraic equations resulting from eq. (18). This set of fuzzy equations can be expressed in matrix form as

$$\mathbf{H}_\lambda \tilde{\mathbf{U}}_\lambda = \mathbf{G}_\lambda \tilde{\mathbf{Q}}_\lambda + \tilde{\mathbf{V}}_\lambda, \tag{19}$$

where $\mathbf{H}_\lambda$ and $\mathbf{G}_\lambda$ are two NxN non-fuzzy matrices and $\tilde{\mathbf{U}}_\lambda, \tilde{\mathbf{Q}}_\lambda, \tilde{\mathbf{V}}_\lambda$ are fuzzy vectors of length N, $\forall\ \lambda \in ]0,1]$. Notice that $N_1$ fuzzy values of $\tilde{u}_\lambda$ and $N_2$ fuzzy values of $\tilde{q}_\lambda$ are known on $\Gamma_1$ and $\Gamma_2$ respectively, hence there are only N fuzzy unknowns in the system of equations (19). One has to rearrange the system to obtain a standard system of fuzzy algebraic equations

$$\tilde{\mathbf{A}}_\lambda \tilde{\mathbf{X}}_\lambda = \tilde{\mathbf{F}}_\lambda, \forall 0 \le \lambda \le 1 \tag{20}$$

where $\tilde{\mathbf{X}}_\lambda$ is a fuzzy (interval) vector of unknown $\lambda$-cuts $\tilde{u}_\lambda$ and $\tilde{q}_\lambda$ fuzzy boundary values. Eq.(20) can generate different types of interval solutions i.e. we can obtain different kinds of fuzzy solutions, i.e. approximations of the "exact" conditional fuzzy solution $\tilde{r}_2(x|\Gamma), x \in \Gamma \in M$ .
If all operations in eq.(20) are formally understood in the sense of interval operations we get a solution, called further an ***Algebraic Interval Solution (AIS)*** $\Sigma_A(\tilde{\mathbf{A}}_\lambda, \tilde{\mathbf{F}}_\lambda)$ [28,30]. Sometimes it's more convenient to treat eq.(20)

as a set of real systems of N algebraic equations. Therefore, it should be better to consider rather a set of solutions of the following form

$$\left\{\mathbf{A}_\lambda \mathbf{X}_\lambda = \mathbf{F}_\lambda : \mathbf{A}_\lambda \in \tilde{\mathbf{A}}_\lambda,\ \mathbf{F}_\lambda \in \tilde{\mathbf{F}}_\lambda \right\} \quad \forall\, 0 \le \lambda \le 1 \tag{21}$$

The ***United Solution Set (USS)*** of eq.(24) is defined as [28,30]

$$\Sigma_U\left(\tilde{\mathbf{A}}_\lambda, \tilde{\mathbf{F}}_\lambda\right) = \Sigma_{\exists\exists}\left(\tilde{\mathbf{A}}_\lambda, \tilde{\mathbf{F}}_\lambda\right) = \left\{\mathbf{X}_\lambda : \exists \mathbf{A}_\lambda \in \tilde{\mathbf{A}}_\lambda,\ \exists \mathbf{F}_\lambda \in \tilde{\mathbf{F}}_\lambda,\ \mathbf{A}_\lambda \mathbf{X}_\lambda = \mathbf{F}_\lambda\right\} =$$
$$= \left\{\mathbf{X}_\lambda : \mathbf{A}_\lambda \mathbf{X}_\lambda = \mathbf{F}_\lambda : \mathbf{A}_\lambda \in \tilde{\mathbf{A}}_\lambda,\ \mathbf{F}_\lambda \in \tilde{\mathbf{F}}_\lambda \right\} = \left\{\mathbf{X}_\lambda : \tilde{\mathbf{A}}_\lambda \mathbf{X}_\lambda \cap \tilde{\mathbf{F}}_\lambda \neq \mathbf{0}\right\} \quad \forall\, 0 \le \lambda \le 1 \tag{22}$$

USS is't an interval vector, but has very complicated form. Usually it is approximated from above by some interval vector.

Assume from now, that both families: AIS's and USS's satisfy conditions (A1-A3). Then we get, that the family of USS's generates the ***Conditional United Fuzzy Solution Set (CUFSS)*** $\tilde{u}_U(x|\Gamma), x \in \Gamma \in M$ in the following way

$$\mu(y; \tilde{u}_U(x|\Gamma)) := \sup\left\{\lambda : y \in \Sigma_U\left(\tilde{\mathbf{A}}_\lambda, \tilde{\mathbf{F}}_\lambda\right)\right\} \quad x \in \Gamma \in M \ , y \in R^N \tag{23}$$

The ***United Fuzzy Solution Set-(UFSS)*** $\tilde{u}_U(x), x \in \tilde{\Gamma}$ is thus defined according to formula (3) as follows

$$\mu(y; \tilde{u}_U(x)) := \sup_{\Gamma \in M} \left(\mu(\Gamma; \tilde{\Gamma}) \wedge \mu(y; \tilde{u}_U(x|\Gamma))\right), \quad y \in R^N \tag{24}$$

Similarly we get, that the family of AIS's generates the ***Conditional Algebraic Fuzzy Solution (CAFS)*** $\tilde{u}_A(x|\Gamma), x \in \Gamma \in M$ in the following way

$$\mu(y; \tilde{u}_A(x|\Gamma)) := \sup\left\{\lambda : y \in \Sigma_A\left(\tilde{\mathbf{A}}_\lambda, \tilde{\mathbf{F}}_\lambda\right)\right\} \quad x \in \Gamma \in M \ , y \in R^N \tag{25}$$

The ***Algebraic Fuzzy Solution (AFS)*** $\tilde{u}_A(x), x \in \tilde{\Gamma}$ is thus defined similarly to eq.(24)

$$\mu(y; \tilde{u}_A(x)) := \sup_{\Gamma \in M} \left(\mu(\Gamma; \tilde{\Gamma}) \wedge \mu(y; \tilde{u}_A(x|\Gamma))\right), \quad y \in R^N \tag{26}$$

Further we omit the parameter $\Gamma$, since it is fixed for further considerations. Assume from now that no $\mathbf{A} \in \tilde{\mathbf{A}}_\lambda$ is singular $\forall\, 0 \le \lambda \le 1$. We wish to obtain the set of upper approximations $\tilde{\mathbf{X}}_\lambda$ of set-solutions $\Sigma_A\left(\tilde{\mathbf{A}}_\lambda, \tilde{\mathbf{F}}_\lambda\right)$ and $\Sigma_U\left(\tilde{\mathbf{A}}_\lambda, \tilde{\mathbf{F}}_\lambda\right)$. We now try to solve eq. (24) for the corresponding $x_{i\lambda}^-$ and $x_{i\lambda}^+$, i=1,2,...,N, $0 \le \lambda \le 1$, and hope they are the $\lambda$-cuts of fuzzy numbers $\tilde{x}_i$, i=1,2,...,N. In any case, assume that this method does produce fuzzy numbers $\tilde{x}_i$, i=1,2,...,N. Thus we can assume, that the family $\left\{\tilde{\mathbf{X}}_\lambda\right\}$ generates the fuzzy function with the membership function

$$\mu(\mathbf{x}; \tilde{\mathbf{X}}) := \min_{1 \le i \le N} \left\{\mu(x_i; \tilde{x}_i)\right\} \ \mathbf{x} = [x_i] \in R^N, \Gamma \in M \tag{27}$$

Since $\Sigma_A(\tilde{\mathbf{A}}_\lambda, \tilde{\mathbf{F}}_\lambda) \subseteq \tilde{\mathbf{X}}_\lambda$ and $\Sigma_U(\tilde{\mathbf{A}}_\lambda, \tilde{\mathbf{F}}_\lambda) \subseteq \tilde{\mathbf{X}}_\lambda$ we get that $\tilde{u}_U \subseteq \tilde{\mathbf{X}}$ and $\tilde{u}_A \subseteq \tilde{\mathbf{X}}$ i.e. the obtained approximation is less accurate than both AFS and UFSS.

Many authors [25,28] have discussed methods for computing an interval vector $\tilde{\mathbf{X}}_\lambda$ containing AFS and UFSS. The exact calculation of $\tilde{\mathbf{X}}_\lambda$ is for multidimensional problems very difficult. Naturally, the smallest $\tilde{\mathbf{X}}_\lambda$ is of interest. We call $\tilde{\mathbf{X}}(\Gamma)$ an ***Upper Approximate Conditional Fuzzy Solution (UACFS)*** of FBEM for arbitrary $\Gamma \in \mathrm{M}$ .

The ***Upper Approximate Fuzzy Solution (UAFS)*** $\tilde{\mathbf{X}}(\mathbf{x}), \mathbf{x} \in \tilde{\Gamma}$ is thus defined similarly to eq.(24)

$$\mu(\mathbf{y}; \tilde{\mathbf{X}}(\mathbf{x})) := \sup_{\Gamma \in \mathrm{M}} \left(\mu(\Gamma; \tilde{\Gamma}) \wedge \mu(\mathbf{y}; \tilde{\mathbf{X}}(\mathbf{x}|\Gamma))\right), \ \mathbf{y} \in \mathrm{R}^N, \mathbf{x} \in \mathrm{R}^N \quad (28)$$

# 6. NUMERICAL RESULTS

## 6.1 Trapezoidal Boundary Conditions

The following example illustrates how the presented methods work for fuzzy values of boundary conditions. Analyse a simple potential problem. Consider the case of a square close domain of the type shown in fig. 1, where the boundary has been discretized into 12 fuzzy constant elements with 5 internal points, cf. [4,13,14,16,17,35].

It is assumed, that boundary conditions $u_0$ and $q_0$ in the considered potential problem are fuzzy functions with trapezoidal membership functions. Such membership functions can be characterised as the ordered quadruple equivalent to the characteristic points of the trapezoid. Numerical values of the considered boundary values are given in details at fig. 1.

Since only boundary conditions are of the fuzzy character, the exact fuzzy solution has membership functions also of the trapezoidal shape. Results are presented at fig. 2. Values of the membership functions of the solution in the internal points have trapezoidal character too. The detailed values are omitted for simplicity cf. [33,34].

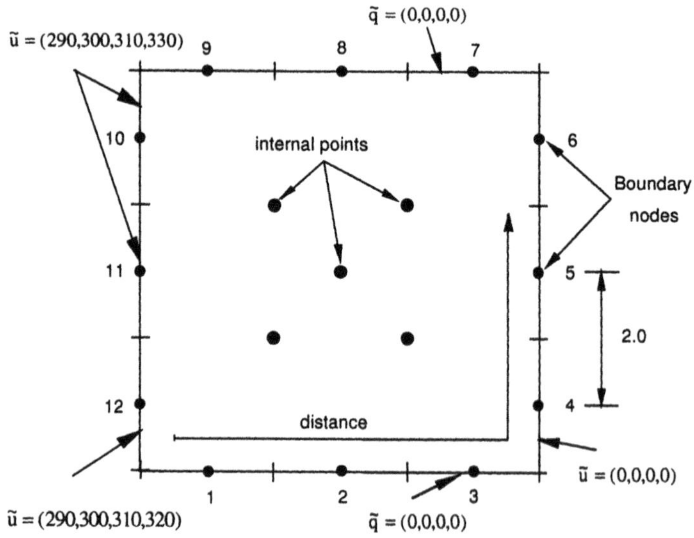


Figure 1. Fuzzy potential problem with fuzzy trapezoidal boundary conditions

## 6.2 Potential Problem In The Fuzzy Domain

Now consider the same potential problem. Assume, that boundary functions are of interval character. This assumption is made for simplicity of calculations only. The methodology of calculations allow to consider any shapes of membership functions. Boundary conditions $u_0$ i $q_0$ are independent with respect to boundary fluctuations and $\tilde{\xi}(x) \equiv 0, \forall x \in \tilde{\Omega}$.
Numerical values for boundary conditions are: $u_{4,5,6} = 0$, $u_{10} = [290,310]$, $u_{11} = [295,305]$, $u_{12} = [295,315]$, $q_{1,2,3,7,8,9} = 0$. Additionally assume, that the considered domain is fuzzy - its boundary is a tetragon with apexes which can take uncertain positions. Let coordinates ($x_i, y_i, i = 1,2,3,4$) be known with accuracy ($\pm\tau_i, \pm\sigma_i, i = 1,2,3,4$) respectively see fig.3. In such a way, the considered domain is the fuzzy function of 8 fuzzy parameters.
At first we analyse the appropriate conditional solutions with respect to fuzzy boundary. We know, that each conditional solution is the interval function. Denote this solution $\tilde{u}(x|\Gamma) = [u^-(x), u^+(x)], x \in \Gamma \in M$. If all conditional solutions are known, we use max-min formula to obtain the global membership functions for interesting solutions - boundary or internal.

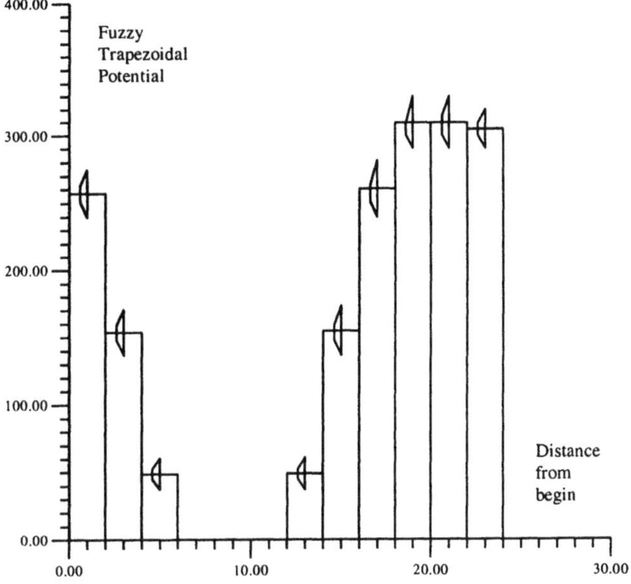


Figure 2. Trapezoidal membership functions of boundary solutions

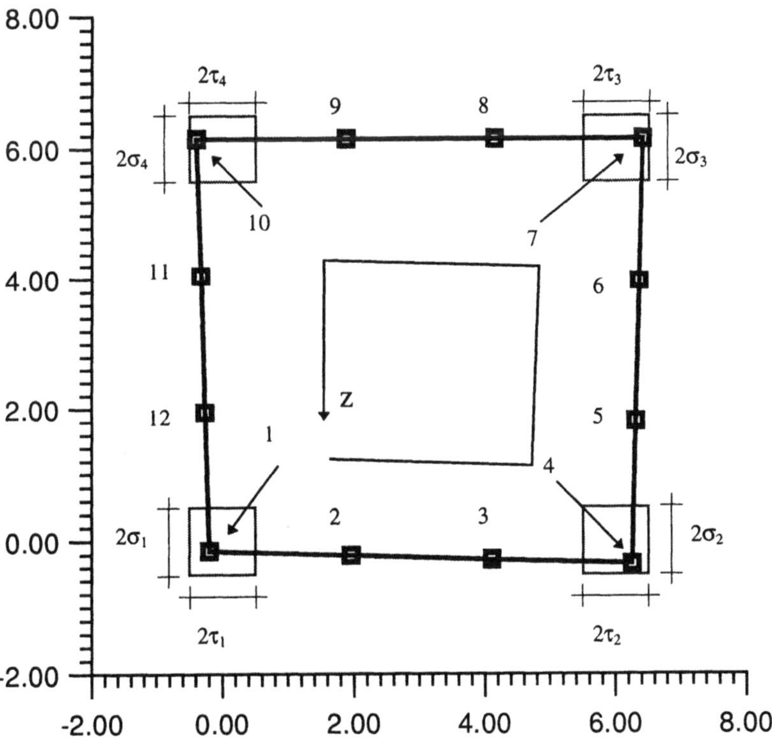


Figure 3. Fuzzy domain of boundary problem

At fig. 4 the global boundary solution for the potential function is given. For comparison one conditional solution is presented for the particular case of the domain signed on fig. 4. There is no special problems to analyse the potential problem in fully complicated form, with all fuzzy elements - internal sources, boundary conditions and the shape of boundary. It only the problem of computational complexity.

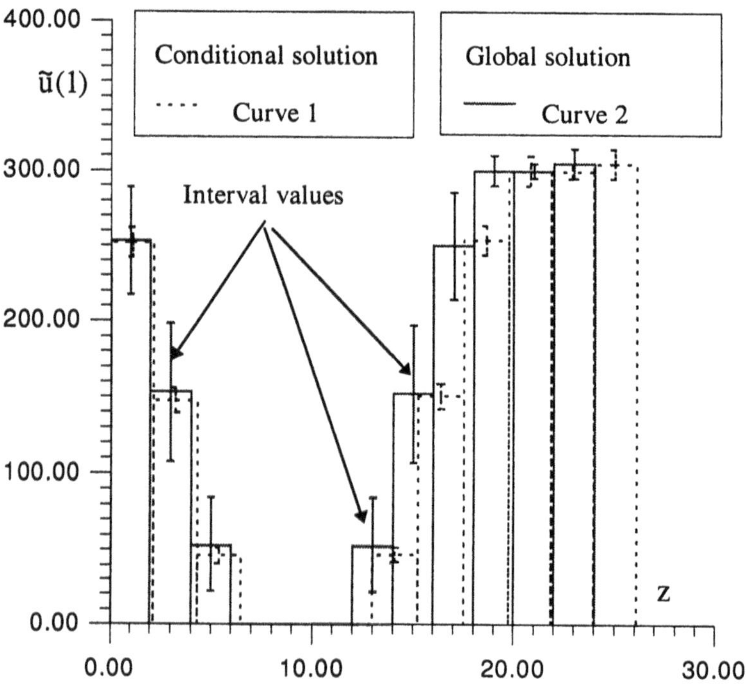


Figure. 4. Global and conditional fuzzy solutions

## 7. CONCLUSIONS

This paper is a continuation of earlier works and summarise our knowledge about analysis with use of boundary element method in fuzzy formulation. It was concerned with the new theoretical and computational methodology of the fuzzy analysis to boundary element method in potential theory. Applications were presented to potential problems with boundary conditions which are not sharply given and are characterised by fuzzy functions of interval type and trapezoidal-type membership functions.

A major conclusion is that fuzzy sets can be effectively used to estimate system uncertainty in boundary problems and random uncertainty can be calculated with a new technique called FBEM.

# 8. REFERENCES

1. Alefeld G., Herzberger J. (1983) Introduction to Interval Computations, Academic Press, New York
2. Aumann R., J. (1965) Integrals of Set-Valued Functions, J.Math.Anal.Appl. **12**, 1-12
3. Bauch H., Jahn K.,U., Oelschlagel, D., Susse, H., Wiebigke, V. (1987) Intervalmathematik, BSG B.G. Teubner Verlagsgeselschaft, Berlin
4. Brebbia C.,A., Dominguez J. (1989) Boundary Elements - An Introductory Course, Comp. Mechanics Publ., Southampton, Boston
5. Brebbia C., Telles J.,C.,F., Wrobel L.,C. (1984) Boundary Element Techniques - Theory and Applications in Engineering, Springer-Verlag, Berlin, Heidelberg, New York, Tokyo
6. Buckley J.,J. (1992) Solving Fuzzy Equations in Economics and Finance, Fuzzy Sets and Systems **48**, 289-296
7. Buckley J.,J. (1992) Solving Fuzzy Equations, Fuzzy Sets and Systems **50**, 1-14
8. Buckley J.,J., Qu Y. (1990) Solving Linear and Quadratic Fuzzy Equations, Fuzzy Sets and Systems **38**, 43-59
9. Buckley J.,J., Qu Y. (1990) On Using $\alpha$-Cuts To Evaluate Fuzzy Equations, Fuzzy Sets and Systems **38**, 309-312
10. Buckley J.,J., Qu Y. (1991) Solving Fuzzy Equations: A New Solution Concept. Fuzzy Sets and Systems **39**, 291-301
11. Buckley J.,J., Qu Y. (1991) Solving Systems of Linear Fuzzy Equations. Fuzzy Sets and Systems **43**, 33-43
12. Burczyński T. (1995) Metoda elementów brzegowych, WN-T, Warszawa
13. Burczyński T., Skrzypczyk J. (1995) The Fuzzy Boundary Element Method: A New Solution Concept. Proc. of XII Polish Conference on COMPUTER METHODS IN MECHANICS, Warsaw - Zegrze, Poland, 9-13 May, 65-66
14. Burczyński T., Skrzypczyk J. (1996) Stochastic And Fuzzy Aspects Of The Boundary Element Method, International Conference on Uncertain Structures: Analytical, Numer. and Experimental Methods, Cruise Ship in the Western Caribbean, March 3-10
15. Burczyński T., Skrzypczyk J. (1996) The Fuzzy Boundary Element Method: A New Methodology. Sci. Fasc. of Silesian Tech. Univ., ser. Civil Eng. **83**, Gliwice, 25-42
16. Burczyński T., Skrzypczyk J. (1997) Fuzzy Aspects of The Boundary Element Method. Engineering Analysis with Boundary Elements, Special Issue: Stochastic Boundary Element Methods **19**, 209-216
17. Burczyński T., Skrzypczyk J. (1997) The Boundary Element Method For Fuzzy Systems. Proc. of the IASTED International Conference on Modelling, Simulation and Optimization, Singapore, August 11-14, 24-27
18. Dubois D., Prade H. (1979) Fuzzy Real Algebra: Some Results. Fuzzy Sets and Systems **2**, 327-348
19. Dubois D., Prade H. (1988) An Approach to Computational Processing of Uncertainty, Plenum Press, New York, London

20. Felbin C. (1992) Finite Dimensional Fuzzy Normed Linear Space, Fuzzy Sets and Systems **48**, 239-248
21. Guang-Quan Z. (1991) Fuzzy Continuous Function and Its Properties, Fuzzy Sets and Systems **43**, 159-171
22. Kaleva O. (1987) Fuzzy Differential Equations, Fuzzy Sets and Systems **24**, 301-317
23. Kaleva O. (1990) The Cauchy Problem For Fuzzy Differential Equations, Fuzzy Sets and Systems **35**, 389-396
24. Mikhlin S.,G. (1986) Singular Integral Operators, Akademie-Verlag, Berlin
25. Moore R.,E. (1966) Interval Analysis, Prentice Hall, Englewood Cliffs
26. Nanda S. (1989) On Integration of Fuzzy Mappings, Fuzzy Sets and Systems **32**,95-101
27. Negoita C.,V., Ralescu D.,A. (1975) Applications of Fuzzy Sets to System Analysis, Ed. Technica, Birkhauser Verlag, Basel und Stuttgart
28. Neumaier A. (1990) Interval Methods for Systems of Equations, Cambridge University Press, Cambridge, New York, Port Chester, Melbourne, Sydney
29. Nguyen H.,T. (1978) A Note on the Extension Principle for Fuzzy Sets, J. of Math. Anal. Appl. **64**, 369-380
30. Shary S. P. (1996) Algebraic Approach to the Interval Linear Static Identification, Tolerance, and Control problems, or One more Application of Kaucher Arithmetic, Reliable Comp. **2**, 3-33
31. Skrzypczyk J. (1996) On Fuzzy Singular Integration. Sci. Fasc. of Silesian Tech. Univ. ser. Civil Eng. **83**, Gliwice, 121-130
32. Skrzypczyk J. (1997) Zadanie brzegowe teorii potencja³u w obszarze rozmytym. Materiały Konferencji naukowej z okazji 70-lecia urodzin Józefa Głomba pt. Wybrane problemy naukowo - badawcze mostownictwa i budownictwa, Gliwice 17 czerwca 1997, Wyd. Politechniki Śląskiej, Gliwice, 243-251
33. Skrzypczyk J., Burczyński T. (1997) The Fuzzy Boundary Element Method. Proc. of XIII Polish Conference on COMPUTER METHODS IN MECHANICS, Vol. 4, Poznań, Poland, 5-8 May, 1195-1202
34. Skrzypczyk J., Burczyński T. (1997) Fuzzy Aspects of The Boundary Element Method in Elastostatics. Proc. of XXXVI Sympozjon "Modelowanie w Mechanice", Wisła 16-20 luty 1997, Zeszyty Naukowe Katedry Mechaniki Stosowanej **3**, 25-30
35. Skrzypczyk J., Burczyński T. (1998) Metoda elementów brzegowych dla obszarów o kształcie rozmytym, Materiały XXVII Sympozjonu "Modelowanie w Mechanice", Wisła 9-13 lutego 1998, Zeszyty Naukowe Katedry Mechaniki Stosowanej **6**, 317-322.
36. Skrzypczyk J., Burczyński T. (1998) Fuzzy Boundary Element Method in the Analysis of Uncertain Systems, J. of Theoretical and Applied Mechanics **36**, 493-512
37. Skrzypczyk J., Burczyński T. (1998) Boundary Element Methods For Fuzzy Domain, Proc. of IABEM International Symposium on Boundary Element Methods, Ecole Polytechnique, Palaiseau (Paris), France, May 26-29
38. Skrzypczyk J., Witek H. (2000) Solution Interpretation in Fuzzy Boundary Problems – New Concepts, Proc. of XXXVI Sympozjon "Modelowanie w Mechanice", Wisła 14-18 luty 2000, Zeszyty Naukowe Katedry Mechaniki Stosowanej, 285-288

# A BOUNDARY INTEGRAL METHOD FOR TRANSIENT HEAT DIFFUSION IN ISOTROPIC NON-HOMOGENEOUS MEDIA

A. Stefanescu,(1) A. Kassab(2)
*Institute of Computational Engineering (ICE)*
*Department of Mechanical, Materials, and Aerospace Engineering*
*University of Central Florida, Orlando, Florida 32816-2450*
*EMail: (1)as93163@pegasus.cc.ucf.edu, (2)kassab@mail.ucf.edu*

**Abstract** This paper presents a Laplace transform-based BEM method for transient heat conduction in heterogeneous media. The method uses the generalized boundary integral method developed by Kassab and Divo[1] as a point of departure. The BIE is modified to solve the resulting modified variable coefficient Helmholtz problem in Laplace domain. Subsequently, the BEM solution is thereafter inverted to the time-domain using the Stehfest transform. Numerical examples for variable conductivity are presented and show agreement between BEM-predicted temperatures and analytical solutions.

**Key Words:** BEM, heat diffusion, non-homogeneous media.

## 1. Introduction

Heat diffusion problems in isotropic or anisotropic heterogeneous media have posed much difficulty when applying the Boundary Element Method (BEM). The Green free space solutions which are essential in the BEM are not known in general for general non-homogeneous media. Notable contributions to this problem are due to Shaw and his colleagues [2-3] and to Clements and his colleagues[4-5]. Recently Kassab and Divo[1] proposed a method to find a generalized fundamental solution, which is individually tailored to each particular case of variable conductivity. Using this generalized method it is possible to derive an integral equation in terms of boundary integrals only.

*T. Burczynski (ed.), IUTAM/IACM/IABEM Symposium on Advanced Mathematical and Computational Mechanics Aspects of the Boundary Element Method,* 365–375.

The Laplace transform BEM was first proposed by Rizzo and Shippy[6] and is adopted in the present work to transform the governing diffusion equation to Laplace domain. The generalized method of Kassab and Divo[1] is then implemented to solve the resulting modified variable coefficient Helmholtz equation resulting in an integral equation which contains boundary integrals as well as a domain integral. The domain integral is approximated by so-called radially symmetric expansion functions, which in turn are chosen to be the divergence of certain (vector-) functions. This allows the transformation of the domain integrals into boundary integrals. The numerical solution in Laplace space is non-transient and depends on the transform parameter. Upon applying the inverse transformation, one obtains the solution at a particular point in space as a function of time. The inversion is performed numerically.

## 2. Brief review of the generalized boundary element method for steady-state non-homogenous heat conduction

Steady state heat conduction in heterogeneous media is described by the following governing equation.

$$\nabla \cdot \left[ k(x) \, \nabla T(x) \right] = 0 \tag{1}$$

The temperature distribution $T(x)$ as well as the conductivity $k(x)$ are both functions of space, and here $x$ generically refers to space (1-D, 2-D, or 3-D). Following the usual procedure in boundary integral methods, the governing equation is multiplied by a so called fundamental solution $E(x, \xi)$ and the partial differential equation is integrated over the domain of interest $\Omega$. Using the Green's first and/or second identities, the following integral equation is derived

$$\int_{\Omega} T(x) \left[ \nabla \cdot k(x) \, \nabla E(x, \xi) \right] \, d\Omega + \oint_{\Gamma} \left( E(x, \xi) \, k(x) \frac{\partial T(x)}{\partial n} - T(x) \, k(x) \frac{\partial E(x, \xi)}{\partial n} \right) d\Gamma = 0 \tag{2}$$

In this notation the domain is denoted by $\Omega$, the boundary is denoted by $\Gamma$, and $\frac{\partial}{\partial n}$ represents the directional derivative normal to the domain boundary. Due to the variable conductivity the fundamental solution is no longer defined to be Green's free space solution for the heat conduction equation perturbed by the Dirac Delta function $\delta$, since this is not generally available. Kassab and Divo[1] introduce instead a generalized forcing function, defined by the following properties:

The Laplace transform BEM was first proposed by Rizzo and Shippy[6] and is adopted in the present work to transform the governing diffusion equation to Laplace domain. The generalized method of Kassab and Divo[1] is then implemented to solve the resulting modified variable coefficient Helmholtz equation resulting in an integral equation which contains boundary integrals as well as a domain integral. The domain integral is approximated by so-called radially symmetric expansion functions, which in turn are chosen to be the divergence of certain (vector-) functions. This allows the transformation of the domain integrals into boundary integrals. The numerical solution in Laplace space is non-transient and depends on the transform parameter. Upon applying the inverse transformation, one obtains the solution at a particular point in space as a function of time. The inversion is performed numerically.

## 2. Brief review of the generalized boundary element method for steady-state non-homogenous heat conduction

Steady state heat conduction in heterogeneous media is described by the following governing equation.

$$\nabla \cdot \left[k(x)\, \nabla T(x)\right] = 0 \tag{1}$$

The temperature distribution $T(x)$ as well as the conductivity $k(x)$ are both functions of space, and here $x$ generically refers to space (1-D, 2-D, or 3-D). Following the usual procedure in boundary integral methods, the governing equation is multiplied by a so called fundamental solution $E(x,\xi)$ and the partial differential equation is integrated over the domain of interest $\Omega$. Using the Green's first and/or second identities, the following integral equation is derived

$$\int_{\Omega} T(x)\, \left[\nabla \cdot k(x)\, \nabla E(x,\xi)\right]\, d\Omega + \oint_{\Gamma} \left( E(x,\xi)\, k(x) \frac{\partial T(x)}{\partial n} - T(x)\, k(x) \frac{\partial E(x,\xi)}{\partial n} \right) d\Gamma = 0 \tag{2}$$

In this notation the domain is denoted by $\Omega$, the boundary is denoted by $\Gamma$, and $\frac{\partial}{\partial n}$ represents the directional derivative normal to the domain boundary. Due to the variable conductivity the fundamental solution is no longer defined to be Green's free space solution for the heat conduction equation perturbed by the Dirac Delta function $\delta$, since this is not generally available. Kassab and Divo[1] introduce instead a generalized forcing function, defined by the following properties:

$$\nabla \cdot k(x)\, \nabla \overline{T}(x,s) - \rho\, c(x) s\, \overline{T}(x,s) = -\rho\, c(x)\, T(x,0) \tag{7}$$

where the term on the RHS can be neglected by simply re-defining the dependent variable as the temperature excess, $T(x,t) - T(x,0)$. Multiplying the above modified variable coefficient Helmholtz equation by the test function $E(x,\xi)$, integrating over the domain $\Omega$ and using Green's second identity, we arrive at

$$\int_{\Omega} \overline{T}(x,s)\, [\nabla \cdot k(x)\, \nabla E(x,\xi)]\, d\Omega - \int_{\Omega} \rho\, c(x)\, s\, \overline{T}(x,s)\, E(x,\xi)\, d\Omega$$
$$+ \oint_{\Gamma} \left( E(x,\xi)\, k(x) \frac{\partial \overline{T}(x,s)}{\partial n} - \overline{T}(x,s)\, k(x) \frac{\partial E(x,\xi)}{\partial n} \right) d\Gamma = 0 \tag{8}$$

The first domain integral can readily be eliminated by using the non-homogeneous fundamental solution for the steady state problem, see eqn (5), as the test function thereby eliminating the first domain integral. It is important to notice, however, that the main difficulty to be addressed in this approach arises from the second domain integral and is due to the fact that the function $E(x,\xi)$ is singular at the collocation point. In order to address this problem, the second domain integral can be manipulated to arrive at the following:

$$\int_{\Omega} \underbrace{\rho\, c(x)\, s\, E(x,\xi)}_{G(x,\xi)}\, \overline{T}(x)\, d\Omega = \tag{9}$$
$$\overline{T}(\xi) \underbrace{\int_{\Omega} G(x,\xi)\, d\Omega}_{B(\xi)} + \int_{\Omega} G(x,\xi) \left[ \overline{T}(x) - \overline{T}(\xi) \right] d\Omega$$

where the domain integral $B(\xi)$ depends on the source point. This integral is weakly singular as the integrand is logarithmically singular in 2-D and $1/r$ singular in 3-D. Naturally, we want to maintain the boundary-only feature of the method and seek to transform $B(\xi)$ into a boundary integral. The fundamental solution is radially symmetric, and if we require that it be equal to the divergence of a radially symmetric vector function $\vec{U}(r)$, then

$$B(\xi) = \int_{\Omega} \rho\, c\, s\, E(r,\xi)\, d\Omega = \oint \rho\, c\, s\, \vec{U}(r) \cdot \vec{n}\, d\Gamma \tag{10}$$

where in the case of constant conductivity

$$\vec{U}(r) = \left\{ \frac{1}{8\pi\, k} \right\} \cdot \vec{e}_r \tag{11}$$

and in the case of variable conductivity

$$\vec{U}(r) = \left\{ \frac{1}{3} + C_1 \frac{r}{3} arctan(C_1 r) + \frac{1}{6C_1^2} \frac{ln(1 + C_1^2 r^2)}{r^2} \right\} C_2 \cdot \vec{e}_r \quad (12)$$

where $C_1$ and $C_2$ are constants containing the expansion coefficients of the thermal conductivity which is modeled as quadratic in 3-D. Furthermore, the second domain integral in eqn (9) can be shown to exist. Neglecting the $(\theta, \phi)$-dependence of $\overline{T}$ close to $\xi$, we expand $\overline{T}(r)$ in a Taylor series about the location of the source point $\xi$,

$$\overline{T}(r) = \overline{T}(\xi) + \frac{1}{1!} \frac{\partial \overline{T}(\xi)}{\partial r} \epsilon + \frac{1}{2!} \frac{\partial^2 \overline{T}(\xi)}{\partial r^2} \epsilon^2 + \ldots + R(r) \quad (13)$$

where $\epsilon$ denotes the distance from the source point, and since the term containing the fundamental solution

$$G(r, \xi) \propto ln\epsilon \quad (14)$$

we obtain an estimate for the integral

$$\int_{\Omega} G(r, \xi) \left[ \overline{T}(r) - \overline{T}(\xi) \right] d\Omega \propto \int_{\Omega} \left( \epsilon \, ln\epsilon + \epsilon^2 \, ln\epsilon + \ldots \right) d\Omega \quad (15)$$

Since $(\epsilon, \epsilon^2, \ldots)$ go faster towards zero than $ln\epsilon$ goes towards infinity, the above integral is well behaved, and its existence is established.

With this, the second area integral in eqn (9) can be evaluated approximately with the aid of a series expansion

$$\epsilon'(\xi^i) = \int_{\Omega} G(x, \xi) \left[ \overline{T}(x) - \overline{T}(\xi) \right] d\Omega = \int_{\Omega} \sum_{j=1}^{N} \alpha_j f_j \, d\Omega \quad (16)$$

where the functions $f_j$ are chosen to be the divergence of radially symmetric vector-functions

$$f_j = \nabla \cdot \vec{u}_j \quad (17)$$

so that using the Green-Gauss Theorem we can transform the domain integral into a boundary integral. In order to determine the coefficients $\alpha_j$ we collocate the above expansion at the $N$ boundary nodes and obtain, for example when the source point is at node $i$

$$\underbrace{G(x_k, \xi) \left[ \overline{T}(x_k) - \overline{T}(\xi^i) \right]}_{h_k^i} = \sum_{j=1}^{N} \alpha_j^i f_j(x_k) \quad (18)$$

or in vector form,

$$\vec{h}^i = \boldsymbol{F}\,\vec{\alpha}^i \tag{19}$$

where $\boldsymbol{F}$ is the interpolation matrix, and $\vec{h}^i$ is a vector whose components contain the values of the function we want to approximate. In order to express the coefficients $\alpha^i_j$ in terms of the temperature, we invert the matrix $\boldsymbol{F}$

$$\vec{\alpha}^i = \boldsymbol{F}^{-1}\,\vec{h}^i \tag{20}$$

With this, we are able to evaluate $\epsilon'(\xi^i)$ using the results from above

$$\epsilon'(\xi^i) = \sum_{j=1}^{N} \alpha^i_j \underbrace{\oint_\Gamma \vec{u}_j \cdot \vec{n}\, d\Gamma}_{v_j} = \underbrace{(\vec{v}^T \boldsymbol{F}^{-1})}_{\vec{w}} \cdot \vec{h}^i = \vec{w} \cdot \vec{h}^i \tag{21}$$

Notice that the $i$-th entry in the above equation is zero. With this the expression for $\epsilon'(\xi^i)$ is given by

$$\epsilon'(\xi^i) = \sum_{\substack{j=1 \\ j \neq i}}^{N} w_j\, G(x_j, \xi)\, \overline{T}(x_j) - \overline{T}(\xi^i) \sum_{\substack{j=1 \\ j \neq i}}^{N} w_j\, G(x_j, \xi^i) \tag{22}$$

so that the initial domain integral is finally approximated by

$$\int_\Omega G(x,\xi)\, \overline{T}(x)\, d\Omega = \overline{T}(\xi^i) \Big( B(\xi) - \sum_{\substack{j=1 \\ j \neq i}}^{N} w_j\, G(x_j, \xi^i) \Big) \tag{23}$$

$$+ \sum_{\substack{j=1 \\ j \neq i}}^{N} w_j\, G(x_j, \xi)\, \overline{T}(x_j)$$

Introducing the above into the integral equation for the temperature in Laplace transform space, collecting terms of $T(\xi)$ and finally discretizing the boundary integrals by introducing boundary elements, leads to the following equation at each source point $i$ (interior or boundary point)

$$\overline{T}(\xi^i)\Big[A(\xi^i)+B(\xi^i)-\sum_{\substack{j=1\\ j\neq i}}^{N} w_j\,G(x_j,\xi^i)\Big]$$
$$+\sum_{\substack{j=1\\ j\neq i}}^{N} w_j\,G(x_j,\xi^i)\,\overline{T}(x_j)+\epsilon(\xi^i)+\sum_{j=1}^{N}\oint_{\Gamma}\overline{T}(x,s)\,k(x)\frac{\partial E(x,\xi^i)}{\partial n}d\Gamma$$
$$=\sum_{j=1}^{N}\oint_{\Gamma}E(x,\xi^i)\,k(x)\frac{\partial \overline{T}(x,s)}{\partial n}\,d\Gamma \tag{24}$$

As can be seen above, in addition to the amplification factor $A(\xi)$ inherent in the steady state method of Kassab and Divo[1], the diagonal element in the kernel-matrix will contain an additional amplification factor $B(\xi)$ and a sum. The amplification factor $A(\xi)$ can be evaluated through recourse to the so-called equipotential theorem, since we can make the argument that for very large times we should retain the boundary integral equation for steady state, and all the transient contributions should vanish. If in this case we further assume a constant temperature throughout the field, the heat flux is zero everywhere. Consequently, the diagonal element of the kernel matrix can be shown to have the negative value of the sum of its off-diagonal row-elements. The second amplification factor term $B(\xi)$ is evaluated using eqn. (10) with Gauss-type quadrature integration. Collocating the above equation at all the boundary points ($i=1,\ldots,N$), the equation is reduced to an algebraic set of the standard form

$$[H]\{\overline{T}(s)\}=[G]\{\overline{q}(s)\} \tag{25}$$

which is solved by standard methods given a value of the transform parameter, $s$.

## 4. Internal Points

As has been discussed in the previous section, domain integrals are approximated by a series expansion. Accuracy of the approximation depends on the number of sample points used, as well as their location in the domain of interest. It should be emphasized here that the accuracy of the approximation determines to a very large extent the accuracy of the present BEM as a whole. An extensive study has been undertaken in order to elucidate this effect, and the present authors found that it is of crucial importance to place sample points not only on the boundary but also in the interior of the domain ( $\rightarrow$ internal points). Doing this, however, does not alter the boundary-only feature of the method at all.

It is very important to place the sample points in such a way that regions with (expected) high gradients of the solution are being resolved accurately. A sufficiently large number of points will usually provide for a good representation of the integrand. This aspect is demonstrated in Fig. 1a-1c. This two dimensional example shows the approximation of a sinus-like function with radial basis functions, using sixteen sample points on the boundary. As a comparison, no internal points were used in Fig. 1a, three internal points were used in Fig. 1b and 25 internal points in Fig. 1c.

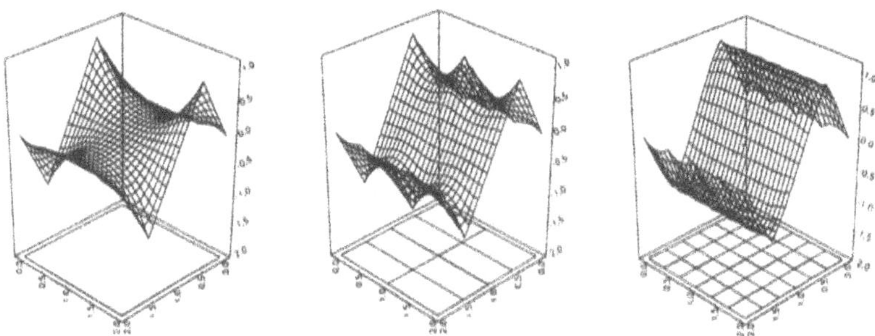

*Figure 1a:* no internal points. *Figure 1b:*3 internal points. *Figure 1c:* 16 internal points.

Clearly, the inclusion of a few internal points increases the quality of the approximation dramatically. For realistic problems, it turned out to be necessary to use a number of internal points equal or larger than the number of boundary elements. In the numerical computation presented here, the internal points were uniformly distributed throughout the interior domain.

## 5. Numerical Inversion of the Laplace Transform

Since we are dealing with a diffusion problem, it is not necessary to employ complex variable-based inversion methods. Indeed, the Stehfest Transform [7] is know to provide a very suitable means of inverting BEM-predicted temperatures back to the time-domain for the case of constant conductivity. Consequently, the Stehfest Transform method is adopted in this work. The inversion algorithm is given by:

$$f(t) \approx ln(\frac{2}{t})\sum_{n=1}^{N} K_n\, F\left(n\, ln(\frac{2}{t})\right)$$

$$K_n = (\text{-}1)^{n+N/2} \sum_{k=[(n+1)/2]}^{min(n,N/2)} \frac{k^{N/2}(2k)!}{(N/2-k)!\,k!\,(k-1)!\,(n-k)!(2k-n)!} \tag{26}$$

where N is even. Here $F()$ represents the Laplace Transform. The samples are computed for a number of time instants at which the solution is sought. Here the Laplace transform parameter "s" is real and explicitly provided by Eq. (26). From the numerical point of view, attention must be given to the magnitude of the transformation parameter "$s$". Since physical time $t$ and the transformation parameter $s$ are related to each other by the Laplace transform as ($s \Leftrightarrow 1/\sqrt{t}$), the transformation parameter will become very large values at short times leading to poorly conditioning of the BEM equations. This problem can be alleviated by scaling the right-hand-side of the BEM equations with the transformation parameter $s$, such that the RHS is always of the same order of magnitude as the LHS, that is, the algebraic set of equations is modified to

$$([H_{ss}] + s\,[H_{tr}])\,\{\overline{T}(s)\} = s\,[G_{ss}]\{\frac{1}{s}\overline{q}(s)\} \tag{27}$$

where $[H_{ss}]$, $[G_{ss}]$ and $[H_{tr}]$ are the influence coefficients (of the steady state and transient part respectively) which depend only on the geometry.

## 6. Numerical Results

We will present three examples for validation of the numerical algorithm. In the first example, we test our approach to approximating the domain integral and the inversion process. We consider transient heat conduction in a cube with homogeneous conductivity. The cube was assumed to be of side length L=0.1 m, and to have the material properties of copper ($\alpha$=1.165 x $10^{-4}$ $\frac{\mathrm{m}^2}{\mathrm{s}}$). The boundary was discretized using 96 boundary elements. A uniform temperature of T=0 °C throughout the domain of interest was chosen as initial condition. Using the exact solution

$$T(x,t) = T_0 + \sum_{n=1}^{\infty}\frac{2\,\mathrm{T_0}}{\lambda_n}[(-1)^n - 1]\,sin(\lambda_n x)e^{-\alpha\lambda_n^2 t} \tag{28}$$

first kind boundary conditions ($T_0$=10 °C) were imposed on two opposite faces, while second kind boundary conditions were imposed on all other faces (adiabatic in this case). Comparison between numerical and analytical solution for temperature along the mid-line of the cube in Fig. 2 show excellent agreement between the two.

In the second example we consider a block with a fillet and re-entrant corner, as depicted in Fig. 3 . The surface was discretized with 205 boundary elements, and 258 internal points were uniformly distributed in the interior domain. The material properties are the same as in the previous example, except for the conductivity which has a two-dimensional variation:

$$k(y,z) = 401 + 800\,y + 400\,z + 400\,yz + 400\,y^2 + 100\,z^2 \tag{29}$$

The following temperature distribution is a solution to the steady state heat conduction equation for the above variable conductivity:

$$T(y,z) = 30\,y - 60\,z \tag{30}$$

This solution was used to impose 1st- and 2nd kind boundary condition as shown in Fig. 3. The relative error, was defined as the difference between computed and analytical temperature, normalized by the maximum temperature difference occurring in the field (% relative error in $T = 100$ $|(T_{num} - T_{exact})/\triangle T_{max}||$), and was < 3 % in the interior domain. The computed solution is shown in the interior in two different planes parallel to the $(y, z)$ - plane in Fig. 4a-4b. The temperature is recovered accurately even in geometrically difficult regions such as close to the reentrant corner.

As a last example we draw attention to transient heat transfer in the cube used in the first example, this time with heterogeneous conductivity and assumed material properties of (density $\rho$=1, heat capacity $c_p$=1). The following solution to the transient heat conduction equation was used to impose 1st- and 2nd kind boundary conditions:

$$k(z) = 1 + 1.5\,z + 0.5\,z^2 \quad \text{and} \quad T(y, z) = (2z + 3)\,e^{t/(\rho c_p)} \qquad (31)$$

The initial condition for the block is then T=$[2z + 3]$ °C. The solution for the temperature along the middle line of the cube is given in Fig. 5, and the maximal relative error was less then 5.2 %. This concludes the comparison between numerical and analytical solutions.

## 7. Conclusions

In this work we present a BEM method for transient heat conduction in heterogeneous media. Using the Laplace transform, temperature solutions for arbitrary variations of conductivity can be obtained without time-marching. The computation effort is hence the same, regardless of the time at which the solution is sought. The numerical results show very good agreement with analytical solutions.

## 8. References

[1] Kassab, A.J. and Divo, E., 'A Generalized Boundary Integral Formulation for Diffusion Problems in Inhomogeneous Media', Chapter 2, in *Advances in Boundary Elements:Numerical and Mathematical Aspects*, Golberg, M.A., (ed.), Computational Mechanics, Boston, 1998, pp.37-76.

[2] Shaw, R.P., and Gipson, G.S., 'A BIE Formulation of Linearly Layered Potential Problems,' *Engineering Analysis*, 1996, Vol. 16, pp. 1-3.

[3] Shaw, R.P., Manolis, G., and Gipson, G.S., 'The 2-D Free-Space Green's Function and BIE for a Poisson Equation with Linearly Varying Conductivity in Two Directions,' *BETECH XI: Proc. 10th Int. Conf. on Boundary Element Technology*, Ertekin, R.C., Brebbia, C.A., Tanaka, M., and Shaw, R.P., (eds.), Computational Mechanics, Boston, 1996, pp. 327-334.

[4] Abdullah, H., Clements, D., and Ang, W.T, 'A Boundary Element Method for the Solution of a Class of Heat Conduction Problems for Non-homogeneous Media,' *Engineering Analysis*, Vol. 11, 1993, pp. 313-318.

[5] Clements, D.L., 'Fundamental Solutions for Second Order Linear Elliptic Partial Differential Equations,' *Computational Mechanics*, Vol. 22, No. 1, 1998, pp. 26-31.
[6] Rizzo, F. and Shippy, D. J., 'A Method of Solution For Certain Problems of Transient Heat Conduction', *AIAA J.*, 1970, 8, pp. 2004-2009.
[7] Stehfest, H., *Comm. ACM 13*, pp. 47-49 and pp. 624, 1970.

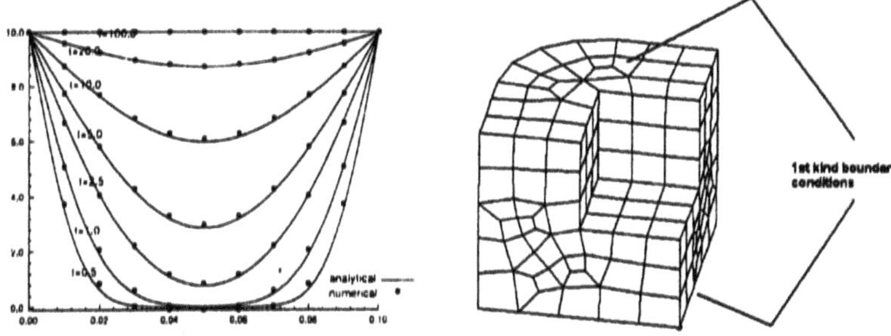


*Figure 2:* Plot of time history of mid-line. *Figure 3:* Cube with fillet and corner removed.

*Figure 4a*: Solution at $\frac{x}{L}$=15% *Figure 4b:* Solution at $\frac{x}{L}$=60%

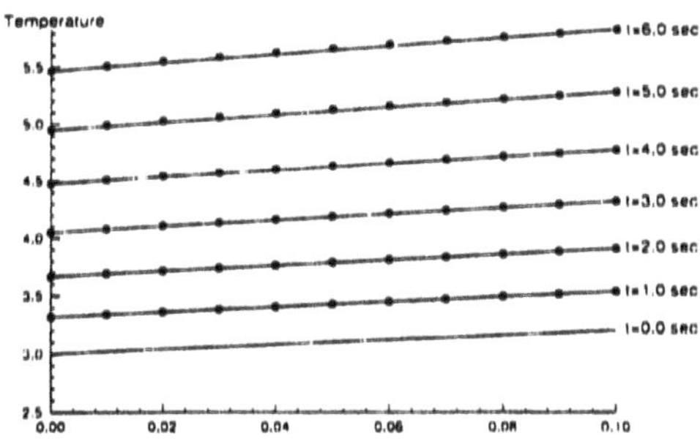


*Figure 5* : Temperature along mid-line of the block.

# A COMPARATIVE STUDY OF THREE SYSTEMS OF BOUNDARY INTEGRAL EQUATIONS IN THE POTENTIAL THEORY

Roman Vodička
*Faculty of Mechanical Engineering, Technical University of Košice*
*Letná 9, 041 87 Slovakia*
vodicka@tuke.sk

Vladislav Mantič
*Escuela Superior de Ingenieros, Universidad de Sevilla*
*Camino de los Descubrimientos s/n, Sevilla, 41092 Spain*
mantic@esi.us.es

**Keywords:** Boundary integral equation; The first kind symmetric BIE system; The second kind BIE system; Boundary element method; Direct BEM; Symmetric Galerkin BEM

**Abstract** A numerical study of Laplace equation solution by the direct BEM in 2D is presented. The following three basic systems of BIEs: the conventional BIE of potential, the first kind symmetric BIE system and the second kind BIE system, are compared. Error behaviour along the boundary and error convergence in $L_2$ and maximum norms for $h$-refinements of BEM meshes are studied and discussed.

## 1. INTRODUCTION

From many different Boundary Integral Equations (BIEs) deduced in the framework of potential theory, elasticity, acoustic, etc., only a few of them are actually applied to numerical solution of engineering problems by the Boundary Element Method (BEM). Several theoretical and numerical comparative studies of different BIEs have already been carried out [4, 3, 14, 20, 23]. Nevertheless, in the opinion of the present

*T. Burczynski (ed.), IUTAM/IACM/IABEM Symposium on Advanced Mathematical and Computational Mechanics Aspects of the Boundary Element Method*, 377–393.

authors, there are still gaps in understanding their relative performance and suitability for engineering applications.

Consider the two basic BIEs of the potential theory for the so-called Direct BEM. First, the conventional BIE, denoted hereinafter as $u$-BIE, which is obtained by performing the limit-to-the boundary of the Green Boundary Integral Representation (BIR) of potential in the interior of the domain. Second, the normal flux BIE, denoted hereinafter as $\partial_n u$-BIE, which is obtained by performing the limit-to-the boundary of the BIR of normal flux in the interior of the domain. Applying one of these BIEs on the whole boundary, or one BIE on the Dirichlet part and the other BIE on the Neumann part of the boundary, the four basic systems of BIEs of the so called Direct BEM can be formulated in a natural manner [23]. The aim of this work is to contribute to comparison and understanding of error behaviour of these basic systems of BIEs.

The first BIE system studied here is given by $u$-BIE. The other two systems studied here are obtained by a combination of $u$-BIE and $\partial_n u$-BIE, the choice of BIE applied at a point being dependent on the kind of the boundary condition prescribed at this point. This choice is complementary in both coupled systems of BIEs. In this way, the first kind symmetric BIE system is obtained when $u$-BIE is applied on Dirichlet part and $\partial_n u$-BIE on Neumann part of the boundary, and vice-versa in obtaining the second kind BIE system. The BIE system obtained applying $\partial_n u$-BIE on the whole boundary does not present any apparent advantage over the three remaining BIE systems and, therefore, it is not studied here.

The above described three basic BIE systems have very different properties as far as their suitability for application in practice is concerned, e.g. complexity of their implementations in a BEM code, symmetry, conditioning of the resulting linear system, etc. It is remarkable that each of these three basic systems presents some advantages over two remaining.

In this paper a numerical comparative study of BEM solutions of the above described three basic BIE systems in two dimensions is presented. In particular, error behaviour along the boundary and error convergence in $L_2$ and maximum norms for a (quasi)uniform $h$-refinement of BEM meshes are presented for two mixed Boundary Value Problems (BVPs). These BVPs have been selected from a more extensive series of BVPs analyzed in [27]. Other theoretical and numerical comparative studies of the basic BIE systems in elasticity were presented in [13, 23, 25, 26].

# 2. THREE BASIC BIE SYSTEMS OF THE DIRECT BEM

## 2.1. PRELIMINARIES

Let $\Omega \subset R^2$ be a domain with a Lipschitz piecewise smooth boundary $\Gamma$. Let $\Gamma_u$ and $\Gamma_q$ define Dirichlet and Neumann boundary parts respectively, $\Gamma = \overline{\Gamma}_u \cup \overline{\Gamma}_q$ (an overline means the closure of a set), $\Gamma_u \cap \Gamma_q = \emptyset$. Consider a mixed BVP for Laplace equation for potential function $u$:

$$\Delta u = 0 \quad \text{in } \Omega \tag{1}$$

$$u|_{\Gamma_u} = \bar{u} \qquad q|_{\Gamma_q} = \partial_n u|_{\Gamma_q} = \bar{q} \tag{2}$$

In definition of the normal flux $q$ appears the unit outward normal $n$ to $\Gamma$. Dirichlet boundary condition $\bar{u}$ is supposed to be extended in a smooth way to $\Gamma$. Support of $\bar{u}$ is denoted as $\Gamma_{\bar{u}} \subset \Gamma$. Potential solution $u$ of (1-2) can be written as $u = \bar{u} + \mathring{u}$, where obviously supp $\mathring{u} \subset \Gamma_q$

Let the fundamental solution of Laplace equation in $R^2$ be defined as $U(x,y) = -(2\pi)^{-1} \ln r$, $r = \|y - x\|$. The following singular integral operators can be defined for sufficiently smooth functions $u$ and $q$:

$$\begin{aligned}
(V_{\alpha\beta} q)(x) &= \int_{\Gamma_\beta} U(x,y) q(y) \mathrm{d}\Gamma(y) && (3)\\
(K_{\alpha\beta} u)(x) &= \int_{\Gamma_\beta} \partial_{n_y} U(x,y) u(y) \mathrm{d}\Gamma(y) && (4)\\
\left(K'_{\alpha\beta} q\right)(x) &= \int_{\Gamma_\beta} \partial_{n_x} U(x,y) q(y) \mathrm{d}\Gamma(y) && (5)\\
(D_{\alpha\beta} u)(x) &= -\int_{\Gamma_\beta} \partial^2_{n_x n_y} U(x,y) u(y) \mathrm{d}\Gamma(y) && (6)
\end{aligned}$$

where $x \in \Gamma_\alpha$ with exception of boundary corners, $\alpha, \beta = u, \bar{u}, q$. Note that for $\alpha = \beta$ the hyper-singular integral in (6) has to be defined in the sense of the finite part.

It can be shown that some generalizations (in the distributional sense) of the above integral operators defined almost everywhere (a.e.), which coincide with these operators for sufficiently smooth functions, are continuous in the following Sobolev functional spaces [5, 6, 7]:

$$\begin{matrix}
V_{\alpha\beta} : \tilde{H}^{-\frac{1}{2}}(\Gamma_\beta) \to H^{\frac{1}{2}}(\Gamma_\alpha) & K_{\alpha\beta} : \tilde{H}^{\frac{1}{2}}(\Gamma_\beta) \to H^{\frac{1}{2}}(\Gamma_\alpha) \\
K'_{\alpha\beta} : \tilde{H}^{-\frac{1}{2}}(\Gamma_\beta) \to H^{-\frac{1}{2}}(\Gamma_\alpha) & D_{\alpha\beta} : \tilde{H}^{\frac{1}{2}}(\Gamma_\beta) \to H^{-\frac{1}{2}}(\Gamma_\alpha)
\end{matrix} \tag{7}$$

where

$$H^{\frac{1}{2}}(\Gamma_\alpha) = \left\{ u|_{\Gamma_\alpha}, u \in H^{\frac{1}{2}}(\Gamma) \right\} \quad \tilde{H}^{\frac{1}{2}}(\Gamma_\alpha) = \left\{ u \in H^{\frac{1}{2}}(\Gamma_\alpha), \tilde{u} \in H^{\frac{1}{2}}(\Gamma) \right\} \tag{8}$$

$\tilde{u}$ means continuation of $u$ by 0 outside $\Gamma_\alpha$, $\tilde{u} = u$ on $\Gamma_\alpha$, $\tilde{u} = 0$ on $\Gamma \setminus \Gamma_\alpha$

$$H^{-\frac{1}{2}}(\Gamma_\alpha) = \left(\tilde{H}^{\frac{1}{2}}(\Gamma_\alpha)\right)' \quad \text{(dual space)} \qquad \tilde{H}^{-\frac{1}{2}}(\Gamma_\alpha) = \left(H^{\frac{1}{2}}(\Gamma_\alpha)\right)' \quad (9)$$

## 2.2. CONVENTIONAL U-BIE SYSTEM. COLLOCATION BEM

Consider $u$-BIE in the following form, which holds a.e. for $x \in \Gamma$ (e.g., for sufficiently smooth functions it holds with exception of corner points):

$$\frac{1}{2}u(x) + \int_\Gamma \partial_{n_y} U(x,y)u(y)\mathrm{d}\Gamma_y - \int_\Gamma U(x,y)q(y)\mathrm{d}\Gamma_y = 0 \qquad (10)$$

By partition of boundary to $\Gamma_u$ and $\Gamma_q$, application of boundary conditions (2) and definitions (3-4), $u$-BIE system for unknowns $\mathring{u}\,|_{\Gamma_q}$ and $q|_{\Gamma_u}$ can be written as follows:

$$\begin{pmatrix} \frac{1}{2} + K_{qq} & -V_{qu} \\ -K_{uq} & V_{uu} \end{pmatrix} \begin{pmatrix} \mathring{u}\,|_{\Gamma_q} \\ q|_{\Gamma_u} \end{pmatrix} = \begin{pmatrix} V_{qq} & -K_{q\bar{u}} \\ -V_{uq} & \frac{1}{2} + K_{u\bar{u}} \end{pmatrix} \begin{pmatrix} \bar{q} \\ \bar{u} \end{pmatrix} \qquad (11)$$

This system was introduced in [11], and its discretization by collocation approach and constant boundary elements was developed in [21]. It is remarkable that the main stream of BEM development till present date has been related with solution of (11) or its analogy in elasticity [17].

Main advantages of a Collocation BEM applied to (11) are:

- An easy theoretical formulation and implementation of BEM discretization through a collocation procedure.

- Reliable results guaranteed by a large and successful experience with its application to solution of engineering and physical problems.

Its main disadvantages are:

- The fact that it is insufficient for an efficient solution of screen and crack problems.

- Its mixed character (first and second kind BIEs on Dirichlet and Neumann boundary parts respectively), which makes more difficult, first, its mathematical analysis, and second, application of iterative methods to solution of the linear algebraic system.

- Difficult mathematical analysis of convergence of collocation procedure by Lagrange boundary elements.

Convergence proofs for solution of (11) by a collocation approach were given in [18] in case of a smooth $\Gamma$ and constant (and also linear) elements, and in [7] for Dirichlet and Neumann problems in case of non-smooth $\Gamma$ and linear elements.

## 2.3. THE FIRST KIND SYMMETRIC BIE SYSTEM. SYMMETRIC GALERKIN BEM

Consider $\partial_n u$-BIE in the following form:

$$\frac{1}{2}q(x) - \int_\Gamma \partial_{n_x} U(x,y)q(y)\mathrm{d}\Gamma_y + \int_\Gamma \partial_{n_x}\partial_{n_y} U(x,y)u(y)\mathrm{d}\Gamma_y = 0 \quad (12)$$

at $x \in \Gamma_q$, and $u$-BIE (10) at $x \in \Gamma_u$, with exception of corner points. By partition of boundary, application of boundary conditions (2) and definitions (3-6), the first kind BIE system can be written as:

$$\begin{pmatrix} -D_{qq} & -K'_{qu} \\ -K_{uq} & V_{uu} \end{pmatrix} \begin{pmatrix} \mathring{u}|_{\Gamma_q} \\ q|_{\Gamma_u} \end{pmatrix} = \begin{pmatrix} -\frac{1}{2} + K'_{qq} & D_{q\bar{u}} \\ -V_{uq} & \frac{1}{2} + K_{u\bar{u}} \end{pmatrix} \begin{pmatrix} \bar{q} \\ \bar{u} \end{pmatrix} \quad (13)$$

Discretization of (13) by a Symmetric Galerkin BEM (SGBEM) was introduced in [19] and also in [9, 22]. For a recent review on development and applications of SGBEM, see [2].

Main advantages of SGBEM are:

- Symmetric linear system of algebraic equations with positive and negative definite diagonal blocks, which provides an effective solution by direct as well as iterative methods.
- It is a natural approach, with reasonable continuity requirements on BEM approximations of variables, for solution of screen and crack problems.
- Mathematical analysis of convergence is relatively easy and guarantees nice convergence properties of this approach, see, e.g., [4, 3, 10] and references therein.
- Solution evaluated at interior points of the domain by a BIR using as the integral density a SGBEM solution on the boundary has a higher convergence order than that obtained using a Collocation BEM solution of $u$-BIE [4, 8].

Its main disadvantages are:

- Theoretical formulation and computational implementation is more complicated than in Collocation BEM applied to $u$-BIE: double integration over the boundary, discretization of the hyper-singular $\partial_n u$-BIE.
- Condition number of the linear system blows up with a higher rate than for $u$-BIE, see [4] for a theoretical explanation and [24] for numerical results.
- Non-unique solution of (13) in case of a multi-connected domain $\Omega$ when Neumann condition is prescribed on the boundary of a hole. In this case two solutions of (13) differ by a constant on the hole boundary.

An approach for removing this non-uniqueness was presented in [10], see also in [27].

## 2.4. THE SECOND KIND BIE SYSTEM. COLLOCATION BEM

Considering $u$-BIE (10) on $\Gamma_q$ and $\partial_n u$-BIR (12) on $\Gamma_u$ the following second kind BIE system can be derived:

$$\begin{pmatrix} \frac{1}{2}+K_{qq} & -V_{qu} \\ -D_{uq} & \frac{1}{2}-K'_{uu} \end{pmatrix} \begin{pmatrix} \mathring{u}|_{\Gamma_q} \\ q|_{\Gamma_u} \end{pmatrix} = \begin{pmatrix} V_{qq} & -K_{q\bar{u}} \\ K'_{uq} & D_{u\bar{u}} \end{pmatrix} \begin{pmatrix} \bar{q} \\ \bar{u} \end{pmatrix} \tag{14}$$

This system was introduced in [12]. It is closely related to the classical second kind BIEs of potential theory which, defining simple layer density as $\tau$ and double layer density as $\mu$, can be written for mixed BVP (1-2) in the following way:

$$\begin{pmatrix} \frac{1}{2}+K'_{qq} & D_{qu} \\ V_{uq} & \frac{1}{2}-K_{uu} \end{pmatrix} \begin{pmatrix} \tau|_{\Gamma_q} \\ \mu|_{\Gamma_u} \end{pmatrix} = \begin{pmatrix} \bar{q} \\ \bar{u} \end{pmatrix} \tag{15}$$

Equation (15) was formulated in the case of elasticity in [1, 16, 23], Mazya and co-workers extensively studied such systems for both potential theory and elasticity, for a review see [14].

Main advantages of a Collocation BEM applied to (14) are:

- Very low condition number and nice spectral properties of the linear system of algebraic equations [12, 16], which provides an effective solution by the iterative methods based on Krylov subspaces [24].

- Multi-domain decomposition method can be implemented in a very effective way, when no singular integrals on interfaces (associated to diagonal blocks in the linear system of algebraic equations) are computed [13].

Its main disadvantages are:

- Computational implementation is more complicated than for $u$-BIE system, because of discretization of $\partial_n u$-BIE.

- Non-unique solution of (14) in case of a multi-connected domain $\Omega$ when Dirichlet condition is prescribed on the boundary of a hole. In this case two solutions of (14) differ by a function from the nullspace of the integral operator $1/2 - K'_{uu}$ restricted to the hole boundary, considering the normal vector directed inwards the hole. An approach for removing this non-uniqueness was presented in [27].

- A way of an efficient application to screen and crack problems is unclear.

- Difficult mathematical analysis of convergence of collocation procedure by Lagrange boundary elements.

# 3. NUMERICAL EXAMPLES

## 3.1. COMPUTATIONAL IMPLEMENTATION OF BEM

Boundary has been discretized by straight boundary elements. For approximation of potential $u$ and normal flux $q$ a consistent linear-constant approximation, introduced in [15], have been used. Potential has been approximated by a piecewise linear continuous function $u_h$, and normal flux by a piecewise constant function $q_h$.

A collocation approach has been used for discretization of $u$-BIE system (11) and the second kind BIE system (14). Collocation points have been defined by element end-points at Neumann boundary part, and by element centres at Dirichlet boundary part. Galerkin method has been used for discretization of the first kind BIE system (13).

Consistent boundary elements, with different approximations of $u$ and $q$, have been used in order to obtain a simple and similar way of discretization, with the same number of algebraic equations, in all formulations. It has to be stressed that the $\partial_n u$-BIE cannot be collocated at element end-points because of the presence of the hypersingular kernel.

All integrals, regular and singular, have been computed analytically [24].

## 3.2. EVALUATION OF NUMERICAL RESULTS

In order to show error behaviour along the boundary in the following numerical tests, absolute errors, $u_h - u$ and $q_h - q$ have been ploted.

The following two norms have been used for error convergence analysis. First, $L_2(\Gamma)$ norm, denoted as $\|\cdot\|_{L_2}$ is computed over the original boundary using Gaussian quadrature with 10 points at each boundary segment defined by two nodes. Second, a discretized maximum norm $\|f\|_{\max} = \max_{x \in S} |f(x)|$, where $S$ is a set of element end-points or centres for potential or normal flux respectively. The following relative errors $\|u_h - u\|/\|u\|$ and $\|q_h - q\|/\|q\|$ are used in log-log plots of convergence. Average convergence values are computed by a least square procedure from these plots. Taking into account approximations of $u$ and $q$ applied here, a-priori estimates of error imply that linear and quadratic convergence rates of error can be expected for $q$ and $u$ respectively.

## 3.3. EXAMPLE 1: ELLIPTIC DOMAIN

Dirichlet, Neumann and mixed BVPs have been solved on an elliptic domain, Fig. 1, with the following sufficiently general and smooth analytic solution

$$u = e^{x+1}\cos(y-1) + e^{x^2-(y+1)^2}\sin(2x(y+1)) + (x-2)^2 - 3(x-2)y^2 \quad (16)$$

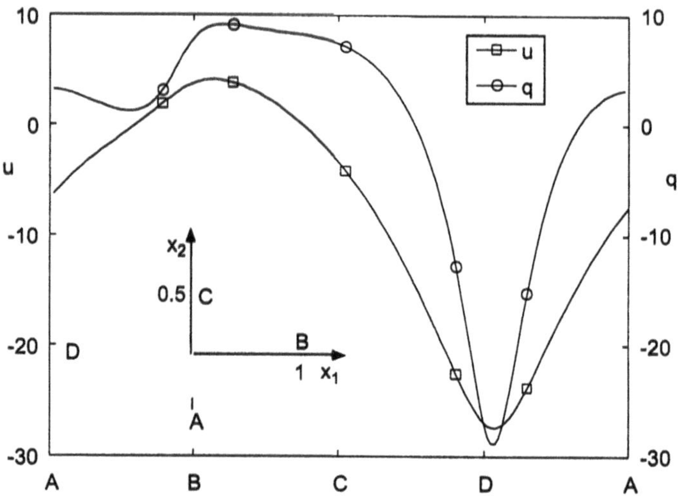


*Figure 1* Elliptic domain. Partition of the boundary for the mixed BVP. Analytic solution of potential $u$ and normal flux $q$.

Quasiuniform boundary element meshes have been defined by uniform partitions of the angle parameter $\theta$ in the natural parameterization of the ellipse $(x_1, x_2) = (\cos\theta, 0.5\sin\theta)$. Meshes with 12, 36, 60, 84 and 108 elements have been used.

Distributions of absolute errors obtained for a sequence of three boundary element discretizations of the mixed BVP are presented in Fig. 2. Relative error norms for all discretizations of the mixed BVP are plotted in log-log scale in Fig. 3. Average convergence rates for Dirichlet, Neumann and mixed BVPs are presented in Tab. I. Convergence behaviours of $u$-BIE and the first kind BIE systems obtained are very similar. Nevertheless, in general, convergence rates and relative error values are slightly better for the first kind BIE system, an exception being the case of maximum norm of error in $q$ in Neumann problem. Convergence rates obtained by these two BIE systems suit well expected convergence rates. An exception is the case of maximum norm in $q$ where a quadratic superconvergence is observed.

On the other hand, convergence behaviour of the second kind BIE system is in general worse than in the remaining two BIE systems. In particular, convergence rates of $u$ in the mixed BVP, about 1.25, are

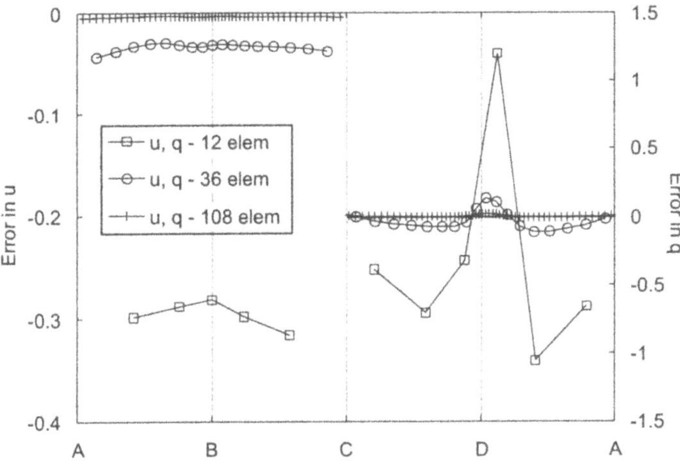


*Figure 2* Absolute errors for the mixed BVP in elliptic domain. *a)* $u$-BIE system.

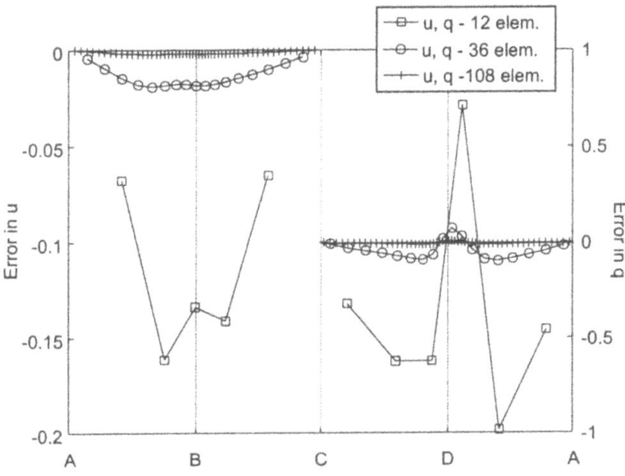


*Figure 2* *b)* The first kind BIE system.

lower than the expected quadratic convergence, and no superconvergence is observed in maximum norm of $q$. Additionally, in the mixed BVP, small divergent peaks in error of $q$ can be observed at points where the change of boundary condition type takes place, see Fig. 2c. These peaks have a negative effect on convergence behaviour of the second kind BIE system in maximum norm, see Fig. 3b.

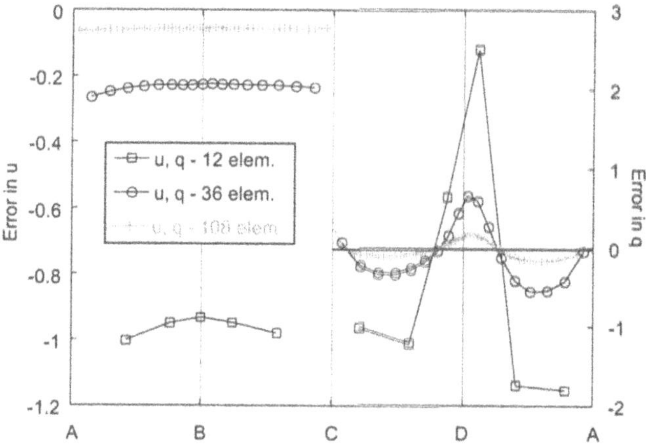


*Figure 2* *c)* The second kind BIE system.

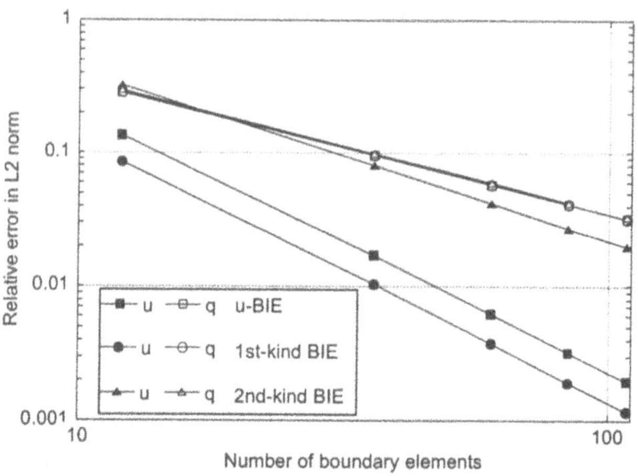


*Figure 3* Convergence of the relative error for the mixed BVP in elliptic domain. *a)* $L_2$ norm.

## 3.4. EXAMPLE 2: TORSION OF A RECTANGULAR BEAM

Torsion problem of a rectangular beam can be solved as a Neumann problem for a warping function $u$ which verifies Laplace equation. Applying symmetry conditions along the symmetry axes of the beam section, a mixed BVP defined over a quarter of beam section can be ob-

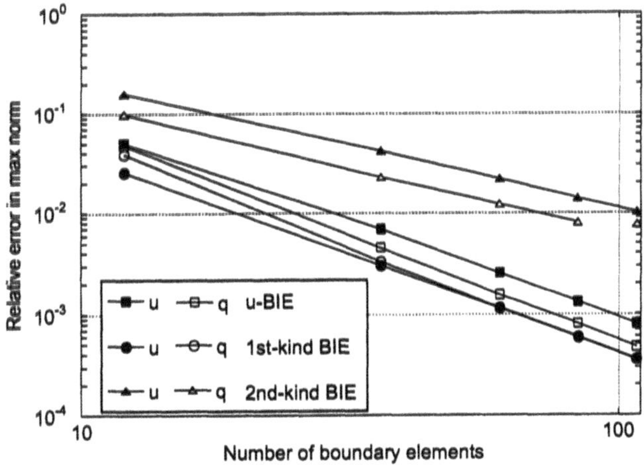


*Figure 3* *b)* Maximum norm.

*Table 1* Average convergence rates. Elliptic domain.

| | Dirichlet | | Neumann | | mixed BVP | | | |
|---|---|---|---|---|---|---|---|---|
| | $L_2$ | $Max$ | $L_2$ | $Max$ | $L_2$ | | $Max$ | |
| | $q$ | $q$ | $u$ | $u$ | $u$ | $q$ | $u$ | $q$ |
| $u$-BIE | 0.99 | 2.09 | 2.01 | 2.21 | 1.92 | 0.99 | 1.88 | 2.09 |
| 1st kind BIE | 0.99 | 2.12 | 2.04 | 1.80 | 1.95 | 0.99 | 1.94 | 2.13 |
| 2nd kind BIE | 1.00 | 1.24 | 2.01 | 2.21 | 1.26 | 0.99 | 1.25 | 1.20 |

tained, Fig. 4. Analytical solution of this mixed BVP problem, which is studied here, can be expressed using a series expansion. The above mixed BVP has been solved using uniform meshes with 48, 96, 144, 192, 240 and 288 boundary elements. Distributions of absolute errors obtained for two coarsest meshes are presented in Fig. 5. Relative error norms for all meshes are plotted in log-log scale in Fig. 6. Average convergence rates are presented in Tab. 2.

Convergence rates of error suit well expected convergence rates for all BIE systems, the only exception being slightly less values, about 1.7, obtained for maximum norm of error in $u$. It is remarkable that, the first kind BIE system gives on the one hand the best values of error in $u$ in $L_2$ norm, and on the other hand the worst values of errors in both $u$ and $q$ in maximum norm. Another outstanding observation is that the second kind BIE system gives similar error norm values as $u$-BIE

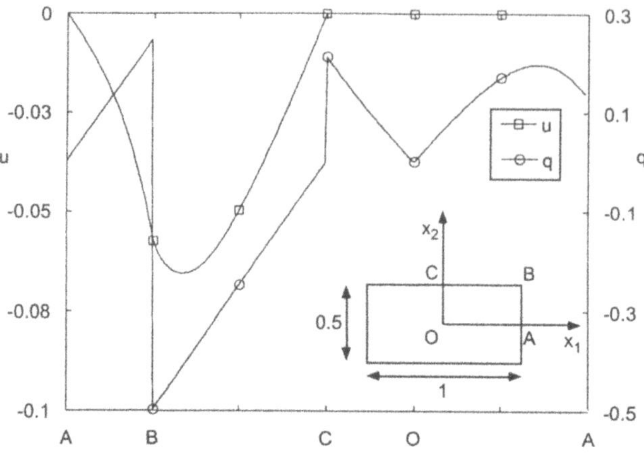


*Figure 4* Mixed BVP for torsion of a rectangular beam. Analytic solution of potential $u$ and normal flux $q$.

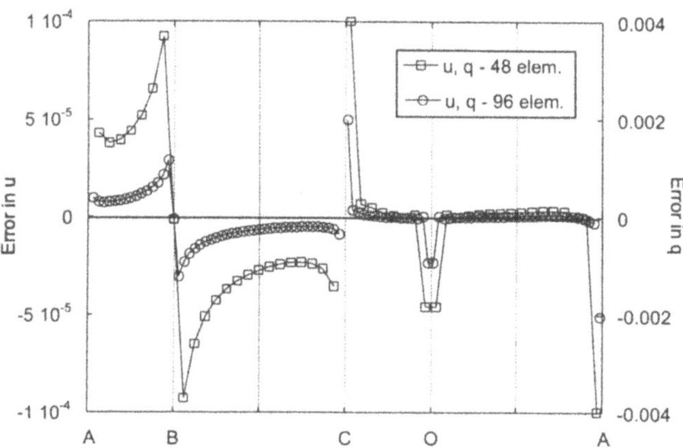


*Figure 5* Absolute errors for the mixed BVP for torsion of a rectangular beam. *a)* $u$-BIE system.

system, the only exception being the case of maximum norm of error in $q$, where the second kind BIE system gives clearly the best values.

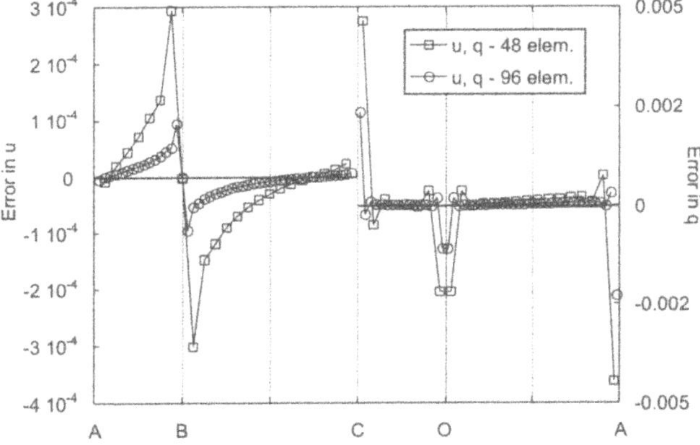


*Figure 5 b)* The first kind BIE system.

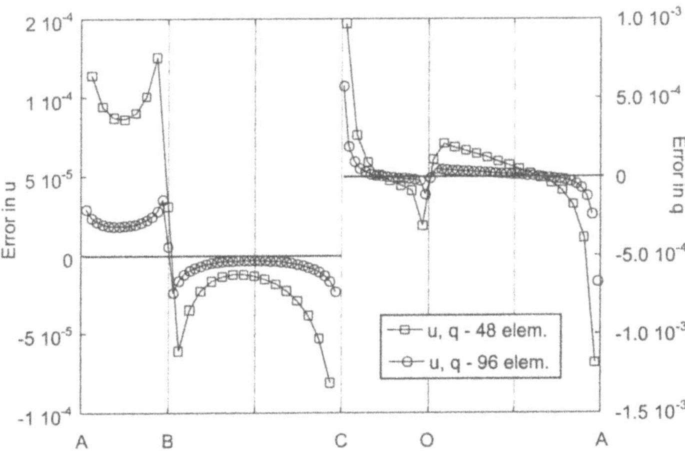


*Figure 5 c)* The second kind BIE system.

# 4. CONCLUDING REMARKS

Although only results of a limited series of numerical tests have been presented here, some interesting conclusions about characteristic features of error behaviour and convergence of each basic BIE system stud-

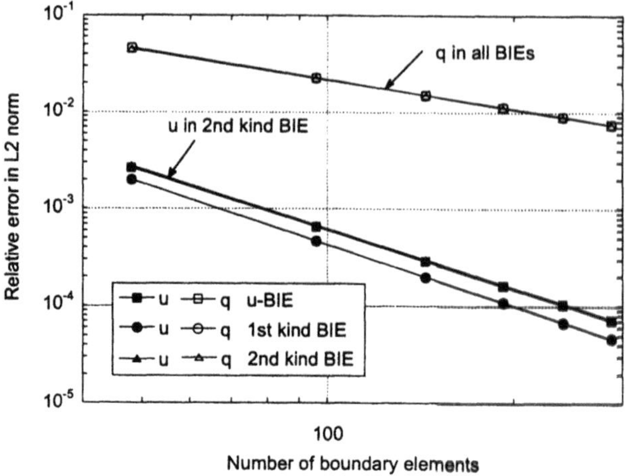


*Figure 6* Convergence of the relative error for the mixed BVP for torsion of a rectangular beam. *a)* $L_2$ norm.

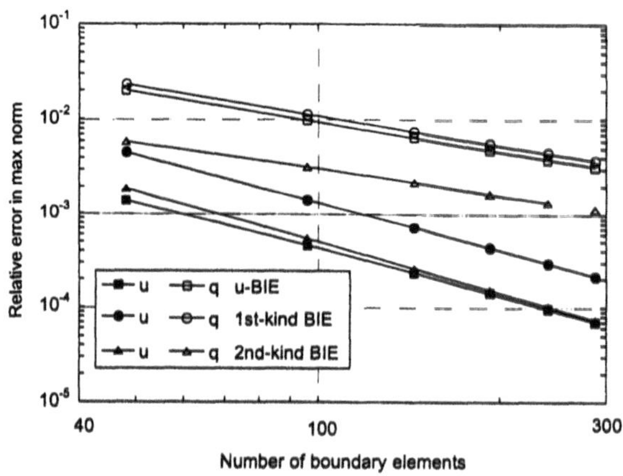


*Figure 6* *b)* Maximum norm.

ied here can be made. It should be pointed out that these conclusions agree with the results of a more extensive series of numerical tests in [27].

Results of $u$-BIE system show that it is a reliable approach, which never has been the worst, but sometimes it can be the best one. This is certainly a satisfactory conclusion for BEM community because this approach represents the main stream in BEM development and applications at present.

*Table 2* Average convergence rates. Mixed BVP for torsion of a rectangular beam.

| | mixed BVP | | | |
|---|---|---|---|---|
| | $L_2$ | | $Max$ | |
| | $u$ | $q$ | $u$ | $q$ |
| $u$-BIE | 2.01 | 1.00 | 1.67 | 1.02 |
| 1st kind BIE | 2.09 | 1.01 | 1.71 | 1.02 |
| 2nd kind BIE | 2.02 | 0.99 | 1.82 | 0.92 |

When errors are evaluated in the integral $L_2$ norm, then the SGBEM solution of the first kind BIE system gives clearly the lowest values of error in $u$ and the same values of error in $q$ as the remaining two approaches. Nevertheless, sometimes SGBEM can give the worst value of error in the maximum norm. This fact could represent an important disadvantage of this approach for engineering applications, considering, e.g., that maximum values of stresses are applied in failure criteria.

The results of the second kind BIE system show that it is the less reliable approach, though sometimes it can give the lowest error values.

## Acknowledgments

The work of the second author was partially supported by the Spanish Ministerio de Educación y Cultura under grant PB98-1118.

## References

[1] Altiero, N.J. and Gavazza, S.D. On a Unified Boundary-Integral Equation Method. *J. Elasticity*,10(1):1–9, 1980.

[2] Bonnet, M., Maier, G. and Polizzotto, C. Symmetric Galerkin Boundary Element Method. *Applied Mechanics Review*, 51(11):669–704, 1998.

[3] Chen, G. and Zhou, J. *Boundary Element Method.* Academic Press, London, 1992.

[4] Costabel, M. Principles of Boundary Element Methods. *Computer Physics Reports*, 6:243–274, 1987.

[5] Costabel, M. Boundary Integral Operators on Lipschitz Domains: Elementary Results. *SIAM J. Math. Anal.*, 19(3):613–626, 1988.

[6] Costabel, M. and Stephan, E. Boundary Integral Equations for Mixed Boundary value Problems in Polygonal Domains and Galerkin Approximations. In W. Fiszdon and K. Wilmanski (eds.),

*Mathematical Models and Methods in Mechanics 1981*, Warszaw, pp. 175–251. Banach Center Publications Vol. 15, Polish Scientific Publishers, 1985.

[7] Costabel, M. and Stephan, E. On the Convergence of Collocation Methods for Boundary Integral equations on Polygons. *Mathematics of Computation*, 49(180):461–478, 1987.

[8] Hartmann, F. *Introduction to Boundary Elements.* Springer-Berlag, Berlin, 1989.

[9] Hartmann, F., Katz, C. and Protopsalis, B. Boundary Elements and Symmetry. *Ingenieur-Archiv*, 55:440–449, 1985.

[10] Hsiao, G.C. and Wendland, W.L. On a Boundary Integral Method for Some Exterior Problems in Elasticity. *Proc. Tbilisi Univ., UDK 539.3, Math. Mech. Astron.*, 257:31–60, 1985.

[11] Jaswon, M. A. Integral Equation Methods in Potential Theory. I. *Proc. Roy. Soc.*, 275(A):23–32, 1963.

[12] Khutorianskii, N. M. Boundary Integral and Integro-Differential Equations of the Second Kind for a Mixed Boundary Value Problem of the Theory of Elasticity (in Russian). *Applied Problems of Strength and Plasticity*, 3–13, 1981.

[13] Mantič, V. *Computational Implementation of Boundary Element Method with Multilevel Substructuring* (in Slovak). PhD thesis, Technical University of Košice 1992.

[14] Maz'ya,V.G. *Boundary Integral Equations. Encyclopaedia of Mathematical Sciences.* Springer-Berlag, Berlin, 1991.

[15] Nappi, A. Boundary Element Elastic Analysis on the Basis of a Consistent Formulation. *Appl. Math. Modelling*, 13(4):2340–241, 1989.

[16] Rieder, G. Extremal Principles, Iteration and BEM for Mixed Problems. In C.A. Brebbia and A. Chaudouet-Miranda (eds.), *Boundary Elements in Mechanical and Electrical Engineering,*, pp. 531–541. Computational Mechanics Publications, Southampton and Springer-Verlag, Berlin, 1990.

[17] Rizzo, F.J. An Integral Equation Approach to Boundary Value Problems of Classical Elastostatics. *Quarterly of Applied Mathematics*, 25:83–95, 1967.

[18] Schmidt, G. and Strese, H. The Convergence of a Direct BEM for the Plane Mixed Boundary Value Problem of the Laplacian. *Numer. Math.*, 54:145–165, 1988.

[19] Sirtori, S. General Stress Analysis Method by Means of Integral Equations and Boundary Elements. *Meccanica*, 14:210–218, 1979.

[20] Sirtori, S., Maier, G., Novati, G. and Miccoli, S. A Galerkin Symmetric Boundary Element Method in Elasticity. Formulation and Implementation. *Int. J. Numer. Meth. Eng.*, 35:255–282, 1992.

[21] Symm, G. T. Integral Equation Methods in Potential Theory. II. *Proc. Roy. Soc.*, 275(A):33–46, 1963.

[22] Tsybenko, A. S. and Lavrikov, S. A. Direct Symmetric Formulation of the Boundary Integral Equation Method (in Russian). *Applied Problems of Strength and Plasticity*, 11:43–50, 1984.

[23] Ugodchikov, A. G. and Khutorianskii, N. M. *Boundary Element Method in the Mechanics of a Deformable Solid Body* (in Russian). University of Kazan Publishing, Kazan, 1986.

[24] Vodička, R. *Iterative Methods for Linear System Solution in BEM* (in Slovak). Diploma thesis, Czech Technical University of Prague, 1993.

[25] Vodička, R. *Coupled Formulations of Boundary Integral Equations for Solving Contact Problems of Elasticity* (in Slovak). PhD thesis, Technical University of Košice 1999.

[26] Vodička, R. Coupled Formulations of Boundary Integral Equations for Solving Contact Problems of Elasticity. *Engng. Anal. Boundary Elements*, 24:407–426, 2000.

[27] Vodička, R. and Mantič, V. *A Comparison of Three Basic Formulations of BIE in the Potential Theory.* Technical Report, Dept. of Applied Mechanics, Technical University of Košice and Elasticity and Strength of Materials Group, University of Seville, December 1995.

# HP-ADAPTIVE BE MODELING OF THE HUMAN EAR ACOUSTICS

Timothy Walsh, Leszek Demkowicz
Texas Institute for Computational and Applied Mathematics
The University of Texas at Austin
Austin, TX 78712

**Abstract** The paper presents an implementation of an *hp*-adaptive BEM for exterior scattering problems motivated with the modeling of human auditory system acoustics. We focus on two practical issues connected with the *hp* discretizations: geometry representation, and numerical integration of 'almost singular' functions. Numerical examples illustrate the discussed implementation issues.

**Keywords:** Boundary element methods, *hp* adaptivity, auditory system, geometry reconstruction, singular integration

## 1. MOTIVATION

With regard to the human hearing system, a general question one may ask is: "How do the shape of the pinna and ear canal, wave propagation properties of the middle ear, cochlea characteristics, and auditory nerve, influence the electrical signal that reaches the brain, given a location and intensity of a source?" Although many acoustical and electrophysical experimental devices have been developed to measure/characterize the response of the outer, middle, and inner ears [22, 25, 24], many of the associated procedures are time consuming and expensive, and are not readily amendable to parametric studies of the effects of various geometrical features on the overall response. Numerical simulations of the auditory system provide the advantage that parametric studies can be performed in a much more efficient manner. This can be useful for studying the effects of small geometrical perterbations on the auditory response, as well as in the design process for technologies such as hearing aids, hearing protection devices, and virtual acoustical simulators.

*T. Burczynski (ed.), IUTAM/IACM/IABEM Symposium on Advanced Mathematical and Computational Mechanics Aspects of the Boundary Element Method,* 395–408.

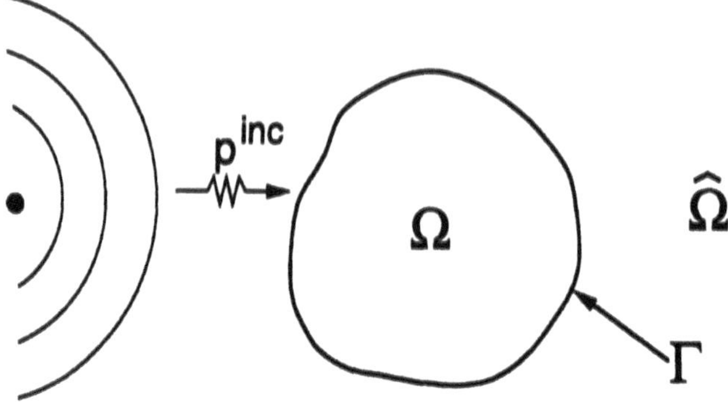


*Figure 1.1* **The model scattering problem in $\mathbb{R}^3$**

In this initial phase of our work, we restrict attention to the outer (external) ear, though we bear in mind that the outer ear model could eventually be coupled to a middle/inner ear model in a simulation of the overall auditory system. In the case of the outer ear, numerical studies on the subject are quite sparse [19, 20, 21], and this provides motivation for the investigations.

## 2. BURTON-MILLER INTEGRAL EQUATION FORMULATION

For exterior acoustic scattering problems, the two widely used classical formulations are the Helmholtz and hypersingular integral formulations. However, neither one is fully equivalent to the original Helmholtz differential equation with Sommerfeld radiation condition, over the entire range of frequencies. Both exhibit the ficticious frequencies, where the integral formulations contain spurious modes, and even for simulation frequencies that are numerically close to these, the approximations become unstable [1].

In 1971, Burton and Miller [23] presented a new formulation that consisted of a complex combination of the Helmholtz and hypersingular integral equations, and proved that this was equivalent to the original problem.

The problem of interest is shown schematically in Figure 1.1. Given an incident wave $p^{inc}$, and a bounded domain $\Omega$ with a boundary $\Gamma$, we wish to find a scattered pressure $p^s$ satisfying the Helmholtz differential equation

$$\Delta p^s + k^2 p^s = 0 \text{ in } \hat{\Omega} = \mathbb{R}^3 - \Omega, \tag{1.1}$$

Sommerfeld radiation condition

$$\left|\frac{\partial p^s}{\partial R} - ikp^s\right| = O\left(\frac{1}{R^2}\right) \text{ for } R \to \infty, \tag{1.2}$$

and rigid boundary condition on $\Gamma$

$$\frac{\partial(p^s + p^{inc})}{\partial n} = 0 \text{ on } \Gamma. \tag{1.3}$$

The Helmholtz equation with Sommerfeld radiation condition can be replaced with the Burton Miller integral equation, which can then be cast into a variational formulation, see [1] for details. The Burton Miller integral equation is as follows.
Find $p$ on $\Gamma$ such that

$$\begin{aligned}\alpha\left[\frac{1}{2}p(x) - \int_\Gamma \frac{\partial \Phi}{\partial n_y}(x,y)p(y)\,dy + \int_\Gamma \Phi(x,y)\frac{\partial p}{\partial n_y}(y)\,dy\right] \\ +(1-\alpha)\frac{i}{k}\left[\frac{1}{2}\frac{\partial p}{\partial n_x} + \int_\Gamma \frac{\partial \Phi}{\partial n_x}(x,y)\frac{\partial p}{\partial n_y}(y)\,dy + \int_\Gamma \frac{\partial^2 \Phi}{\partial n_x \partial n_y}(x,y)p(y)\,dy\right] \\ = \alpha p^{inc} + (1-\alpha)\frac{i}{k}\frac{\partial p^{inc}}{\partial n_x} \quad \text{on } \Gamma\end{aligned} \tag{1.4}$$

where

- $\Phi$ is the fundamental solution of the Helmholtz equation
- $p(x)$ is the pressure on $\Gamma$
- $k$ is the wave number
- $0 < \alpha < 1$ determines the relative contributions of the Helmholtz and hypersingular integral equations

For the rigid scattering problems, $\frac{\partial p}{\partial n} = 0$, and the corresponding weak formulation reads as follows:

Find total pressure $p = p^{inc} + p^s \in H^{\frac{1}{2}}(\Gamma)$ such that

$$d(p,q) = l(q) \quad \forall q \in H^{\frac{1}{2}}(\Gamma) \tag{1.5}$$

where the sesquilinear and antilinear forms are defined as follows

$$d(p,q) \;=\; \frac{1}{2}\int_\Gamma p(x)\bar{q}(x)\,dx - \int_\Gamma\int_\Gamma \frac{\partial \phi}{\partial n_y}(x,y)p(y)\bar{q}(x)\,dydx$$

$$
\begin{aligned}
& +\frac{i}{k}\int_\Gamma\int_\Gamma \Phi(x,y)\mathrm{rot}p(y)\mathrm{rot}\bar{q}(x)\,dydx \\
& -ik\int_\Gamma\int_\Gamma \Phi(x,y)n_x n_y p(y)\bar{q}(x)\,dydx \\
l(q) \quad = \quad & \int_\Gamma p^{inc}(x)\bar{q}(x)\,dx + \frac{i}{k}\int_\Gamma \frac{\partial p^{inc}}{\partial n_x}(x)\bar{q}(x)\,dx
\end{aligned}
$$

It has been shown [10, 9] that the inf-sup (BB) constant governing the formulation decreases slowly and monotonically with wave number k. For $k \to 0$, the coupling coefficients have to be modified [9]. Other references on the weak formulation of boundary integral equations can be found in [2, 3, 4, 5]

# 3. *HP* BOUNDARY ELEMENT DISCRETIZATIONS

The present work is a continuation of the earlier work at TICAM on *hp*-adaptive boundary elements for the solution of exterior and structure-fluid interaction problems in underwater acoustics [12, 1, 6, 11, 13]. We follow the original strategy [12, 1] with the $L^2$-residual driving adaptive refinements. For the moment, we use only the $h$-adaptive strategy with fixed order of approximation $p$, although $p$ can vary in a global manner. Figure 1.2 illustrates a typical result for the problem of scattering of a plane wave on a rigid sphere. The three curves, corresponding to the *actual approximation error*, the *best approximation error* [1], and the $L^2$-residual, asymptotically converge to each other. This is predicted by the convergence theory for the Helmholtz formulation [12, 9, 8], and is observed in practice for the Burton-Miller formulation as well.

The present boundary element code has been implemented within a new, FORTRAN 90 implementation of a two-dimensional *hp*-adaptive package described in [14]. There have been two practical problems precluding the development of the *hp*-adaptive boundary elements. The first one deals with the higher order representation of complex geometries. Most of the existing commercial mesh generators rely on the use of lower order elements and produce meshes that are prohibitive in size from the point of view of BEM. Moreover, *hp*-adaptivity calls for a more precise geometry representation that allows not only for generation of higher order meshes but also for updating the geometry degrees-of-freedom (d.o.f.) during both $h$ and $p$ mesh refinements. For the problem at hand, the geometry representation must be constructed from MRI scans, and this is the first practical issue that we address in this paper.

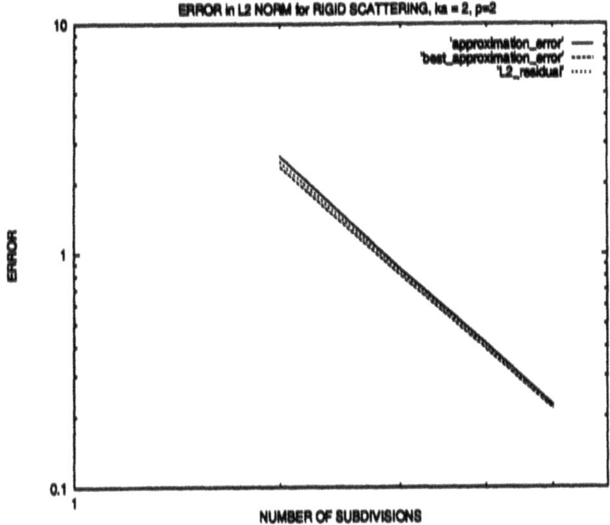


*Figure 1.2* Convergence results for rigid scattering on a sphere.

The second issue that we address in this contribution deals with the numerical integration of 'almost singular functions'. By virtue of the BE discretization, the double integral in (1.5) is split over finite elements. For each pair of elements $K_1, K_2$, the corresponding stiffness matrix is defined using formula (1.5) with the two integrals extended over elements $K_1, K_2$, and solution $p$ and test function $q$ replaced with the elements' shape fucntions. Typically, we use standard Gaussian integration for the outer integral over element $K_1$, with the quadrature order corresponding to order of approximation $p$. If elements $K_2$ and $K_1$ coincide, then for each integration point used to approximate the outer integral, the inner integral is (weakly) singular. Here we use the standard Duffy's coordinates for the integration. For two different elements, formally the integrand is regular (bounded), and standard Gaussian quadrature can be used if the two elements are sufficiently distant. If, however, the outer integration point is *close* to inner integral element $K_2$, then the functions become numerically singular and standard Gaussian quadrature has to be replaced with expensive adaptive integration rules. This will occur for higher order meshes and higher order Gaussian quadratures for the outer integral, where the integration points will approach the boundary of the element, steepening the integrand and introducing an 'almost' singularity.

# 4. GEOMETRIC RECONSTRUCTION USING TRIANGULAR A-PATCHES

The surface triangulation of the head/ear used in this study, obtained from MRI scans, consists of a set of vertices and element connectivities. Such a geometrical representation is found to be incompatible with *hp* adaptive acoustic simulations for many reasons. First, scannings provide no surface parametrization to govern the placement of new geometry degrees of freedom, and consequently neither $h$ nor $p$ refinements would improve the resolution of the geometry. Stated another way, with such a mesh the geometrical error will dominate the approximation error after several levels of adaptive refinements. Second, for wave propagation problems, a $C^0$ approximation of the surface may be insufficient as it allows artificial scattering along element edges for high enough frequency. This, in turn, will cause elements to cluster around the original element edges, increasing computational cost. Replacing these areas with a $C^1$ representation will reduce such artificial scattering and the associated increased computational costs.

In light of the deficiencies listed above, a three-step approach has been adopted, following the work of Bajaj, Xu et al [17], [18]. First, the vertex normals are estimated on the original fine mesh, using the algorithms described in [17]. Second, to eliminate the need for unrefinements, the original triangular mesh is decimated adaptively using a vertex-normal based decimation scheme [17]. This involves vertex deletion, retriangulation, and retention of the fine mesh normals. Finally, an A-patch, $C^1$ surface reconstruction is generated on the decimated mesh. Figure 1.3 shows an example of the effects of the reconstruction. The first mesh is the original linear triangulation of the head/ear, obtained from an MRI scan. The succeeding meshes have increasing orders of shape function approximation to the A-patches. The smoothness increases with polynomial order, even though the number and distribution of elements does not change. As usual, in the BE calculations, the A-patch is replaced by its *hp*-interpolant [14].

Figure 1.4 shows a more dramatic example in the case of adaptivity. Here the left column shows a sequence of adaptive meshes formed during a scattering simulation for a 20-sided polygon, in the case of no geometry reconstruction. The right column shows the same situation, only this time with geometry reconstruction. In the latter case, *both* the geometrical and the solution approximation improve with refinements, whereas in the left column the geometrical error stays fixed. Further, without reconstruction, elements cluster around the sharp boundaries, attempting to resolve the artificial singularities.

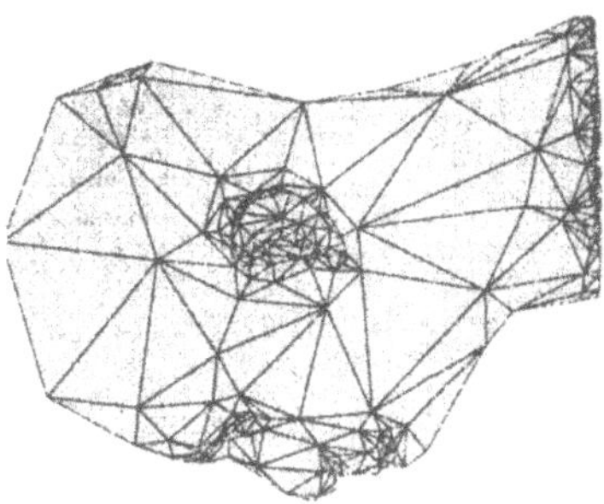

original linear mesh

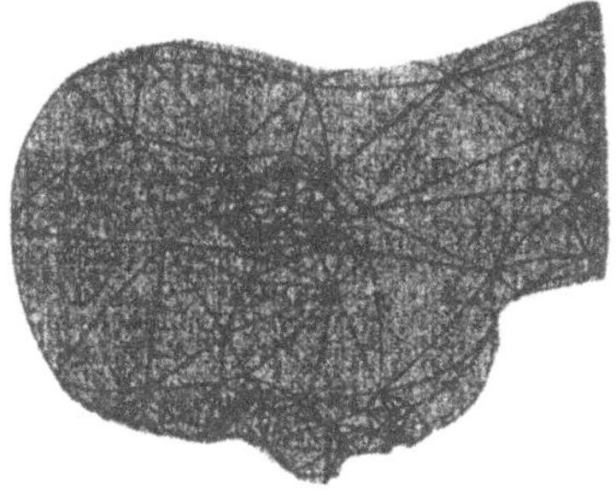

quadratic elements

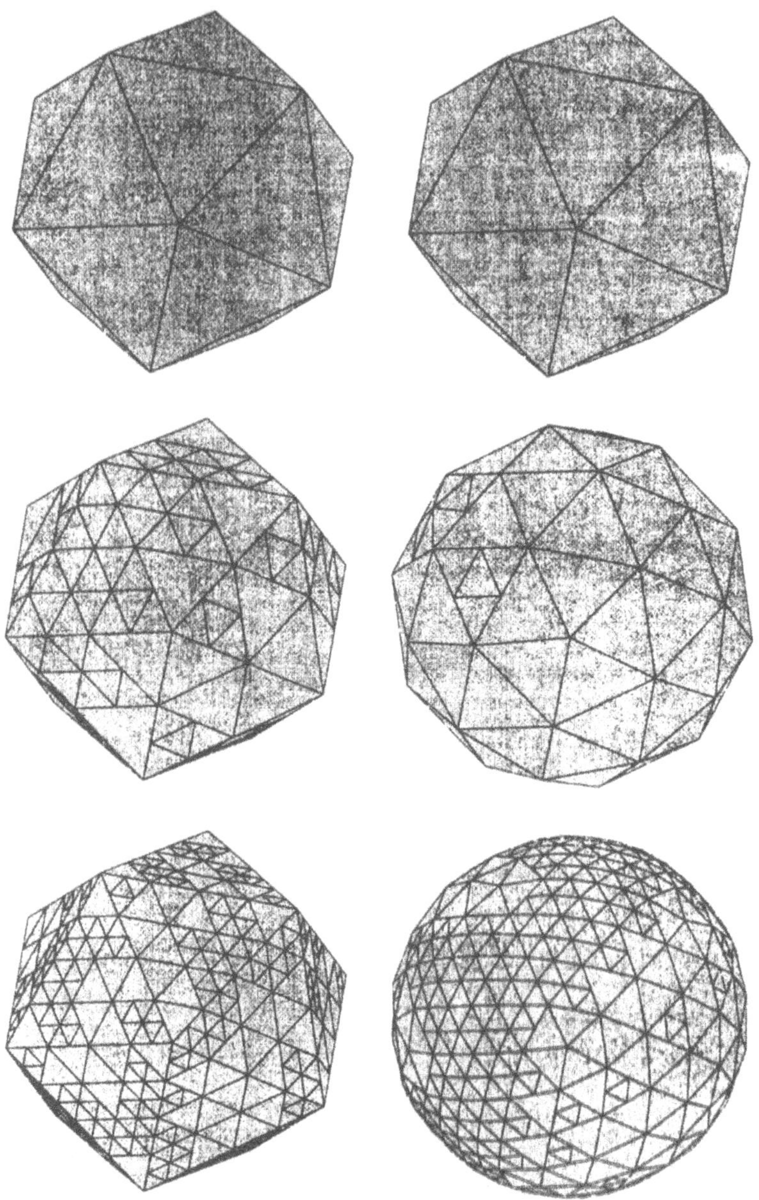

(a) Without geometry reconstruction (b) With geometry reconstruction

*Figure 1.4* Acoustical scattering from a rigid polygon using $h$ adaptive meshes, with (right) and without (left) geometry reconstruction

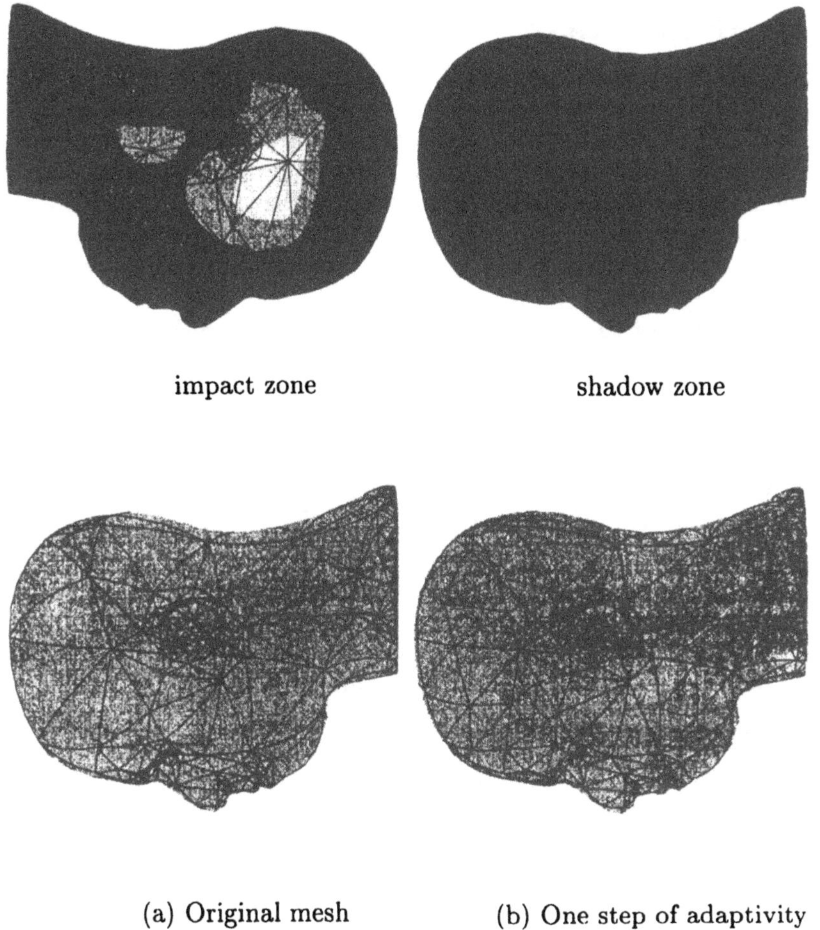


(a) Original mesh (b) One step of adaptivity

*Figure 1.5* Acoustical scattering from the head/ear with reconstructed geometry, after one step of $h$ adaptivity.

Figure 1.5 shows an example of the BEM computation of scattering of a plane wave on the human head/ear. In this case the geometry was reconstructed with quadratic elements. The real component of the pressure distribution shows a similar pattern as that obtained on the sphere, with distortion owing to the complex shape. After estimating the $L^2$ residual, the mesh has been subjected to a single $h$-refinement with new elements added at various locations around the head, where the residual was large.

# 5. INTEGRATION OF ALMOST SINGULAR INTEGRALS

Singular integrations are the major computational expense in BEM codes. They appear in formation of the linear system, as well as in $L^2$ residual error estimation for adaptivity [1]. In fact, the cost of error estimation is of the same order as for formation of the stiffness matrix, which is already the dominant computational task in Galerkin BEM [6]. Thus, the large meshes involved in the head/ear problem as well as the need for error estimation and adaptivity necessitate fast methods for singular integration.

For the case of the *almost* singular integration, which consumes the majority of the computational time, the recent emergence of exponentially convergent *hp* integration schemes [15] has provided an opportunity to greatly extend the applicability of the Galerkin BEM to larger problems. The scheme is based on a geometrically graded integration mesh with order of quadrature decreasing linearly towards the singularity. The advantages of such a scheme are as follows

- integration mesh constructed a-priori using known location of singularity
- geometrically graded integration mesh guarantees a fast (exponential) convergence in terms of the number of integration points
- no convergence checks are necessary

More precisely, the following theorem can be found in [15].
Given:

- $f \in \mathfrak{B}_\beta(\Omega)$ [2], analytic in $\Omega$ but singular at origin, for some $0 < \beta < 1$
- quadrature degree vector $\mathbf{p}$ with $p_j = max\{2, \lfloor j\mu \rfloor + 1\}, \quad 2 \leq j \leq n+1$, where $\mu$ is the slope of the quadrature (in practice $\mu = 1$),

we have

**Theorem (Schwab)**
*For every $0 < \sigma < 1$ and linear degree vector $\mathbf{p}$,*
*there exist $b, C > 0$ independent of $N$ such that*

$$| \int_K f(x)dx - Q_\sigma^{n,\mathbf{p}} f | \leq Cexp(-bN^{\frac{1}{r+1}}).$$

Here $r$ is the dimension of the region of integration, and N is the number of integrand evaluations. ∎

Using this theorem, we have implemented an *hp*-integration algorithm that is used when the distance between elements $K_1, K_2$ is of the order of the element diameters. For two distant elements, $K_1, K_2$, based on results of Karafiat [7], we use the standard Gaussian quadrature. These ideas have been combined into an *hp* integration algorithm for triangles and implemented in our new boundary element code. Timing studies of the entire solution procedure have been conducted, in one case using the *hp* integration and in the other using the bisection approach given in [1]. For the head/ear mesh with 3000 linear elements, our new algorithm shows a speedup factor of 8. Thus, these methods show great promise in terms of extending the applicability of Galerkin BEM to solve practical problems.

# 6. CONCLUSIONS

In the paper we have presented a new implementation of an *hp*-adaptive boundary element method for the modeling of the human auditory system. The method is based on the Galerkin form of the Burton-Miller formulation that guarantess a uniformly stable performance in the range of wave numbers of interest. In this preliminary work we have focused on two practical implementation problems, necessary for the overall success of the method: a geometry representation based on MRI scans that supports *hp* discretizations, and the numerical integration of 'almost singular' integrands.

# References

[1] L. Demkowicz, A. Karafiat and J. T. Oden. Solution of Elastic Scattering Problems in Linear Acoustics using *hp* Boundary Element Method. *Computer Methods in Applied Mechanics and Engineering*, vol. 101, 1992, pp. 251-282.

[2] Nedelec, J. C., Approximation par potential de double couche de probléme de Neumann extérieur, *C. R. Acad. Sc. Paris* t. 286, Série A, (16 janvier 1978), pp. 103-106.

[3] Hamdi, M.A., Une formulation variationelle par équations intégrales pour la résolution de l'equation de Helmholtz avec des conditions aux limites mextes, *C.R. Acad. Sci. Paris Ser. II* 292 , 1981, pp. 17-20.

[4] Bamberger, A., and Ha-Duong, T., Formulation variationelle espace-temps pour le calcul par potentiels retardé de la diffraction d'onde acoustique, *Math. Meth. in Appl. Sc.*, 8, 1986, pp. 405-435.

[5] Bamberger, A., and Ha-Duong, T., Formulation variationelle pour le calcul de la diffraction d'une onde acoustique par une surface rigide, *Math. Meth. in Appl. Sc.*, 8, 1986, pp. 598-608.

[6] P. Geng, J.T. Oden, and R.A. van de Geijn, Massively Parallel Computation for Acoustical Scattering Problems using Boundary Element Methods, *Journal of Sound and Vibration*, 1996, 191(1), pp. 145-165

[7] Karafiat, A., Habilitation Thesis, Politechnika Krakowska, Krakow, 1996.

[8] Demkowicz, L., Asymptotic convergence in finite and boundary element methods: Part 1: Theoretical Results , *Computers Math. Applic*, Vol. 27, No. 12, 1994, pp.69-84.

[9] Demkowicz, L., Asymptotic convergence in finite and boundary element methods: Part 2: The LBB constant for rigid and elastic scattering problems, *Computers Math. Applic*, Vol. 28, No. 6, 1994, pp.93-109.

[10] Amini, S., On the choice of the coupling parameter in boundary integral formulations of the exterior acoustic problem, *Appl. Anal.*, 35, 1990, pp.75-92.

[11] Geng, P., Parallel *hp* methods for coupled BE/FE analusis of structural acoustics problems, Doctoral dissertation, The University of Texas at Austin, 1994.

[12] Demkowicz, L., Oden, J., Ainsworth, M., and Geng, P., Solution of elastic scattering problems in linear acoustics using *hp* boundary element methods, *J Comp Appl Math*, 36, 1991, pp.29-63.

[13] Chang, Y., Scattering of acoustic waves on viscoelastic structures modeled by means of *hp* adaptive BE/FE methods, Doctoral dissertation, The University of Texas at Austin, 1996.

[14] Demkowicz, L., Gerdes, K., Schwab, C., Bajer, A., and Walsh, T., HP90: A general and flexible Fortran 90 *hp*-FE code *Computing and Visualization in Science*, 1, 1998, pp. 145-163.

[15] Schwab, C., Variable order composite quadrature of singular and nearly singular integrals, *Computing*, 53, 1994, pp. 173-194.

[16] Sauter, S., Schwab, C., Quadrature for *hp* Galerkin BEM in $\Re^3$, Research Report No. 96-02, Seminar fur Angewandte Mathematik, March 1996.

[17] Bajaj, C., Xu, G., Smooth adaptive reconstruction of free-form fat surfaces, submitted to *ACM Transactions on Graphics*.

[18] Bajaj, C., Chen, J., and Xu, G., Modeling with cubic A- patches, *ACM Transactions on Graphics*, Vol. 14, No. 2, April 1995, pp. 103-133.

[19] Ciskowski, R., Applications in bioacoustics, In *Boundary element method in acoustics*, Ciskowski, R., and Brebbia, C., (Ed.), Computational Mechanics Publications, 1991.

[20] Kahana, Y., Nelson, P., Petyt, M., and Choi, S., Numerical modeling of the transfer functions of a dummy-head and of the external ear, *AES* $16^{th}$ *International Conference on Spatial Sound Reproduction*, Finland, 1999.

[21] Katz, B., Measurement and calculation of individual head related transfer functions using a boundary element model including the measurement and effect of skin and hair impedance, Doctoral dissertation, The Pennsylvania State University, May 1998.

[22] Sandlin, R., Handbood of Hearing Aid Amplification, Vol. 1, Singular Publishing, 1995.

[23] Burton, A., and Miller, G., The Application of Integral Equation Methods to the Numerical Solution of Some Exterior Boundary-Value Problems, *Proc. Roy. Soc.*, London Ser., A323, 1971, pp.201-210.

[24] Crocker, M., (Ed), *Handbook of Acoustics*, John Wiley & Sons 1995.

[25] Yost, W.A., Fundamentals of Hearing: An Introduction, ($3^{rd}$ edition), Academic Press, Inc. San Diego, CA, 1994.

# QUADRATIC AND SINGULAR BOUNDARY ELEMENTS FOR CRACKS AND NOTCHES IN THREE DIMENSIONS

J O Watson
Department of Mining Engineering
University of New South Wales
Sydney 2052, Australia

## 1. Introduction

The purpose of the work described here is to provide a convenient and accurate means of analysis of stress near cracks and notches in three dimensions. The techniques employed are those that have already been verified in the development of Hermitian cubic and singular elements for plane strain (1). The displacement and traction interpolates over elements on faces of a crack include singular shape functions which, as the crack root is approached, asymptotically vary in the same way as the theoretical modes of crack opening displacement determined by Williams (2). These shape functions are multiplied by stress intensity factors just as smoothly varying (isoparametric) functions are multiplied by displacements and tractions at nodes. Construction of the discrete system of equations in terms of unknown nodal values of displacement, traction and stress intensity factors generally is by nodal collocation. The integral equations are taken over a modified boundary which is constructed such that the collocation point lies outside the domain enclosed by that boundary. At collocation points on the faces of a crack, hypersingular integral equations are obtained by differentiation of the singular integral equation in the normal direction. Finally, the boundary integral is taken to be the sum of an explicitly computed component which is evaluated by Gaussian quadrature, and an implicitly computed component which is calculated by consideration of trial displacements (referred to by Lutz et al (3) as isolated displacement fields) of subdomains to either side of a crack, as previously done for plane strain (4).

At present there is quadratic variation of element geometry and the smoothly varying components of displacement and traction, but it is intended eventually to introduce Hermitian cubic and singular elements similar to those developed for plane strain, as such elements offer improved functional variation over each element in return for practically no global increase in the number of degrees of freedom.

*T. Burczynski (ed.), IUTAM/IACM/IABEM Symposium on Advanced Mathematical and Computational Mechanics Aspects of the Boundary Element Method*, 409–418.

## 2. Boundary elements

Surfaces of cracks are modelled by coincident pairs of eight- and nine-node quadrilateral elements, and six-node triangular elements. The variations of cartesian coordinates, traction and smoothly varying component of displacement are quadratic with respect to intrinsic coordinates. The general form of variation of displacement over an element $S_b$ on a surface of a crack is

$$u_j(\xi,\eta) = \sum_{v=1}^{n} M_v(\xi,\eta)\, u_j(b,v) + \sum_{k=1}^{3} \Psi^u_{jk}(\xi,\eta)\, K_k(y_0) \qquad (1)$$

where $M_v$ are quadratic shape functions of intrinsic coordinates $(\xi,\eta)$, $u_j(b,v)$ are displacements at node $v$, $\Psi^u_{jk}(\xi,\eta)$ are singular shape functions and $K_k$ are stress intensity factors at the point $y_0$ of the crack root nearest to the point $y(\xi,\eta)$ of $S_b$ as shown in Figure 1. The stress intensity factors $K_k(y_0)$ are taken to vary quadratically over each element side on a crack root (so in general $u_j(\xi,\eta)$ depends upon stress intensity factors at three nodes), and

$$\Psi^u_{jk}(\xi,\eta) = \psi^u_{jk}(\xi,\eta) - \sum_{v=1}^{n} M_v(\xi,\eta)\, \psi^u_{jk}(b,v) \qquad (2)$$

where $\psi^u_{jk}(\xi,\eta)$ are displacements according to Williams (2) in mode $k$ of a crack the root of which is straight and tangential to the actual crack at $y_0$, and $\psi^u_{jk}(b,v)$ are values of $\psi^u_{jk}(\xi,\eta)$ at node $v$ of $S_b$. The effect of the second term on the right side of equation (2) is to ensure that $\Psi^u_{jk}(\xi,\eta)$ equals zero at each node of $S_b$. To improve accuracy, singular shape functions are continued for more than one element away from the crack root.

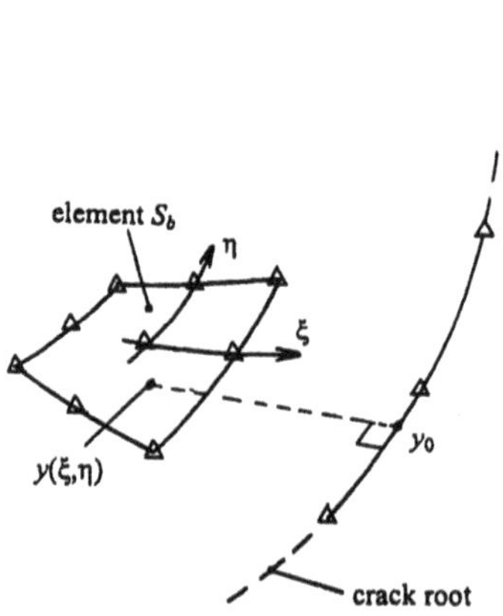


Figure 1 Definition of singular shape function

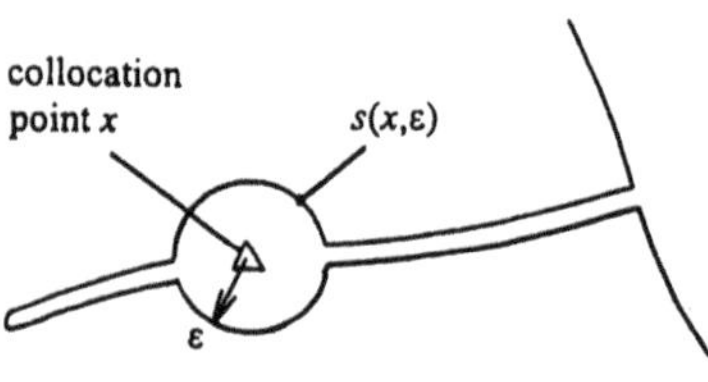


(a) collocation point on crack

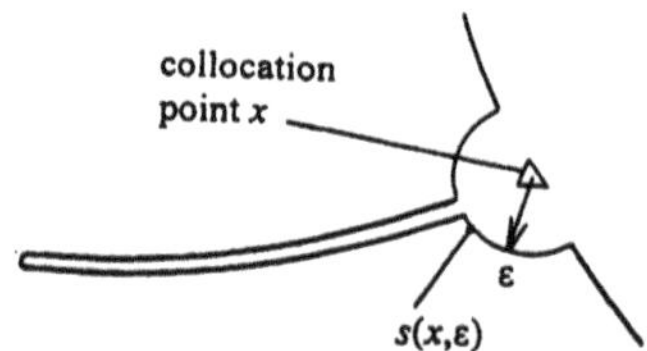


(b) collocation point at crack-surface junction

Figure 2 Regions of exclusion

Compared with shifting of side nodes of elements adjacent to the crack root to quarter points, the introduction of singular shape functions has the advantage that there is no loss of capacity to model the smoothly varying component of displacement. For notch roots, account can be taken of the different eigenvalues (powers of distance from the crack root in expressions for displacement) in each of the three modes. Finally, stress intensity factors are computed at the same time as displacements, so there is no need for postprocessing such as consideration of crack opening displacement.

## 3. Integral equations

The singular boundary integral equation (or zero order equation) is that of the direct formulation, obtained by substitution of the unknown displacement field and the Kelvin solution for load at the collocation point into Betti's theorem (5). Because the Kelvin solution is not twice continuously differentiable at the load point, the integral equation is taken over the boundary $S - S(x,\varepsilon) + s(x,\varepsilon)$ of a domain consisting of that part of the elastic body which lies outside a region of exclusion as shown in Figure 2, in the limit as a characteristic dimension $\varepsilon$ of that region tends to zero. The hypersingular boundary integral equation (or first order equation) is the derivative of the zero order equation in the direction normal to the crack: the differentiation is considered to be carried out before $\varepsilon$ tends to zero, so that differentiation under the integral sign is legitimate. The resulting zero and first order equations are

$$\lim_{\varepsilon \to 0} \int_{S - S(x,\varepsilon) + s(x,\varepsilon)} [T_{ij}(x,y)\, u_j(y) - U_{ij}(x,y)\, t_j(y)]\, dS_y = 0 \tag{3}$$

and

$$\lim_{\varepsilon \to 0} \int_{S - S(x,\varepsilon) + s(x,\varepsilon)} [V_{ij}(x,y)\, u_j(y) - W_{ij}(x,y)\, t_j(y)]\, dS_y = 0 \tag{4}$$

where for example

$$V_{ij}(x,y) = n_s(x)\, \partial T_{ij}(x,y)/\partial x_s \tag{5}$$

Equations (3) and (4) remain valid if the region of exclusion is not spherical, and even if it changes in shape as its size tends to zero. The collocation point $x$ may be at an edge or corner of the surface, and in particular it may be at a point at which a crack meets the surface as shown in Figure 2(b).

In the implementation of Hermitian cubic and singular elements for plane strain (1,4), zero, first and second order equations were collocated at all nodes of cracks, including nodes at crack roots. As is explained in a later section of this paper, it is not at present convenient to collocate the first order equation at a crack root. In the scheme adopted here, only the zero order equation is collocated at the crack root, and the first order equation is collocated at points such as $(\xi,\eta) = (0.5,1.0)$, $(0.5,-1.0)$ and $(0.5,0.0)$ of a quadrilateral element the side $\xi = 1.0$ of which lies on a crack root, as illustrated in Figure 3. These points are referred to as auxiliary collocation points.

The most commonly proposed methods of computation of coefficients of discrete systems of equations approximating to hypersingular integral equations require regularisation (6) or the evaluation of Hadamard finite parts (7). In the approach taken here, the boundary integral is written as the sum of an explicitly computed component $H^1_i(\varepsilon)$ or $I^1_i(\varepsilon)$ and an implicitly computed component $H^2_i(\varepsilon)$ or $I^2_i(\varepsilon)$. Equations (3) and (4) become

$$\lim_{\varepsilon \to 0} [H^1_i(\varepsilon) + H^2_i(\varepsilon)] = 0 \tag{6}$$

and

$$\lim_{\varepsilon \to 0} [I^1_i(\varepsilon) + I^2_i(\varepsilon)] = 0 \tag{7}$$

respectively, where

$$H^2_i(\varepsilon) = \sum_{\substack{b=1 \\ x \in S_b}}^{p} \int_{S^\varepsilon_b} T_{ij}(x,y)\, M^c_0(\xi,\eta)\, u_j(x)\, dS_y + \int_{s(x,\varepsilon)} [T_{ij}(x,y)\, u_j(y) - U_{ij}(x,y)\, t_j(y)]\, dS_y \tag{8}$$

and

$$I^2_i(\varepsilon) = \sum_{\substack{b=1 \\ x \in S_b}}^{p} \int_{S^\varepsilon_b} [V_{ij}(x,y)\, \{M^c_0(\xi,\eta)\, u_j(x) + N\xi(\xi,\eta)\, [\partial u_j/\partial\xi]_x + N\eta(\xi,\eta)\, [\partial u_j/\partial\eta]_x\} + W_{ij}(x,y)\, M^c_0(\xi,\eta)\, t^b_j(x)]\, dS_y$$

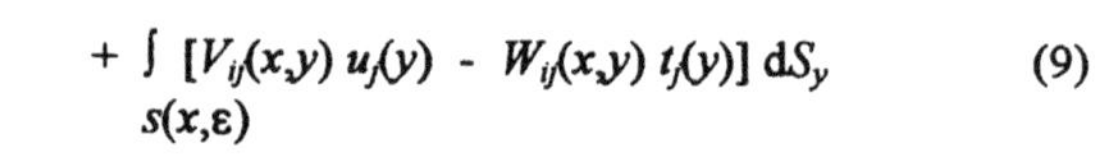

$$+ \int_{s(x,\varepsilon)} [V_{ij}(x,y)\, u_j(y) - W_{ij}(x,y)\, t_j(y)]\, dS_y \tag{9}$$

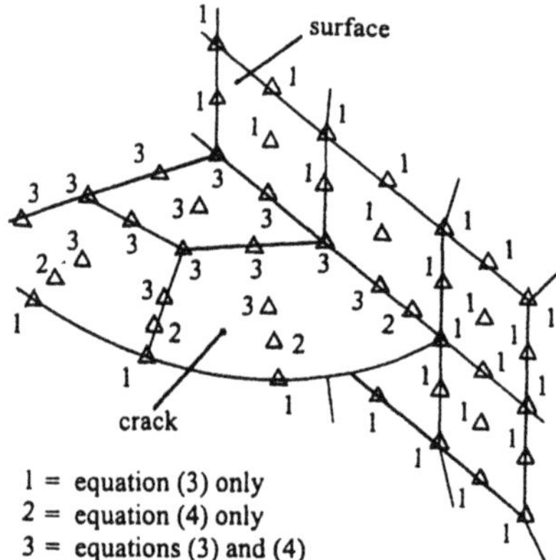


Figure 3 Collocation points

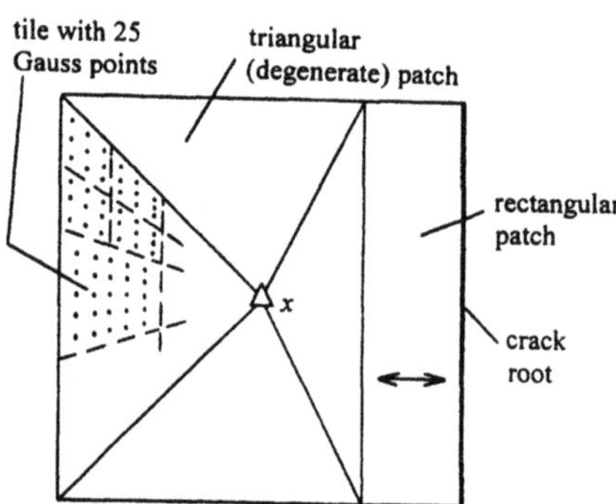


Figure 4 Typical integration scheme, $x \in S_b$

In equations (8) and (9), $S^{\varepsilon}_{b}$ is that part of $S_b$ which lies outside the region of exclusion, $t_j(y)$ is traction at $y$, $t^{b}_{j}(x)$ is traction on $S_b$ at $x$, and $M'_0$, $M''_0$, $N\xi$ and $N\eta$ are auxiliary shape functions with the following properties.

1. $1 - M'_0(\xi,\eta) \in O(r)$ where $r$ is distance from $x$ to $y(\xi,\eta)$.
2. $1 - M''_0(\xi,\eta) \in O(r^2)$
3. $N\xi(\xi,\eta) \in O(r)$ and $N\eta(\xi,\eta) \in O(r)$
4. $1 - \partial N\xi/\partial\xi \in O(r)$ and $\partial N\xi/\partial\eta \in O(r)$
5. $\partial N\eta/\partial\xi \in O(r)$ and $1 - \partial N\eta/\partial\eta \in O(r)$

## 4. Explicitly computed integrals

The explicitly computed components, the forms of which may be inferred from equations (3) to (9), are only weakly singular at $x$ and so are evaluated by Gaussian quadrature formulae. Elements which include the collocation point are divided into patches as illustrated for an element adjacent to a crack root in Figure 4, and each patch is further subdivided into tiles according to the same quadtree scheme as is used to integrate over elements which do not include the collocation point. There are between two and five Gauss points in each direction of a tile, the arrangement of tiles and numbers of points each way being determined according to estimated upper bounds of error in terms of the $2m$th derivative of the integrand where $m$ is the number of Gauss points (5,8). The double-ended arrow shown in the patch adjacent to the crack root in Figure 4 indicates a non-linear transformation of an intrinsic coordinate in the integration of products of kernels and singular shape functions (9).

Integrations over elements which include the collocation point give rise to coefficients of $u_j(x)$, $[\partial u_j/\partial\xi]_x$ and $[\partial u_j/\partial\eta]_x$. If $x$ is an auxiliary collocation point, it is necessary to eliminate coefficients of $u_j(x)$ by substitution in terms of nodal displacements and stress intensity factors according to equation (1). Whether $x$ is an auxiliary collocation point or a node, it is always necessary to substitute for $[\partial u_j/\partial\xi]_x$ and $[\partial u_j/\partial\eta]_x$. For example, to eliminate $[\partial u_j/\partial\xi]_x$ we write

$$[\partial u_j/\partial\xi]_x = \sum_{v=1}^{n} [\partial M_v/\partial\xi]_x\, u_j(b,v) + \sum_{k=1}^{3} (\Psi^{u}_{jk}(\xi_0,\eta_0)\, [\partial K_k/\partial\xi]_x + K_k(x_0)\, [\partial\Psi^{u}_{jk}/\partial\xi]_x) \quad (10)$$

where $(\xi_0,\eta_0)$ are the intrinsic coordinates of $x$, and $x_0$ is the point on the crack root nearest to $x$. The derivatives of $K_k$ and $\Psi^{u}_{jk}$ are evaluated by differencing at $x$.

## 5. Implicitly computed integrals

By writing a Taylor series expansion for displacement at the collocation point and using Hooke's law to make certain substitutions, it can be shown that for $x$ on a crack

$$\lim_{\varepsilon\to 0} H^2_i(\varepsilon) = \sum_{x=x^+,x^-} a^0_{i}\beta(x)\, u\beta(x) \qquad (11)$$

and

$$\lim_{\varepsilon\to 0} H^2_i(\varepsilon) = \sum_{x=x^+,x^-} (a^1_{i}\beta(x)\, u\beta(x) + b^{1A}_{i}\beta(x)[\partial u\beta/\partial x_A]_x$$
$$+ b^{1B}_{i}\beta(x)[\partial u\beta/\partial x_B]_x + c^1_{i}\beta(x)\, t\beta(x)) \qquad (12)$$

where $x^+$ and $x^-$ are points on the upper and lower surface of the crack, the subscript $\beta$ denotes a component tangential or normal to the crack as shown in Figure 5, and $t\beta(x)$ is traction on the plane tangential to the crack at $x$. The coefficients $a^0_{i}\beta$, $a^1_{i}\beta$ *etc* are determined by consideration of the trial displacement fields (9) $u\beta_l$ listed in Table 1, not for the entire elastic body but for subdomains which lie to each side of the crack as shown in Figure 6. This is permissible because the implicitly computed components depend only upon the geometry of elements which include the collocation point (the 'tent floor'), which is the same for a subdomain as for the entire elastic body (4).

Equations (6) and (7) are written for each subdomain in turn. Explicitly computed integrals $H^1_{il}$ and $I^1_{il}$ corresponding to the displacement fields $u\beta_l$ are computed by Gaussian quadrature over the surface (the tent, and tent floor), and coefficients appearing in equations (11) and (12) are then computed from $H^2_{il}$ (which equals $-H^1_{il}$)and $I^2_{il}$ (which equals $-I^1_{il}$) as indicated in Table 2. Finally, if $x$ is an auxiliary collocation point then coefficients of $u\beta(x)$ are eliminated by substitution according to equation (1) and, whether $x$ is an auxiliary collocation point or a node, $[\partial u\beta/\partial x_A]_x$ and $[\partial u\beta/\partial x_B]_x$ are eliminated in the same way as is $[\partial u_i/\partial \xi]_x$ according to equation (10). These eliminations give rise to leading diagonal coefficients of stress intensity factors.

In the implementation of Hermitian cubic and singular elements for plane strain (1,4), first and second order equations are collocated at the crack root, and implicitly computed

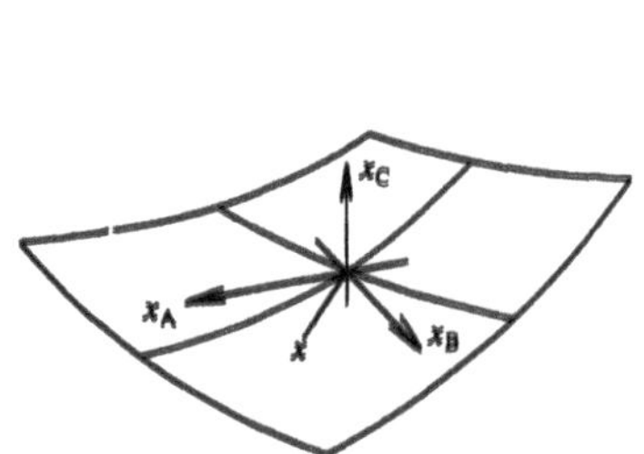

Figure 5 Local axes tangential and normal to crack

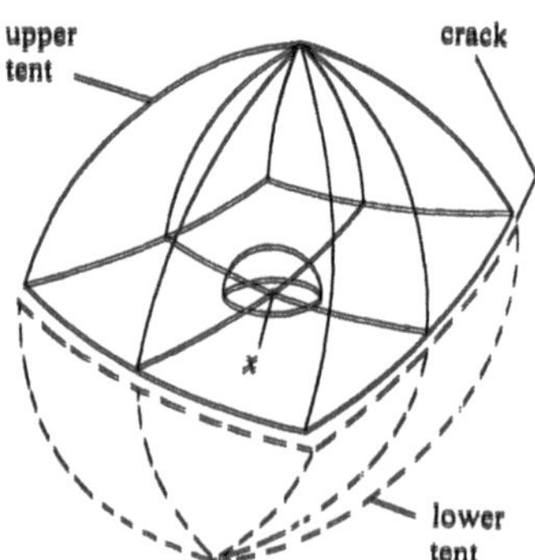

Figure 6 Subdomains to each side of crack

Table 1 Trial displacement fields

| $l$ | $u_{Al}$ | $u_{Bl}$ | $u_{Cl}$ |
|---|---|---|---|
| 1 | 1 | 0 | 0 |
| 2 | 0 | 1 | 0 |
| 3 | 0 | 0 | 1 |
| 4 | $x_A$ | $-\nu x_B$ | $-\nu x_C$ |
| 5 | $-\nu x_A$ | $x_B$ | $-\nu x_C$ |
| 6 | $-\nu x_A$ | $-\nu x_B$ | $x_C$ |
| 7 | $x_B$ | $x_A$ | 0 |
| 8 | 0 | $x_C$ | $x_B$ |
| 9 | $x_C$ | 0 | $x_A$ |
| 10 | $-x_B$ | $x_A$ | 0 |
| 11 | 0 | $-x_C$ | $x_B$ |
| 12 | $x_C$ | 0 | $-x_A$ |

Table 2 Implicitly computed coefficients

| coefficient | value |
|---|---|
| $a^0{}_{i\beta}(x)$ | $H^2{}_{i\beta}$ |
| $a^1{}_{i\beta}(x)$ | $I^2{}_{i\beta}$ |
| $b^{1A}{}_{iA}(x)$ | $[I^2{}_{i4} + \nu I^2{}_{i5}]/(1 - \nu^2)$ |
| $b^{1A}{}_{iB}(x)$ | $[I^2{}_{i7} + I^2{}_{i10}]/2$ |
| $b^{1A}{}_{iC}(x)$ | $-I^2{}_{i12}$ |
| $b^{1B}{}_{iA}(x)$ | $[I^2{}_{i7} - I^2{}_{i10}]/2$ |
| $b^{1B}{}_{iB}(x)$ | $[I^2{}_{i5} + \nu I^2{}_{i4}]/(1 - \nu^2)$ |
| $b^{1B}{}_{iC}(x)$ | $I^2{}_{i11}$ |
| $c^1{}_{iA}(x)$ | $[I^2{}_{i9} + I^2{}_{i12}](1 + \nu)/E$ |
| $c^1{}_{iB}(x)$ | $[I^2{}_{i8} - I^2{}_{i11}](1 + \nu)/E$ |
| $c^1{}_{iC}(x)$ | $[I^2{}_{i6} + \nu(I^2{}_{i4} + I^2{}_{i5})/(1 - \nu)]$ |

integrals are evaluated by consideration of trial displacement fields of a single subdomain which (in two dimensions) encloses the crack root. The trial displacement fields taken in the computation of leading diagonal coefficients of stress intensity factors are the crack opening displacements for a planar crack, denoted $\psi^u{}_{jk}$ in equation (2). In three dimensional analysis the crack root generally is curved, so the displacements $\psi^u{}_{jk}$ determined by Williams (2) can no longer be taken as trial displacements: hence the need for auxiliary collocation points.

## 6. Buried elliptical crack

The crack, with ratio of lengths of axes 2.0, lies in an infinite space of elastic material with Poisson's ratio 0.3 subjected to unit uniaxial stress at infinity. The solution is taken as the sum of two components (5), one of which is displacement due to unit pressure inside the crack. The coarse and fine boundary element meshes consist of 28 and 112 nine node quadrilaterals per face respectively. The coarse mesh is shown in Figure 7: in the fine mesh, there are four elements in place of each element of the coarse mesh. Computed stress intensity factors at each node on the crack root are shown in Table 3. Results are generally too high. In the analysis of cracks in plane strain, it is found that computed results are similarly too high if only dominant modes of crack opening displacement are taken into account, so it is likely that accuracy can be significantly improved by the inclusion of subdominant modes (1).

Table 3 Computed Mode 1 stress intensity factors, buried elliptical crack

| β (degrees) | coarse mesh $K_I$ (MPa.mm$^{0.5}$) | coarse mesh error (%) | fine mesh $K_I$ (MPa.mm$^{0.5}$) | fine mesh error (%) |
|---|---|---|---|---|
| 0.000 | 1.059 | +2.3 | 1.041 | +0.6 |
| 5.625 | | | 1.045 | +0.3 |
| 11.250 | 1.080 | +1.6 | 1.073 | +0.9 |
| 16.875 | | | 1.095 | 0.0 |
| 22.500 | 1.121 | -1.1 | 1.131 | -0.2 |
| 28.125 | | | 1.188 | +1.0 |
| 33.750 | 1.252 | +2.7 | 1.233 | +1.1 |
| 39.375 | | | 1.279 | +1.4 |
| 45.000 | 1.334 | +2.5 | 1.312 | +0.8 |
| 50.625 | | | 1.354 | +1.2 |
| 56.250 | 1.400 | +2.2 | 1.383 | +0.9 |
| 61.875 | | | 1.415 | +1.2 |
| 67.500 | 1.453 | +2.2 | 1.435 | +0.9 |
| 73.125 | | | 1.458 | +1.3 |
| 78.750 | 1.484 | +2.1 | 1.467 | +1.0 |
| 84.375 | | | 1.479 | +1.2 |
| 90.000 | 1.499 | +2.4 | 1.479 | +1.1 |

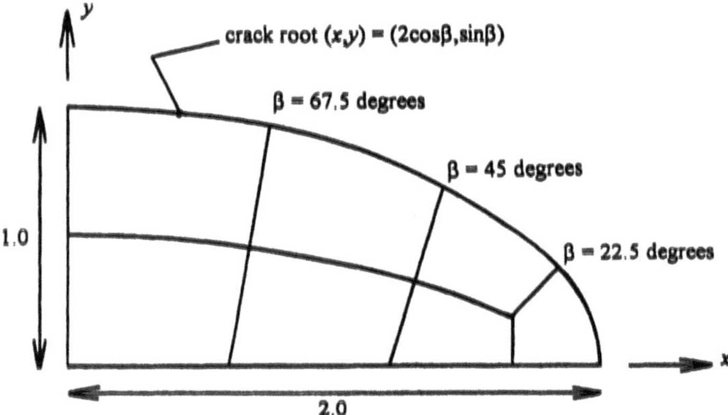


Figure 7 One quarter of coarse mesh, buried elliptical crack

## 7. Semi-elliptical surface crack

The crack lies in a halfspace of elastic material of Poisson's ratio 0.3, subjected to uniaxial stress at infinity. The solution is again taken as the sum of two components. The coarse and fine meshes consist of nine node quadrilateral and infinite elements (5), with 14 and 56 elements per crack face respectively. The coarse mesh is shown in Figure 8: in the fine mesh, there are four elements in place of each element of the coarse mesh on the faces of the crack. The computed deformed shape of the crack is shown in Figure 9, and in Figure 10 are shown computed stress intensity factors at each node on the crack root, and the variation of stress intensity factor according to Murakami (10). As expected, computed results show improbable oscillations near the surface because the shape functions over elements adjacent to an end of the crack root are unable to approximate well the dimpling of the surface that occurs if Poisson's ratio is not zero. Results for Poisson's ratio equal to zero (not shown here) are free of such oscillations.

## 8. Conclusion

It is demonstrated that values of stress intensity factors of sufficient accuracy for engineering purposes can be computed using fairly coarse meshes of quadratic and singular boundary elements. It is probable that precision can be improved by the inclusion of subdominant modes of crack opening displacement and further improvements of detail including Hermitian cubic crack root geometry.

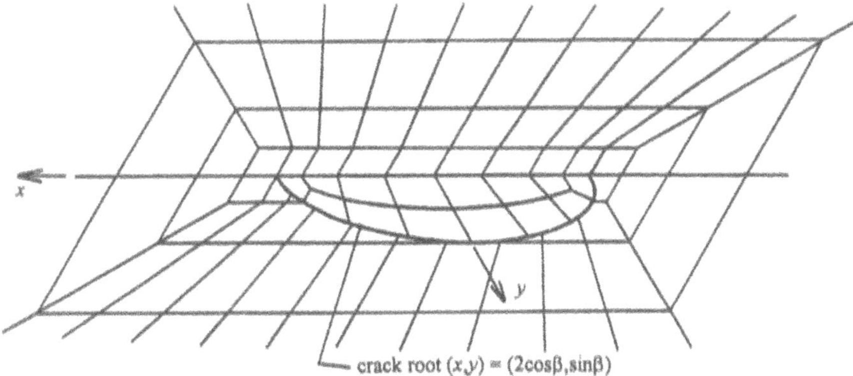


**Figure 8** Coarse mesh, semi-elliptical surface crack

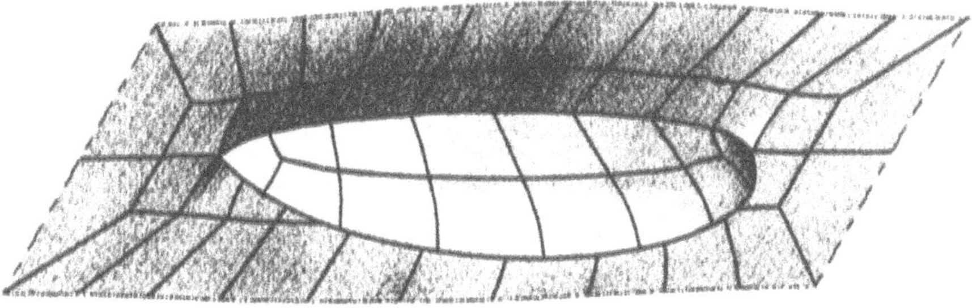

**Figure 9** CADFIX plot of computed displacements of surface crack

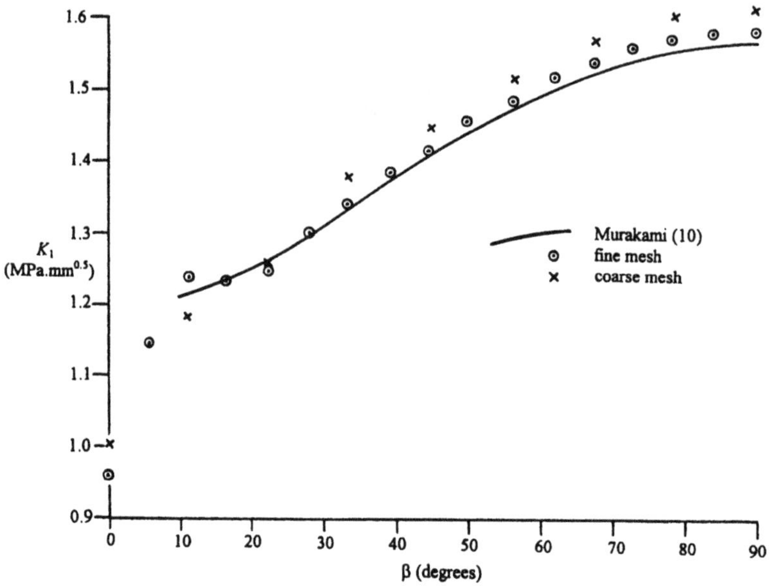


Figure 10 Mode 1 stress intensity factor, semi-elliptical surface crack

## 9. Acknowledgement

The author is grateful to FEGS Ltd of Cambridge, UK for permission to use the graphics package CADFIX to produce Figures 8 and 9.

## 10. References

1. Watson, J.O. (1995) Singular boundary elements for the analysis of cracks in plane strain, *I. J. Num. Meth. Eng.* **38**, 2389-2411.
2. Williams, M.L. (1952) Stress singularities resulting from various boundary conditions in angular corners of plates in extension, *J. Appl. Mech. ASME* **74**, 526-528.
3. Lutz, E., Ingraffea, A.R. and Gray, L.J. (1992) Use of 'simple solutions' for boundary integral methods in elasticity and fracture analysis, *I. J. Num. Meth. Eng.* **35**, 1737-1751.
4. Watson, J.O. (1986) Hermitian cubic and singular elements for plane strain, in P.K. Banerjee and J.O. Watson (eds.) *Developments in Boundary Element Methods – 4*, Elsevier Applied Science Publishers, London, 1-28.
5. Beer, G. and Watson, J.O. (1992) *Introduction to Finite and Boundary Element Methods for Engineers*, Wiley, Chichester.
6. Gallego, R. and Dominguez, J. (1996) Hypersingular BEM for transient elastodynamics, *I. J. Num. Meth. Eng.* **39**, 1681-1705.
7. Gray, L.J., Martha, L.F. and Ingraffea, A.R. (1990) Hypersingular integrals in boundary element fracture analysis, *I. J. Num. Meth. Eng.* **29**, 1135-1158.
8. Stroud, A.H. and Secrest, D. (1966) *Gaussian Quadrature Formulas*, Prentice-Hall, Englewood Cliffs, New Jersey.
9. Watson, J.O. (1982) Hermitian cubic boundary elements for plane problems of fracture mechanics, *Res Mech.* **4**, 23-42.
10. Murakami, Y. (1987) *Stress Intensity Factors Handbook*, Pergamon, Oxford.

# DYNAMIC CRACK ANALYSIS BY HYPERSINGULAR AND NON-HYPERSINGULAR TIME-DOMAIN BEM

Ch. Zhang
*Department of Civil Engineering, Hochschule Zittau/Görlitz, University of Applied Sciences, D-02763 Zittau, Germany*
c.zhang@hs-zigr.de

A. Savaidis
*Department of Civil Engineering, TU Darmstadt, D-64289 Darmstadt, Germany*

**Keywords:** Boundary element method, boundary integral equations, dynamic crack analysis, dynamic fracture mechanics

**Abstract** Transient elastodynamic crack analysis by time-domain boundary element method (BEM) is presented. Time-dependent traction boundary integral equations (BIEs) are used for this purpose. Both the hypersingular and the non-hypersingular formulations are dealt with. The time-domain traction BIEs are solved numerically by a time-stepping scheme. The efficiency and the accuracy of the time-domain BEM for transient elastodynamic crack analysis are demonstrated by a numerical example. Advantages and drawbacks of the hypersingular and the non-hypersingular BEM formulations are compared and discussed.

## 1. INTRODUCTION

Time-domain boundary element method (BEM) or time-domain boundary integral equation method (BIEM) provides an efficient and accurate numerical tool for transient elastodynamic crack analysis. This is due to its semi-analytical nature, high accuracy, reduced problem dimension, and simple pre- and post-processing for input and output data. The classical displacement-based BEM alone leads to a degenerate formulation and is therefore not suited for crack analysis. To avoid this difficulty, several BEM formulations are suggested in the past: the crack Green's function method [1], the multi-domain method [2]-[3], the traction BEM

*T. Burczynski (ed.), IUTAM/IACM/IABEM Symposium on Advanced Mathematical and Computational Mechanics Aspects of the Boundary Element Method*, 419–428.

([1], [7]-[11]), and the dual BEM [4]. Among these methods, the traction BEM is, to the opinion of the authors, the most promising method for crack analysis.

Two different traction BEM formulations exist in the literature, namely, the hypersingular and the non-hypersingular BEM formulations. Hypersingular traction BIEs can be directly obtained by using Betti-Rayleigh reciprocal theorem and Hooke's law. The arising hypersingular integrals are understood as Hadamard finite-part integrals. A regularization of the hypersingular traction BIEs leads to non-hypersingular traction BIEs with Cauchy principal value integrals. Non-hypersingular time-domain traction BIEs can also be derived immediately without regularization by using a two-state conservation integral of elastodynamics ([7]-[9]). For an unbounded domain, the hypersingular traction BIEs contain only the crack-opening-displacements (CODs) as unknown quantities ([10]-[11]), while both the spatial and the temporal derivatives of the crack-opening-displacements occur in the non-hypersingular traction BIEs.

A comparative study on both hypersingular and non-hypersingular time-domain traction BEM formulations used previously by the authors ([7]-[11]) is presented in this paper. A time-stepping scheme for solving the hypersingular and the non-hypersingular time-domain BIEs is applied. The scheme uses a collocation method with a linear shape-function in time and two different spatial shape-functions. Time-integrations arising in the computation of the system matrices are carried out analytically in the method. A constant spatial shape-function is used away from crack-tips, while a square-root crack-tip shape-function is adopted at crack-tips for describing the proper local behavior of the crack-opening-displacements at crack-tips. This makes an accurate COD-based calculation of the dynamic stress intensity factors feasible. Analytical and numerical methods for evaluating the hypersingular and the non-hypersingular integrals arising in the time-domain traction BEM are applied. A numerical example is given to show that both the hypersingular and the non-hypersingular time-domain traction BEM are very efficient and accurate for transient elastodynamic crack analysis. Advantages and drawbacks of the hypersingular and the non-hypersingular traction BEM formulations are compared and discussed.

## 2. PROBLEM STATEMENT

We consider an isotropic, homogeneous, and linearly elastic solid contaning a finite crack in two-dimensional plane strain or plane stress. The cracked solid satisfies the equations of motion

$$\sigma_{ij,j} = \rho \ddot{u}_i, \tag{1}$$

the Hooke's law

$$\sigma_{ij} = E_{ijkl} u_{k,l}, \tag{2}$$

the initial conditions

$$u_i(\mathbf{x},t) = \dot{u}_i(\mathbf{x},t) = 0\,, \qquad \text{for} \quad t \le 0, \tag{3}$$

and the boundary conditions

$$f_i(\mathbf{x},t) = 0\,, \qquad \mathbf{x} \in \Gamma_c, \tag{4}$$

$$f_i(\mathbf{x},t) = f_i^*(\mathbf{x},t)\,, \qquad \mathbf{x} \in \Gamma_\sigma, \tag{5}$$

$$u_i(\mathbf{x},t) = u_i^*(\mathbf{x},t)\,, \qquad \mathbf{x} \in \Gamma_u. \tag{6}$$

Here, body forces are not considered, $f_i(\mathbf{x},t) = \sigma_{ij}(\mathbf{x},t)n_j$ denotes the traction components, $\sigma_{ij}$ and $u_i$ denote the stress and the displacement components, $\rho$ is the mass density, $E_{ijkl}$ is the elasticity tensor, $n_j$ is the unit normal vector, $\Gamma_c = \Gamma_c^+ + \Gamma_c^-$ with $\Gamma_c^\pm$ being the upper and the lower crack-faces, $\Gamma_\sigma$ represents the extrenal boundary where the stresses or the tractions are given, and $\Gamma_u$ represents the external boundary where the displacements are prescribed, respectively. A comma after a quantity designates the spatial derivatives, while superscript dots stand for the temporal derivatives of the quantity. The conventional summation rule over double indices is applied, with $i,j,k = 1,2$ for two-dimensional plane strain or plane stress.

## 3. HYPERSINGULAR TIME-DOMAIN BIES

By using Betti-Rayleigh reciprocal theorem, a representation integral for the displacement field $u_k(\mathbf{x},t)$ can be obtained as

$$u_k(\mathbf{x},t) = \int\limits_\Gamma \left(u_{ik}^G * \sigma_{ij} - \sigma_{ijk}^G * u_i\right) n_j ds, \quad \mathbf{x} \in \Omega, \tag{7}$$

where $u_{ik}^G$ and $\sigma_{ijk}^G$ are the displacement and the stress Green's functions, an asterisk $*$ denotes Riemann convolution, and $\Omega$ is the interior domain under consideration enclosed by the boundaries $\Gamma = \Gamma_\sigma + \Gamma_u + \Gamma_c^\pm$.

As is well-known that Eq. (7) cannot be directly applied to crack analysis since the usual limiting process $\mathbf{x} \to \Gamma_c^\pm$ for deriving displacement BIEs leads in this case to a degenerate BIE formulation. This difficulty can be avoided by using traction BIEs, which can be obtained by substituting Eq. (7) into Hooke's law (2), letting $\mathbf{x} \to \Gamma$, considering the stress continuity condition on the crack-faces, and invoking the boundary conditions (4)-(6). The result is

$$\delta(\mathbf{x}) f_p(\mathbf{x},t) = n_q(\mathbf{x}) \int\limits_{\Gamma_\sigma + \Gamma_u} \left(U_{ipq}^G * f_i - T_{ipq}^G * u_i\right) ds$$

$$+ \quad n_q(\mathbf{x}) \int\limits_{\Gamma_c^+} T^G_{ipq} * \Delta u_i ds, \qquad \mathbf{x} \in \Gamma, \tag{8}$$

where $\Delta u_i(\mathbf{y}, \tau)$ denotes the crack-opening-displacements (COD) defined by

$$\Delta u_i(\mathbf{y}, \tau) = u_i(\mathbf{y} \in \Gamma_c^+, \tau) - u_i(\mathbf{y} \in \Gamma_c^-, \tau). \tag{9}$$

and

$$\delta(\mathbf{x}) = \begin{cases} \frac{1}{2}, & \mathbf{x} \in \Gamma_\sigma + \Gamma_u, \\ 1, & \mathbf{x} \in \Gamma_c^+, \end{cases} \tag{10}$$

$$U^G_{ipq} = -E_{pqkl} u^G_{ik,l} = -\sigma^G_{pqi}; \qquad T^G_{ipq} = -E_{pqkl} \sigma^G_{ijk,l} n_j. \tag{11}$$

Unfortunately, the traction BIEs (7) are hypersingular since in 2-D case $T^G_{ipq}$ behaves as

$$T^G_{ipq}(\mathbf{x}, t; \mathbf{y}, \tau) \propto \frac{1}{|\mathbf{x} - \mathbf{y}|^2}, \qquad \text{for} \quad \mathbf{x} \to \mathbf{y}. \tag{12}$$

Here, the hypersingular integrals are understood as Hadamard "finite-part" integrals. The hypersingular traction BIEs (7) can be solved numerically by different methods such as the Galerkin method, the regularization method or the direct method. A direct solution method is applied in this paper.

## 4. NON-HYPERSINGULAR TIME-DOMAIN BIES

Non-hypersingular time-domain traction BIEs can be obtained by regularization techniques. Most of them use partial integration to reduce the hypersingular integrals in (7) to strongly singular or Chauchy principal value integrals which also appear in the classical displacement BIEs. Another straightforward way to obtain non-hypersingular traction BIEs uses an elastodynamic conservation law or the $J_D$-integral which gives rise to the following representation integral for the displacement gradients $u_{k,l}(\mathbf{x}, t)$ ([7]-[9])

$$u_{k,l}(\mathbf{x}, t) = \int\limits_{\Gamma} \left( \epsilon_{rst} \epsilon_{rlj} \sigma^G_{ijk} * u_{i,t} n_s + \rho u^G_{ik} * \ddot{u}_i n_l - u^G_{ik,l} * f_i \right) ds, \qquad \mathbf{x} \in \Omega, \tag{13}$$

where $\epsilon_{rst}$ is the permutation tensor.

By substituting Eq. (12) into Hooke's law (2), taking the limiting process $\mathbf{x} \to \Gamma$, considering the stress continuity condition on the crack-faces, and using the boundary conditions Eqs. (4)-(6), non-hypersingular time-domain traction BIEs can be obtained as ([7]-[9])

$$\begin{aligned} f_p(\mathbf{x}, t) &= E_{pqkl} n_q(\mathbf{x}) \times \\ &\left[ \int\limits_{\Gamma_\sigma + \Gamma_u} \left( \epsilon_{rst}\epsilon_{rlj}\sigma^G_{ijk} * u_{i,t} n_s + \rho u^G_{ik} * \ddot{u}_i n_l - u^G_{ik,l} * f_i \right) ds \right. \\ &\left. - \int\limits_{\Gamma_c^+} \left( \epsilon_{rst}\epsilon_{rlj}\sigma^G_{ijk} * \Delta u_{i,t} n_s + \rho u^G_{ik} * \Delta\ddot{u}_i n_l \right) ds \right] , \quad \mathbf{x} \in \Gamma . \end{aligned} \tag{14}$$

Since no spatial derivatives of the stress Green's functions are contained in Eq. (14), the traction BIEs (14) are non-hypersingular with the exception at crack-tips and they possess the usual strong-singularities arising in the classical displacement BIEs. The strongly singular integrals in Eq. (14) are Cauchy principal value integrals. Unlike the hypersingular traction BIEs (7), the non-hypersingular BIEs (14) contain spatial and temporal derivatives of the displacements and the crack-opening-displacements. At a propagating crack-tip $\Delta\ddot{u}_i$ is singular since

$$\Delta\ddot{u}_i \propto \frac{\dot{a}^2(\tau)}{\mid \mathbf{a}(\tau) - \mathbf{y} \mid^{3/2}} , \qquad \mathbf{y} \to \mathbf{a}(\tau), \tag{15}$$

where $\mathbf{a}$ is a vector describing the current position of a propagating crack-tip and $\dot{a}$ is the corresponding propagation velocity of a crack-tip. Consequently, the non-hypersingular traction BIEs (14) may be numerically disadvantageous for treating propagating cracks.

## 5. NUMERICAL SOLUTION PROCEDURE

Both the hypersingular and the non-hypersingular time-domain traction BIEs can be solved numerically by a time-stepping scheme as presented in this section. In what follows, the essential steps of the time-stepping scheme will be given.

The crack-face $\Gamma_c^+$, the external boundary $\Gamma_\sigma + \Gamma_u$, and the time variable $t$ are discretized as

$$\Gamma = \sum_{e=1}^{E} \Gamma_e ; \qquad t = \sum_{n=1}^{N} n\Delta t, \tag{16}$$

where $\Gamma_e$ denotes the length of the $e$-th element, $E = E_1 + E_2$ is the total element-number with $E_1$ being the element-number of the crack-face $\Gamma_c^+$

and $E_2$ being the element-number of the external boundary $\Gamma_\sigma + \Gamma_u$, and $\Delta t$ represents the time-step, respectively. The unknown boundary quantities are approximated by the following interpolation functions

$$\Delta u_i(\mathbf{y},\tau) = \sum_{e=1}^{E_1}\sum_{n=1}^{N} \mu_{e(\Delta u)}(\mathbf{y})\eta^n_{(u)}(\tau)(\Delta u_i)^n_e , \tag{17}$$

$$u_i(\mathbf{y},\tau) = \sum_{e=1}^{E_2}\sum_{n=1}^{N} \mu_{e(u)}(\mathbf{y})\eta^n_{(u)}(\tau)(u_i)^n_e , \tag{18}$$

$$f_i(\mathbf{y},\tau) = \sum_{e=1}^{E_2}\sum_{n=1}^{N} \mu_{e(f)}(\mathbf{y})\eta^n_{(f)}(\tau)(f_i)^n_e , \tag{19}$$

where $\mu_{e(\cdot)}(\mathbf{y})$ and $\eta^n_{(\cdot)}(\tau)$ are the spatial and the temporal shape-functions. The spatial shape-function can be expressed as

$$\mu_{e(\cdot)}(\mathbf{y}) = g_{(\cdot)}(\mathbf{y})H(\mathbf{y}) = \begin{cases} g_{(\cdot)}(\mathbf{y}), & \mathbf{y} \in \Gamma_e , \\ 0, & \mathbf{y} \notin \Gamma_e , \end{cases} \tag{20}$$

where the function $g_{(\cdot)}(\mathbf{y})$ describes the spatial variation of the unknown boundary quantities within the $e$-th element and $H(\mathbf{y})$ is the Heaviside function. The temporal variation of the unknown boundary data is approximated by a linear shape-function as

$$\eta^n_{(u)}(\tau) = \eta^n_{(f)}(\tau) = \begin{cases} 1 - \dfrac{|\tau - n\Delta t|}{\Delta t}, & |\tau - n\Delta t| \leq \Delta t , \\ 0, & \text{otherwise} . \end{cases} \tag{21}$$

Substitution of Eqs. (17)-(19) into the hypersingular BIEs (7) or the non-hypersingular BIEs (14) results in a system of linear algebraic equations

$$\mathbf{A}^1\mathbf{v}^q = \mathbf{B}^1\mathbf{f}^q + \sum_{s=1}^{q-1}\left(\mathbf{B}^{q-s+1}\mathbf{f}^s - \mathbf{A}^{q-s+1}\mathbf{v}^s\right) , \tag{22}$$

where $\mathbf{A}^s$ and $\mathbf{B}^s$ are the system matrices, $\mathbf{v}^s$ is a vector containing the crack-opening-displacements on the crack-face and the displacements on the external boundary, and $\mathbf{f}^s$ is a vector containing the traction components on discrete boundary-points, respectively. By invoking the boundary conditions (4)-(6), Eq. (22) can be recast into the following form

$$\bar{\mathbf{A}}^1\mathbf{z}^q = \bar{\mathbf{B}}^1\mathbf{r}^q + \sum_{s=1}^{q-1}\left(\mathbf{B}^{q-s+1}\mathbf{f}^s - \mathbf{A}^{q-s+1}\mathbf{v}^s\right) , \tag{23}$$

in which $\bar{\mathbf{A}}^1$ and $\bar{\mathbf{B}}^1$ are the reordered system matrices from $\mathbf{A}^1$ and $\mathbf{B}^1$ according to the boundary conditions, $\mathbf{z}^q$ is a vector containing the unknown boundary quantities, and $\mathbf{r}^q$ is a vector containing the prescribed boundary data. From Eq. (23) the following time-stepping scheme is obtained

$$\mathbf{z}^q = (\bar{\mathbf{A}}^1)^{-1}\left[\bar{\mathbf{B}}^1\mathbf{r}^q + \sum_{s=1}^{q-1}\left(\mathbf{B}^{q-s+1}\mathbf{f}^s - \mathbf{A}^{q-s+1}\mathbf{v}^s\right)\right], \qquad (q = 1, 2, ..., N), \tag{24}$$

where $(\bar{\mathbf{A}}^1)^{-1}$ is the inverse of the system matrix $\bar{\mathbf{A}}^1$ at the first time-step. At each time-step, only two new system matrices $\mathbf{A}^q$ and $\mathbf{B}^q$ have to be calculated. By using the time-steping scheme (23) the unknown boundary data can be determined step by step.

With the linear temporal shape-function (21), time-integrations arising in the evaluation of the system matrices $\mathbf{A}^s$ and $\mathbf{B}^s$ can be carried out analytically ([10]-[11]). For simplicity, two different spatial shape-functions are adopted to approximate the spatial variation of the crack-opening-displacements. For elements away from crack-tips a constante shape-function is applied, while for elements behind the crack-tips a special shape-function ("crack-tip element") is adopted. In this manner, the local behavior of the crack-opening-displacements at crack-tips can be properly described. In addition, the crack-tip-element is also used as a "transition-element" for the second element behind the crack-tips. The use of the so-called "transition-element" improves the accuracy of the computed dynamic stress intensity factors. On the external boundary of the cracked body, both the displacements and the tractions are approximated by a constant spatial shape-function. In the case of regular elements, i.e. $\mathbf{x}_d \notin \Gamma_e$, spatial integrations arising in the system matrices can be carried out numerically by using standard Gaussian quadrature, since the corresponding integrals are regular. For singular elements, i.e. $\mathbf{x}_d \in \Gamma_e$, both hyper- and strong-singularities occur in the hypersingular BEM based on Eqs. (7), while only strong-singularities arise in the non-hypersingular BEM based on Eqs. (14). The hypersingular and strongly singular integrals can be evaluated analytically and numerically ([10]-[11]). Also, there are integrable singularities at the wave-fronts, but they provide no difficulties by using standard Gaussian quadrature.

# 6. NUMERICAL EXAMPLE AND CONCLUSIONS

As an example we consider a rectangular plate with an inclined crack in two-dimensional plane strain as shown Fig. 1. The inclination angle of the crack with respect to the horizontal plate-boundary is 45°. The plate is subjected to an impact loading. The material and the geometrical data used in the numerical calculations are: Poisson's ratio $\nu$=0.3; shear modulus $\mu$=76923MPa; mass density $\rho$=5000$kg/m^3$; plate length $H$=60mm; plate width $B$=30mm; and crack length $2a$=14.14mm. The crack-face is discretized by using 6 elements, the horizontal plate-boundaries are discretized by using 24 elements, and 48 elements are used for the vertical plate-boundaries. The time-step is taken as $\Delta t = (B/12)/c_L = 0.17035\mu s$.

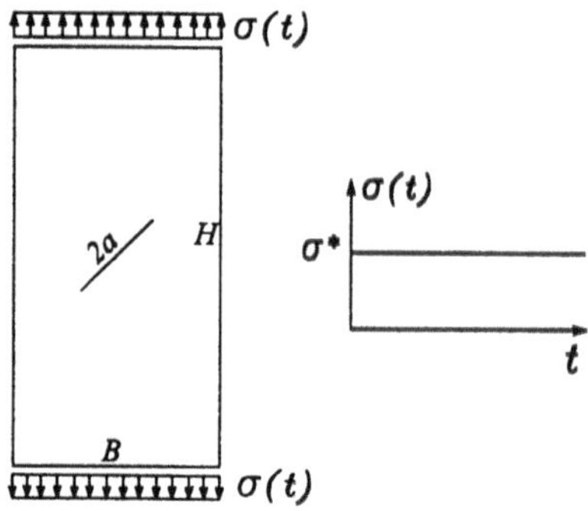


*Figure 1* A rectangular plate with an inclined crack

The normalized dynamic stress intensity factors obtained by using the hypersingular time-domain traction BEM are shown in Figs. 2 and 3. For comparison, the numerical results obtained by Murti and Valliappan [6] via a finite element method, by Dominguez and Gallego [3] via a time-domain displacement BEM in conjunction with a domain-substructuring, and by Fedelinski [5] via a Laplace-transform dual BEM are also given in Figs. 2 and 3. Fedelinski [5] used 40 quadratic elements for the disretization of the plate: 24 for the external boundaries and 8 for each of the crack-faces. He used quarter-point-elements and 50 Laplace-parameters to obtain the dynamic stress intensity factors. In the small-time region, the present numerical results agree very well with those obtained by other authors. For the normalized Mode-II dynamic stress intensity factor $\bar{K}_{II}(t)$, the agreement between the present numerical results and those of other authors is still very good in the large-time region. By contrast, some discrepancies in the normalized Mode-I dynamic stress intensity factor $\bar{K}_I(t)$ are noted between different methods for large $t$. In this case, the present numerical results agree very well with the results of Murti and Valliapan [6] by a finite element method. With the same discretization and time-step, the non-hypersingular time-domain traction BEM provided nearly the same numerical results, but they are not given here to avoid possible confusions in Figs. 2 and 3.

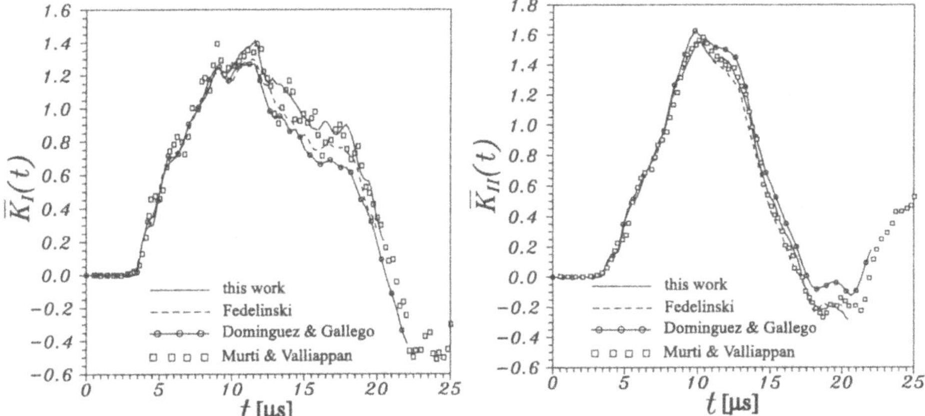


*Figure 2* Normalized $\bar{K}_I(t)$-factor

*Figure 3* Normalized $\bar{K}_{II}(t)$-factor

*Table 1* A comparison of hypersingular and non-hypersingular BEM

| **Issues** | **Hypersingular** | **Non-Hypersingular** |
| --- | --- | --- |
| *Derivation by* | Betti & Hooke | Regularization |
| | | $J_D$-integral |
| *Green's functions used* | $u^G_{ik,l}$; $\sigma^G_{ijk,l}$ | $u^G_{ik}$; $u^G_{ik,l}$; $\sigma^G_{ijk}$ |
| *Highest singularity* | $1/r^2$ (2-D) | $1/r$ (2-D) |
| | $1/r^3$ (3-D) | $1/r^2$ (3-D) |
| *Unknowns* | $u_i$; $\Delta u_i$; $f_i$ | $u_{i,j}$; $\Delta u_{i,j}$; $\ddot{u}_i$; $\Delta\ddot{u}_i$; $f_i$ |
| *Continuity requirement* | higher | lower |
| *Time-integration* | analytically | analytically |
| *Spatial integration* | numerically | numerically |
| *Diagonal dominance* | better | worse |
| *Condition number* | larger | smaller |
| *Adaptivity* | worse | better |
| *Choice of time-step* | crucial | crucial |

In Tab. 1, both the hypersingular and the non-hypersingular time-domain traction BEM are compared. As conclusions it can be said that both methods are very accurate and efficient for transient elastodynamic crack analysis as shown in this paper for 2-D case. A common drawback is that the choice of the time-step is in both methods crucial. The stability and the quality of the numerical results could be strongly influenced by the used time-step. A too small time-step causes an instability of the numerical scheme, while a too large time-step leads to a numerical damping of the results. Due to this disadvantage, the resolution of the numerical results at rapidly propagating crack-tips is in both methods yet not satisfactory and it requires further improvements.

## References

[1] Cruse, T. A.: *Boundary Element Analysis in Computational Fracture Mechanics.* Kluwer Academic Publishers, Boston, Mass., 1988.

[2] Dominguez, J.: *Boundary Elements in Dynamics.* Computaional Mechanics Publications, Southampton, UK, 1993.

[3] Dominguez, J. and Gallego, R.: Time Domain Boundary Element Method for Dynamic Stress Intensity Factor Computations. *Int. J. Numer. Meth. Engng.* **33** 635-647 (1992).

[4] Fedelinski, P., Aliabadi, M. H. and Rooke, D. P.: The Dual Boundary Element Method in Dynamic Fracture Mechanics. *Engng Anal. Bound. Elem.* **12** 203-210 (1993).

[5] Fedelinski, P.: Private Communication, 1999.

[6] Murti, V. and Valliappan, S.: The Use of Quarter Point Element in Dynamic Crack Analysis. *Eng. Fract. Mech.* **23** 585-614 (1986).

[7] Zhang, Ch.: A Novel Derivation of Non-Hypersingular Time-Domain BIEs for Transient Elastodynamic Crack Analysis. *Int. J. Solids Struct.* **28** 267-281 (1991).

[8] Zhang, Ch. and Gross, D.: A Non-Hypersingular Time-Domain BIEM for Transient Elastodynamic Crack Analysis. *Int. J. Numer. Meth. Engng.* **36** 2997-3017 (1993).

[9] Zhang, Ch. and Gross, D.: *On Wave Propagation in Elastic Solids With Cracks.* Computational Mechanics Publications, Southampton, UK, 1997.

[10] Zhang, Ch. and Savaidis, A.: A Hypersingular BEM for Dynamic Crack Analysis. *ZAMM Z. Angew. Math. Mech.*, **79**, (1999).

[11] Zhang, Ch. and Savaidis, A.: Time-Domain BEM for Dynamic Crack Analysis. *Mathematics and Computers in Simulation*, **50** 351-362 (1999).

GPSR Compliance
The European Union's (EU) General Product Safety Regulation (GPSR) is a set of rules that requires consumer products to be safe and our obligations to ensure this.

If you have any concerns about our products, you can contact us on

ProductSafety@springernature.com

In case Publisher is established outside the EU, the EU authorized representative is:

Springer Nature Customer Service Center GmbH
Europaplatz 3
69115 Heidelberg, Germany

www.ingramcontent.com/pod-product-compliance
Ingram Content Group UK Ltd.
Pitfield, Milton Keynes, MK11 3LW, UK
UKHW021859190726
13853UKWH00003B/1337

* 9 7 8 9 4 0 1 5 9 7 9 4 4 *